THEORY OF CRITICAL PHENOMENA IN FINITE-SIZE SYSTEMS

Scaling and Quantum Effects

Series in
Modern
Condensed
Matter
Physics

Vol. 9

THEORY OF CRITICAL PHENOMENA IN FINITE-SIZE SYSTEMS

Scaling and Quantum Effects

Jordan G. Brankov

Institue of Mechanics, Bulgarian Academy of sciences

Daniel M. Danchev

Institue of Mechanics, Bulgarian Academy of sciences

Nicholai S. Tonchev

*G. Nadjakov Institue of Solid State Physics,
Bulgarian Academy of Sciences*

World Scientific
Singapore • New Jersey • London • Hong Kong

Published by

World Scientific Publishing Co. Pte. Ltd.

P O Box 128, Farrer Road, Singapore 912805

USA office: Suite 1B, 1060 Main Street, River Edge, NJ 07661

UK office: 57 Shelton Street, Covent Garden, London WC2H 9HE

Library of Congress Cataloging-in-Publication Data
Brankov, Iordan.
    Theory of critical phenomena in finite-size systems : scaling and
quantum effects / Jordan G. Brankov, Daniel M. Danchev, Nicholai S.
Tonchev.
        p.   cm. -- (Series on modern condensed matter physics ; vol. 9)
    Includes bibliographical references and index.
    ISBN 9810239254
    1. Critical phenomena (Physics) 2. Finite size scaling (Physics)
I. Danchev, Daniel M. II. Tonchev, Nicholai S. III. Title.
IV. Series: Series in modern condensed matter physics : vol. 9.
QC173.4.C74B68   000
530.4'74--dc21   99-33063
                                                  CIP

British Library Cataloguing-in-Publication Data
A catalogue record for this book is available from the British Library.

Printed in Singapore by World Scientific Printers

To our children

Milena   and   Valentin,
Marin,
Ivailo   and   Rossen

Preface

It is hard to find a field in theoretical physics which can compete with the theory of phase transitions and critical phenomena with respect to the number of written textbooks, reviews and monographs. In our opinion, writing a new book can be justified only by treating in a different manner new trends existing in the area. What do we have in mind in our case? There are just a few books on critical phenomena in systems with confined geometry: a collection of reprints [Cardy, ed. (1988)], a collection of reviews [Privman, ed. (1990)], and the monograph [Krech (1994)] on the Casimir effect. Indeed, in some modern texbooks on critical phenomena one can find special chapters devoted to this topic, see, e.g., [Cardy (1996)], [Domb (1996)], [Zinn-Justin (1996)], [Henkel (1999)]. Against the background of the numerous papers that appear annually, the gap in the monographic literature on the subject is obvious. The present book attempts to partially fill up this gap. We hope also to give our modest contribution in spreading the scaling ideas for fruitful interpretation and analysis of phase transitions in classical and quantum systems of finite volume.

It is a well known fact that the volume is an irrelevant parameter for the local properties of a macroscopic system and, therefore, can be chosen arbitrary large. The conventional statistical mechanical theory studies abstract systems, consisting of infinitely many particles in an infinite volume, due to the essential simplifications that occur in their description. Moreover, it becomes possible to describe phase transitions mathematically in terms of discontinuous or singular behavior of some thermodynamic functions. In constructing the above, so called thermodynamic limit [Van Hove (1949)], [Fisher (1964)], one has to keep constant values of some intensive quantities. For example, in the canonical Gibbs ensemble the increase in the number of particles N has to be accompanied by a proportional increase in the volume V, so that the density $\rho = N/V$ be constant. Any intensive quantity a_V of a finite system can be written in the form $a_V = a_\infty + \delta a_V$, where a_∞ is the bulk value and δa_V is a finite-size correction which tends to zero as $V \to \infty$. The finite-size correction δa_V contains a more detailed information about the shape of the system and the boundary conditions. Usually, the correction term becomes essential under rather special conditions, e.g., in the vicinity of a second-order phase transition. When the relevant thermodynamic parameters approach a critical point, the correlation length ξ of an infinite system will grow unboundedly. Therefore, if we apply the theory of infinite (bulk) systems for the description of a large but finite system, very close to the critical point the bulk correlation length will become comparable with the smallest size of the system and deviations from the real critical behaviour will set in.

In the beginning of the 70's, M. Fisher and his followers developed a comprehensive theory of finite-size scaling. As in the bulk case, this theory of critical behaviour of finite systems is based on renormalization group properties and offers a universal description depending, in addition, on the shape of the system and the type of boundary conditions.

During the last two decades, the study of finite-size effects has undergone an extensive development and gained still growing importance for the theory of phase transitions and critical phenomena. The factors stimulating the interest in these studies may be classified in three groups.

1. On the one side, singularities in intensive thermodynamic functions may appear only in the limit of an infinitely large system. On the other side, all the experimental observations pertain to finite samples. As mentioned above, finite-size effects necessarily become essential near a critical point. This fact, combined with the traditional difficulties in the description of strongly interacting many-particle systems, poses a serious challenge to the theory. The study of the universal features of finite-size effects, which arise due to large-scale collective behaviour (highly correlated classical or quantum fluctuations), is a subject of the modern theory of finite-size scaling. The latter theory reveals the intimate mechanism of how the critical singularities build up in the thermodynamic limit, as well as the role of boundary and shape effects.

2. The interpretation of experimental results on finite-size effects in real systems is not a simple task for several reasons. First of all, they are de-

tectable only in samples of a rather small size (in all, or at least one spatial dimension), and are generally mixed up with strong effects due to gravity, impurities, or other inhomogeneities. In practice, the divergence of the correlation length is limited also by the finite temperature resolution. Nevertheless, the high precision of some modern experimental techniques makes finite-size effects accessible. Correlation lengths of the order of hundreds of nanometers have been achieved. A number of successful measurements have been performed and substantial progress has been made in overcoming most of the standing problems. There are various types of experimental results, for which the theory of finite-size effects seems relevant. Such are the specific heat and superfluid density measurements near the superfluid ("lambda") transition of liquid helium in small pores and thin films [Gasparini and Rhee (1992)], [Nissen et. al. (1993)]; or the direct measurements of magnetization in nanoscale ferrimagnetic particles [Tang et. al. (1991)], [Kulkarni et. al. (1994)]. In some binary fluids (2.6-lutidine and water) demixed between narrowly spaced plates [Scheiber et. al. (1979)], shifts in the critical temperature T_c as large as 100 mK have been observed. Magnetic insulators, such as transition metals difluorides, can be epitaxially grown in very thin films. The latter show both finite-size shift in the critical point and rounding of the thermodynamic singularities. It is an experimental observation that the specific heat as a function of temperature shows one or two maxima, depending on the layer thickness [Lederman et. al. (1993a)], [Lederman et. al. (1993b)], in quantitative agreement with the theory of finite-size scaling. Highly precise measurements of magnetic phase transitions in some ultrathin Fe and Ni films exhibit thickness dependence of T_c and crossovers between different regimes which reveal important aspects of the universality hypothesis and finite-size scaling theory [Durr et. al. (1989)], [Li and Baberschke (1992)]. Some layered superconductors (including high-temperature superconductors) also exhibit a pronounced size effect in T_c which has been discussed in [Michielsen et. al. (1991)], [Schneider (1991)].

3. Along with the development of highly productive computer systems, the numerical methods of modelling took a prominent place between the laboratory experiment and theory in the study of phase transitions and critical phenomena. Since, due to technical limitations, one works with rather small systems consisting of, say, up to 10^6 particles, the finite-size effects in the numerical data are essential. It is the finite-size scaling theory

which is the most reliable tool for extrapolation to the thermodynamic limit
of data obtained by Monte Carlo and Molecular Dynamics simulations,
as well as by transfer-matrix calculations, or exact diagonalization of the
Hamiltonian for small lattice systems.

The aim of the present book is to familiarize the reader with the rich
collection of ideas, methods and results relevant to the theory of critical
phenomena in systems with confined geometry. At that the authors believe
in the instructive role of the simple models approach towards the better
understanding of complex physical systems. Thus, by following a tradition
which exists from the very beginning of the theory of phase transition, we
put the accent on the derivation of rigorous and exact results. We have
confined ourselves to the investigation of few exactly solved models which
allow one to derive analytical expressions for the physical quantities of
interest, preferably at any space dimensionality.

Here it is in place to emphasize that the thermodynamic limit leads to
great simplifications in the analytical results. In the case of finite systems,
the derivation of exact results (valid for any finite number of particles) is
a much more complicated task. Hence, the number of models subject to
rigorous finite-size analysis is extremely limited. For example, even in the
case of ideal gases, see [Ziff et. al. (1977)], one has to avoid taking the usual
limit (as $V \to \infty$)

$$\frac{1}{V} \sum_{\mathbf{k}} \to \frac{1}{(2\pi)^d} \int \mathrm{d}^d\mathbf{k},$$

where d is the space dimensionality, and the sum runs over the set of normal
modes which depends on the shape of the system and the boundary conditions. Obviously, the starting expressions one usually has at hand for finite
systems are very cumbersome. Moreover, one has to derive their asymptotic
form in special regimes, when the values of the temperature and the relevant external fields tend to the critical point together with $L \to \infty$, so that
the ratio L/ξ remains of the order of unity. Such a finite-size scaling analysis requires specific analytical techniques, the description of which takes
a due place in the present book. Some of the calculations are performed
rather circumstantially, so that the reader could master the mathematical
tools, the details of which are usually lacking in the specialized papers on
the subject.

Of course, the selection of the material in a monograph depends on the
taste and professional interests of its authors. The present case is not an

exception. Attention is paid to some specific problems and mathematical tools characteristic of the case of long-range interactions with power-law decay. Another example is the probabilistic point of view on finite-size scaling which makes close connection with the concept of limit Gibbs states. Of special interest to the reader could be the investigation of systems in which the quantum fluctuations play an essential role together with the thermal ones. The extension of the finite-size scaling theory to such systems is based on the observation that the Gibbs weight $\exp(-\beta\mathcal{H})$, where $\beta = (kT)^{-1}$ is the inverse temperature, and $\mathcal{H}$ is the Hamiltonian, is formally identical with the time evolution operator $\exp(it\mathcal{H}/h)$ upon replacement of time t by the imaginary quantity $ih\beta$. Hence, the partition function of a quantum system looks like a classical partition function with an additional dimension, except that this extra dimension is of finite extent βh in units of time. When the temperature goes to zero and the quantum effects become important, the size in this "imaginary time direction" tends to infinity. This observation provides an useful interpretation of quantum effects as finite-size effects.

Confined critical systems, due to the presence of strong fluctuations, exhibit appearance of long-range forces between the walls - a phenomenon which is the direct analogue of the well-known Casimir effect in electromagnetism. There are different model approaches for obtaining the universal scaling functions that govern the Casimir forces. In this book a detailed, pedagogical account is given of the exact results obtained in the spherical approximation for both classical and quantum systems. Finally, we present a survey of the theoretical results known for other models, make comments and give reference to experimental results.

Unfortunately, many important topics are omitted due to the lack of space and expertize on the part of the authors.

The exposition of the main issues is given in a self-contained form which presumes the reader's knowledge in the framework of standard courses on the theory of phase transitions and critical phenomena. The authors experience in the preparation of one-semester courses for students at the Catholic University of Leuven (JGB) and the University of Wuppertal (DMD) is taken into account.

The authors are greatly indebted to Professor Yu Lu for initiating the writing of this book and for his permenent encouragement. We are very greatful to Professors S. Dietrich, J. Rudnick, V. A. Zagrebnov, the Doctors H. Chamati, E. Korutcheva and M. Krech for the exchange of information, critical reading and commenting on parts of the text. Two of the authors (D.M.D. and N.S.T.) would like to thank the International Center for Theoretical Physics, Trieste, for the hospitality during their visits in 1998 and 1999, when a part of this book was written.

The partial support of the Bulgarian National Fund for Scientific Research, Projects F-608 and MM-603, is greatfully acknowledged.

Contents

Chapter 1

Overview of Critical Phenomena in Bulk Systems

1.1 Preliminaries

Condensed matter physics studies systems with a huge number of degrees of freedom. For a sensible description of such systems one needs the principles and methods of statistical mechanics. The main aim of statistical mechanics is to predict the observable characteristics of macroscopic systems on the basis of our knowledge about their structure, constituent elements and the microscopic interactions among them. Thus, statistical mechanics can be thought of as a bridge between the macroscopic and the microscopic properties of systems. In the present book we are interested only in systems at thermal equilibrium, the statistical properties of which are described by one of the Gibbs ensembles. The basic quantities used in equilibrium statistical mechanics can be classified into several groups. The first group consists of quantities which are average values of functions of the microscopic state of a many-body mechanical system. For example, such is the internal energy U of a system. Other quantities, like the entropy S, have a pure statistical origin and cannot be represented as linear functionals of the probability distribution of the microscopic states. A third group of quantities comprises the parameters of the Gibbs distribution, such as the temperature T in the canonical and grand canonical ensembles, or the chemical potential μ in the latter ensemble. Their physical meaning is related to the interaction of the system with the environment. Finally, real systems are considered to exist in a finite region of space, the most important measure of which is the volume V. All these quantities characterize the macroscopic state of a given system and constitute the class of thermodynamic parameters.

1

In thermodynamics the choice of the set of independent external parameters is a matter of convenience. For example, from the Gibbs equation for a simple fluid

$$dU = T\,dS - p\,dV, \tag{1.1}$$

where p is the pressure, it follows that the internal energy is a thermodynamic potential of the system when it is expressed as a function of the entropy and the volume, $U = U(S, V)$. Since the entropy is not a directly measurable quantity, in some applications one may wish to consider as external parameters the temperature T and the volume V. By the *Legendre transformation* of U,

$$d(U - TS) = -S\,dT - p\,dV, \tag{1.2}$$

one obtains the thermodynamic potential known as *Helmholtz free energy*

$$A(V, T) = U - TS. \tag{1.3}$$

By another Legendre transformation of U,

$$d(U - TS + pV) = -S\,dT + V\,dp, \tag{1.4}$$

one obtains the *Gibbs free energy*

$$F(p, T) = U - TS + pV. \tag{1.5}$$

In this book we consider mainly magnetic systems. In such a case we ignore possible changes in the volume of the magnetic system and adopt the fruitful analogy between a simple fluid and a simple magnet, based on the following correspondence rules for the variables (see, e.g., [Stanley (1971)])

$$p \leftrightarrow H, \qquad V \leftrightarrow -M, \tag{1.6}$$

where H is the external magnetic field and M is the magnetization. Therefore, we call $A(M, T)$ the Helmholtz free energy, and $F(H, T)$ the Gibbs free energy. The corresponding full differentials are given by

$$\begin{aligned} dA &= -S\,dT + H\,dM, \\ dF &= -S\,dT - M\,dH, \end{aligned} \tag{1.7}$$

Note that we have omitted the dependence of the free energies $A(M, T)$ and $F(H, T)$ on the fixed (by assumption) volume V of the system.

How can one calculate the thermodynamic potentials of a given system, provided its microstructure, as well as the physical laws governing the behaviour of its elements are known? The answer to this question was given in 1902 by J.W. Gibbs [Gibbs (reprinted, 1960)]. In the next section we briefly sketch the Principles of Statistical Mechanics due to Gibbs, by using nowadays terminology.

1.2 Principles of Statistical Mechanics

Equilibrium statistical mechanics is based on statistical postulates which relate the microscopic properties of a system of many interacting particles to its thermodynamic functions, as well as to more subtle observable characteristics as correlation functions, boundary and finite-size effects. To state these postulates we need first a description of the set of all possible microscopic states of a given system. This set is defined in a different way for classical and quantum systems; it may have continuous and/or discrete components, depending on the theoretical model used.

Classical continuous systems

Let us start with a classical mechanical system of N identical and pointwise particles, labeled by $i = 1, \ldots, N$, which move in a finite region Λ of the d-dimensional Euclidean space R^d. Assume for simplicity that the particles have only translational degrees of freedom. The mechanical state of such a system is described by a point ω in the continuous *phase space* $\Omega_{\Lambda,N} = \Lambda^N \times R^{dN}$ of all the allowed coordinates $x_i \in \Lambda$ and momenta $p_i \in R^d$ of the particles. If the particles have hard cores, one has to exclude from the *configuration space* Λ^N all the configurations in which the hard cores of at least two particles overlap. Let the given classical system be described by the Hamiltonian

$$\mathcal{H}^{cl.}_{\Lambda,N}(\omega) = \sum_{i=1}^{N} \frac{p_i^2}{2m} + U(x_1, \ldots, x_N), \tag{1.8}$$

defined for all $\omega \in \Omega_{\Lambda,N}$. The first term in the right-hand side describes the kinetic energy, and the function $U(x_1, \ldots, x_N)$ is the total potential energy of the particles. According to the postulate for the so-called *canonical Gibbs ensemble*, the microscopic state ω of the system in thermal equilibrium with a heat bath at temperature T, is a random variable on the phase space $\Omega_{\Lambda,N}$ with probability density $P_{\Lambda,N,T}(\omega)$ given by

$$P^{cl.}_{\Lambda,N,T}(\omega) = \frac{\exp[-\beta \mathcal{H}^{cl.}_{\Lambda,N}(\omega)]}{N!\, Z^{cl.}(\Lambda,N,T)}. \qquad (1.9)$$

Here $\beta = 1/(k_B T)$, where k_B is the *Boltzmann constant*, $k_B \simeq 1.38 \times 10^{-23}\,JK^{-1}$. The probability density is defined with respect to the infinitesimal volume of the phase space

$$d\omega = \prod_{i=1}^{N}(\,dp_i\,dx_i), \qquad (1.10)$$

which is invariant under the time evolution of the system. The constant factor $1/N!$ is included to identify configurations which differ only by a permutation of the particle labels. This factor ensures additivity of the entropy, otherwise it is inessential and can be omitted. Note that the quantity $dp_i\,dx_i$ has the dimension of action; one can pass to a dimensionless volume of phase space by the change $dp_i\,dx_i \to dp_i\,dx_i/h^3$, where h is the Plank constant, $h \simeq 6.63 \times 10^{-34}\,J\,sec.$

The normalization constant $Z^{cl.}(\Lambda,N,T)$ of the probability density is called the *canonical partition function* of the system. Explicitly it reads

$$Z^{cl.}(\Lambda,N,T) = \frac{1}{N!}\int_{\Omega_{\Lambda,N}} d\omega\,\exp[-\beta \mathcal{H}^{cl.}_{\Lambda,N}(\omega)]. \qquad (1.11)$$

For the sake of simplicity, in Eqs. (1.9) and (1.11) we have omitted the explicit dependence on other fixed external parameters. It is important to note that for classical systems the contribution of the kinetic energy term into the canonical partition function singles out in the form of the universal factor

$$\prod_{i=1}^{N}\int_{R^d} dp_i\,\exp[-\beta p_i^2/(2m)] = (2\pi m/\beta)^{dN/2}. \qquad (1.12)$$

The phase space of a classical system with variable number of identical particles, say, due to a contact with reservoir of particles, is defined as the direct sum $\Omega_\Lambda = \oplus_{n=0}^{\infty}\Omega_{\Lambda,n}$, where $\Omega_{\Lambda,0}$ is a one-point space of measure 1. The so-called *grand canonical Gibbs ensemble* describes systems in equilibrium with both a heat bath at temperature T, and a reservoir of particles with chemical potential μ. The probability density $P^{cl.}_{\Lambda,\mu,T}(\omega)$ on Ω_Λ is given by

$$P^{cl.}_{\Lambda,\mu,T}(\omega) = \frac{\exp[-\beta\mathcal{H}_\Lambda(\omega) + \beta\mu n(\omega)]}{n(\omega)! \, \Xi^{cl.}(\Lambda,\mu,T)}. \tag{1.13}$$

Here $\mathcal{H}_\Lambda(\omega)$ equals the N-particle Hamiltonian (1.8) and $n(\omega) = N$ whenever $\omega \in \Omega_{\Lambda,N}$, with the convention that $\mathcal{H}_\Lambda(\omega) = 0$ for $\omega \in \Omega_{\Lambda,0}$. The partition function $\Xi^{cl.}(\Lambda,\mu,T)$ of the grand canonical ensemble has the explicit form

$$\Xi^{cl.}(\Lambda,\mu,T) = 1 + \sum_{n=1}^{\infty} \exp(\beta\mu n) Z^{cl.}(\Lambda,n,T). \tag{1.14}$$

Quantum continuous systems

In the case of a quantum system with fixed number N of identical particles, the Hamiltonian in coordinate representation is given by the operator

$$\mathcal{H}^{q.}_{\Lambda,N} = \sum_{i=1}^{N} \left(-\frac{1}{2m}\Delta_i\right) + U(x_1,\ldots,x_N), \tag{1.15}$$

where Δ_i is the Laplace operator acting on the coordinates x_i of the i-th particle. By taking into account the boundary condition imposed on the boundary $\partial\Lambda$ of the region Λ, the Hamiltonian is defined as a self-adjoint operator acting on the Hilbert space $L^2(\Lambda^N)$ of square-integrable functions of the coordinates $x_i \in \Lambda$, $i = 1,\ldots,N$, of all the particles. In the considered case of a finite region Λ, the spectrum $\{E_\alpha(\Lambda,N),\ \alpha = 1,2,\ldots\}$ of the Hamiltonian consists of discrete eigenvalues with a finite degeneracy. The corresponding eigenfunctions obey the quantum statistics of indistinguishable particles with respect to permutations of the arguments x_i: they are symmetric functions for *bosons* (Bose-Einstein statistics) and anti-symmetric for *fermions* (Fermi-Dirac statistics). The correspondingly symmetrized Hilbert space will be denoted by $L^2_\epsilon(\Lambda^N)$, where $\epsilon = +$ for Bose-Einstein statistics and $\epsilon = -$ for Fermi-Dirac statistics.

Let α label the (properly symmetrized) eigenfunctions of the Hamiltonian (1.15), so that the eigenvalues are repeated in the list $\{E_\alpha(\Lambda,N),\ \alpha = 1,2,\ldots\}$ as many times as their degeneracy is. In the quantum canonical ensemble the probability of finding the system in eigenstate α equals

$$P^{q\cdot}_{\Lambda,N,T}(\alpha) = \frac{\exp[-\beta E_\alpha(\Lambda,N)]}{Z^{q\cdot}(\Lambda,N,T)}. \qquad (1.16)$$

The quantum canonical partition function is

$$Z^{q\cdot}(\Lambda,N,T) = \operatorname{Tr}\exp[-\beta\mathcal{H}^{q\cdot}_{\Lambda,N}] = \sum_\alpha \exp[-\beta E_\alpha(\Lambda,N)]. \qquad (1.17)$$

Here Tr denotes the *trace* of an operator, i.e., the sum of its diagonal elements in any orthonormal basis of the Hilbert space $L^2_\epsilon(\Lambda^N)$. The trace value is independent of the choice of the basis. As in the case of classical systems, the possible dependence on other fixed external parameters is omitted. Note an important distinction from the classical case: for quantum systems the contribution of the kinetic energy term in the canonical partition function does not separate out from the one of the configurational degrees of freedom.

The Hamiltonian $\mathcal{H}^{q\cdot}_\Lambda$ of a quantum system with variable number of particles is defined on the *Fock space* $\oplus_{n=0}^\infty L^2_\epsilon(\Lambda^n)$, which is a direct sum of n-particle symmetrized Hilbert spaces. On each n-particle subspace, $n \geq 1$, the Hamiltonian is defined by Eq. (1.15), and, by convention, it is set to zero for $n = 0$. In the quantum grand canonical ensemble the probability of finding the system to contain n particles and be in the eigenstate α is

$$P^{q\cdot}_{\Lambda,\mu,T}(\alpha,n) = \frac{\exp[-\beta E_\alpha(\Lambda,n) + \beta\mu n]}{\Xi^{q\cdot}(\Lambda,\mu,T)}. \qquad (1.18)$$

Correspondingly, the quantum grand canonical partition function equals

$$\begin{aligned}
\Xi^{q\cdot}(\Lambda,\mu,T) &= \operatorname{Tr}\exp[-\beta\mathcal{H}^{q\cdot}_\Lambda + \beta\mu\mathcal{N}] \\
&= 1 + \sum_{n=1}^\infty \exp(\beta\mu n)Z^q(\Lambda,n,T). \qquad (1.19)
\end{aligned}$$

In the right-hand side of the first equality the trace is taken over the above mentioned Fock space; in each subspace $L^2_\epsilon(\Lambda^n)$ the particle-number operator $\mathcal{N}$ equals n times the unit operator.

For details on the mathematical foundations of statistical mechanics of quantum continuous systems we refer the reader to [Petrina (1995)].

Classical lattice systems

So far we have considered classical and quantum systems with continuous configuration space. There are two other important classes of classical and quantum model systems called *lattice systems*. Lattice systems are idealized models of crystals in which the translational degrees of freedom of the particles are completely ignored. In the traditional approach followed here, the model is defined initially on a finite subset Λ of an infinite regular lattice. For the sake of simplicity, we consider cubic lattices of dimensionality d, the sites of which have integer coordinates; such an infinite lattice is denoted by Z^d. With each lattice site $i \in Z^d$ a classical or quantum dynamical variable is associated, say S_i, which may be a scalar or a vector with finite number of components. For simplicity, S_i are considered as identical (for all i) classical variables or quantum operators.

In *classical lattice systems*, each single-site variable S_i may take values in a finite or continuous set X_i. A *configuration* ω of the system in a finite region Λ one calls any set of the form $\omega = \{S_i \in X_i : i \in \Lambda\}$. Since the particles are identical, all X_i are identical copies of one set X. Therefore, the space of all configurations of the finite system is $\Omega_\Lambda = X^N$, where $N = |\Lambda|$ is the number of sites in Λ. For uniformity of terminology with the continuous case, we call classical Hamiltonian of the lattice system any reasonable potential energy function

$$\mathcal{H}_\Lambda^{cl.latt.}(\omega) = U(\omega), \tag{1.20}$$

defined for all configurations $\omega \in X^N$. Note that the above definition of the Hamiltonian (energy) function neglects completely the interaction of the degrees of freedom in the region Λ with those in the environment $Z^d \setminus \Lambda$; this case corresponds to *open* or *zero-Dirichlet* boundary conditions. Other types of boundary conditions will be considered in Chapter 7 on the example of concrete models with specified interaction potentials. In accordance with Eq. (1.9), one can define the probability density in the canonical Gibbs ensemble as

$$P_{\Lambda,T}^{cl.latt.}(\omega) = \frac{\exp[-\beta \mathcal{H}_\Lambda^{cl.latt.}(\omega)]}{Z^{cl.latt.}(\Lambda, T)}. \tag{1.21}$$

This expression differs from the corresponding one for continuous systems in two ways. First, the factor $N!$ in the denominator is absent, due to the fact that the lattice sites are distinguished by their coordinates. Second, the density of particles in the system (the ratio $\rho = N/V$) is not any more an independent variable, since $N = |\Lambda| \equiv V$ plays the role of the volume (absolutely rigid lattice). Thus N becomes a redundant variable which is omitted from the list of arguments. Having in mind these changes, the canonical partition function in (1.21) equals

$$Z^{cl.latt.}(\Lambda, T) = \int_{\Omega_\Lambda} d\omega \exp[-\beta \mathcal{H}_\Lambda^{cl.latt.}(\omega)]. \tag{1.22}$$

Note that in the case when the single-site configuration spaces $X_i = X$ are finite, the confiquration space of the whole system is finite too, and Eq. (1.21) defines simply the probability of finding a given configuration ω. Then the integral in the definition (1.22) of the partition function has to be understood as a sum over the configuration space Ω_Λ. In such a case, one of the single-particle states in X can be interpreted as the absence of a particle (i.e., an empty site), while the remaining states in X can be thought of as particles with a finite number of different properties. Then one can extend the above definitions in an obvious way, to define a grand canonical ensemble for classical lattice systems.

Quantum lattice systems

In quantum lattice systems a quantum operator $\mathcal{S}_i$, acting on a finite-dimensional Hilbert space X_i, is associated with each site i of the lattice Z^d. For a finite subset Λ of the lattice, the Hamiltonian $\mathcal{H}_\Lambda^{q.latt.}$ is a self-adjoint operator on the finite-dimensional Hilbert space $\otimes_{i \in \Lambda} X_i$. In the quantum canonical Gibbs ensemble the probability of finding the system in an eigenstate α of the quantum Hamiltonian, having the eigenvalue $E_\Lambda^{q.latt.}(\alpha)$, equals

$$P_{\Lambda,T}^{q.latt.}(\alpha) = \frac{\exp[-\beta E_\Lambda^{q.latt.}(\alpha)]}{Z^{q.latt.}(\Lambda, T)}. \tag{1.23}$$

Here the canonical partition function is given by

$$Z^{q.latt.}(\Lambda, T) = \mathrm{Tr}\,\exp(-\beta \mathcal{H}_\Lambda^{q.latt.}) = \sum_\alpha \frac{\exp[-\beta E_\Lambda^{q.latt.}(\alpha)]}{Z^{q.latt.}(\Lambda, T)}. \tag{1.24}$$

The trace is taken over the finite-dimensional Hilbert space $\otimes_{i \in \Lambda} X_i$.

One can consider also the case of quantum lattice systems with variable number of particles (grand canonical ensemble). Since the corresponding definitions can easily be obtained from the ones for quantum continuous systems, their explicit formulation is omitted. We note only that if the system is a quantum lattice gas, its "kinetic" energy can be modelled by a finite-difference operator.

According to the principles of statistical mechanics, the partition function for each of the Gibbs ensembles (canonical, grand canonical, and others not mentioned here) is related to a particular thermodynamic potential. For example, the statistical definition of the Helmholtz free energy for a continuous system (either classical or quantum) is given in terms of the canonical partition function by the equation

$$A(\Lambda, N, T) = -\beta^{-1} \ln Z(\Lambda, N, T). \tag{1.25}$$

In the case of a lattice system of n-component vector-spins (classical variables or quantum operators) $\mathbf{S}_i$, $i \in Z^d$, placed in an inhomogeneous external magnetic field which is represented the set of vectors $\{\mathbf{H}_i, i \in \Lambda\}$, the total Hamiltonian has the form

$$\mathcal{H}_\Lambda^{latt.}(\{\mathbf{H}_i, i \in \Lambda\}) = \mathcal{H}_{\Lambda, int}^{latt.} - \sum_{i \in \Lambda} \mathbf{H}_i \cdot \mathbf{S}_i. \tag{1.26}$$

Here the first term in the right-hand side describes the spin-spin interaction energy, and the second term represents the additional energy due to the interaction of the spins with the external field. Note that the corresponding Gibbs free energy

$$F^{latt.}(\Lambda, \{\mathbf{H}_i, i \in \Lambda\}, T) = -\beta^{-1} \ln Z^{latt.}(\Lambda, \{\mathbf{H}_i, i \in \Lambda\}, T), \tag{1.27}$$

is not a purely thermodynamic quantity, since it depends on a macroscopic number of external parameters (the local fields $\mathbf{H}_i$); the number of these parameters becomes infinite in the *thermodynamic limit*, see Section 1.3. That is why, one usually considers the case when the local fields depend on just a few parameters. The simplest example is provided by the uniform magnetic field, when $\mathbf{H}_i = \mathbf{H}$ for all $i \in \Lambda$.

The grand canonical partition function for a continuous system (classical or quantum) defines another thermodynamic potential, which is simply the volume $V = |\Lambda|$ of the system times the pressure p applied to it by the environment:

$$Vp(\Lambda, \mu, T) = \beta^{-1} \ln \Xi(\Lambda, \mu, T) \tag{1.28}$$

(note the absence of the minus sign in the right-hand side).

It is in place here to remark that, apart from the definition of the configuration space and the exchange of arguments $\mu \leftrightarrow N$, the grand canonical partition function $\Xi(\Lambda, \mu, T)$ differs formally from the canonical one $Z(\Lambda, N, T)$ only in the replacement of the Hamiltonian $\mathcal{H}$ by $\mathcal{H} - \mu N$. Moreover, the statistical definitions of the corresponding thermodynamic potentials are analogous, compare (1.28) with (1.25). Due to the above analogy, in some of the works specialized on applications of statistical mechanics to problems of solid state physics, the thermodynamic potential $-Vp$ is also called "free energy" of the system. When no ambiguity arises, we will also make use of this loose terminology in the remainder of the book.

In order to avoid unnecessary repetition of general expressions, it is convenient to unify notations for classical and quantum systems. To this end we will use the more concise definitions appropriate for quantum systems, with the convention that for classical systems one has to make the replacement

$$\mathrm{Tr}\mathcal{A} = \sum_s \mathcal{A}(s) \quad \to \quad \int_{\Omega_\Lambda} d\omega \mathcal{A}(\omega). \qquad (1.29)$$

Here it is understood that: (1) In the canonical Gibbs ensemble $\mathcal{A}$ is an operator acting on the N-particle symmetrized Hilbert space, the sum over s is taken over an orthonormal basis in that space, and $\mathcal{A}(s)$ is the diagonal element of the operator in the chosen basis. When the basis is formed by the eigenfunctions of the Hamiltonian, the label s coincides with α used in Eqs. (1.16) and (1.17), or (1.23) and (1.24). The integral in the right-hand side of (1.29) is taken over the space of microstates of the N-particle classical system, and $d\omega$ is the infinitesimal volume of that space; in the case of continuous systems the factor $1/N!$ is absorbed in the volume element. The quantity $\mathcal{A}(\omega)$ represents the classical function of state corresponding to the quantum operator $\mathcal{A}$. (2) In the grand canonical Gibbs ensemble $\mathcal{A}$ is an operator acting on the symmetrized Fock space and the trace is taken over an orthonormal basis in that space. Therefore, the label s is actually a double one, $s = (n, \alpha)$, so that the outer sum runs over the n-particle subspaces of the Fock space, and the inner sum over α is taken over an orthonormal basis in the n-particle Hilbert subspace. The diagonal elements of the operator following the trace symbol are denoted again by $\mathcal{A}(s)$. When the basis in the n-particle subspace is formed by the eigenfunctions of the n-particle Hamiltonian, the above notation is consistent with Eqs. (1.18) and

(1.19). The integral in the right-hand side of (1.29) symbolizes summation over all the n-particle phase spaces and integration over each n-particle phase space of the classical system; $d\omega$ is the infinitesimal volume of the spaces. The quantity $\mathcal{A}(\omega)$ represents the classical function of state which corresponds to the quantum operator $\mathcal{A}$.

Statistical definition of thermodynamic functions

Now we pass to the consideration of average values and fluctuations in the Gibbs canonical and grand canonical ensembles. By using the appropriate probability $P(s)$, see Eqs. (1.21), (1.23), the average (expectation) value of any function $\mathcal{A}(s)$ of the microstate s of the system can be calculated as

$$A \equiv \langle \mathcal{A} \rangle = \sum_s P(s)\mathcal{A}(s). \tag{1.30}$$

In particular, the internal energy U in the canonical ensemble is defined as the average value of the Hamiltonian $\mathcal{H}$. It can be calculated by using the equation

$$U \equiv \langle \mathcal{H} \rangle = Z^{-1} \sum_s E(s) \exp[-\beta E(s)]. \tag{1.31}$$

From the definition of the Gibbs free energy, see Eq. (1.27), it follows that the internal energy of a lattice magnetic system can be written also as

$$U = \frac{\partial(\beta F)}{\partial \beta} = F - T\frac{\partial F}{\partial T}. \tag{1.32}$$

More information about the statistical properties of random variables, such as the energy in the canonical ensemble, is given by their moments of higher degree. The initial moment of first degree is simply the average value. Among the moments of higher degree most important is the *variance* (central moment of degree 2)

$$\langle \mathcal{H}^2 \rangle - \langle \mathcal{H} \rangle^2 = -\frac{\partial U}{\partial \beta} = k_B T^2 \frac{\partial U}{\partial T}. \tag{1.33}$$

Obviously, the variance provides a measure for the fluctuations of the given quantity, in our case the energy. By using the thermodynamic definition of the *heat capacity at constant volume* C_V,

$$C_V = \frac{\partial U}{\partial T}, \tag{1.34}$$

one obtains the important relationships

$$C_V = (k_B T^2)^{-1} \left(\langle \mathcal{H}^2 \rangle - \langle \mathcal{H} \rangle^2 \right) = -\frac{\partial^2}{\partial \beta^2}(\beta F). \qquad (1.35)$$

Hence, the thermodynamic condition for thermal stability of the system, namely $C_V > 0$, implies that βF is a *strictly concave* (in other words, *convex upwards*) function of the parameter β. Note that sometimes, in a loose terminology, the dimensionless function βF is also called a "free energy" or "reduced free energy".

In the grand canonical ensemble the number of particles in a region Λ of volume $V = |\Lambda|$ is a random variable $\mathcal{N}$ with average value

$$N \equiv \langle \mathcal{N} \rangle = V \frac{\partial p}{\partial \mu}. \qquad (1.36)$$

In thermodynamics.the pressure is an intensive parameter independent of the volume, i.e. $p = p(\mu, T)$, and the derivative $\partial p / \partial \mu$ equals the average density of particles $\rho(\mu, T) \equiv N(V, \mu, T)/V$. The Gibbs differential equation for the grand canonical potential reads

$$\mathrm{d}(pV) = S\,\mathrm{d}T + p\,\mathrm{d}V + N\,\mathrm{d}\mu. \qquad (1.37)$$

Consider now the variance of the number of particles in the system. Taking into account Eq. (1.36) and the definition of the probability distribution in the grand canonical ensemble, one can write

$$\langle \mathcal{N}^2 \rangle - \langle \mathcal{N} \rangle^2 = V k_B T \frac{\partial^2 p}{\partial \mu^2}. \qquad (1.38)$$

From the positivity of the left-hand side it follow that the pressure, hence the grand canonical potential Vp, is a *strictly convex* function of the chemical potential μ. Furthermore, in thermodynamics one can change the independent variables and regard both μ and p as functions of ρ and T (see, e.g. [Huang (1987)]). Thus one can show that at constant temperature

$$\left(\frac{\partial^2 p}{\partial \mu^2} \right)_T = \rho^2 \kappa_T, \qquad (1.39)$$

where κ_T is the *isothermal compressibility,,* defined thermodynamically as

$$\kappa_T = -\frac{1}{V} \left(\frac{\partial V}{\partial p} \right)_T. \qquad (1.40)$$

Therefore, the thermodynamic condition for mechanical stability of the system, namely $\kappa_T > 0$, is closely related to the convexity of the pressure as a function of μ.

In a similar way one can show also that the positivity of the variance of the random variable $\mathcal{H} - \mu\mathcal{N}$ leads to the the convexity of βp (and $V\beta p$) as a function of β.

According to thermodynamics, the observed properties of macroscopic systems have to be either extensive (proportional to the volume), or intensive (independent of the volume). In either case they should not depend on the shape of the spatial region Λ occupied by the system, provided "too long-range" interactions are excluded. However, the above mentioned features are far from being obvious from the general statistical-mechanical definitions. Fortunately, they have been shown to hold for certain classes of model systems in the so-called *thermodynamic limit*.

1.3   Thermodynamic limit

The concept of the thermodynamic limit plays a crucial role in statistical mechanics. From a physical point of view, it is related to the fact that one considers macroscopic systems consisting of a huge number of particles, say, of the order of 10^{19} - 10^{23}. Experience has shown that the gross features of such systems, taken per unit volume or per particle, are practically independent of the system size.

To be more specific, let us consider the case of the canonical Gibbs ensemble. Rigorous studies of model systems of interacting particles have established that the extensivity property of the Helmholtz free energy (1.25) holds only asymptotically, when the volume $|\Lambda|$ of the system and the number of particles in it tend to infinity simultaneously, so that the particle density $\rho = N/|\Lambda|$ has a finite limit. Naturally, all the other external parameters, such as temperature T, external fields, etc, are kept fixed. Then, under certain general regularity conditions (to be mentioned below) imposed on the shape of Λ and on the interaction Hamiltonian, the asymptotic form of the Helmholtz free energy is

$$A(\Lambda, N, T) = |\Lambda|\, a(\rho, T) + o(|\Lambda|). \tag{1.41}$$

Here $o(|\Lambda|)$ denotes terms such that $o(|\Lambda|)/|\Lambda| \to 0$ as $|\Lambda| \to \infty$. On general physical grounds one expects that these terms incorporate the effects due to the interaction of the system with its surrounding (the so called *boundary effects*). Evidently, the $o(|\Lambda|)$ terms may depend on the shape of Λ and on the specific boundary conditions imposed at its boundary $\partial\Lambda$. From a mathematical point of view, Eq. (1.41) implies that the leading asymptotic form of the free energy, $A^{as\cdot}(|\Lambda|, N, T) \equiv |\Lambda|\,a(\rho, T)$, is a *homogeneous function* of the natural extensive variables $|\Lambda|$ and N of degree -1. This means that for any $\lambda > 0$

$$A^{as\cdot}(\lambda|\Lambda|, \lambda N, T) = \lambda^{-1} A^{as\cdot}(|\Lambda|, N, T). \tag{1.42}$$

By choosing $\lambda = |\Lambda|^{-1}$, one obtains that $a(\rho, T) = A^{as\cdot}(1, \rho, T)$.

Alternatively, one has to prove the existence of the Helmholtz free energy density $a(\rho, T)$ defined by the limit

$$a(\rho, T) = \lim_{\substack{|\Lambda| \to \infty, N \to \infty \\ N/|\Lambda| = \rho}} |\Lambda|^{-1} A(\Lambda, N, T). \tag{1.43}$$

For a shorthand notation of the above limit we will use the symbol t-lim. If the thermodynamic limit exists, it demonstrates one of the radical simplifications that occur: the *shape-independence* of the free energy density.

Obviously, for a correct definition of the limit in Eq. (1.43) one has to impose some reasonable restrictions on the admissible shapes of the region Λ as $|\Lambda| \to \infty$. Basically, one has to require that Λ expands rather uniformly in all space dimensions, and that its surface remains fairly smooth. For our purposes it suffices to assume that in the case of a continuous system Λ is a parallelepiped $\times_{\nu=1}^{d}[0, L_\nu]$ with edges of linear size L_ν. For lattice systems Λ is a subset of the d-dimensional lattice Z^d of the form $\times_{\nu=1}^{d}\{1, \ldots, L_\nu\}$, where L_ν are integers. Then, by "Λ tends to infinity" in d-dimensions we will mean that $L_\nu \to \infty$, $\nu = 1, \ldots, d$, at fixed nonzero shape ratios $l_\nu = L_\nu/L_d$. The resulting geometry will be denoted by the symbol ∞^d. We shall consider also the case when "Λ tends to infinity in d'-dimensions", where $d' < d$. By that we mean fixed finite values of $L_1, \ldots, L_{d-d'}$, while $L_\nu \to \infty$ for $\nu = d - d' + 1, \ldots, d$ at fixed nonzero shape ratios. This case will be called $L^{d-d'} \times \infty^{d'}$ geometry. For more general conditions on the shape of the region we refer the reader to [Van Hove (1949)].

The analysis of the integrand in the classical canonical partition function (1.11) shows that a finite thermodynamic limit for the free energy density exists only if the attractive part of the total potential energy grows not faster than N as $N \to \infty$, i.e.,

$$U(x_1, \ldots, x_N) \geq -BN, \qquad (1.44)$$

for all N and some constant $B > 0$. This is the so called *stability condition*. It imposes restrictions on the rate of asymptotic decrease of the attractive potential with the increase of the inter-particle distance. To avoid configurations with infinite potential energy, the repulsive part of the potential has to obey some restrictions too: on the one hand, it should not allow the particles to collapse on each other, on the other hand it should also decay sufficiently fast at large inter-particle separations.

Some of the above conditions are trivially fulfilled for lattice spin systems. Consider for example a quantum spin model with a pairwise spin-spin interaction. Let the Hamiltonian of the system, placed in uniform magnetic field $\mathbf{H}$, be

$$\mathcal{H}_\Lambda(\mathbf{H}) = -\frac{1}{2} \sum_{i,j \in \Lambda} \sum_{\alpha=1}^{3} J_{ij}^\alpha S_i^\alpha S_j^\alpha - \sum_{i \in \Lambda} \mathbf{H} \cdot \mathbf{S}_i, \qquad (1.45)$$

where S_i^α is the α-component of the quantum spin operator $\mathbf{S}_i$ at site i, $J_{ij}^\alpha = J^\alpha(|\mathbf{r}_i - \mathbf{r}_j|)$ is the spin-spin coupling between the α-components of the spins at position vectors $\mathbf{r}_i$ and $\mathbf{r}_j$. Then, the existence of the thermodynamic limit for the Gibbs free energy density

$$f(\mathbf{H}, T) = \lim_{|\Lambda| \to \infty} |\Lambda|^{-1} F(\Lambda, \mathbf{H}, T) \qquad (1.46)$$

has been proven under the condition (see [Ruelle (1969)])

$$\lim_{|\Lambda| \to \infty} \sum_{i \in \Lambda} |J^\alpha(|\mathbf{r}_i|)| < \infty, \qquad \alpha = 1, 2, 3. \qquad (1.47)$$

The extensivity problem for the thermodynamic potential $|\Lambda| p(\Lambda, \mu, T)$ in the grand canonical ensemble seems simpler due to the presence of just one extensive parameter, the volume $V = |\Lambda|$. The thermodynamic limit $V \to \infty$, to be denoted also by t-lim, is taken at fixed chemical potential μ, temperature T, and, eventually, other external parameters. However, in the case of non-positive chemical potentials, additional care should be taken to exclude collapse of the system into states with infinite density. This can be avoided by inclusion of hard-core repulsion or sufficiently strong repulsive potential at small inter-particle distances.

1.4   Equivalence of Gibbs ensembles

One of the conceptually important consequences of the thermodynamic limit (if it exists) is the *equivalence of the Gibbs ensembles*. For example, from Eqs. (1.38) and (1.39) it follows that the relative fluctuations of the number of particles in the grand canonical ensemble,

$$\frac{(\langle \mathcal{N}^2 \rangle - \langle \mathcal{N} \rangle^2)^{1/2}}{\langle \mathcal{N} \rangle} = \left(\frac{k_B T \kappa_T}{V} \right)^{1/2}, \tag{1.48}$$

vanishes in the thermodynamic limit as $O(V^{-1/2})$, provided κ_T is finite. This condition is violated only at the critical point of a liquid-gas phase transition, where the infinite (in the thermodynamic limit) value of the isothermal compressibility (1.40) indicates the occurrence of *anomalously large fluctuations*. Such anomalous fluctuations of the particle density, known as *critical opalescence*, are indeed observed in light-scattering experiments. The different Gibbs ensembles are equivalent in the thermodynamic limit in the sense that the densities of the corresponding thermodynamic potentials are related by a Legendre transformation. The equivalence of the canonical and grand canonical ensembles for a simple fluid means that the pressure $p(\mu, T)$ and the density of the Helmholtz free energy $a(\rho, T)$ are related by the transformations, (see, e.g., [Glimm and Jaffe (1981)])

$$p(\mu, T) = \sup_{\rho} \left[\mu\rho - a(\rho, T) \right]$$
$$a(\rho, T) = \sup_{\mu} \left[\mu\rho - p(\mu, T) \right]. \tag{1.49}$$

In the case of magnetic systems, a Legendre transformation relates the Helmholtz free energy density $a(m, T)$ and the Gibbs free energy density $f(H, T)$ as follows

$$f(H, T) = \inf_{m} \left[-Hm + a(m, T) \right]$$
$$a(m, T) = \sup_{H} \left[Hm + f(H, T) \right]. \tag{1.50}$$

The thermodynamic equivalence allows one to use the ensemble (usually the grand canonical one) in which the thermodynamic potential is most easily evaluated. After that, one can pass to the desired variables by using the appropriate Legendre transformation.

1.5 Phase transitions

Generally speaking, the thermodynamic potentials for "well-behaved" finite systems are real-analytic functions of the external parameters. This can be explicitly demonstrated by the example of lattice models with finite single-site space X and finite-range interparticle interactions. However, the densities of the thermodynamic potentials may exhibit singularities in the thermodynamic limit. It is believed that these densities are continuous functions which depend piecewise analytically on the thermodynamic parameters. The loci of singular points determine the *phase diagram* of the system in the space of thermodynamic parameters. The regions of analytic behavior are called *thermodynamic phases*. The singularities are interpreted as changes in the phase structure of the system, i.e. as *phase transitions*. According to the Erenfest classification, when some first derivative of a given thermodynamic potential is a discontinuous function of the temperature, or some other external parameter, one speaks about *first order phase transition* with respect to that parameter. A line of first order phase transitions is called a *coexistence line*, since two different phases can coexist on it. Such a line may terminate at an end-point called a *critical point*. The critical point is a point in the space of thermodynamic variables at which the differences between two coexisting phases disappear. On crossing a critical point the system exhibits a *second order (or continuous) phase transition*, since the first derivatives of the thermodynamic potential are continuous, while the second derivatives are not (usually diverge). One of the most fascinating problems in statistical mechanics is the study of the singular behavior of thermodynamic functions at a critical point.

The phase diagram of a simple ferromagnet is shown in Fig. 1.1. At the line $H = 0$, $T \geq 0$, different types of phases can be found: two low-temperature $(0 \leq T < T_c)$ ordered phases, and a high-temperature $(T > T_c)$ disordered phase. These phases are characterized by different values of a thermodynamical variable called *order parameter*. It is convenient to introduce the order parameter so that it vanishes at the critical point and remains zero everywhere in the disordered phase. The order parameter of a simple ferromagnetic system is the *spontaneous magnetization* per particle (or per unit volume) defined as

$$m_0(T) = \lim_{H \to 0^+} m(T, H). \tag{1.51}$$

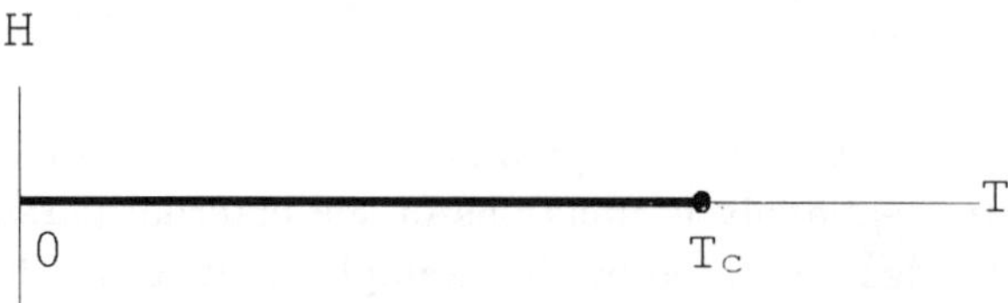

Fig. 1.1   The phase diagram of a simple ferromagnetic system. There is a cut along the T axis from 0 to T_c. The magnetization density m is an analytic function of both T and H at all points except those on the cut; it is discontinuous across the cut. The cut is a line of first order phase transitions and its end-point is the *critical point*; $m(T, H)$ is singular at this point.

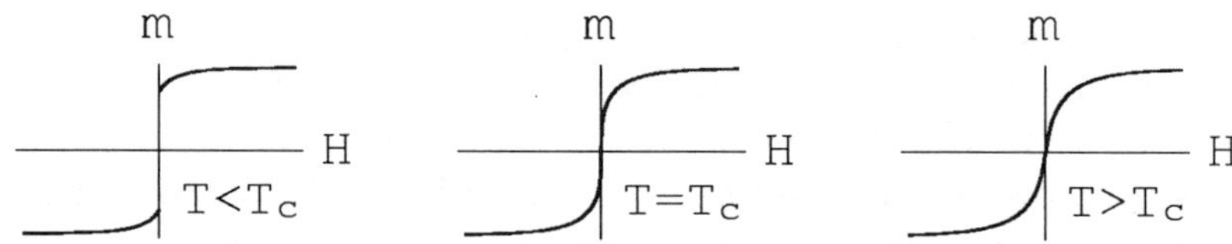

Fig. 1.2   The magnetization density $m(T, H)$ at: a) $T < T_c$, b) $T = T_c$ and c) $T > T_c$.

where $m(T, H)$ is the magnetization density in uniform magnetic field H, at temperature T. At temperatures below the *critical temperature* T_c, the order parameter $m_0(T)$ is nonzero and the m versus H plot is discontinuous at $H = 0$. This can be seen by taking the limit $H \to 0^-$ (i.e., for vanishing negative values of H). Then one obtains a phase with opposite sign of the spontaneous magnetization. Therefore, on the segment $H = 0, 0 \le T < T_c$, there are two coexisting ordered phases which differ by the sign of the order parameter. The magnetic field H "breaks the symmetry" between these two phases and stabilizes one of them at the expense of the other. At $T = T_c$ the spontaneous magnetization m_0 vanishes and $m(T, H)$ becomes a continuous function of H with infinite slope (initial susceptibility) at $H = 0$. At $T > T_c$ the magnetization density $m(T, H)$ remains a continuous function of H and becomes analytical at $H = 0$. The above observations are conveniently summarized in Fig. 1.2. The behavior of the spontaneous magnetization $m_0(T)$ is illustrated in Fig. 1.3.

1.5.1 *Order parameter*

The existence of an order parameter is related to the presence of *long-range order* (LRO) in the thermodynamic limit. A way of looking at this is to consider the *two-point total correlation function*

$$\Gamma(\mathbf{r}_i, \mathbf{r}_j; H, T) \equiv \langle S(\mathbf{r}_i) S(\mathbf{r}_j) \rangle, \tag{1.52}$$

where $S(\mathbf{r}_i)$ and $S(\mathbf{r}_j)$ are the spin variables (operators) at position vectors $\mathbf{r}_i$ and $\mathbf{r}_j$, respectively. If the system is translational invariant, Γ depends only on the difference vector $\mathbf{r}_i - \mathbf{r}_j$. If, in addition, the system is isotropic, Γ is a function of the distance $r_{ij} = |\mathbf{r}_i - \mathbf{r}_j|$ between the sites i and j. We shall generally assume that the system under consideration has the above properties and $\Gamma = \Gamma(r_{ij}, H, T)$. In each of the pure ordered phases, which can be found below the critical temperature and in the absence of a magnetic field, the limit $r \to \infty$ exists and equals

$$\Gamma(r = \infty; H = 0, T < T_c) \equiv \lim_{r \to \infty} \langle S(\mathbf{0}) S(\mathbf{r}) \rangle = [m_0(T)]^2. \tag{1.53}$$

The non-zero value of the above limit implies the existence of a *long-range order* (LRO). In the general case one expects, although there is no rigorous proof, that the limiting value of the two-point total correlation function equals the squared value of the order parameter, see also Section 9.1.1.

In the ferromagnetic example, sketched in the previous subsection, it was simple enough to identify the order parameter. There are cases when this problem needs a more elaborate consideration. Roughly speaking, the

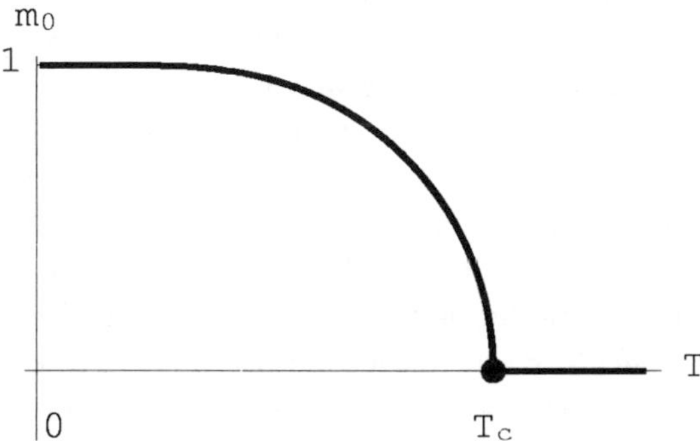

Fig. 1.3   The spontaneous magnetization density $m_0(T)$. It is identically zero above T_c.

order parameter should characterize the type of order which spontaneously arises below T_c. In the liquid-gas phase transition in a simple fluid the order parameter is the density difference between the liquid and gas phases; in the superconducting transition it is the energy gap between the ground state and the spectrum of single-particle excitations, etc.

In the discussion of the ferromagnetic phase transition we have tacitly assumed that the magnetization can point in two directions only, i.e., that there is a single easy-axis of magnetization. Hence, one obtains a rather trivial up-down symmetry, which is precisely the case of the Ising-type models. Generally speaking, the order parameter can be a scalar, as in our ferromagnetic example, but it can also be a vector, tensor, etc. While in what follows most of the considerations will be quite general, for the sake of concreteness, we will use the notations for ferromagnetic n-vector models on the d-dimensional cubic lattice Z^d. We assume that at each lattice site $i \in \Lambda \subset Z^d$ (where Λ contains a finite number of sites $N = |\Lambda|$) there is a n-dimensional classical vector of unit length $\mathbf{S}_i$, with components S_i^α, $\alpha = 1, \ldots, n$. Let the spin-spin interaction be described by the Hamiltonian

$$\mathcal{H} = -J \sum_{\langle i,j \rangle} \sum_{\alpha=1}^{n} S_i^\alpha S_j^\alpha, \quad J > 0, \tag{1.54}$$

where the first sum is taken over the pairs $\langle i, j \rangle$ of nearest-neighbor sites of Λ. This Hamiltonian is rotationally invariant in the n-dimensional spin space, where from comes the name *$O(n)$-symmetric models*. For $n = 1, 2, 3$ the model corresponds to the (classical) Ising, XY and Heisenberg models, respectively. Systems with $n \geq 2$, like the XY and Heisenberg models, are said to possess a *continuous symmetry*. The interaction with an external magnetic field breaks this symmetry and establishes a preferred direction for the spin alignment. If one first takes the thermodynamic limit, and then reduces the external field to zero, the system may exhibit spontaneous magnetization pointing in the initial direction of the field. It has become customary to refer to the dimensionality n of the order parameter as *symmetry index*. Under periodic boundary conditions the limit $n \to \infty$ leads to the *spherical model* (SM) introduced in [Berlin and Kac (1952)], see Section 3.2. It has be shown that other values of n may also have a physical meaning. For example, the formal limit $n \to 0$ leads to the self-avoiding walk (SAW) problem which is applicable to polymers.

The concept of symmetry is of fundamental importance and plays a crucial role in the theory of phase transitions and critical phenomena. This fact was first realized by L. D. Landau, see [Landau (reprinted, 1965)]. Later on, it has been rigorously proven [Mermin and Wagner (1966)] that spontaneous LRO cannot exist in two-dimensional systems with short-range interactions if at $n \geq 2$. This is in contrast to the Ising model ($n = 1$), where spontaneous LRO exists at $d = 2$ but does not exist at $d = 1$. *Lower critical dimension* for a given model is called the highest space dimension at which spontaneous LRO cannot exist. Furthermore, even for $d = 3$, where a phase transition occurs for all values of n, the critical behavior depends essentially on the symmetry index.

In the modern theory of phase transitions a general definition of an order parameter, associated with a given limit Gibbs state, can be given by following the construction suggested in [Fisher (1971)]. Consider a system of N particles confined to a finite space region $\Lambda \subset R^d$. Let $\psi(\mathbf{r})$ be a microscopic variable (operator) associated with the space point $\mathbf{r} \in \Lambda$ and let $K(\mathbf{r}_1, \ldots, \mathbf{r}_n; \mathbf{r})$, $n \leq N$, be a function defined for all $\mathbf{r}_i$, $\mathbf{r} \in R^d$, which rapidly decreases to zero as $|\mathbf{r}_i - \mathbf{r}| \to \infty$ for all $i = 1, \ldots, n$. Then a general *local observable of order n* can be defined as

$$\Psi_\Lambda(\mathbf{r}) = \int_{\Lambda^n} \left(\prod_{i=1}^n d\mathbf{r}_i \right) K(\mathbf{r}_1, \ldots, \mathbf{r}_n; \mathbf{r}) \, \psi(\mathbf{r}_1) \cdots \psi(\mathbf{r}_n). \qquad (1.55)$$

If the Hamiltonian $\mathcal{H}_\Lambda$ is a linear functional of the local observables $\Psi_\Lambda(\mathbf{r})$, $\mathbf{r} \in \Lambda$, having the form

$$\mathcal{H}_\Lambda(h) = \mathcal{H}_\Lambda - h \int_\Lambda d\mathbf{r} \, \Psi_\Lambda(\mathbf{r}), \qquad (1.56)$$

then h is called a *field conjugate to the density*

$$\psi_\Lambda(T, h) = |\Lambda|^{-1} \int_\Lambda d\mathbf{r} \, \langle \Psi_\Lambda(\mathbf{r}) \rangle_{\Lambda, h}. \qquad (1.57)$$

Here $\langle \cdots \rangle_{\Lambda, h}$ denotes the average value with respect to the finite-volume Gibbs probability distribution for the Hamiltonian (1.56). To identify the order parameter, one has to find first the lowest-order local observable $\Psi_\Lambda(\mathbf{r})$ such that the two-point *net (or connected) correlation function*

$$G_{\Psi\Psi}(\mathbf{r}; T, h) =$$
$$t\text{-}\lim \left[\langle \Psi_\Lambda(\mathbf{r}') \Psi_\Lambda(\mathbf{r}' + \mathbf{r}) \rangle_{\Lambda, h} - \langle \Psi_\Lambda(\mathbf{r}') \rangle_{\Lambda, h} \langle \Psi_\Lambda(\mathbf{r}' + \mathbf{r}) \rangle_{\Lambda, h} \right], \quad (1.58)$$

taken in the thermodynamic limit, has the slowest decay with $|\mathbf{r}| \to \infty$ at the critical point $T = T_c$, $h = h_c$.

Let us take into account that the bulk Gibbs free energy density for a system with the Hamiltonian (1.56) is given by

$$f(T, h) = -\beta^{-1}\, t\text{-lim}\, |\Lambda|^{-1} \ln \mathrm{Tr} \exp[-\beta \mathcal{H}_\Lambda(h)]. \tag{1.59}$$

Then, in view of the general fluctuation-dissipation relationship, see, e.g. Chapter 7 in [Huang (1987)],

$$\chi_T(T, h) := -\frac{\partial^2 f(T, h)}{\partial h^2} = \beta \int_{R^d} d\mathbf{r}\, G_{\Psi\Psi}(\mathbf{r}; T, h), \tag{1.60}$$

the slowest decay of (1.58) implies that the bulk isothermal susceptibility $\chi_T(T, h)$ is the most strongly divergent susceptibility as the critical point is approached. The bulk density of the above specified local observable, averaged over the Gibbs ensemble and over the volume of the system,

$$\psi(T, h) := t\text{-lim}\, |\Lambda|^{-1} \int_\Lambda d\mathbf{r} \langle \Psi_\Lambda(\mathbf{r}) \rangle_{\Lambda, h}, \tag{1.61}$$

taken in the limit $h \to h_c$, plays the role of a long-range order parameter at $T < T_c$. It is important to note that at $T < T_c$ the thermodynamic limit generally does not commute with the limit $h \to h_c$.

1.5.2   *Critical exponents*

From thermodynamic point of view, a critical point is characterized by the type of singularity of the thermodynamic potential density and its derivatives with respect to the *relevant variables*. In the case of a simple ferromagnet the relevant variables are the temperature and the magnetic field. Considering critical behavior, it is convenient to replace T by $t = (T - T_c)/T_c$. Then the thermodynamic functions have singularities at $H = t = 0$. The most common case is a branching point singularity, when the expansion of the thermodynamic functions in the neighborhood of the critical point contains terms proportional to non-integer powers of the t and/or H. The exponents of the leading-order singular terms as $t \to 0$ and $H \to 0$ are called *critical exponents*. They give a quantitative description of the critical-point singularity and are inherent to the physical quantities under consideration.

Now we pass to the definition of the critical exponents which describe the critical behavior of the most important thermodynamic functions. Note that they have a customary notation in terms of Greek letters. Let us start with the magnetization. On approaching the critical temperature from below along the line $H = 0$, the spontaneous magnetization per spin vanishes as

$$m_0(T) \simeq A_m\,|t|^\beta, \qquad t = T/T_c - 1 \to 0^+. \tag{1.62}$$

The behavior of the magnetization at the critical temperature as a function of H is characterized by the critical exponent δ,

$$m(T_c, H) \simeq A_H\,|H|^{1/\delta}\mathrm{sign}(H), \qquad H \to 0. \tag{1.63}$$

For the magnetic susceptibility per spin, which is defined as

$$\chi(T, H) = \frac{\partial m(T, H)}{\partial H}, \tag{1.64}$$

one has

$$\chi(T, 0) \simeq \begin{cases} A_\chi^+\, t^{-\gamma}, & t \to 0^+ \\ A_\chi^-\, |t|^{-\gamma'}, & t \to 0^-. \end{cases} \tag{1.65}$$

For the specific heat capacity per spin, which is defined as

$$c(T, H) = -T\frac{\partial^2 f(T, H)}{\partial T^2}, \tag{1.66}$$

the corresponding critical behavior at $H = 0$ is

$$c(T, 0) \simeq \begin{cases} A_c^+\, t^{-\alpha}, & t \to 0^+ \\ A_c^-\, |t|^{-\alpha'}, & t \to 0^-. \end{cases} \tag{1.67}$$

If, as happens in mean field theories, the specific heat is discontinuous at $t = 0$, then one formally sets $\alpha = 0$. If it diverges like $-\ln|t|$, as in the $d = 2$ Ising model, one can set again $\alpha = 0$, since

$$-\ln x = \lim_{\alpha \to 0}(x^{-\alpha} - 1)/\alpha. \tag{1.68}$$

Both theory and experiment suggest that in most of the cases $\alpha = \alpha'$. It is also possible that the specific heat capacity remains finite at T_c but its graph has a cusp. In the case of a symmetric cusp

$$c(T, 0) \simeq C_c - D|t|^{-\alpha_s}, \qquad |t| \to 0, \tag{1.69}$$

where $-1 < \alpha_s < 0$. Then $dc(T)/dT$ diverges with an exponent $1 + \alpha_s$, and one says that the singular part of $c(T, 0)$ has the exponent α_s. The subscript "s" is often omitted, as in Table 1.1. The exponents α, β, γ and δ, defined so far, describe the critical behavior of thermodynamic quantities.

Of equal importance is the critical behavior of the net correlation function in the thermodynamic limit

$$G(\mathbf{r}_i, \mathbf{r}_j; T, H) \equiv \langle S(\mathbf{r}_i)S(\mathbf{r}_j)\rangle - \langle S(\mathbf{r}_i)\rangle\langle S(\mathbf{r}_j)\rangle =$$
$$= -k_B T \text{ t-}\lim \left.\frac{\partial^2 F(\Lambda, T, \{H_i, i \in \Lambda\})}{\partial H_i \partial H_j}\right|_{\{H_i = H\}} \tag{1.70}$$

The last equality follows by differentiation of the (pseudo) free energy (1.27) in the presence of local magnetic fields. After taking the derivatives, one may set $H_i = H$ for all $i \in \Lambda$, and then pass to the thermodynamic limit. On physical grounds, supported by rigorous studies of spin models, it is expected that the net correlation function in a *pure thermodynamic phase* always vanishes as $|\mathbf{r}_i - \mathbf{r}_j| \to \infty$, see Section 9.1.1. The rate of its asymptotic decay with the distance is often used to define the correlation length of the system, especially, in pure ordered phases. For translation invariant and isotropic systems with short-range interactions, it is expected that the asymptotic behavior of $G(r, T, 0)$ close to the critical temperature is

$$G(r; T, 0) \simeq \begin{cases} A_G^+ \, r^{-\tau} \exp[-r/\xi(T,0)], & r \to \infty, T > T_c \\ A_G^- \, r^{-\tau'} \exp[-r/\xi(T,0)], & r \to \infty, T < T_c. \end{cases} \tag{1.71}$$

Here $\xi(T, 0)$ defines the zero-field correlation length. From a physical point of view, $\xi(T, H)$ measures the distance over which the spins are significantly correlated. Note that the critical exponents τ and τ' in the power-law prefactor can be different; for example, in the $d = 2$ Ising model one has, $\tau = 1/2$ and and $\tau' = 2$ [McCoy and Wu (1973)].

As the critical point is approached, $\xi(T, 0)$ diverges as

$$\xi(T, 0) \simeq \begin{cases} A_\xi^+ \, t^{-\nu}, & t \to 0^+ \\ A_\xi^- \, |t|^{-\nu'}, & t \to 0^-. \end{cases} \tag{1.72}$$

Exponent	α	β	γ	δ	ν	η
Mean field	0	1/2	1	3	1/2	0
3d theory						
$n = 0$ (SAW)	$0.24^{a)}$	$0.30^{a)}$	$1.16^{a)}$		$0.59^{a)}$	
$n = 1$ (Ising)	0.119	$0.326^{b)}$	$1.239^{b)}$	4.80	$0.627^{b)}$	0.024
$n = 2$ (XY model)	$-0.01^{a)}$	$0.35^{a)}$	$1.32^{a)}$		$0.67^{a)}$	$0.04^{a)}$
$n = 3$ (Heisenberg)	$-0.08^{a)}$	$0.38^{a)}$	$1.38^{a)}$	$4.63^{a)}$	$0.715^{a)}$	$0.07^{a)}$
$n = \infty$ (SM)	-1	1/2	2	5	1	0
2d theory						
$n = 1$ (Ising)	0	1/8	7/4	15	1	1/4
Experiment						
Xe ($n = 1$)	< 0.2	0.35	1.3	4.2	≈ 0.57	0.1
Binary fluid ($n = 1$)	0.113	0.322	1.239	4.85	0.625	0.017
β-brass ($n = 1$)	0.05	0.305	1.25		0.65	0.08
^{4}He ($n = 2$)	-0.014	0.34	1.33	3.95	0.672	0.021
Fe ($n = 3$)	-0.03	0.37	1.33	4.3	0.69	0.07
Ni ($n = 3$)	0.04	0.358	1.33	4.29	0.64	0.041

Table 1.1. Representative examples of critical exponents from theory and experiment. The data with a superscript $a)$ are adopted from [Chaikin and Lubensky (1995)]; exponents without superscript are determined by accepting formally the validity of the corresponding scaling relations between them, see Section 1.5.4, or are exact, as the results for the $d = 2$ Ising model and $d = 3$ spherical models; the data with superscript $b)$ are from [Ferrenberg and Landau (1991)]. The experimental data are adopted from [Ahlers (1980)], [Anders and Stierstadt (1981)], [Bally et. al. (1968)], [Chang et. al. (1979)], [Cohen and Carver (1977)], [Collins et. al. (1969)], [Heller (1967)], [Hiroyoshi et.al. (1980)], [Kobeissi (1981)], [Kumar et. al. (1983)], [Rocker et al. (1971)], [Rowlinson and Winton (1982)], [Soeffge (1980)], [Suter and Hohenemser (1978)], [Vygovskiy and Yergin (1972)]. The data for a binary fluid are: α and β – for methanol-hexane, while γ, ν and η are for a mixture of trimethylpentane and nitroethane. The exponents δ for a binary fluid, α for β-brass, and γ and η for helium are calculated from the scaling relations. The data for helium pertain to the He I - He II transition.

Therefore, *at $T = T_c$* the asymptotic form of the correlation function changes qualitatively from exponential to power-law decay,

$$G(r; T_c, 0) \simeq A_G^c \, r^{-(d-2+\eta)}, \tag{1.73}$$

where d is the dimensionality of the system. The exponent of r is written in accordance with tradition: it defines the critical exponent η.

From the above examples it is easy to deduce, that in all but some exceptional cases one writes $X(T) \sim |t|^{x_\pm}$ for a given quantity X if the limit

$$\lim_{t \to \pm 0} \frac{\ln X(t)}{\ln |t|} = x_\pm \tag{1.74}$$

exists. In most cases $x_- = x_+$.

A representative table of critical exponents, obtained from both theory and experiment, is given in Table 1.1.

Looking at the critical exponents, some questions arise immediately: *(i)* Which features of the system actually determine the critical exponents? *(ii)* Are all these exponents independent of each other? We postpone the answers to Section 1.5.4.

1.5.3   *The Landau theory*

In 1937 L.D. Landau developed a phenomenological theory of phase transitions, see [Landau (reprinted, 1965)], by generalizing the ideas due to W.L. Bragg and E.J. Williams (1934) in the theory of binary alloys. His theory deals with macroscopic (coarse-grained) quantities and is supposed to hold near a critical point, where the order parameter is small. Here we will give a sketch of a simplified version of Landau's theory, by assuming the existence of a homogeneous scalar order parameter ψ. This version reproduces the results of the *mean-field theory*, known also from the Van der Waals model of a simple fluid (1873) and the Weiss model of a ferromagnet (1907).

According to the Landau theory, the important features close to the critical point can be determined from the expansion of the thermodynamic potential in a power series of the order parameter ψ. Let us assume that, in the absence of a symmetry-breaking field h, the expansion of the Helmholtz free energy density $a(\psi, T)$ has the form

$$a(\psi, T) = c_0(T) + c_2(T)\psi^2 + c_4(T)\psi^4 + \dots, \tag{1.75}$$

where the phenomenological coefficients c_p, $p = 0, 2, 4, \dots$ are real-analytic functions of the temperature T. The critical temperature T_c is defined by the conditions

$$c_2(T) = 0, \qquad c_4(T) > 0. \tag{1.76}$$

The different phases correspond to the minima of $a(\psi, T)$ with respect to the order parameter ψ.

Close to the critical point, the equation of state

$$h = \frac{\partial}{\partial \psi} a(\psi, T) \simeq 2c_2(T)\psi + 4c_4(T)\psi^3, \tag{1.77}$$

takes the form

$$h \simeq 4b\,\psi\left(\frac{a}{2b}t + \psi^2\right). \tag{1.78}$$

A remarkable feature of this (classical) equation of state is that the term in the parentheses is a homogeneous function of degree 1 in the variables t and ψ^2. At the critical temperature, $t = 0$, the equation of state yields

$$\psi(T_c, h) \simeq (4b)^{-1/3} \operatorname{sign}(h)|h|^{1/3}, \qquad |h| \to 0, \tag{1.79}$$

hence $\delta = 3$. The other critical exponents can readily be found to take the values $\alpha = \alpha' = 0$, $\beta = 1/2$, and $\gamma = \gamma' = 1$.

Somewhat more elaborate considerations, based on Landau's theory for spatially inhomogeneous (coarse-grained) order parameter $\psi(\mathbf{r})$, perturbed by a "test field" $h(\mathbf{r})$ which is localized at the origin $\mathbf{r} = \mathbf{0}$, allow one to obtain (at $d \geq 2$) the critical exponents for the correlation function, $\eta = 0$, and the correlation length, $\nu = 1/2$, see e.g. [Huang (1987)].

A remarkable feature of the mean-field critical exponents, obtained above by using the Landau theory, is their *universality*, that is, their independence of the phenomenological parameters. This result suggests that the critical exponents are not sensitive to the details of the microscopic Hamiltonian. Moreover, the mean-field exponents (called also "classical" exponents) are independent of the space dimensionality d too. Their values, however, are inconsistent with precise experimental data, with rigorous results for exactly solvable models (see Table 1.1), as well as with *Renormalization Group* predictions, see Section 1.6. It turns out that there exists a space dimension $d = d_u$, called *upper critical dimension*, such that the exact exponents coincide with the corresponding mean-field ones for all $d \geq d_u$. The value of d_u may depend on the type of the inter-particle interaction.

1.5.4 *Universality and scaling laws*

The universality hypothesis answers the question what features of the system determine the values of its critical exponents. It can be stated as follows [Kadanoff (1971)]:

"All phase transition problems can be divided into a small number of different classes depending upon the dimensionality of the system and the symmetries of the ordered state. Within each class, all phase transitions have identical behavior in the critical region, only the names of thermodynamic variables are changed."

As to almost any rule, there are exceptions to this rule too. The famous example is the eight-vertex model solved by [Baxter (1982)] – it has exponents that vary continuously with the parameters of the partition function. This feature violates the universality hypothesis, but it is nowadays generally believed that such violations occur only in very special cases.[*]

According to the hypothesis of universality the main features that determine which *universality class* a given system belongs to, hence, the values of its critical exponents, are the dimensionality d, the symmetry index n, and the range of the interaction potential. Let us suppose that the Hamiltonian of the system $\mathcal{H}$ can be written as $\mathcal{H} = \mathcal{H}_0 + \lambda \mathcal{H}_1$, where $\mathcal{H}_0$ has some symmetry which $\mathcal{H}_1$ has not. Then the critical exponents will depend on λ only in so far that they will have one value for $\lambda = 0$ and another for $\lambda \neq 0$. On the other hand, if $\mathcal{H}_0$ is some simple Hamiltonian and $\mathcal{H}_1$ has the same symmetry and dimensionality as $\mathcal{H}_0$, then one can simplify the initial system up to the one described by $\mathcal{H}_0$ without loosing any information about the critical exponents. This is the most important practical advantage of the universality hypothesis. On these grounds one believes, for example, that carbon dioxide, xenon, and $d = 3$ Ising model should

---

[*]In the renormalization group language this happens when the Hamiltonian contains marginal operators; for details consult [Kadanoff (1971)], [Kadanoff and Wegner (1971)], [Baxter (1972)].

have the same critical exponents, as it has indeed been checked out (within experimental error-bars) in experiments [Hocken and Moldover (1976)].

The universality hypothesis is commonly stated together with the hypothesis of scaling for the thermodynamic functions close to the critical point.[†] Widom (1965), and Domb and Hunter (1965) conjectured that certain thermodynamic functions are homogeneous functions of their variables around T_c. More explicitly, Griffiths (1967) suggested that the equation of state for a ferromagnet can be written in the scaling form

$$h = m|m|^{\delta-1}h_s(t|m|^{-1/\beta}), \tag{1.80}$$

i.e., as a homogeneous function of t and $m^{1/\beta}$. According to the universality hypothesis the dimensionless *scaling function* $h_s(x)$ is the same for all systems within a given universality class. Here we will not go into details about the properties of $h_s(x)$ and the relations between the critical exponents (such relations are called *scaling laws*) that follows from Eq. (1.80). Instead, we shall formulate the scaling hypotheses for the free energy and the two-point correlation function, wherefrom one obtains all standard s-caling relations. Note that it is possible to motivate the scaling hypotheses via the renormalization group methods, see Section 1.6.5. Here we adopt the simplest approach to postulate them.

Let the bulk free energy density $f(T, h)$ be split into singular, f_{sing}, and nonsingular (regular), f_{ns}, parts,

$$f(T, h) = f_{\text{sing}}(T, h) + f_{\text{ns}}(T, h), \tag{1.81}$$

where the singular part is responsible for the critical behavior. The scaling hypothesis concerns the singular part of the free energy density f_{sing}. It is supported by the renormalization group theory, see Section 1.6, which predicts that in the case of a simple ferromagnet the asymptotic behavior of $f_{\text{sing}}(T, h)$ as $(T, h) \to (T_c, 0)$ has the following form

$$\beta f_{\text{sing}}(T, h) \simeq A_1 |t|^{2-\alpha} W^{\pm}(A_2 h|t|^{-\Delta}), \tag{1.82}$$

where $\pm$ refers to $t > 0$ and $t < 0$, respectively. The scaling functions $W^{\pm}$ and the exponents α and $\Delta = \beta + \gamma$ are the same for all systems in a given universality class. Within a given universality class, the lattice structure,

---

[†]In fact, these are independent assumptions – one of them may hold, while the other fails – as in the case of the eight-vertex model when universality fails but scaling still holds [Baxter (1982)].

coupling constants, etc., may vary, but all such variations are summarized in the values of the nonuniversal metric factors A_1 and A_2.

Another fundamental hypothesis concerns the scaling form of the net pair correlation function, defined by Eq. (1.70), in the neighborhood of a critical point, namely

$$G(\mathbf{r}; T, h) \simeq D_1 r^{-(d-2+\eta)} X_G^{\pm}(\mathbf{r}/\xi; D_2 h|t|^{-\Delta}), \tag{1.83}$$

where $X^{\pm}$ are universal scaling functions. One can take this postulate as a basic one and, by using the *fluctuation-dissipation relationship*

$$\chi(T, h = 0) \equiv \chi_0(t) = \beta \sum_{\mathbf{r}} G(\mathbf{r}; T, 0), \tag{1.84}$$

derive the scaling form of the free energy density (1.82) with

$$\alpha = 2(1 - \Delta + \nu) - \eta\nu,$$

where Δ is the so-called "gap exponent" [Huang (1987)]. This form of scaling is called *three-exponent scaling*, since *all* the bulk critical exponents can be expressed in terms of the basic triplet η, ν and Δ. For example, from Eqs. (1.82) and (1.83) it follows that

$$\alpha + 2\beta + \gamma = 2, \quad \beta + \gamma = \Delta, \quad (2 - \eta)\nu = \gamma.$$

Such relations are called *scaling relations between the critical exponents*.

Note that given the scaling form of the pair correlation function (1.83), via the fluctuation-dissipation relationship (1.84) one can link the metric factors D_1 and D_2 to A_1 and A_2. Indeed, when $\xi \gg 1$ one can replace the summation over $\mathbf{r}$ in the right-hand side of Eq. (1.84) by integration over $\mathbf{x} = \mathbf{r}/\xi$, and obtain a scaling expression for the susceptibility in the form

$$k_B T \, \chi(T, h) \simeq D_1 \xi^{2-\eta} X_\chi^{\pm}\left(D_2 h|t|^{-\Delta}\right), \tag{1.85}$$

where $X_\chi^{\pm}$ are new universal functions. By assuming that the total correlation function (1.52) has a scaling form similar to that of the net correlation function,

$$\Gamma(\mathbf{r}; T, h) \simeq D_1 r^{-(d-2+\eta)} X_\Gamma^{\pm}\left(\mathbf{r}/\xi; D_2 h|t|^{-\Delta}\right), \tag{1.86}$$

where $X_\Gamma^{\pm}$ are universal functions, and requiring that $\Gamma(\mathbf{r}, T, h)$ reduces to $m^2(T, h)$ in the limit $r \to \infty$, we obtain

$$m(T, h) \simeq D_1^{1/2} \xi^{-(d-2+\eta)/2} X_m^{\pm}\left(D_2 h|t|^{-\Delta}\right) \tag{1.87}$$

On the other hand, by differentiating the scaling form of the free energy density (1.82) with respect to the magnetic field variable h, one obtains successively

$$m(T, h) \simeq A_1 A_2 |t|^\beta W_1^\pm \left(A_2 h |t|^{-\Delta}\right) \tag{1.88}$$

with $\beta = 2 - \alpha - \Delta$, and

$$k_B T\, \chi(T, h) \simeq A_1 A_2^2 |t|^{-\gamma} W_2^\pm \left(A_2 h |t|^{-\Delta}\right) \tag{1.89}$$

with $\gamma = 2\Delta + \alpha - 2$, and $W_{1,2}^\pm$ being universal functions.

The scaling expressions for $m(T, h)$ and $\chi(T, h)$, derived above in two different ways, can be compared by taking into account the critical behavior (1.72) of the correlation length ξ. Thus one obtains that the scaling relation $\gamma = (2 - \eta)\nu$ and the *hyperscaling* relation $2\beta = (d - 2 + \eta)\nu$ are fulfilled. Then, by setting

$$W_1^\pm(x) \equiv Q_1 X_m^\pm(Q_2 x), \qquad W_2^\pm(x) \equiv Q_3 X_\chi^\pm(Q_2 x), \tag{1.90}$$

where Q_1, Q_2 and Q_3 are universal constants, one obtains the equalities

$$A_2 \;=\; D_2 Q_2 \tag{1.91}$$

$$A_1 A_2 \;=\; Q_1 \left[D_1 \left(A_\xi^\pm\right)^{-(d-2+\eta)} \right]^{1/2} \tag{1.92}$$

$$A_1 A_2^2 \;=\; Q_3 D_1 \left(A_\xi^\pm\right)^{2-\eta} \tag{1.93}$$

From Eq. (1.92) it follows that $A_1^2 A_2^2 = D_1 (A_\xi^\pm)^{2-\eta-d} Q_1^2$, and from Eq. (1.93) one derives

$$A_1 = Q \left(A_\xi^\pm\right)^{-d} \tag{1.94}$$

with $Q = Q_1^2/Q_3$. The above equation yields the hyperuniversality relationship, see [Stauffer et. al. (1972)],

$$\lim_{T \to T_c, h \to 0} [\beta f_{\text{sing}}(T, h) \xi^d(T, h)] = Q = \text{universal constant}\,, \tag{1.95}$$

which represents the *hyperuniversality hypothesis* in the form of *two-scale factor universality*. Obviously, Eq. (1.95) yields the *hyperscaling relation*

$$2 - \alpha = d\nu. \tag{1.96}$$

Here we mention that the finite-size scaling theory, based on the two-scale factor universality, predicts that the singular part of the free energy density of a system in a finite volume L^d *at the critical point* is a universal quantity, see Section 4.3.

The universality of critical exponents and certain critical amplitude ratios is a central concept in the modern theory of critical phenomena. For example, Eq. (1.91) expresses D_2 by A_2 and shows that they are simply proportional to an universal constant. In a similar way from (1.92) and (1.94) one can express D_1 via A_1 and A_2:

$$D_1 = \left(A_2 A_1^{\Delta/d\nu} \right)^2 \times \text{ universal constant.}$$

The universality of scaling functions naturally leads to a variety of universal critical amplitudes and amplitude combinations which characterize a given universality class as much as the critical exponents do. A review on the available theoretical and experimental data on this topic can be found in [Privman et. al. (1991)].

Finally, it should be emphasized that the three-exponent scaling relations do not depend on the dimensionality d and are generally valid. On the other hand, it is well known that above the upper critical dimension d_u, *hyperscaling does not hold.*[‡]

As a concrete illustration of the scaling hypothesis for the correlation function, we give the corresponding results for the $d = 2$ Ising model, where

$$G(\mathbf{r}; T, 0) \simeq r^{-1/4} X^{\pm}(r/\xi)$$

with

$$X^+(x) \sim x^{-1/4} e^{-x}, \qquad X^-(x) \sim x^{-7/4} e^{-x}, \qquad x \to \infty.$$

One may pose the question about the restrictions on the relations between critical exponents that can be derived from general thermodynamic considerations. It turns out that this approach leads to inequalities between the exponents that become equalities at the critical point. Let us give some examples.

- Inequality of Rushbrooke: $\alpha' + 2\beta + y' \geq 2$;
- First inequality of Griffiths: $\alpha' + \beta (\delta + 1) \geq 2$.

[‡]For Ising-type and $O(n)$, $n \geq 2$, models with short range interactions $d_u = 4$.

These two inequalities are derived from thermodynamic stability conditions and, therefore, are considered as rigorous. By introducing some plausible assumptions, one obtains

- Second inequality of Griffiths: $\gamma(\delta + 1) \geq (2 - \alpha)(\delta - 1)$;
- Inequality of Fisher: $(2 - \eta)\nu \geq \gamma$;
- Inequality of Josephson: $d\nu \geq 2 - \alpha$;
- Inequality of Buckingham – Gunton: $d\frac{\delta-1}{\delta+1} \geq 2 - \eta$.

Upon assuming the scaling hypotheses, all these inequalities actually turn into equalities.

1.6   Basic facts from the renormalization group theory

The renormalization group (RG) methods have a big amount of important applications in different fields of statistical physics. A comprehensive review of the nature, origins and stand of RG ideas and methods in statistical physics and condensed matter can be found in [Fisher (1998)]. There are many other books, e.g., [Ma (1976)], [Ravandal (1976)], [Yukhnovskii (1987)], [Parisi (1988)], [Goldenfeld (1992)], [Binney et. al. (1992)], [Cardy (1996)], and [Zinn-Justin (1996)], in which the RG theory is adapted to the reader's background training and interests on different levels of detail. The RG methods can be roughly divided into two groups: real-space and field-theoretical. The former are closer in spirit to the original Kadanov's idea which carries out most of the arguments in configuration space, while the latter are based on perturbation analysis in momentum space. Here we present briefly some basic facts from the RG theory which are related to the universal properties of critical phenomena.

1.6.1   *The Ginsburg-Landau model*

An essential feature of the RG methods is the adequate treatment of the fact that critical fluctuations in a system undergoing phase transition are associated with *long-wavelength* fluctuations in momentum space or *long-range* correlations in real space. This means that the integration out of excitations with large wavevectors, or block averaging of degrees of freedom on a short-distance scale in real space, leaves critical features of the system unchanged. It is the idea to work with an effective ("coarse-grained")

Hamiltonian, instead of the initial microscopic Hamiltonian, that makes successful the after-following attempts to go beyond mean field results, or results obtained for a limited number of exactly solved models. A basic model which disregards the short-range detail of the system is the Ginsburg-Landau (GL) model. The effective Hamiltonian of the GL model is defined by the expression

$$\mathcal{L}_{eff}[M] =$$
$$\int_{\Omega} \mathrm{d}^d \mathbf{r} \left[\frac{1}{2} b_1 [\nabla M_\Lambda(\mathbf{r})]^2 + \frac{b_2}{2} t M_\Lambda^2(\mathbf{r}) + \frac{1}{4!} b_3 M_\Lambda^4(\mathbf{r}) - h(\mathbf{r}) M_\Lambda(\mathbf{r}) \right] , \quad (1.97)$$

where $M_\Lambda(\mathbf{r})$ is the average magnetization of the spins in a block around the point $\mathbf{r}$. The size Λ^{-1} of the block, over which the spins are averaged, must be much larger than the lattice spacing a, but much less than the correlation length $\xi(T)$. Indeed, only in this case the final results of calculations with (1.97) do not depend on the choice of Λ, which indicates that they are meaningful. Without going into detail, we shall only note that the above procedure, called *coarse-graining* procedure, must guarantee that $M_\Lambda(\mathbf{r})$ is a sufficiently smooth function of $\mathbf{r}$, having no Fourier components with wavenumbers greater than Λ.

Let us briefly comment on the meaning of the different terms in the effective Hamiltonian (1.97). The term $\frac{1}{2} b_1 [\nabla M_\Lambda(\mathbf{r})]^2$ describes a ferromagnetic interaction. The strength of the correlations in the system depends on the magnitude of $b_1 > 0$, which depends on the range of the microscopic interaction between the spins. The only temperature dependent term in (1.97) is the second one: b_2 is a positive constant and $t = (T - T_c)/T_c$, where T_c is the critical (mean field) temperature. The temperature independent coefficient b_3 is positive. There are different ways of more or less systematical derivation of an effective coarse-grained Hamiltonian from a given microscopic one, which we do not consider here.

The probability density for the field (local magnetization) in the Ginzburg-Landau model is by definition

$$P[M] \propto \exp\{-\beta \mathcal{L}_{eff}[M]\}, \quad (1.98)$$

and the partition function is given by

$$Z_{GL} = \int \mathcal{D}[M] P[M], \quad (1.99)$$

where $\int \mathcal{D}[M]$ denotes a functional integral over the configuration space.

To avoid the problems connected with the mathematical definition of the above expressions (which are not directly related to the physics they describe, but are of rather special interest), and due to some physical reasons, it is useful to consider the lattice counterpart of model (1.97). Let us formally consider a field $\{M_i\}$ defined only at the sites of a lattice with lattice spacing a. Then, instead of (1.97) we may consider the effective lattice Hamiltonian of the GL model

$$\mathcal{L}^a_{eff}[M] = a^d \left[\sum_{ij} \frac{1}{2a^2} J_{i,j}(M_i - M_j)^2 + \sum_i \left(\frac{b_2}{2} t M_i^2 + \frac{b_3}{4!} M_i^4 + h_i M_i \right) \right]. \tag{1.100}$$

The introduction of a lattice reduces the problem to a finite number of degrees of freedom, $M_i, i = 1, 2, \ldots, N$, instead of the continuous field $M(\mathbf{r})$, interacting via the nearest-neighbor couplings J_{ij}. This makes the mathematical structure of the theory much more simple. If M_i is a slowly varying function of the lattice site, then Eq. (1.100) reduces to

$$\begin{aligned} \mathcal{L}^a_{eff}[M] &= \int d^d\mathbf{r} \left\{ \frac{1}{2} b_1 [\nabla M(\mathbf{r})]^2 + \frac{b_2}{2} t M(\mathbf{r})^2 + \frac{1}{4!} b_3 M(\mathbf{r})^4 \right. \\ &\quad - \left. h(\mathbf{r}) M(\mathbf{r}) + O(a^2 (\nabla^2 M)^2) \right\}. \end{aligned} \tag{1.101}$$

The above expression is obtained by using the finite-difference derivative

$$\frac{M(\mathbf{r}_i + a\mathbf{x}) - M(\mathbf{r}_i)}{a} \to \nabla M(\mathbf{r}). \tag{1.102}$$

In the *continuum limit* $a \to 0$, the number of degrees of freedom (denoted by the integer N in the lattice case) tends to infinity, so that the volume of the system $V = Na^d$ remains constant. This means that in a system of finite size there are now infinitely many (in fact, a continuum of) degrees of freedom $M(\mathbf{r})$. When the lattice spacing $a \to 0$ the last term in the right-hand side of Eq. (1.101) can be neglected.

It is convenient to absorb the factor a^d in Eq. (1.100) by introducing a new field $Q_i = a^{(d-2)/2} M_i$ and redefining the constants

$$A = b_2 t a^2, \qquad B = b_3 a^{4-d}, \qquad H_i = a^{(d+2)/2} h_i. \tag{1.103}$$

Then, instead of Eq. (1.100), we obtain

$$\mathcal{L}_{eff} = \sum_{i,j} \frac{1}{2} J_{ij}(Q_i - Q_j)^2 + \frac{A}{2}\sum_i Q_i^2 + \frac{B}{4!}\sum_i Q_i^4 + \sum_i H_i Q_i. \quad (1.104)$$

A reduced, exactly solved quantum version of the model (1.104), called the *reduced Q^4 model*, will be considered in Chapter 3.

Now we continue our study of the continuous model (1.101). It is convenient to absorb the inverse temperature β in the "Hamiltonian" $\mathcal{H}_{eff}\{\phi\}$ and to suitably redefine the field $M(\mathbf{r})$ and the constants of the model (hereafter $h(\mathbf{r}) \equiv 0$):

$$\mathcal{H}_{eff}\{\phi\} := \beta \mathcal{L}_{eff}[M] = \int \mathrm{d}^d\mathbf{r} \left[\frac{1}{2}(\nabla\phi)^2 + \frac{1}{2}r_0\phi^2 + \frac{1}{4!}u_0\phi^4 \right], \quad (1.105)$$

where

$$\phi := (\beta b_1)^{1/2} M(\mathbf{r}); \qquad r_0 := \frac{b_2 t}{b_1}; \qquad u_0 := \frac{b_3}{\beta b_1^2}. \quad (1.106)$$

The complications that arise when one attempts at developing a perturbation theory in u_0 can be illustrated easily by the *dimensional analysis* of the partition function

$$Z_{GL} = \int \mathcal{D}[\phi] \mathrm{e}^{-\mathcal{H}_{eff}\{\phi\}}, \quad (1.107)$$

The physical requirement that the argument of the exponential in Eq. (1.107) must be dimensionless implies that in units of length $L \sim \Lambda^{-1}$

$$[\phi] = 1 - d/2, \qquad [r_0] = -2, \qquad [u_o] = d - 4, \quad (1.108)$$

where the brackets [...] stand for "dimensionality of". Hence it follows that if a perturbation scheme for calculation of Eq. (1.107) is constructed, then the only dimensionless parameter that appears is

$$\bar{u}_0 = u_0 r_0^{(d-4)/2} \sim u_0 t^{(d-4)}. \quad (1.109)$$

Thus, if $d > 4$ one can expect that as $T \to T_c^+$, our perturbation expansion will yield critical exponents that have accurate values. If $d < 4$, then $\bar{u}_0$ goes to infinity at the transition point, and a strong-coupling regime takes place that can make a perturbation approach meaningless.

1.6.2 *The Gaussian approximation*

If we drop the first term in the right-hand side of Eq. (1.105), we recover the mean field Landau theory considered in Section 1.5.3; if we set $u_0 = 0$, we obtain the so called *continuous Gaussian model.* The latter case is of basic interest, firstly, because of its mathematical tractability, i.e. the functional integration in the partition function (1.107) can be performed exactly, and, secondly, the best that can be done in the $u_0 \neq 0$ case is the evaluation of Z_{GL} in terms of an infinite series of Gaussian functional integrals.

Let us note that the properties of the lattice Gaussian model in the case of finite-size systems with different boundary conditions are considered in Section 7.1.

Since the calculation of the partition function for the continuous Gaussian model in the presence of external field,

$$Z_G(h_0) = \int \mathcal{D}[\phi] \exp\left\{ -\int \mathrm{d}^d \mathbf{r} \left[\frac{1}{2}(\nabla\phi)^2 + \frac{1}{2}r_0\phi^2 - h_0\phi \right] \right\}, \qquad (1.110)$$

can be found in many references, e.g., [Binney et. al. (1992)] and [Goldenfeld (1992)], here we shall only sketch the basic points and give the final results. As it was pointed out above, of physical interest for the theory of critical phenomena are fields smoothly varying on spatial scales less than Λ^{-1}. Such are the functions $\phi \equiv \phi(\mathbf{r})$ which can be represented as

$$\phi(\mathbf{r}) = \frac{1}{(2\pi)^d} \int_{k=0}^{\Lambda} \mathrm{d}^d\mathbf{k}\, \hat{\phi}(\mathbf{k})\, \mathrm{e}^{i\mathbf{k}\cdot\mathbf{r}}, \qquad (1.111)$$

where Λ naturally appears as a *cut-off* for the wave numbers $\mathbf{k}$. The usefulness of Eq. (1.111) is twofold: it makes the functional integral (1.110) well defined, and breaks it up into a product of ordinary integrals over the different components of $\hat{\phi}(\mathbf{k})$. The idea is to consider the system in a box of size L, under periodic boundary conditions, and instead of (1.111) to use the representation

$$\phi(\mathbf{r}) = \frac{1}{L^d} \sum_{|\mathbf{k}|<\Lambda} \hat{\phi}(\mathbf{k})\, \mathrm{e}^{i\mathbf{k}\cdot\mathbf{r}}. \qquad (1.112)$$

Note that Eq. (1.112) becomes a Fourier transformation in the limit $\Lambda \to \infty$. A similar relation for the external field $h_0 \equiv h_0(\mathbf{r})$ is also needed. After that we can formally perform the corresponding Gaussian integrals. The final result is

$$Z_G[\hat{h}_0] = Z_G[0]\exp\left[\frac{1}{2}\int d^d\mathbf{r_1}d^d\mathbf{r_2}h_0(\mathbf{r_1})G_0(\mathbf{r_1}-\mathbf{r_2})h_0(\mathbf{r_2})\right], \qquad (1.113)$$

where, retaining only the leading-order asymptotic dependence on L,

$$Z_G[0] = \exp\left[-\frac{1}{2(2\pi)^d}L^d\int_0^\Lambda d^d\mathbf{k}\ln(k^2+r)\right] \qquad (1.114)$$

and

$$\int d^d\mathbf{r_1}d^d\mathbf{r_2}h_0(\mathbf{r_1})G_0(\mathbf{r_1}-\mathbf{r_2})h_0(\mathbf{r_2}) = \frac{1}{(2\pi^d)}\int_0^\Lambda d^d\mathbf{k}\frac{\hat{h}_0(\mathbf{k})\hat{h}_0(-\mathbf{k})}{k^2+r_0}.$$
$$(1.115)$$

By differentiating Eq. (1.113) with respect to the field h_0, one can calculate *all* the correlation functions of the field ϕ, for example,

$$\langle\phi(\mathbf{r_1})\rangle = \int d^d\mathbf{y}\,G_0(\mathbf{r_1}-\mathbf{y})h_0(\mathbf{y}), \qquad (1.116)$$

and

$$\langle\phi(\mathbf{r_1})\phi(\mathbf{r_2})\rangle - \langle\phi(\mathbf{r_1})\rangle\langle\phi(\mathbf{r_2})\rangle = G_0(\mathbf{r_1}-\mathbf{r_2}). \qquad (1.117)$$

In the following $G_0(\mathbf{r_1}-\mathbf{r_2})$ will be called the *bare propagator*; its Fourier transform is

$$\hat{G}_0(k) = \frac{1}{k^2+r_0}. \qquad (1.118)$$

The zero-field partition function $Z_G[0]$ strongly depends on the value of Λ, since the integral in Eq. (1.114) is divergent as $k\to\infty$. On the other hand, the h_0-dependent part in Eq. (1.113) does not depend strongly on Λ, since we consider smooth functions $h_0(\mathbf{r})$, the Fourier transform of which has a cut-off at large moments, see Eq. (1.115). Since we are interested only in relationships between different macroscopic quantities, it is desirable to eliminate any microscopic information (e.g., b_1, b_2, b_3, Λ) from the final results. In the case of the Gaussian model this is possible to be done by simply letting $\Lambda\to\infty$, with a reasonable interpretation, see, e.g., [Parisi (1988)] or [Binney et. al. (1992)]. Problems arise if the ϕ^4 term persists in the effective Hamiltonian. Then, because of new divergences, one needs a special renormalization procedure. We shall come back to this point later.

Let us consider the heat capacity per unit volume c of the Gaussian model. By using Eqs. (1.34) and (1.114) one obtains

$$c = a_1 \int_0^\Lambda \frac{d^d\mathbf{k}}{k^2 + r_0} + a_2 \int_0^\Lambda \frac{d^d\mathbf{k}}{(k^2 + r_0)^2}, \qquad (1.119)$$

where a_1 and a_2 are some constants. If we let $\Lambda \to \infty$, a simple analysis of the above integrals shows that

$$c \sim \begin{cases} t^{-(2-d/2)}, & \text{if} \quad d < 4 \\ \text{finite}, & \text{if} \quad d > 4 \end{cases}. \qquad (1.120)$$

Therefore, in the Gaussion approximation the spatial fluctuations have changed the exponent α from the mean field value $\alpha = 0$, to $\alpha = 2 - d/2$. It is easy to show that all other exponents preserve their mean field values.

1.6.3 *Perturbation expansion*

The aim of this subsection is very limited: just to give some ideas and basic information about problems that are closely related to the main topics of the book.

There exists a formal perturbation expansion scheme for calculation of the free energy and correlation functions of the Ginzburg-Landau model. It leads to a representation of functional integrals by an infinite series of diagrams called *Feynman diagrams*. The latter are recipes, or mnemonic rules, which tell one how to construct the terms of different order in the coupling constant u_0, and show how to assign the corresponding algebraic expression to each of these terms. Here we omit the details of this basic part of every perturbation theory and refer the reader to textbooks on field theory, e.g., [Parisi (1988)]), [Binney et. al. (1992)] for introduction to the subject.

Let us comment on the sensitivity of the theory to the cut-off parameter Λ. In calculating the specific heat in the Gaussion approximation one can set $\Lambda \to \infty$ without any problems. However, this is not the case when the coupling constant $u_0 \neq 0$. If one makes an expansion in powers of u_0, for example, in the calculation of correlation functions, its coefficients are represented in terms of integrals given by the Feynman rules. If the dimensionality of the system is sufficiently high, all these integrals diverge and a meaningful limit does not exist. This fact is due to the behavior of the integrands at large wavevectors and is known as *ultraviolet divergence*.

Indeed, it is very hard to extract some interesting information from an infinite series of different diagrams, unless some tricks or special techniques are used. Firstly, one can show that in the case when the spatial dimension is less or equal to the upper critical dimension $d_u = 4$, it is possible to choose three new coupling constants in the effective Hamiltonian (1.105), which absorb all the divergences. The expansion in terms of these new renormalized variables should be Λ-insensitive. Secondly, a simple possible way of using the perturbation expansion consists in selecting, among the many diagrams with different topology, sets of diagrams that form a geometric series. Such type of selections are laid in the basis of many approximate methods that are popular in different areas of condensed matter physics, like the Hartree-Fock approximation, random-phase approximation, etc.

In practice, it is more convenient to work with Feynman rules in momentum space. The perturbation series for the correlation function in momentum space $\hat{G}(k)$ can be represented by the following infinite series of terms

$$
\begin{aligned}
\hat{G}(k) &= \hat{G}_0(k) + (-u_0/2)\hat{G}_0(k)A\hat{G}_0(k) + \cdots \\
&= \hat{G}_0(k)\frac{1}{1 + (u_0/2)A\hat{G}_0(k)},
\end{aligned}
\tag{1.121}
$$

where $\hat{G}_0(k)$ is the Fourier transform of the bare propagator, see Eq. (1.118), and

$$
A \equiv A(r_0, \Lambda) = \frac{1}{(2\pi)^d} \int_0^\Lambda \frac{\mathrm{d}^d\mathbf{q}}{q^2 + r_0}.
\tag{1.122}
$$

Note that we are considering only connected Feynman diagrams, as it is prescribed by analisis of the series. Two comments are in place here: the series of diagrams we have chosen among many others is a very special one; the quantity A is momentum independent, and obviously divergent in the limit $\Lambda \to \infty$ for every $d \geq 2$. Let us ignore these facts for a while. From Eq. (1.122) one obtains the result

$$
[\hat{G}(k)]^{-1} = k^2 + r_0 + \frac{u_0}{2}A(r_0, \Lambda).
\tag{1.123}
$$

The quantity $[\hat{G}(k)]^{-1}$, called *two-point vertex function*, is of particular physical importance; its value at $k = 0$ is related to the susceptibility of the system. Here it was obtained, using the terminology of the Feynman diagrams, in one loop approximation.

One can see that many other diagrams contribute to the two-point vertex function. Summing up the group of diagrams of the same kind, we can write the formal result

$$[\hat{G}(k)]^{-1} = k^2 + r^0 - \Sigma(k, r_0, u_0, \Lambda), \tag{1.124}$$

where $\Sigma(k, r_0, u_0, \Lambda)$ is the sum of the so-called *1-particle irreducible self-energy* diagrams. To give an idea of how to remove the divergences that appear in integrals like (1.122), let us add and subtract the term $(1/2)r_{0c}\phi^2$ in the Hamiltonian (1.105). Then, instead of Eq. (1.124) we obtain

$$[\hat{G}(k)]^{-1} = k^2 + r - [\Sigma(k, r, u_0, \Lambda) - r_{0c}], \tag{1.125}$$

where $r = r_0 - r_{0c}$. The critical temperature is now given by the condition

$$r_{0c} = \Sigma(0, 0, u_0, \Lambda), \tag{1.126}$$

hence it depends on the coupling constant u_0 and the cut-off parameter Λ. After substitution of Eq. (1.126) into Eq. (1.125), the latter takes the form

$$[\hat{G}(k)]^{-1} = k^2 + r - [\Sigma(k, r, u_0, \Lambda) - \Sigma(0, 0, u_0, \Lambda)]. \tag{1.127}$$

Now we have an algorithm for obtaining different approximations in powers of u_0. The simplest one is up to first order in u_0, when the self-energy Σ is given by the one-loop diagram as in Eq. (1.122), with r_0 replaced by r. By inserting $-(u_0)/2A(r, \Lambda)$ in Eq.(1.127), we obtain finally

$$
\begin{aligned}
[\hat{G}(k)]^{-1} &= k^2 + r + \frac{u_0}{2(2\pi)^d} \int_0^\Lambda \mathrm{d}p\, p^{d-1} \left(\frac{1}{p^2 + r} - \frac{1}{p^2} \right) \\
&= k^2 + r \left[1 - \frac{u_0}{2(2\pi)^d} \int_0^\Lambda \mathrm{d}p \frac{p^{d-3}}{p^2 + r} \right].
\end{aligned}
\tag{1.128}
$$

The last integral is convergent in the limit $\Lambda \to \infty$ if $2 < d < 4$.

Next we calculate the critical exponent γ. From Eq. (1.128) we obtain the shifted temperature

$$t = r + \frac{u_0 r^{\frac{d-2}{2}}}{2(2\pi)^d} \int_0^{\Lambda/\sqrt{r}} \mathrm{d}p \frac{p^{d-3}}{p^2 + 1}. \tag{1.129}$$

There are no problems with ultraviolet divergence of the last integral, but *infrared divergence* arise in the limit $r \to 0$ at dimensions $d < 4$. However, at $d > 4$ we simply have $t \sim r$, and from $r = t^\gamma$ the mean field result

$\gamma = 1$ follows. A more subtle analysis of the integral in Eq. (1.129) when $2 < d < 4$ yields for the temperature shift (see, e.g., [Ravandal (1976)])

$$t = r \left\{ 1 + \frac{u_0}{(4\pi)^2 \varepsilon} \left[\frac{\pi\varepsilon/2}{\sin(\pi\varepsilon/2)} \left(\frac{\Lambda}{\sqrt{r}} \right)^\varepsilon - 1 \right] \right\}, \qquad (1.130)$$

where $\varepsilon = 4 - d$. One can see that the perturbation term becomes arbitrary large as the critical point is approached, $r \to 0^+$, and our perturbation theory breaks down. In the limit $\varepsilon \to 0^+$ the above equation reduces to

$$t = r \left[1 + \frac{g_0}{(4\pi)^2} \ln \frac{\Lambda}{r^{1/2}} \right], \qquad (1.131)$$

where $g_0 = u_0/\Lambda^\varepsilon$. How to avoid the infrared divergence is a cornerstone problem of the field theory of critical phenomena. There exist two ways of working around this problem: the so called *momentum-shell RG* (see, e.g., [Goldenfeld (1992)]) and the *field-theoretical approach* (see, e.g., [Zinn-Justin (1996)]).

Let us remark that if one considers a temperature region not very close to the critical point, i.e. when r is small but $r \neq 0$ (we do not go into the detail of what that exactly means), it is possible to set the upper limit of the integral in Eq. (1.129) equal to infinity, and thus to obtain

$$t \approx r + f(u_0, d) r^{(d-2)/2}. \qquad (1.132)$$

Since the second term in Eq. (1.132) is dominant when $2 < d < 4$, from $r = t^\gamma$ the non-mean field result $\gamma = 2/(d-2)$ follows. However, in this approximation there is no momentum dependence of the self-energy (1.124), and the critical exponent $\eta = 0$ remains the mean field one. In spite of the fact that the selection of diagrams in the Hartree-Fock approximation looks rather arbitrary, and the assumption to consider Eq. (1.132) is quite formal, there exist models that give exactly the above results for the critical exponents. The problem can be clarified by using the techniques of the $1/n$ expansion, n being the number of field components. It is possible to check that when $n \to \infty$ in the framework of a n-component version of the model (1.105), the only surviving diagrams are those included in Eq. (1.121), see, e.g., [Ravandal (1976)] and [Parisi (1988)].

1.6.4 *The Renonormalization Group: Generalities*

All the different RG approaches exploit the idea of mapping, under some rules, the initial parameters which define the physical model onto some other parameters. The mapping emphasizes those aspects of the problem that are of basic interest. In critical phenomena, where the long-distance effects are of interest, this can be done through some kind of coarse-graining of the short-distance degrees of freedom, of the form that we have discussed earlier in the context of the Ginzburg-Landau theory. For example, by using the Kadanoff-cell procedure (see, e.g., [Goldenfeld (1992)] and [Stanley (1999)]), we can define the *block spins*

$$M_I \equiv \frac{1}{m_b b^d} \sum_{i \in I} M_i, \tag{1.133}$$

where m_b is the average magnetization of the spins M_i in a block of linear size ba. Under this mapping the block spins have the same magnitude as the original spins M_i.

Let us now make the assumption that the block spins interact with each other and with an effective external field in the same manner, as the original spins do. This assumption implies that the parameters $\{K\}$ of the model have to be redefined in a way that preserves the form of the effective Hamiltonian. The transformation of the parameters can be put in the form

$$\{K'\} = \mathcal{R}_b(\{K\}) \tag{1.134}$$

The function $\mathcal{R}_b(\{K\})$ is termed *renormalization group transformation*; it depends on the length rescaling parameter factor b and describes how the parameters $\{K\}$ change after rescaling of space and local operators. An exact and very illuminating example of this transformation in the case of one dimensional percolation is described in [Stanley (1999)]. Let us note that the lattice spacing between the original spins is a, while the lattice spacing between the block spins is ba. After the transformation (1.134) the new correlation length $\xi'(\{K'\})$ is smaller than the original one $\xi(\{K\})$ by a factor of b,

$$\xi'(\{K'\}) = b^{-1}\xi(\{K\}), \tag{1.135}$$

which means that the system is further away from the critical point than the original system. Thus, the effect of a succession of such transformations is to take the system away from criticality. However, if the system is initially *at* its critical point, i.e. $\xi = \infty$, then by Eq. (1.135) ξ' is also infinite. A *necessary condition* of this to occur is $K' = K$. The value of K for which

$$\{K^*\} = \mathcal{R}_b(\{K^*\}) \tag{1.136}$$

is called *critical fixed point* $\{K^*\}$ of the transformation $\mathcal{R}_b(\{K\})$. The fixed point is physically significant, because the system at it remains invariant under changes of length scale. In general, several fixed points may exist. For a given fixed point $\{K^*\}$ one can define an *attraction basin*, or *critical manifold*, which is the set of all points in $\{K\}$-space such that after infinite number of iterations of $\mathcal{R}_b$ they "flow" to, and reach $\{K^*\}$. It can be easily shown that all points in the basin of attraction of a critical fixed point have infinite correlation length. This fact explains the phenomenon *universality of the critical behavior.*

It is important to characterize the stability of any fixed point. In general, the map (1.134) is nonlinear and its analysis is very difficult. However, this problem can be studied by linearizing the RG equations around the fixed point $\{K^*\}$, where $\mathcal{R}_b(\{K\})$ is assumed to be differentiable. Then the linearized RG transformation around the point $\{K\} = \{K^*\}$ reads

$$K'_a - K^*_a \simeq \sum_b T_{ab}(K'_b - K^*_b), \tag{1.137}$$

where

$$T_{ab} = \partial \mathcal{R}^a_b(\{K\})/\partial K_b \big|_{\{K\}=\{K^*\}} . \tag{1.138}$$

Since the matrix T_{ab} is not symmetric in general, there is no reason for its left and right eigenvectors to be equal. Denoting the eigenvalues of T_{ab} by λ^i (assumed to be real), and its left eigenvectors by $\{\phi^i\}$, we can write

$$\sum_a \phi^i_a T_{ab} = \lambda^i \phi^i_b. \tag{1.139}$$

Let us define the *scaling variables* (or *scaling fields*) $\{u_i\}$ as coordinates of the deviation of $\{K\}$ from $\{K^*\}$ along the direction of the eigenvectors $\{\phi^i\}$:

$$u_i := \sum_a \phi^i_a(K_a - K^*_a), \tag{1.140}$$

Under the RG transformation they transform without mixing with each other,

$$
\begin{aligned}
u_i' &= \sum_a \phi_a^i (K_a' - K_a^*) = \sum_{ab} \phi_a^i T_{ab}(K_b - K_b^*) \\
&= \sum_b \lambda^i \phi_b^i (K_b - K_b^*) = \lambda^i u_i.
\end{aligned}
\tag{1.141}
$$

It is convenient to introduce the *renormalization group eigenvalues* y_i by the relation $\lambda^i = b^{y_i}$. For a given fixed point one distinguishes three types of scaling variables:

(i) if $y_i > 0$, u_i is said to be *relevant*,

(ii) if $y_i < 0$, u_i is said to be *irrelevant*,

(iii) if $y_i = 0$, u_i is said to be *marginal*.

The type of the y_i's shows how the repeated RG iterations change the scaling variables u_i: they grow in case (i), shrink in case (ii), and do not change in case (iii). The relevant scaling variables and the corresponding RG eigenvalues are related to RG flows away from the fixed point $\{K^*\}$. They define the local stability of the fixed point. The number of relevant operators is equal to the number of physical parameters that is necessary to fix in order to place the system on the critical surface. The number of the relevant eigenvalues is equal to the co-dimension of the critical surface; the latter is defined as the difference between the dimensionalities of the space of coupling constants and the critical surface.

The irrelevant eigenvalues correspond to flow into the fixed point.

The marginal eigenvalues are associated with logarithmic corrections to scaling; they are important near the upper and lower critical dimensions of the system.

Here we emphasize that there is no true small parameter in the RG equations. Hence, only approximate methods based on artificial small parameters, are available for their solution. Such a small parameter is $\varepsilon = d_u - d$, provided one can regard the spatial dimension d as a continuous variable. The idea is to deduce the universal properties of the system related to one fixed point (nontrivial) from those related to another fixed point (trivial). Suppose that there is a perturbation theory around the trivial fixed point. Then, if ε determines how close are the two fixed points, and if ε can be done arbitrary small, then the nontrivial fixed point may become accessible by the perturbation theory. Since we shall not use such approximate methods in this book, we refer the reader for details on the Wilson - Fisher

ε-expansion to the textbooks, e.g., [Ma (1976)], [Ravandal (1976)], [Goldenfeld (1992)], [Binney et. al. (1992)].

1.6.5 *Scaling variables and critical exponents*

To illustrate how the critical exponents can be obtained from the eigenvalues y_i, let us consider the universality class of the short-range Ising model. In this case there are two relevant scaling variables (fields): thermal u_t, with eigenvalue y_t, and magnetic u_h, with eigenvalue y_h. In addition, there is an infinite number of irrelevant variables $u_i,...$ The latter will not be considered at the moment. The values of the scaling variables u_t, u_h depend analytically on the deviation (t, h) from the critical point $(0, 0)$. Close to the critical point we can assume

$$u_t = u_t^1 t + \mathcal{O}(t^2, h^2), \qquad u_h = u_h^1 h + \mathcal{O}(th). \tag{1.142}$$

Since the RG transformations preserve the functional form of the singular part of the free energy, with parameters K' instead of K, we have the equation $N' f_s(K') = N f_s(K)$, where $N' = b^{-d} N$ is the total number of blocks. Hence, we find $f_s(K) = b^{-d} f_s(K')$. After n RG steps close to the fixed point, we obtain in terms of the scaling variables (1.142),

$$f_s(u_t, u_h) = b^{-n} f_s(b^{y_t} u_t, b^{y_h} u_h) = b^{-nd} f_s(b^{ny_t} u_t^1 t, b^{ny_h} u_h^1 h). \tag{1.143}$$

Now, because of the arbitrary value of the rescaling factor b, we can choose $b = |t|^{-(1/y_t n)}$, and from Eq. (1.143) we obtain the scaling result

$$f_s(t, h) = |t|^{d/y_t} \Phi_f(h|t|^{-y_h/y_t}), \tag{1.144}$$

where the scaling function Φ_f depends on the particular system only through the scale factors u_t^1 and u_h^1. By using the definition of the critical exponents, see Section 1.5.2, and after the appropriate differentiation of Eq.(1.144), we obtain

$$\alpha = 2 - \frac{d}{y_t}, \quad \beta = \frac{d - y_h}{y_t}, \quad \gamma = \frac{2y_h - d}{y_t}, \quad \delta = \frac{y_h}{d - y_h}. \tag{1.145}$$

We see that these four exponents are expressed in terms of the two RG eigenvalues y_t and y_h. This means that two relations between them exist, for example

$$\alpha + 2\beta + \gamma = 2, \qquad \gamma = \beta(\delta - 1). \tag{1.146}$$

Analogous consideration of the spin-spin correlation function yields the scaling relation

$$G(r,t) = \frac{1}{r^{2(d-y_h)}} \Phi_G(r|t|^{1/y_t}), \tag{1.147}$$

where, for the sake of simplicity, we have set $h = 0$. For large r the correlation function $G(r,t)$ decays asymptotically as $e^{-r/\xi}$, and from Eq.(1.147) we obtain for the correlation length $\xi \propto |t|^{-1/y_t}$, so that $\nu = 1/y_t$. At the critical point the asymptotic form of the correlation function is $\propto r^{-(d-2+\eta)}$, hence $2(d - y_h) = d - 2 + \eta$. We see that the exponents ν and η of the correlation function can be expressed in terms of y_t and y_h, and that they are related to the thermodynamic exponents (1.145), through the scaling relations

$$\alpha = 2 - d\nu \qquad \text{and} \qquad \gamma = \nu(2 - \eta). \tag{1.148}$$

Let us note that the scaling relations (1.146) and (1.148) follow from the Kadanoff's block-spin transformation. Here we have demonstrated that the critical exponents are actually related to the RG eigenvalues, which provides a means of their calculation in the framework of a specific model.

Let us now suppose that, in addition to the relevant variables u_t and u_h, there is an irrelevant scaling variable u_3. Unlike the case of relevant variables only, c.f. Eq. (1.142), we have to assume now

$$u_3 = u_3^o + u_3^1 + \mathcal{O}(h^2). \tag{1.149}$$

Repeating the above consideration, instead of Eq.(1.144) we obtain

$$f_s(t, h) = |t|^{d/y_t} \Phi_f \left(\frac{h}{|t|^{y_h/y_t}}, \frac{u_3^0}{|t|^{y_3/y_t}} \right). \tag{1.150}$$

Since u_3 is irrelevant, its associated eigenvalue y_3 is negative. Then, on approaching the critical point, we see that

$$z = u_3|t|^{-|y_3|/y_t} \to 0. \tag{1.151}$$

If the scaling function $\Phi_f(x, z)$ is an analytical function of its arguments, it can be expanded in a Taylor series in z,

$$\Phi_f(x, z) = \Phi_f^0(x, 0) + \Phi_f^1(x, 0)z + \dots. \tag{1.152}$$

In this case one can see that the effect of an irrelevant variable is to generate corrections to the leading scaling term.

The fact that u_3 is irrelevant, however, does not exclude the possibility of $\Phi_f(x, z)$ being a nonanalytic function of z. In the latter case the variable z is called *dangerous irrelevant variable*. In many cases such a nonanalyticity can be established by explicit calculations, and it provides the clue to the explanation of the breakdown of some scaling relations between critical exponents, failure of finite-size scaling under some circumstances, etc. The concept of dangerous irrelevant variables has been developed by M.E. Fisher in [Fisher (1983)]. Let us see how it works when $\Phi_f(x, z)$ has a simple power law-divergence of the form

$$\Phi_f(x, z) = z^{-\mu} W(x), \qquad \mu > 0. \tag{1.153}$$

By substitution of Eq.(1.153) into Eq.(1.150) one obtains

$$f_s(t, h) \sim |t|^{(d-\mu y_3)/y_t} W(h|t|^{-y_h/y_t}). \tag{1.154}$$

In terms of the standard thermodynamic exponents the above behavior is interpreted as breakdown of the hyperscaling relation Eq.(1.148), if still assuming $\nu = 1/y_t$, since $2 - \alpha = (d - \mu y_3)\nu$. A singularity of the type (1.153) is not exceptional at all: as it has been pointed out in [Fisher (1983)] and [Shapiro and Rudnick (1986)], it occurs in the mean field regime of the spherical and continuous ϕ^4 models. The issue is discussed further in Chapter 6, where it is shown how finite-size scaling at and above the upper critical dimension is modified by the presence of a dangerous irrelevant variable.

Chapter 2

The Approximating Hamiltonian Method

In this chapter we present a concise review of the Approximating Hamiltonian Method (AHM) which allows one to solve exactly in the thermodynamic limit some special classes of quantum and classical models. The models are defined initially in a finite region Λ of the d-dimensional space R^d or Z^d. Generally speaking, they describe quantum or classical systems containing some mean-field type interaction between paricles, or quasi-particles, which has equal strength, independent of the interparticle distance (the so-called equivalent-neighbors interaction) and inversely proportional to tne total number of particles in the system. The general theory of the method aims at proving that if a given model Hamiltonian belongs to a well-defined class, then the thermodynamic limit of the free energy density exists and equals the one corresponding to a simpler approximating Hamiltonian. In most cases one is interested in one-particle approximating Hamiltonians which admit exact calculation of the free energy for *systems in a finite volume*. As a rule, the approximating Hamiltonian depends on a (set of) variational parameter(s), which satisfy a variational principle for the density of the thermodynamic potential. Usually, such systems exhibit mean-field type critical behavior in the thermodynamic limit. However, there exists a variety models, the approximating Hamiltonian of which contains exactly solvable short-range interaction. These models have a non-trivial critical behavior with developed fluctuations, see the example given in Sections 2.3. The exact solvability of such models makes possible the rigorous study of finite-size effects in quantum systems *defined by the approximating Hamiltonian*. The proof of the thermodynamic equivalence of the original and approximating models guarantees that their bulk critical behavior

49

is identical. However, the finite-size scaling functions for the original and approximating systems may be different.

2.1　Background ideas

The Approximating Hamiltonian Method provides a rigorous approach to the study of some classes of statistical mechanical systems in the thermodynamic limit. The method consists of the following interrelated ingredients:

a) Description of classes of model systems which admit a rigorous treatment in terms of a more simple approximating Hamiltonian (AH);

b) Rules according to which the approximating Hamiltonian is constructed from the original one;

c) Mathematical techniques for derivation of bounds which prove the thermodynamic (and, when possible, statistical) equivalence of the approximating and original Hamiltonians;

d) Investigation of the thermodynamic and statistical properties of the system described by the approximating Hamiltonian (to be called approximating system).

For the first time the idea of the AHM has been suggested by N.N. Bogolyubov in his paper on the theory of the weakly non-ideal Bose-gas [Bogolyubov (1947)]. He conjectured there that, under the existence of Bose condensate in the system, the normalized by the sqare root of the volume creation/annihilation operators for particles with zero momentum can be replaced by c-numbers in the model Hamiltonian. The value of these complex-conjugate numbers was determined on the grounds of thermodynamic arguments. Thus, the initial model Hamiltonian was replaced by an approximating (trial) one, which has the advantage to be easilly diagonalizable by a canonical u-v transformation, introduced in the same paper. The furhter development of the AHM took place in the framework of the Bardeen-Cooper-Schrieffer (BCS) reduced model of superconductivity [Bardeen et. al. (1957)]. In 1957 the corresponding approximating Hamiltonian was suggested and diagonalized by means of a u-v canonical transformation [Bogolyubov et. al. (1957)]. Somewhat later N.N. Bogolyubov [Bogolyubov (1960)] rigorously proved that the ground state energy density and the zero-temperature Green functions for the model and approximating Hamiltonians coincide in the thermodynamic limit. The foundations of the modern formulation of the AHM have been laid down by N.N. Bogolyubov Jr., see [Bogolyubov Jr. (1972)] and references therein. The essential generalization to classes of quantum systems with separable interaction, considered at nonzero temperatures, has been achieved by using *the Bogolyubov variational principle* for the free energy density and a special majorization technique based on integration over external sources. Some major restrictions on the applicability of the method, such as the quadratic form of the interaction Hamiltonian and its boundedness in norm, were removed by further extensions of the AHM, see the book [Bogolyubov Jr. et. al. (1981)] and the review article [Bogolyubov Jr. et. al. (1984)].

The systems under consideration are defined for a finite number of particles in a bounded domain Λ, see Section 1.2. The Hamiltonian $\mathcal{H}_\Lambda$ is defined as a self-adjoint operator acting in an appropriate Hilbert space, and the corresponding free energy density $f_\Lambda[\mathcal{H}_\Lambda]$ is assumed to exist. For the sake of simplicity, and when no confusion arises, we do not explicitly distinguish between a Hamiltonian $\mathcal{H}_\Lambda$, describing a system of fixed number of particles N in Λ, and the statistical operator $\mathcal{H}_\Lambda - \mu\mathcal{N}$ in the grand canonical ensemble, where μ is the chemical potential and $\mathcal{N}$ is the particles number operator. In both cases the thermodynamic limit will be denoted by t$-$lim. The norm of a bounded operator A will be denoted by $\|A\|$. The symbol A° stays for both the operator A and its adjoint $A^\dagger$. As usual, $[A, B] = AB - BA$ denotes the commutator.

Each class of models, admitting exact solution by the AHM, requires certain structure of the Hamiltonian and special properties of the operators in terms of which its structure is defined. A common feature is that the interaction Hamiltonian is an extensive function of averaged over the volume of the system local operators. We remind the reader that the space average of a local (or quasilocal) operator $A(x)$, $x \in \Lambda$, over the region $\Lambda \subset \mathbf{R}^d$ is defined by the integral

$$A_\Lambda = |\Lambda|^{-1} \int_\Lambda \mathrm{d}x A(x).$$

In the case when $\Lambda \subset \mathbf{Z}^d$, the corresponding discrete measure $\mathrm{d}x$ induces summation over the sublattice Λ. The operator A_Λ is called an intensive (or normalized) operator. The heuristic rule for construction of the approximating Hamiltonian consists in linearization of the original interaction Hamiltonian in the deviations of such operators from c-numbers. The latter are considered as variational parameters, chosen so as to minimize the contribution of the residual interaction Hamiltonian in the free energy density. The main mathematical tool for setting an upper bound on the difference in the free energy densities for the original and approximating Hamiltonians is *the Bogolyubov variational principle* [Bogolyubov (1947)] and the inequalities following from it. The equivalence of the Bogolyubov and Feynman variational principles, and their relation to other quantum statistical variational principles, is discussed in [Dörre et. al. (1979)].

The Bogolyubov inequalities. Let $\mathcal{H}_\Lambda = \mathcal{H}_\Lambda^{(0)} + \mathcal{H}_\Lambda^{(1)}$, where $\mathcal{H}_\Lambda^{(k)}$, $k = 0, 1$, be self-adjoint operators, $\exp(-\beta\mathcal{H}_\Lambda^{(0)})$ be trace-class*, and let the average values of $\mathcal{H}_\Lambda^{(1)}$ in the Gibbs ensembles for the Hamiltonians $\mathcal{H}_\Lambda$ and $\mathcal{H}_\Lambda^{(0)}$ exist. Then the following inequalities hold

$$|\Lambda|^{-1}\langle\mathcal{H}_\Lambda^{(1)}\rangle_{\mathcal{H}_\Lambda} \leq f_\Lambda[\mathcal{H}_\Lambda] - f_\Lambda[\mathcal{H}_\Lambda^{(0)}] \leq |\Lambda|^{-1}\langle\mathcal{H}_\Lambda^{(1)}\rangle_{\mathcal{H}_\Lambda^{(0)}}, \qquad (2.1)$$

where

$$\langle\cdots\rangle_{\Gamma_\Lambda} = \mathrm{Tr}[\cdots\exp(-\beta\Gamma_\Lambda]/Z[\Gamma_\Lambda] \qquad (2.2)$$

is the average value in the Gibbs ensemble for the Hamiltonian Γ_Λ, and

$$f_\Lambda[\Gamma_\Lambda] = -(\beta|\Lambda|)^{-1}\ln Z[\Gamma_\Lambda] \qquad (2.3)$$

is the density of the corresponding thermodynamic potential.

Now we pass to the formulation of the main results of the AHM for several representative classes of model systems.

2.2 Systems with separable attraction

Historically, the first example of this class is the reduced model Hamiltonian suggested by Bardeen, Cooper and Schrieffer (*BCS model*) in the theory of superconductivity [Bardeen et. al. (1957)]. It is instructive to begin our consideration with this specific example.

---

*An operator $\mathcal{A}$ is called trace-class if $\sum_n |(\varphi_n, \mathcal{A}\varphi_n)| < \infty$ for any othonormal basis $\{\varphi_n\}$ in the Hilbert space.

The Bardeen-Cooper-Schrieffer model

In the second-quantization representation the Hamoltonian of the BCS model has the form

$$\mathcal{H}_\Lambda = \mathcal{T}_\Lambda - g|\Lambda|A_\Lambda A_\Lambda^\dagger, \qquad g > 0, \tag{2.4}$$

where

$$\mathcal{T}_\Lambda = \sum_{\mathbf{k},\sigma}(\varepsilon_\mathbf{k} - \mu)\mathbf{a}_{\mathbf{k},\sigma}^\dagger \mathbf{a}_{\mathbf{k},\sigma} \tag{2.5}$$

is the kinetic energy term, $\mathbf{a}_{\mathbf{k},\sigma}^\dagger$ $(\mathbf{a}_{\mathbf{k},\sigma})$ is the creation (annihilation) operator of an electron in the state with wavevector $\mathbf{k}$ and spin $\sigma = \pm 1$, $\varepsilon_\mathbf{k}$ is the one-electron energy and μ is the chemical potential. The second term in the right-hand side of Eq.(2.4), where

$$A_\Lambda = |\Lambda|^{-1}\sum_{\mathbf{k},\sigma}\lambda(\mathbf{k},\sigma)\mathbf{a}_{\mathbf{k},\sigma}^\dagger \mathbf{a}_{-\mathbf{k},-\sigma}^\dagger, \tag{2.6}$$

describes attractive interaction of strength $g > 0$ between electrons with opposite quasi-momenta and opposite spins. Here $\lambda(\mathbf{k},\sigma)$ is a real-valued cut-off function which restricts the attractive electron-electron interaction to states close to the Fermi surface. For our purposes it is sufficient to assume that $\lambda(\mathbf{k},\sigma)$ satisfies the conditions

$$\lambda(-\mathbf{k},\sigma) = -\lambda(\mathbf{k},\sigma), \qquad |\lambda(\mathbf{k},\sigma)| \le c_1, \qquad |\Lambda|^{-1}\sum_{\mathbf{k},\sigma}|\lambda(\mathbf{k},\sigma)| \le c_2, \tag{2.7}$$

with some constants c_1 and c_2. The corresponding approximating Hamiltonian, $\mathcal{H}_\Lambda^{(0)}(a)$, depends on a complex parameter $a \in \mathbf{C}$ ($a^\dagger$ denotes its complex conjugate) and has the form

$$\mathcal{H}_\Lambda^{(0)}(a) = \mathcal{T}_\Lambda - g|\Lambda|(a^\dagger A_\Lambda + aA_\Lambda^\dagger - aa^\dagger). \tag{2.8}$$

The proper choice of the parameter a will be given below. Now we note that the right-hand side of Eq. (2.8) represents a quadratic in the Fermi creation/annihilation operators form which can be diagonalized by a suitable canonical transformation.

Absolute minimum principle

Note that the model Hamiltonian (2.4) can be written identically as

$$\mathcal{H}_\Lambda = \mathcal{H}_\Lambda^{(0)}(a) + \mathcal{H}_\Lambda^{(1)}(a), \tag{2.9}$$

where the residual Hamiltonian

$$\mathcal{H}_\Lambda^{(1)}(a) = -g|\Lambda|(A_\Lambda - a)(A_\Lambda^\dagger - a^\dagger) \tag{2.10}$$

is quadratic in the deviation of the intensive operators $A_\Lambda^\diamond$ from the complex numbers $a^\diamond$. In view of the non-positive definiteness of $\mathcal{H}_\Lambda^{(1)}(a)$, the application of the Bogolyubov inequality (2.1) yields for all $a \in \mathbf{C}$

$$0 \le f_\Lambda[\mathcal{H}_\Lambda^{(0)}(a)] - f_\Lambda[\mathcal{H}_\Lambda] \le g\langle(A_\Lambda - a)(A_\Lambda^\dagger - a^\dagger)\rangle_{\mathcal{H}_\Lambda}. \tag{2.11}$$

Hence it follows that the best approximation to the free energy density of the model system is given by the *absolute minimum principle*,

$$\min_a f_\Lambda[\mathcal{H}_\Lambda^{(0)}(a)] = f_\Lambda[\mathcal{H}_\Lambda^{(0)}(\bar{a}_\Lambda)]. \tag{2.12}$$

The fact that the absolute minimum of the free energy density $f_\Lambda[\mathcal{H}_\Lambda^{(0)}(a)]$ is attained at a finite value $a = \bar{a}_\Lambda$, which depends on the thermodynamic parameters of the Gibbs ensemble, as well as on the size and shape of the domain Λ, can be readily proved by using the boundedness of the kinetic free energy density and the norm-boundedness of the operator (2.6):

$$|f_\Lambda[\mathcal{T}_\Lambda]| \le M_1, \qquad \|A_\Lambda\| \le M_2. \tag{2.13}$$

Here M_1 and M_2 are constants independent of the volume $|\Lambda|$. The last inequality follows from the norm-boundedness of the Fermi creation and annihilation operators and the third condition in (2.7). Note that the absolute minimum condition (2.12) and the Bogolyubov inequality (2.11) imply the following upper bound on the difference between the approximating and original free energy densities:

$$0 \le \min_a f_\Lambda[\mathcal{H}_\Lambda^{(0)}(a)] - f_\Lambda[\mathcal{H}_\Lambda] \le f_\Lambda[\mathcal{H}_\Lambda^{(0)}(\langle A_\Lambda\rangle_{\mathcal{H}})] - f_\Lambda[\mathcal{H}_\Lambda] \le$$
$$g\langle(A_\Lambda - \langle A_\Lambda\rangle_{\mathcal{H}_\Lambda})(A_\Lambda^\dagger - \langle A_\Lambda^\dagger\rangle_{\mathcal{H}_\Lambda})\rangle_{\mathcal{H}_\Lambda}. \tag{2.14}$$

Here it is in place to mention that the attempt to prove directly that the correlation function in the right-hand side of inequality (2.14) tends to zero as $|\Lambda| \to \infty$ may turn out a rather difficult task. A more efficient way of solving the problem, which avoids the evaluation of the above correlator, is based on the special majorization technique due to N.N. Bogolyubov Jr., see [Bogolyubov Jr. (1972)]. We shall give a sketch of it in order to illustrate the general sufficient conditions for the thermodynamic equivalence.

Majorization technique

One starts by including source terms in the Hamiltonian $\mathcal{H}_\Lambda$ which are linear in the operators $A_\Lambda^\diamond$ and break its gauge invariance,

$$\mathcal{H}_\Lambda(\nu) = \mathcal{H}_\Lambda - |\Lambda|(\nu^\dagger A_\Lambda + \nu A_\Lambda^\dagger). \tag{2.15}$$

Here $\nu^\diamond$ is a complex field conjugate to the operator $A_\Lambda^\diamond$. The same source terms are added to the approximating Hamiltonian, which has already broken gauge invariance, in order to keep the residual Hamiltonian unchanged:

$$\mathcal{H}_\Lambda^{(0)}(a, \nu) = \mathcal{T}_\Lambda - |\Lambda|[(ga^\dagger + \nu^\dagger)A_\Lambda + (ga + \nu)A_\Lambda^\dagger - gaa^\dagger]. \tag{2.16}$$

For the difference between the free energies in the presence of sources,

$$\Delta_\Lambda(\nu) \equiv f_\Lambda[\mathcal{H}_\Lambda^{(0)}(\bar{a}_\Lambda(\nu), \nu)] - f_\Lambda[\mathcal{H}_\Lambda(\nu)], \tag{2.17}$$

inequalities (2.14) yield

$$0 \le \Delta_\Lambda(\nu) \le g\langle \delta A_\Lambda \delta A_\Lambda^\dagger \rangle_{\mathcal{H}_\Lambda(\nu)}, \tag{2.18}$$

where $\delta A_\Lambda^\diamond \equiv A_\Lambda^\diamond - \langle A_\Lambda^\diamond \rangle_{\mathcal{H}_\Lambda(\nu)}$. Our aim now is to majorize the correlation function in the right-hand side of (2.18) in terms of the second derivative of the model free energy density with respect to the complex field ν:

$$\frac{\partial^2}{\partial \nu^\dagger \partial \nu} f_\Lambda[\mathcal{H}_\Lambda(\nu)] = -\beta|\Lambda|(\delta A_\Lambda, \delta A_\Lambda)_{\mathcal{H}_\Lambda(\nu)}. \tag{2.19}$$

Here the $(J_\Lambda, J_\Lambda)_{\mathcal{H}}$ denotes the inner product.

$$(J_\Lambda, J_\Lambda)_{\mathcal{H}_\Lambda} \equiv (\beta Z_\Lambda[\mathcal{H}])^{-1} \int_0^\beta d\tau \, \mathrm{Tr}\left[e^{-(\beta-\tau)\mathcal{H}} J_\Lambda^\dagger e^{-\beta\mathcal{H}} J_\Lambda \right]. \tag{2.20}$$

The following inequality obtained in [Bogolyubov Jr. (1966)] holds irrespectively of the particular definition of the operators A_Λ, $A_\Lambda^\dagger$:

$$\langle \delta A_\Lambda \delta A_\Lambda^\dagger \rangle_{\mathcal{H}_\Lambda(\nu)} \leq (\beta|\Lambda|)^{-1} \left(-\frac{\partial^2}{\partial \nu^\dagger \partial \nu} f_\Lambda[\mathcal{H}_\Lambda(\nu)] \right)$$

$$+ \frac{\{2\langle [\mathcal{H}_\Lambda(\nu), A_\Lambda^\dagger][A_\Lambda, \mathcal{H}_\Lambda(\nu)]\rangle_{\mathcal{H}_\Lambda(\nu)}\}^{1/3}}{|\Lambda|^{2/3}} \left(-\frac{\partial^2}{\partial \nu^\dagger \partial \nu} f_\Lambda[\mathcal{H}_\Lambda(\nu)] \right)^{2/3} . \quad (2.21)$$

An upper bound on the average value $\langle [\mathcal{H}_\Lambda(\nu), A_\Lambda^\dagger][A_\Lambda, \mathcal{H}_\Lambda(\nu)]\rangle_{\mathcal{H}_\Lambda(\nu)}$ can be obtained by impose the additional sufficient conditions

$$\|[A_\Lambda, A_\Lambda^\dagger]\| \leq |\Lambda|^{-1} M_3, \qquad \|[\mathcal{T}_\Lambda, A_\Lambda^\diamond]\| \leq M_4, \qquad (2.22)$$

where M_3 and M_4 are some constants independent of the volume $|\Lambda|$. In the case of the reduced BCS model, the first inequality (2.22) is fulfilled due to the explicit form of the operators involved, see also conditions (2.7). To satisfy the second inequality (2.22) one has to assume in addition

$$|\Lambda|^{-1} \sum_{\mathbf{k},\sigma} |\varepsilon_{\mathbf{k}} \lambda(\mathbf{k},\sigma)| \leq c_3. \qquad (2.23)$$

Thus one can readily show that

$$\|[\mathcal{H}_\Lambda(\nu), A_\Lambda^\diamond]\| \leq M_4 + (gM_2 + |\nu|)M_3 \equiv K(g,\nu), \qquad (2.24)$$

hence, the considered average value is bounded from above by $K^2(g,\nu)$.

The procedure for proving the thermodynamic equivalence of the model and approximating systems exploits essentially the bondedness of the first derivatives of the corresponding free energy densities:

$$\left| \frac{\partial}{\partial \nu} f_\Lambda[\mathcal{H}_\Lambda(\nu)] \right| = \left| \langle A_\Lambda^\dagger \rangle_{\mathcal{H}_\Lambda(\nu)} \right| \leq M_2,$$

$$\left| \frac{\partial}{\partial \nu} f_\Lambda[\mathcal{H}_\Lambda^{(0)}(\bar{a}_\Lambda(\nu),\nu)] \right| = \left| \langle A_\Lambda^\dagger \rangle_{\mathcal{H}_\Lambda^{(0)}(\bar{a}_\Lambda(\nu),\nu)} \right| \leq M_2. \qquad (2.25)$$

The last upper bound holds due to the analyticity of $f_\Lambda[\mathcal{H}_\Lambda^{(0)}(a,\nu)]$ as a function of $a \in \mathbf{C}$ and the fact that it reaches extremun (minimum) at the point $a = \bar{a}_\Lambda(\nu)$. As a next step one considers the inequality

$$\Delta_\Lambda(\nu) \le g(\beta|\Lambda|)^{-1}\left(-\frac{\partial^2}{\partial\nu^\dagger\partial\nu}f_\Lambda[\mathcal{H}_\Lambda(\nu)]\right)$$
$$+2^{1/3}gK^{2/3}(g,\nu)|\Lambda|^{-2/3}\left(-\frac{\partial^2}{\partial\nu^\dagger\partial\nu}f_\Lambda[\mathcal{H}_\Lambda(\nu)]\right)^{2/3}, \qquad (2.26)$$

which follows from (2.18), (2.21) and (2.24), and takes the average value of both sides of it over a domain in the space of the external source field $\nu \in \mathbf{C}$ containing the origin $\nu = 0$. We present here a version of the proof which does not rely upon the fact that both $f_\Lambda[\mathcal{H}_\Lambda^{(0)}(\bar{a}_\Lambda(\nu),\nu)]$ and $f_\Lambda[\mathcal{H}_\Lambda(\nu)]$ depend only on the absolute value $|\nu|$ of the complex field, as is the case of the BCS model. We note that by construction the above free energy densities are real-valued functions of the two real variables $x = \mathrm{Re}(\nu)$ and $y = \mathrm{Im}(\nu)$, i.e.

$$f_\Lambda[\mathcal{H}_\Lambda^{(0)}(\bar{a}_\Lambda(\nu),\nu)] \equiv f_\Lambda^{(0)}(x,y), \qquad f_\Lambda[\mathcal{H}_\Lambda(\nu)] \equiv f_\Lambda(x,y), \qquad (2.27)$$

and, correspondingly, $\Delta_\Lambda(\nu) \equiv \Delta_\Lambda(x,y)$. Let us take the average of inequality (2.26) over the square domain $D = [-\delta \le x \le \delta] \times [-\delta \le y \le \delta]$, where the value of $\delta > 0$ will be chosen below in an optimal way. By the Mean Value Theorem the left-hand side of (2.26) yields $\Delta_\Lambda(\xi,\eta)$, where (ξ,η) is a point in D. Due to the bounds on the first partial derivatives of $\Delta_\Lambda(x,y)$ with respect to x and y, which follow from Eqs. (2.25), one has

$$\Delta_\Lambda(0,0) \equiv f_\Lambda[\mathcal{H}_\Lambda^{(0)}(\bar{a}_\Lambda)] - f_\Lambda[\mathcal{H}_\Lambda] \le \Delta_\Lambda(\xi,\eta) + 8\delta M_2. \qquad (2.28)$$

An upper bound on the averaged over D right-hand side of Eq. (2.26) follows by applying the Hölder inequality, see, e.g., [Marshall and Olkin (1979)], to the second term and using the bounds on the first derivatives of $f_\Lambda(x,y)$:

$$\frac{g}{\beta|\Lambda|}\int_{-\delta}^{\delta}\int_{-\delta}^{\delta}\frac{\mathrm{d}x\mathrm{d}y}{4\delta^2}\left[-\frac{1}{4}\left(\frac{\partial^2}{\partial x^2}+\frac{\partial^2}{\partial y^2}\right)f_\Lambda(x,y)\right]+$$
$$\frac{2^{1/3}gK^{2/3}(g,\delta)}{|\Lambda|^{2/3}}\int_{-\delta}^{\delta}\int_{-\delta}^{\delta}\frac{\mathrm{d}x\mathrm{d}y}{4\delta^2}\left[-\frac{1}{4}\left(\frac{\partial^2}{\partial x^2}+\frac{\partial^2}{\partial y^2}\right)f_\Lambda(x,y)\right]^{2/3} \le$$
$$\frac{gM_2}{\beta|\Lambda|\delta}+\frac{g}{2^{1/3}}\left(\frac{K(g,\delta)M_2}{|\Lambda|\delta}\right)^{2/3} \qquad (2.29)$$

Hence, by combining inequalities (2.28) and (2.29), one sees that the optimal choice of δ is $\delta \propto |\Lambda|^{-2/5}$, which implies

$$0 \leq \Delta_\Lambda(0,0) \equiv f_\Lambda[\mathcal{H}_\Lambda^{(0)}(\bar{a}_\Lambda)] - f_\Lambda[\mathcal{H}_\Lambda] \leq \epsilon_\Lambda \to 0, \quad \text{as } |\Lambda| \to \infty, \quad (2.30)$$

with $\epsilon_\Lambda = O(|\Lambda|^{-2/5})$.

Extension of the model

The above result has been extended to the general class of model Hamiltonians of the form [Bogolyubov Jr. (1972)].

$$\mathcal{H}_\Lambda = \mathcal{T}_\Lambda - |\Lambda| \sum_{k=1}^{n} g_k A_{k,\Lambda} A_{k,\Lambda}^\dagger, \qquad g_k > 0, \qquad (2.31)$$

where $\mathcal{T}_\Lambda$ is a self-adjoint operator which satisfies the first condition (2.13). Let, in addition, the following inequalities hold for some constants, M_i, $i = 2, \ldots, 4$, independent of the volume $|\Lambda|$, and for all $k, l = 1, \ldots, n$:

$$\|A_{k,\Lambda}^\diamond\| \leq M_2, \qquad (2.32)$$

$$\|[A_{k,\Lambda}, A_{l,\Lambda}^\diamond]\| \leq |\Lambda|^{-1} M_3, \qquad \|\mathcal{T}_\Lambda, A_{k,\Lambda}^\diamond]\| \leq M_4. \qquad (2.33)$$

In this case the approximating Hamiltonian has the form

$$\mathcal{H}_\Lambda^{(0)}(a) = \mathcal{T}_\Lambda - |\Lambda| \sum_{k=1}^{n} g_k (a_k A_{k,\Lambda}^\dagger + a_k^\dagger A_{k,\Lambda} - a_k a_k^\dagger), \qquad (2.34)$$

where $a = (a_1, \ldots, a_n) \in \mathbf{C}^n$. The free energy density of the model system with Hamiltonian (2.31) is equivalent in the thermodynamic limit to the approximating free energy density

$$f[\mathcal{H}_\Lambda^{(0)}(\bar{a}_\Lambda)] = \min_{a} f[\mathcal{H}_\Lambda^{(0)}(a)], \qquad (2.35)$$

since

$$0 \leq f_\Lambda[\mathcal{H}_\Lambda^{(0)}(\bar{a}_\Lambda)] - f_\Lambda[\mathcal{H}_\Lambda] \leq \epsilon_\Lambda \to 0, \quad \text{as } |\Lambda| \to \infty. \qquad (2.36)$$

We mention that by using the above described procedure one can obtain a better estimate for ϵ_Λ in (2.36), namely $\epsilon_\Lambda = O(|\Lambda|^{-1/2})$, if instead of the inequalty (2.21) due to Bogolyubov Jr., one uses the second of *the Ginibre inequalities* [Ginibre (1968)]

$$(\delta A_\Lambda, \delta A_\Lambda)_{\mathcal{H}_\Lambda(\nu)} \le \frac{1}{2}\langle \delta A_\Lambda \delta A_\Lambda^\dagger + \delta A_\Lambda^\dagger \delta A_\Lambda \rangle_{\mathcal{H}_\Lambda(\nu)} \le (\delta A_\Lambda, \delta A_\Lambda)_{\mathcal{H}_\Lambda(\nu)}$$

$$+ \frac{1}{2}\{\beta \langle [A_\Lambda, [\mathcal{H}_\Lambda(\nu), A_\Lambda^\dagger]] \rangle_{\mathcal{H}_\Lambda(\nu)}\}^{1/2} (\delta A_\Lambda, \delta A_\Lambda)^{1/2}_{\mathcal{H}_\Lambda(\nu)}. \tag{2.37}$$

However, to obtain an upper bound on $\langle [A_\Lambda, [\mathcal{H}_\Lambda(\nu), A_\Lambda^\dagger]] \rangle_{\mathcal{H}(\nu)}$, one needs two additional sufficient conditions:

$$\||[A_\Lambda^\diamond, [A_\Lambda^\diamond, A_\Lambda]]|\| \le |\Lambda|^{-2} M_5, \quad \||[A_\Lambda^\diamond, [A_\Lambda, \mathcal{T}_\Lambda]] \le |\Lambda|^{-1} M_6, \tag{2.38}$$

where M_5 and M_6 are constants independent of $|\Lambda|$.

The generalization to a Hamiltonian with a finite number n of interaction components is straightforward. An extension to infinite number of components has been given in [Bogolyubov Jr. and Sadovnikov (1975)] under the additional condition that the series $\sum_{k=1}^n g_k$ converges as $n \to \infty$. Stronger result about the existence of an exact solution when the number n grows slower than the volume of the system $|\Lambda|$ has been proven in [Pastur and Shcherbuna (1984)].

Existence of the thermodynamic limit

The existence of the thermodynamic limit for the free energy density is one of the fundamental requirements of thermodynamics, see Section 1.3. So far all the considerations have been carried out for systems in a finite domain Λ. The results stated above ensure only that the difference between the model and approximating free energy densities tends to zero in the thermodynamic limit. The existence of the limit itself for any of these functions has not been established yet; it should be considered as a restriction on the admissible classes of Hamiltonians. For the considered class of models it suffices to require the existence of the thermodynamic limit

$$t-\lim f_\Lambda[\mathcal{H}_\Lambda^{(0)}(a)] = f^{(0)}(a). \tag{2.39}$$

for all finite $a = (a_1, \ldots, a_n)$. Then the thermodynamic limit for the model free energy density exists and equals

$$t-\lim f_\Lambda[\mathcal{H}_\Lambda] = \min_{a}\{t-\lim f_\Lambda[\mathcal{H}_\Lambda^{(0)}(a)]\}. \tag{2.40}$$

The last result is important for the practical applications of the AHM, see [Bogolyubov Jr. et. al. (1981)]. The absolute minimum of the function which represents the thermodynamic limit of the approximating free energy density is reached at points $a_1 = \bar{a}_1, \ldots, a_n = \bar{a}_n$, which satisfy the limiting form of the self-consistency equations

$$a_k = \mathrm{t} - \lim \langle A_{k,\Lambda} \rangle_{\mathcal{H}_\Lambda^{(0)}(\mathrm{a})}, \qquad k = 1, \ldots, n. \tag{2.41}$$

The proof is based on the following lemma.

The Griffiths - Fisher lemma. There is a mathematical lemma which states that if a squence of convex functions converges pointwise to a limit function, then the sequence of its derivatives converges to the derivative of the limit function at the points of its continuous differentiability [Griffiths (1964)]. Here we give the more general result due to [Fisher (1965)]. Let $\{f_n(x)\}$, $x \in I \subset R$ be a sequence of convex functions which converges pointwise to $f_\infty(x)$ as $n \to \infty$. Then the left, $f'(x-0)$, and right, $f'(x+0)$, derivatives at any point $x \in I$ obey the inequalities

$$f'_\infty(x - 0) \leq \liminf_{n\to\infty} f'_n(x - 0) \leq \limsup_{n\to\infty} f'_\infty(x + 0) \leq f'_\infty(x + 0). \tag{2.42}$$

In particular, if all functions $\{f_n(x)\}$ and the limit function $f_\infty(x)$ are differentiable at a point $x_0 \in I$, then

$$\lim_{n\to\infty} f'_n(x_0) = f'_\infty(x_0). \tag{2.43}$$

This lemma is useful in proving the asymptotic closeness of certain average values in the model and approximating systems, see Section 2.5.

2.3   Systems with separable repulsion

The change of the sign of the interaction constant g in the Hamiltonian (2.4) leads to a model with separable repulsion:

$$\mathcal{H}_\Lambda = \mathcal{T}_\Lambda + g|\Lambda| A_\Lambda A_\Lambda^\dagger, \qquad g > 0. \tag{2.44}$$

The general properties (2.13) of the operators $\mathcal{T}_\Lambda$ and $A_\Lambda^\diamond$, respectively, are assumed to hold. It turns out that the class of models with separable repulsion requires some additional properties of these operators which are basically different from the ones in the case of separable attraction. At the same time, it should be emphasized that the second condition (2.13) is sufficient but not necessary: in Section 3.4 we consider an exactly solvable models of the type (2.11) with unbounded operators $A_\Lambda^\diamond$.

By following the heuristic principle of linearization, one constructs the corresponding approximating Hamiltonian,

$$\mathcal{H}_\Lambda^{(0)}(b) = \mathcal{T}_\Lambda + g|\Lambda|(b^\dagger A_\Lambda + b A_\Lambda^\dagger - b b^\dagger), \tag{2.45}$$

which depends on the complex parameter $b \in \mathbf{C}$.

Absolute maximum principle

Since the residual Hamiltonian

$$\mathcal{H}_\Lambda^{(1)}(b) \equiv \mathcal{H}_\Lambda - \mathcal{H}_\Lambda^{(0)}(b) = g|\Lambda|(A_\Lambda - b)(A_\Lambda^\dagger - b^\dagger) \tag{2.46}$$

is now non-positive definite, the application of the Bogolyubov inequality (2.1) yields for all $b \in \mathbf{C}$

$$0 \leq f_\Lambda[\mathcal{H}_\Lambda] - f_\Lambda[\mathcal{H}_\Lambda^{(0)}(b)] \leq g\langle(A_\Lambda - b)(A_\Lambda^\dagger - b^\dagger)\rangle_{\mathcal{H}_\Lambda^{(0)}(b)}. \tag{2.47}$$

Hence, the best approximation for the free energy density of the model system is given by the *absolute maximum principle* [Bogolyubov Jr. (1972)],

$$\max_b f_\Lambda[\mathcal{H}_\Lambda^{(0)}(b)] = f_\Lambda[\mathcal{H}_\Lambda^{(0)}(\bar{b}_\Lambda)]. \tag{2.48}$$

As in the case of attraction, the fact that the absolute maximum of the free energy density $f_\Lambda[\mathcal{H}_\Lambda^{(0)}(b)]$ is attained at a finite value $b = \bar{b}_\Lambda$ can be proved with the aid of conditions (2.13). From the explicit form of the approximating Hamiltonian (2.45) and the analyticity of the corresponding free energy density $f_\Lambda[\mathcal{H}_\Lambda^{(0)}(b)]$ it follows that the point $b = \bar{b}_\Lambda$ is a solution of the equation

$$b = \langle A_\Lambda \rangle_{\mathcal{H}_\Lambda^{(0)}(b)}. \tag{2.49}$$

A characteristic property of the systems with separable repulsion is that the approximating free energy density $f_\Lambda[\mathcal{H}_\Lambda^{(0)}(b)]$ is a strictly convex function of the variables $x = \mathrm{Re}(b)$ and $y = \mathrm{Im}(b)$, therefore Eq. (2.49) has a *unique solution* [Bogolyubov Jr. (1972)]. The absolute maximum condition (2.48) and the Bogolyubov inequality (2.1) imply the following bounds on the difference between the approximating and original free energy densities:

$$g\langle (A_\Lambda - \bar{b}_\Lambda)(A_\Lambda^\dagger - \bar{b}_\Lambda^\dagger)\rangle_{\mathcal{H}_\Lambda} \leq f_\Lambda[\mathcal{H}_\Lambda] - f_\Lambda[\mathcal{H}_\Lambda^{(0)}(\bar{b}_\Lambda)] \leq$$
$$g\langle (A_\Lambda - \bar{b}_\Lambda)(A_\Lambda^\dagger - \bar{b}_\Lambda^\dagger)\rangle_{\mathcal{H}_\Lambda^{(0)}(\bar{b}_\Lambda)}. \tag{2.50}$$

From the second of the above inequalities one obtains the following sufficient condition for thermodynamic equivalence:

$$\langle (A_\Lambda - \langle A_\Lambda\rangle_{\mathcal{H}_\Lambda^{(0)}(\bar{b}_\Lambda)})(A_\Lambda^\dagger - \langle A_\Lambda^\dagger\rangle_{\mathcal{H}_\Lambda^{(0)}(\bar{b})})\rangle_{\mathcal{H}_\Lambda^{(0)}(\bar{b}_\Lambda)} \leq \kappa_\Lambda, \quad \text{t}-\lim \kappa_\Lambda = 0. \tag{2.51}$$

The validity of the above condition can be easily established in the case when $\mathcal{T}_\Lambda$ and A_Λ° are one-particle operators. There are also non-mean-field type models for which one can prove the validity of (2.51). Such an example is given below; another example is provided by the reduced Q^4 model considered in Section 3.4.

Nontrivial example. Consider the two-dimensional square Ising model with 4-spin repulsive interaction (cf [Oitmaa and Barber (1975)]) described by the Hamiltonian (2.44) with

$$\mathcal{T}_\Lambda = -J|\Lambda|A_\Lambda, \qquad A_\Lambda = A_\Lambda^\dagger = |\Lambda|^{-1}\sum_{\langle i,j\rangle}\sigma_i\sigma_j, \qquad J > 0. \tag{2.52}$$

Here $\sigma_i = \pm 1$ is the classical spin variable at site i, the summation runs over all the pairs $\langle i,j\rangle$ of nearest-neighbor sites on a finite square lattice with periodic boundary conditions. Note that the corresponding approximating Hamiltonian (2.45) depends on a real parameter b,

$$\mathcal{H}_\Lambda^{(0)}(b) = -(J - 2gb)|\Lambda|A_\Lambda - |\Lambda|gb^2, \tag{2.53}$$

and describes the classical Ising model with pairwise interaction and renormalized coupling constant. The correlation function (2.51) turns out to be proportional to specific heat capacity c_Λ of the Ising model, additionally divided by the number of particles $|\Lambda|$:

$$\langle A_\Lambda^2\rangle_{\mathcal{H}_\Lambda^{(0)}(\bar{b}_\Lambda)} - \langle A_\Lambda\rangle_{\mathcal{H}_\Lambda^{(0)}(\bar{b}_\Lambda)}^2 = [\beta(J - 2g\bar{b}_\Lambda)]^{-2}c_\Lambda/(k_B|\Lambda|). \tag{2.54}$$

By studying explicitly the well-known thermodynamic limit of the Ising model free energy density, $\text{t}-\lim f_\Lambda[\mathcal{H}_\Lambda^{(0)}(\bar{b}_\Lambda)]$, as a function of the parameter b, one can show that $0 < (J - 2g\bar{b}_\infty) < J$. Hence, from the asymptotic behavior of the specific heat capacity [Ferdinand and Fisher (1969)], $c_\Lambda/k_B = O(\ln|\Lambda|)$, one obtains $\kappa_\Lambda = O(\ln|\Lambda|/|\Lambda|) \to 0$ as $|\Lambda| \to \infty$.

The above example illustrates the fruitful application of finite-size estimates for proving thermodynamic equivalence in the AHM.

A special feature of the systems with separable repulsion is that they admit non-trivial necessary conditions for the thermodynamic equivalence of the model and approximating Hamiltonians. To formulate them, we note that the correlation function in the left-hand side of (2.50) can be written identically as

$$\langle (A_\Lambda - \bar{b}_\Lambda)(A_\Lambda^\dagger - \bar{b}_\Lambda^\dagger) \rangle_{\mathcal{H}_\Lambda} \equiv$$
$$\langle (A_\Lambda - \langle A_\Lambda \rangle_{\mathcal{H}_\Lambda})(A_\Lambda^\dagger - \langle A_\Lambda^\dagger \rangle_{\mathcal{H}_\Lambda}) \rangle_{\mathcal{H}_\Lambda} + |\langle A_\Lambda \rangle_{\mathcal{H}_\Lambda} - \bar{b}_\Lambda|^2. \quad (2.55)$$

Hence, the asymptotic closeness of the free energy densities,

$$f_\Lambda[\mathcal{H}_\Lambda] - f_\Lambda[\mathcal{H}_\Lambda^{(0)}(\bar{b}_\Lambda)] \le \epsilon_\Lambda, \quad (2.56)$$

where $t-\lim \epsilon_\Lambda = 0$, implies the estimates

$$\langle \delta A_\Lambda \delta A_\Lambda^\dagger \rangle_{\mathcal{H}_\Lambda} \le \epsilon_\Lambda, \quad (2.57)$$

where $\delta A_\Lambda \equiv A_\Lambda - \langle A_\Lambda \rangle_{\mathcal{H}}$, and

$$|\langle A_\Lambda \rangle_{\mathcal{H}_\Lambda} - \langle A_\Lambda \rangle_{\mathcal{H}_\Lambda^{(0)}(\bar{b}_\Lambda)}| \le \epsilon_\Lambda^{1/2}. \quad (2.58)$$

Thus, the smallness of the left-hand sides of both (2.57) and (2.58) is necessary for the thermodynamic equivalence of the model and approximating systems. The meaning of condition (2.58) is clear: the average value of the operator $A_\Lambda^\diamond$ must have the same asymptotic behavior in the Gibbs ensembles corresponding to the Hamiltonians $\mathcal{H}_\Lambda$ and $\mathcal{H}_\Lambda^{(0)}(\bar{b}_\Lambda)$. Condition (2.57) can be put in a more transparent form with the aid the first of the Ginibre inequalities (2.37), which yields

$$\langle \delta A_\Lambda \delta A_\Lambda^\dagger \rangle_{\mathcal{H}_\Lambda} \ge (\delta A_\Lambda, \delta A_\Lambda)_{\mathcal{H}_\Lambda} + \frac{1}{2} \langle [A_\Lambda, A_\Lambda^\dagger] \rangle_{\mathcal{H}_\Lambda}. \quad (2.59)$$

By assuming the validity of the first inequality in (2.33), one sees that the absolute value of the second term in the right-hand side of (2.56) is bounded by $|\Lambda|^{-1} M_3/2$. The inner product $(\delta A_\Lambda, \delta A_\Lambda)_{\mathcal{H}}$ is directly related to the initial susceptibility of the system with respect to the external complex field $\nu^\diamond$ conjugate to the operator $A_\Lambda^\diamond$, see Eq. (2.19). Thus, the necessary condition for (2.57) to hold takes the form

$$(\beta|\Lambda|)^{-1}\left(-\frac{\partial^2}{\partial\nu^\dagger\partial\nu}f_\Lambda[\mathcal{H}_\Lambda(\nu)]\right)_{\nu=0}\le\epsilon_\Lambda+|\Lambda|^{-1}M_3/2. \tag{2.60}$$

Extension of the model

In the works [Bogolyubov Jr. (1972)], [Bogolyubov Jr. et. al. (1981)], the above results have been extended to the general class of model Hamiltonians of the form

$$\mathcal{H}_\Lambda=\mathcal{T}_\Lambda+|\Lambda|\sum_{k=1}^{n}g_kA_{k,\Lambda}A^\dagger_{k,\Lambda}, \qquad g_k>0, \tag{2.61}$$

where $\mathcal{T}_\Lambda$ is a self-adjoint operator which satisfies the first condition (2.13). Let, in addition, the inequalities $\|A^\diamond_{k,\Lambda}\|\le M_2$ hold for all $k=1,\dots,n$, with M_2 being a constant independent of the volume $|\Lambda|$. Then the approximating Hamiltonian has the form

$$\mathcal{H}^{(0)}_\Lambda(\mathrm{b})=\mathcal{T}_\Lambda+|\Lambda|\sum_{k=1}^{n}g_k(b_kA^\dagger_{k,\Lambda}+b^\dagger_kA_{k,\Lambda}-b_kb^\dagger_k). \tag{2.62}$$

where $\mathrm{b}=(b_1,\dots,b_n)\in\mathbf{C}^n$. The free energy density of the model system with Hamiltonian (2.61) is equivalent in the thermodynamic limit to the approximating free energy density

$$f[\mathcal{H}^{(0)}_\Lambda(\bar{\mathrm{b}}_\Lambda)]=\max_{\mathrm{b}}f[\mathcal{H}^{(0)}_\Lambda(\mathrm{b})], \tag{2.63}$$

under the additional set of conditions:

$$\langle A_{k,\Lambda}A^\dagger_{k,\Lambda}\rangle_{\mathcal{H}^{(0)}_\Lambda(\bar{\mathrm{b}}_\Lambda)}-\left|\langle A_{k,\Lambda}\rangle_{\mathcal{H}^{(0)}_\Lambda(\bar{\mathrm{b}}_\Lambda)}\right|^2\le\kappa_\Lambda, \qquad k=1,\dots,n, \tag{2.64}$$

where $\mathrm{t-}\lim\kappa_\Lambda=0$. Then

$$0\le f_\Lambda[\mathcal{H}_\Lambda]-f_\Lambda[\mathcal{H}^{(0)}_\Lambda(\bar{\mathrm{b}}_\Lambda)]\le\kappa_\Lambda, \qquad \mathrm{t-}\lim\kappa_\Lambda=0. \tag{2.65}$$

Here $\bar{\mathrm{b}}_\Lambda=\bar{b}_{1,\Lambda},\dots,\bar{b}_{n,\Lambda}$ is the unique solution of the set of equations

$$b_k=\langle A_{k,\Lambda}\rangle_{\mathcal{H}^{(0)}_\Lambda(\mathrm{b})}, \qquad k=1,\dots,n. \tag{2.66}$$

Existence of the thermodynamic limit

In the considered case of separable repulsion it suffices again to require the existence of the thermodynamic limit

$$t-\lim f_\Lambda[\mathcal{H}_\Lambda^{(0)}(b)] = f^{(0)}(b) \qquad (2.67)$$

for all finite (complex) numbers $b = (b_1,\ldots,b_n)$. It has been shown [Bogolyubov Jr. et. al. (1981)] that the finite-volume functions $f_\Lambda[\mathcal{H}_\Lambda^{(0)}(b)]$ are strictly convex with respect to the set of arguments $x_k = \mathrm{Re}(b_k)$, $y_k = \mathrm{Im}(b_k)$, $k = 1,\ldots,n$. Therefore, the limit function (2.67) is also a strictly convex function and the approach of the finite-volume functions to the thermodynamic limit function is uniform in any compact subset of $\mathbf{C}^n$. Hence, there exists the limit

$$t-\lim(\bar{b}_{1,\Lambda},\ldots,\bar{b}_{n,\Lambda}) = \bar{b}, \qquad (2.68)$$

where $\bar{b} = (\bar{b}_1,\ldots,\bar{b}_n)$ is the unique point at which the limit function (2.67) reaches its absolute maximum. Thus, the thermodynamic limit for the model free energy density exists and equals

$$t-\lim f_\Lambda[\mathcal{H}_\Lambda] = \max_{b}\{t-\lim f_\Lambda[\mathcal{H}_\Lambda^{(0)}(b)]\} = t-\lim f_\Lambda[\mathcal{H}_\Lambda^{(0)}(\bar{b})]. \qquad (2.69)$$

This result is useful for practical applications of the AHM.

2.4   Generalization to other types of systems

For the sake of completeness, we briefly mention here some generalizations of the AHM to other classes of model systems, that have also found important applications in solid state physics.

2.4.1   *Systems with attractive and repulsive components*

The model Hamiltonian contains simultaneously attractive and repulsive components of a separable interaction ($g_k > 0, k = 1,\ldots,m$):

$$\mathcal{H}_\Lambda = \mathcal{T}_\Lambda - |\Lambda| \sum_{k=1}^{n} g_k A_{k,\Lambda} A_{k,\Lambda}^\dagger + |\Lambda| \sum_{k=n+1}^{m} g_k A_{k,\Lambda} A_{k,\Lambda}^\dagger, \qquad (2.70)$$

where $\mathcal{T}_\Lambda$ is a self-adjoint operator which satisfies the first condition (2.13), and the operators $A_{k,\Lambda}^\diamond$, $k = 1,\ldots,m$, satisfy conditions (2.32) and (2.33).

The problem can be treated by the AHM with the aid of a generalized version of the *minimax principle* due to [Bogolyubov Jr. (1970)]. According to the linearization principle, the corresponding approximating Hamiltonian has the form

$$\mathcal{H}_\Lambda^{(0)}(\mathrm{a},\mathrm{b}) = \mathcal{T}_\Lambda - |\Lambda| \sum_{k=1}^{n} g_k(a_k A_{k,\Lambda}^\dagger + a_k^\dagger A_{k,\Lambda} - a_k a_k^\dagger) + \tag{2.71}$$

$$|\Lambda| \sum_{k=n+1}^{m} g_k(b_k A_{k,\Lambda}^\dagger + b_k^\dagger A_{k,\Lambda} - b_k b_k^\dagger),$$

where $\mathrm{a} = (a_1, \ldots, a_n) \in \mathbf{C}^n$ and $\mathrm{b} = (b_{n+1}, \ldots, b_m) \in \mathbf{C}^{m-n}$.

A sufficient condition for the thermodynamic equivalence of the model, (2.70), and approximating, (2.71), Hamiltonians is the validity of the set of inequalities ($k = n + 1, \ldots, m$)

$$\langle A_{k,\Lambda} A_{k,\Lambda}^\dagger \rangle_{\mathcal{H}_\Lambda^{(0)}(\mathrm{a},\bar{\mathrm{b}}_\Lambda(\mathrm{a}))} - \left| \langle A_{k,\Lambda} \rangle_{\mathcal{H}_\Lambda^{(0)}(\mathrm{a},\bar{\mathrm{b}}_\Lambda(\mathrm{a}))} \right|^2 \leq \kappa_\Lambda(\mathrm{a}), \tag{2.72}$$

where $\bar{\mathrm{b}}_\Lambda(\mathrm{a}) = \{\bar{b}_{n+1,\Lambda}(\mathrm{a}), \ldots, \bar{b}_{m,\Lambda}(\mathrm{a})\}$ is the (unique) solution of the set of equations

$$b_k = \langle A_{k,\Lambda} \rangle_{\mathcal{H}_\Lambda^{(0)}(\mathrm{a},\mathrm{b})}, \qquad k = n+1, \ldots, m, \tag{2.73}$$

and $\mathrm{t}- \lim \kappa_\Lambda(\mathrm{a}) = 0$ for all finite $\mathrm{a} \in \mathbf{C}^n$. Given also the existence of the pointwise limit

$$\mathrm{t}- \lim f_\Lambda[\mathcal{H}_\Lambda^{(0)}(\mathrm{a},\mathrm{b})] = f^{(0)}(\mathrm{a},\mathrm{b}) \tag{2.74}$$

for all finite $\mathrm{a} \in \mathbf{C}^n$ and $\mathrm{b} \in \mathbf{C}^{m-n}$, it has been proven, see [Bogolyubov Jr. et. al. (1981)], that the thermodynamic limit for the free energy density of the model system with Hamiltonian (2.70) exists and equals

$$\mathrm{t}- \lim f_\Lambda[\mathcal{H}_\Lambda] = \min_{\mathrm{a}} \max_{\mathrm{b}} \{\mathrm{t}- \lim f_\Lambda[\mathcal{H}_\Lambda^{(0)}(\mathrm{a},\mathrm{b})]\}. \tag{2.75}$$

In the case when the limit function (2.74) is twice differentiable with respect to all of its arguments, the minimax principle (2.75) can be reformulated in the practically more convenient form:

$$\min_{\mathrm{a}} \max_{\mathrm{b}} \{\mathrm{t}- \lim f_\Lambda[\mathcal{H}_\Lambda^{(0)}(\mathrm{a},\mathrm{b})]\} = \min_{(\mathrm{a},\mathrm{b})\in \mathbf{S}_\Lambda} \{\mathrm{t}- \lim f_\Lambda[\mathcal{H}_\Lambda^{(0)}(\mathrm{a},\mathrm{b})]\}, \tag{2.76}$$

where $\mathbf{S}_\Lambda$ is the set of solutions of the system of equations

$$\frac{\partial}{\partial a_k}\{\text{t}-\lim f_\Lambda[\mathcal{H}_\Lambda^{(0)}(\text{a},\text{b})]\} = 0, \qquad k = 1,\ldots,n, \tag{2.77}$$

$$\frac{\partial}{\partial b_k}\{\text{t}-\lim f_\Lambda[\mathcal{H}_\Lambda^{(0)}(\text{a},\text{b})]\} = 0, \qquad k = n+1,\ldots,m. \tag{2.78}$$

It is rather staraightforward to generalize the model to the case when the interaction Hamiltonian is a symmetric quadratic form in the operators $A_{k,\Lambda}^\diamond$, $k = 1,\ldots,m$. For examples of physical applications we refer the reader to [Tonchev and Brankov (1980)] and [Chamati and Tonchev (1992)], where the Vonsovskii-Zener model of regularly positioned magnetic impurities has been considered, with and without superconducting BCS interaction, respecively. See also [Gochev and Tonchev (1992)], where the problem of the order parameters in the Lieb-Mattis model of a quantum antiferromagnet has been studied.

2.4.2 *Systems with nonpolynomial interaction*

The AHM has been extended to Hamiltonians with nonpolynomial interaction of the form [den Ouden et. al. (1976)], [Brankov et. al. (1977)], [Brankov et. al. (1979)],

$$\mathcal{H}_\Lambda = \mathcal{T}_\Lambda - |\Lambda|\phi(A_\Lambda), \tag{2.79}$$

where $\mathcal{T}_\Lambda$ and the self-adjoint operator $A_\Lambda = A_\Lambda^\dagger$ satisfy the general conditions (2.13) and (2.22). The operator-valued function $\phi(A_\Lambda)$ is defined by the standard spectral representation, see e.g. [Reed and Simon (1980)],

$$\phi(A_\Lambda) = \int_{-M_2}^{M_2+0} \mathrm{d}E_\lambda \, \phi(\lambda), \tag{2.80}$$

where E_λ is the decomposition of the identity operator generated by A_Λ. We assume that the function $\phi(\cdot)$, defined on the closed interval $I = [-M_2, M_2]$, is twice differentiable and has a bounded second derivative $\phi''(\cdot)$,

$$|\phi''(x)| \leq K, \qquad -M_2 < x < M_2. \tag{2.81}$$

In this case the approximating Hamiltonian is

$$\mathcal{H}_\Lambda^{(0)}(a) = \mathcal{T}_\Lambda - |\Lambda|\phi'(a)(A_\Lambda - a) - |\Lambda|\phi(a), \tag{2.82}$$

where ϕ' is the first derivative of the function ϕ. The additional clustering condition, which is sufficient for the thermodynamic equivalence of the model (2.79) and the approximating (2.82) Hamiltonians, reads

$$\text{t}-\lim\langle(A_\Lambda - \langle A_\Lambda\rangle_{\mathcal{H}_\Lambda^{(0)}(\bar a_\Lambda)})^2\rangle_{\mathcal{H}_\Lambda^{(0)}(\bar a_\Lambda)} = 0, \tag{2.83}$$

for $\bar a_\Lambda$ belonging to the set S_Λ of solutions of the self-consistency equation for the system with approximating Hamiltonian (2.82), see [Brankov et. al. (1977)]:

$$\bar a_\Lambda \in S_\Lambda \equiv \{a \in R : a = \langle A_\Lambda\rangle_{\mathcal{H}_\Lambda^{(0)}(a)}\}. \tag{2.84}$$

Note that in the case of attractive interaction, when $\phi''(a) > 0$ for all $|a| \le M_2$, the clustering condition (2.84) is redundant.

The best approximation to the free energy density of the finite model system is given by the absolute minimum of the function $f_\Lambda[\mathcal{H}_\Lambda^{(0)}(a)]$ *over the set $a \in S_\Lambda$.* The problem of passing to the thermodynamic limit here has been solved by assuming the existence of

$$\text{t}-\lim f_\Lambda[\mathcal{T}_\Lambda - x|\Lambda|A_\Lambda] = F(x), \tag{2.85}$$

for all finite $x \in R$. In [Brankov et. al. (1979)] it has been proved that

$$\text{t}-\lim\{\min_{a\in S_\Lambda} f_\Lambda[\mathcal{H}_\Lambda^{(0)}(a)]\} = \min_{a\in S}\{\text{t}-\lim f_\Lambda[\mathcal{H}_\Lambda^{(0)}(a)]\}, \tag{2.86}$$

where the set S is defined by the inequalities

$$S \equiv \{a \in R : -F'(\phi'(a) - 0) \le a \le -F'(\phi'(a) + 0)\}. \tag{2.87}$$

This rather unexpected generalization of the self-consistency equation has been obtained first in [den Ouden et. al. (1976)], in the case of analytic functions of a finite number of intensive self-adjoint operators $A_{k,\Lambda}$, $k = 1,\dots,n$, which obey certain "short-range" conditions.

2.4.3 *Systems of matter interacting with Boson fields*

Here are two examples of models in solid state physics, which belong to this class: *(a) The Dicke model of superradiance,* solved exactly in [Hepp and Lieb (1973a)], [Hepp and Lieb (1973b)], and by the AHM in [Brankov et. al.

(1975a)]. The model has been generalized to include interactions with both electromagnetic field and phonons [Klemm and Zagrebnov (1977)], [Klemm and Zagrebnov (1978)], and to the case of infinitely many modes of the electromagnetic field [Zagrebnov (1984)]. *(b) The Mattis-Langer model of structural instability* [Mattis and Langer (1970)], solved exactly by the AHM in [Brankov et. al. (1975b)]. The one-dimensional case with countably infinite set of phonon modes has been solved by means of theta-function integration in [Belokolos and Petrina (1984)].

The model Hamiltonian is defined on the tensor product of two Hilbert spaces, one for the subsystem describing matter (e.g., electrons in a solid, spins on a lattice), and the other for the Boson field (lattice vibrations, electromagnetic field). In the second quantization representation the creation, $\mathbf{b}_k^\dagger$, and annihilation, $\mathbf{b}_k$, operators of the Boson field modes (labelled by the subscript k) satisfy the cananical commutation relations

$$[\mathbf{b}_k, \mathbf{b}_l^\dagger] = \delta_{k,l}, \qquad [\mathbf{b}_k, \mathbf{b}_l] = [\mathbf{b}_k^\dagger, \mathbf{b}_l^\dagger] = 0, \qquad (2.88)$$

for all allowed k and l. Since the Boson operators are unbounded, it is not possible to obtain easy bounds on their average values in terms of Hilbert-space norm.

Another characteristic feature of this class of models is that exact solvabilty by the AHM is possible when the interaction with only a finite (or growing slower than $|\Lambda|$, as $|\Lambda| \to \infty$) number of Boson modes is taken into account. The typical model Hamiltonian has the form

$$\mathcal{H}_\Lambda = \mathcal{T}_\Lambda + \sum_{k=1}^{n} \omega_k\, \mathbf{b}_k^\dagger \mathbf{b}_k + |\Lambda|^{1/2} \sum_{k=1}^{n} \lambda_k (\mathbf{b}_k A_{k,\Lambda}^\dagger + \mathbf{b}_k^\dagger A_{k,\Lambda}). \qquad (2.89)$$

Here the operators $\mathcal{T}_\Lambda = \mathcal{T}_\Lambda^\dagger$ and $A_{k,\Lambda}^\diamond$, $k = 1,\ldots,n$, refer to the matter subsystem and satisfy the general conditions (2.13) and (2.32). The second term in the right-hand side of (2.89) describes a finite number of free Boson modes, $k = 1,\ldots,n$, in the space domain Λ. For the sake of simplicity the energies $\omega_k > 0$ and the interaction constants $\lambda_k \in R$ are taken to be independent of the volume $|\Lambda|$. The mathematical definition of the above Hamiltonian is given in [Zagrebnov et. al. (1976)].

Since the contribution of the Boson subsystem to the free energy density vanishes as $|\Lambda| \to \infty$, the thermodynamically equivalent to (2.89) approximating Hamiltonian is the one that describes the "matter" subsystem,

$$\mathcal{H}^{(0)}_{\mathrm{M},\Lambda}(\mathrm{a}) = \mathcal{T}_\Lambda - |\Lambda| \sum_{k=1}^{n} (\lambda_k^2/\omega_k)(a_k A^\dagger_{k,\Lambda} + a_k^\dagger A_{k,\Lambda} - a_k^\dagger a_k). \qquad (2.90)$$

It has been proven that, see [Bogolyubov Jr. et. al. (1984)],

$$|\min_{\mathrm{a}} f[\mathcal{H}^{(0)}_{\mathrm{M},\Lambda}(\mathrm{a})] - f[\mathcal{H}_\Lambda]| \le \epsilon_\Lambda, \qquad (2.91)$$

where $\epsilon_\Lambda \to O$ as $|\Lambda| \to \infty$. By comparison with Eq. (2.34) one finds that $\mathcal{H}^{(0)}_{\mathrm{M},\Lambda}(\mathrm{a})$ is actully the approximating Hamiltonian for a system with effective separable attraction described by the Hamiltonian, cf Eq. (2.31),

$$\mathcal{H}_{\mathrm{M},\Lambda} = \mathcal{T}_\Lambda - |\Lambda| \sum_{k=1}^{n} (\lambda_k^2/\omega_k) A_{k,\Lambda} A^\dagger_{k,\Lambda}. \qquad (2.92)$$

Here the quadratic in the intensive operators $A^\diamond_{k,\Lambda}$ ($k = 1, \dots, n$) attraction is induced by the interaction with a finite number of Boson modes in the original model (2.89). However, the thermodynamic equivalence of the Hamiltonians $\mathcal{H}_{\mathrm{M},\Lambda}$ and $\mathcal{H}^{(0)}_{\mathrm{M},\Lambda}(\bar{\mathrm{a}}_\Lambda)$ has been established in Section 2.2 under the additional contitions (2.33) on the asymptotic behavior of the commutators of the operators $A_{k,\Lambda}$, $A^\diamond_{k,\Lambda}$ and $\mathcal{T}_\Lambda$.

The existence of the thermodynamic limit for the approximating free energy density can be established in complete analogy with the case of separable attraction. Thus one obtains the existence of the limit

$$\mathrm{t-}\lim f_\Lambda[\mathcal{H}_\Lambda] = \min_{\mathrm{a}}\{\mathrm{t-}\lim f[\mathcal{H}^{(0)}_{\mathrm{M},\Lambda}(\mathrm{a})]\}, \qquad (2.93)$$

where the absolute minimum in the right-hand side is attained on the set of solutions of the limiting self-consistency equations

$$a_k = \mathrm{t-}\lim\langle A_{k,\Lambda}\rangle_{\mathcal{H}^{(0)}_{\mathrm{M},\Lambda}(\mathrm{a})}, \qquad k = 1, \dots, n. \qquad (2.94)$$

2.5   Asymptotic closeness of average values

In the preceding sections we have confined ourselves to proving that for some classes of systems the model Hamiltonian and the approximating one are thermodynamically equivalent in the sense that the corresponding free energy densities coincide in the thermodynamic limit. In a broad understanding, however, the concept of thermodynamic equivalence includes the equality of all the observable characteristics of the equilibrium state, such as average values of local operators and correlation functions of any finite order. For instance, in Section 2.2 it is shown that, given a model Hamiltonian $\mathcal{H}_\Lambda$ with separable attraction and the corresponding approximating Hamiltonian $\mathcal{H}_\Lambda^{(0)}(a)$, under certain conditions one has the equality

$$\mathrm{t}-\lim f_\Lambda[\mathcal{H}_\Lambda] = \mathrm{t}-\lim f_\Lambda[\mathcal{H}_\Lambda^{(0)}(\bar{a}_\Lambda)], \qquad (2.95)$$

where

$$f_\Lambda[\mathcal{H}_\Lambda^{(0)}(\bar{a}_\Lambda)] = \min_a f_\Lambda[\mathcal{H}_\Lambda^{(0)}(a)]. \qquad (2.96)$$

Can one then expect that, for any operator Q belonging to the algebra of observables for the considered system,

$$\mathrm{t}-\lim\langle Q\rangle_{\mathcal{H}_\Lambda} = \mathrm{t}-\lim\langle Q\rangle_{\mathcal{H}_\Lambda^{(0)}(\bar{a}_\Lambda)}, \qquad (2.97)$$

provided the limits exist? The non-triviality of the problem can be illustrated by a simple, exactly solved example of a classical spin system.

Example: the Husimi - Temperley model. This spin model belongs to the simplest class of exactly solvable by the AHM statistical mechanical problems. Its characteristic (and unrealistic) feature is that the pairwise interaction between the particles is independent of the distance between them, hence, it is often called *equivalent-neighbors interaction.*. In order to ensure the thermodynamic extensivity of the total interaction energy, the interaction constant is chosen inversely proportional to the number of particles in the system [Husimi (1953)], [Temperley (1954)]. It is instructive to consider systems of this class, since their thermodynamic behaviour reproduces exactly the results of the mean-field theory. The Husimi - Temperley model of ferromagnetism is known also as the Curie - Weiss model. It is defined for a finite collection of $N = |\Lambda|$ Ising spins $\sigma_i \in \{-1, +1\}$, $i \in \Lambda$, described by the Hamiltonian

$$\mathcal{H}_N(h) = -\frac{J}{2N}\sum_{i,j=1}^{N}\sigma_i\sigma_j - h\sum_{i=1}^{N}\sigma_i, \qquad (2.98)$$

where $h \in R$ is an external magnetic field, and $J > 0$ is the interaction constant. Obviously, this Hamiltonian belongs to the class of models with separable attraction considered in Section 2.2, see Eq. (2.4) with

$$T_\Lambda = -h \sum_{i=1}^{N} \sigma_i, \quad A_\Lambda = A_\Lambda^\dagger = \frac{1}{N} \sum_{i=1}^{N} \sigma_i,$$

and $g = J/2$. According to Eq. (2.8), the approximating Namiltonian is

$$\mathcal{H}_N^{(0)}(a, h) = -(Ja + h) \sum_{i=1}^{N} \sigma_i + \frac{1}{2} JNa^2, \tag{2.99}$$

and describes a paramagnetic system of N spins in an effective magnetic field of magnitude $Ja + h$. Consider now the magnetization per spin in the zero-field case $h = 0$. Due to the spin-reversal symmetry of the spin-spin interaction Hamiltonian in Eq. (2.98), one has $N^{-1} \sum_{i=1}^{N} \langle \sigma_i \rangle_{\mathcal{H}_N}(0) = 0$ for every finite N. On the other hand, the free enrgy per spin for the approximating Hamiltonian (2.99) is *independent of the number of spins N* and reads

$$f_N[\mathcal{H}_N^{(0)}(a, h)] = \frac{1}{2} Ja^2 - \beta^{-1} \ln\{2 \coth[\beta(Ja + h)]\}. \tag{2.100}$$

Its absolute minimum is attained at a point $a = \bar{a}(\beta, h)$ which satisfies the equation

$$a = \tanh[\beta(Ja + h)]. \tag{2.101}$$

Thus we obtain that the average magnetization per spin of the approximating system is *independent of the number of spins N* and equals

$$N^{-1} \sum_{i=1}^{N} \langle \sigma_i \rangle_{\mathcal{H}_N^{(0)}(\bar{a}(\beta, h), h)} = \bar{a}(\beta, h). \tag{2.102}$$

It can be readily shown that in the zero-field case, for all temperatures below the critical one, $\beta > \beta_c = 1/J$, the absolute minimum of the function (2.100) is reached at two different points, $\bar{a}(\beta, 0) = \pm m_0(\beta)$, where $m_0(\beta) > 0$ is the spontaneous magnetization. Therefore, the zero-field magnetizations per spin, calculated in the ensembles corresponding to the two thermodynsmically equivalent Hamiltonians, $\mathcal{H}_N$ and $\mathcal{H}_N^{(0)}(\bar{a}(\beta, 0), 0)$, differ by the same value for all system sizes, hence, in the thermodynamic limit as well. The explanation of this fact lies in the *spontaneous symmetry breaking* in the pure thermodynamic phases: the approximating Hamiltonian (2.99) has a broken spin-reversal symmetry even at $h = 0$, and the corresponding Gibbs ensemble yields the spontaneous magnetization of one of the two pure phases. In contrast, the Gibbs ensemble for the initial Hamiltonian (2.98) at $h = 0$ respects the spin-reversal symmetry and yields the zero magnetization of a mixed phase, see also Section 9.1. For the analysis of such situations, and overcomming the arising difficulties, one can use the Bogolyubov concept of quasi-averages [Bogolyubov (1961)].

A quasiaverage value $\langle\langle Q \rangle\rangle$ of an intensive operator Q is the result of first taking the average value for a finite system in the presence of a symmetry breaking external field (source), then passing to the thermodynamic limit, and *after that* switching off the external field, see Eq. (2.111) below. A generalized version of this concept is described in Chapter 9.

Here we establish the closeness of quasiaverages of the type (2.97) in the case when the problem can be reduced to the one of closeness of first derivatives of the corresponding free energy densities, [Bogolyubov Jr. et. al. (1981)], [Bogolyubov Jr. et. al. (1984)]. For the sake of simplicity we consider the case of a self-adjoint operator $Q = Q^\dagger$ the norm of which is bounded uniformly in $|\Lambda|$, i.e., $\|Q\| \le M$ for some constant M. We consider a Hamiltonian $\mathcal{H}_\Lambda$ which satisfies all the conditions for applicability of the AHM, stated in one of the previous subsections. Further, we assume that the auxiliary Hamiltonian

$$\mathcal{H}_\Lambda(h) = \mathcal{H}_\Lambda + h|\Lambda|Q, \tag{2.103}$$

satisfies the same conditions as $\mathcal{H}_\Lambda$ for all finite $h \in R$. Since the free energy density $f_\Lambda[\mathcal{H}_\Lambda(h)]$ is a real-analytic function of the field h for all finite $|\Lambda|$, one has

$$\langle Q \rangle_{\mathcal{H}_\Lambda} = \frac{\partial}{\partial h} f_\Lambda[\mathcal{H}_\Lambda(h)]\bigg|_{h=0}. \tag{2.104}$$

Note that the general case $Q \ne Q^\dagger$ can be treated by introducing two real fields, h_1 and h_2, conjugate to the self-adjoint operators $Q_1 = Q^\dagger + Q$ and $Q_2 = i(Q^\dagger - Q)$. The next natural assumption is the existence of the thermodynamic limit of the free energy density for the auxiliary Hamiltonian (2.103),

$$\text{t} - \lim f_\Lambda[\mathcal{H}_\Lambda(h)] = f_\infty(h), \tag{2.105}$$

at least for all $h \in I$, where I is a neighborhood of the point $h = 0$. We emphasize that the existence of the limit (2.105) does not imply, in general, that the derivative in the right-hand side of Eq. (2.104) converges at all as

$|\Lambda| \to \infty$. Moreover, the limit function (2.105) need not be differentiable at $h = 0$. We know only that $\{f_\Lambda[\mathcal{H}_\Lambda(h)]\}$ is a family of concave in $h \in R$ functions, therefore its limit $f_\infty(h)$ is also concave in $h \in I$. From the Griffiths - Fisher lemma, see Section 2.2, it follows that if the limit function $f_\infty(h)$ is differentiable at the point $h = 0$, then Eq. (2.43) implies the existence of the limit

$$\text{t}-\lim\langle Q\rangle_{\mathcal{H}_\Lambda} = f'_\infty(h)|_{h=0}. \tag{2.106}$$

But if $f_\infty(h)$ is not differentiable at $h = 0$, then inequalities (2.42) imply only that

$$\limsup{}_{|\Lambda|\to\infty}\langle Q\rangle_{\mathcal{H}_\Lambda} \leq f'_\infty(-0), \quad \liminf{}_{|\Lambda|\to\infty}\langle Q\rangle_{\mathcal{H}_\Lambda} \geq f'_\infty(+0). \tag{2.107}$$

In this case one can choose a sequence of points $h_n^+ \in I$, $n = 1, 2, \ldots$, such that $h_n^+ > 0$, $h_n^+ \to 0$, at which the first derivatives of the limit function

$$f'_\infty(h_n^+) = \text{t}-\lim\langle Q\rangle_{\mathcal{H}_\Lambda(h_n^+)} \tag{2.108}$$

exist. Now, from the right-continuity of $f'_\infty(h + 0)$ at $h = 0$ and from the convexity of $f_\infty(h)$, $h \in I$, it follows that the limit

$$\lim_{n\to\infty} \text{t}-\lim\langle Q\rangle_{\mathcal{H}_\Lambda(h_n^+)} = f'_\infty(+0) \tag{2.109}$$

exists. Similarly, it can be shown that there exists a sequence $h_n^- \in I$, $n = 1, 2, \ldots$, such that $h_n^- < 0$, $h_n^- \to 0$, for which

$$\lim_{n\to\infty} \text{t}-\lim\langle Q\rangle_{\mathcal{H}_\Lambda(h_n^-)} = f'_\infty(-0). \tag{2.110}$$

Usually, the thermodynamic limit of the free energy density, $f_\infty(h)$, is a piecewise differentialbe function of the parameters of the Gibbs ensemble. In this case, in a sufficiently small neighborhood of the point of non-differentiability $h = 0$, Eqs. (2.109) and (2.110) hold for any sequences of points $h_n^\pm \to 0$, and thus, the left-hand sides of the above equations converge to the Bogolyubov quasiaverages:

$$\langle\langle Q\rangle\rangle_{\mathcal{H}(\pm 0)} = \lim_{h\to\pm 0} \text{t}-\lim\langle Q\rangle_{\mathcal{H}_\Lambda(h)}. \tag{2.111}$$

From a mathematical point of view, the main result is based on the following lemma, which is an extension of a theorem due to Hadamard (1914).

Lemma [Bogolyubov Jr. and Sadovnikov (1975)]. Let $\Delta_n(x)$, $n = 1, 2, \ldots$, be a sequence of functions that are continuously differentiable on the interval $I = [a, b] \in R$ and have second derivatives $\Delta_n''(x + 0)$ to the right of each point $x \in [a, b)$. Suppose that for all $x \in I$

$$|\Delta_n(x)| \leq \epsilon_n(I) \to 0, \quad \text{as } n \to \infty, \tag{2.112}$$

and that there exists a fixed positive number $D(I)$ such that one of the following two conditions holds for all $x \in [a, b)$:

$$(i) \quad \Delta_n''(x + 0) \leq D(I), \quad \text{or} \quad (ii) \quad \Delta_n''(x + 0) \geq -D(I). \tag{2.113}$$

Then, for all x and n satisfying the inequalities

$$a + l_n \leq x \leq b - l_n, \quad \text{where } l_n = 2[\epsilon_n(I)D(I)]^{1/2}, \tag{2.114}$$

the following upper bound on the first derivative $\Delta_n'(x)$ holds:

$$|\Delta_n'(x)| \leq 2[\epsilon_n(I)D(I)]^{1/2} \to 0, \quad \text{as } n \to \infty. \tag{2.115}$$

Let the Hamiltonian $\mathcal{H}_\Lambda(h)$, for example, contain separable attraction and satisfy all the conditions for applicability of the AHM stated in Section 2.2. Its thermodynamically equivalent approximating Hamiltonian has the form

$$\mathcal{H}_\Lambda^{(0)}(a, h) = \mathcal{H}_\Lambda^{(0)}(a) + h|\Lambda|Q, \tag{2.116}$$

where $\mathcal{H}_\Lambda^{(0)}(a)$ is the approximating Hamiltonian for $\mathcal{H}_\Lambda$, and the parameter a has to be chosen from the absolute minimum principle

$$f_\Lambda[\mathcal{H}_\Lambda^{(0)}(\bar{a}_\Lambda(h), h)] = \min_a f_\Lambda[\mathcal{H}_\Lambda^{(0)}(a, h)]. \tag{2.117}$$

To make use of the lemma, we set

$$\Delta_\Lambda(h) \equiv f_\Lambda[\mathcal{H}_\Lambda(h)] - f_\Lambda[\mathcal{H}_\Lambda^{(0)}(\bar{a}_\Lambda(h), h)]. \tag{2.118}$$

Note that since the point $a = \bar{a}_\Lambda(h)$ depends on the auxiliary field $h \in R$, we cannot establish for finite $|\Lambda|$ the concavity of the function $f_\Lambda[\mathcal{H}_\Lambda^{(0)}(\bar{a}_\Lambda(h), h)]$ with respect to h, although from the thermodynamic equivalence,

$$\text{t}-\lim f_\Lambda[\mathcal{H}_\Lambda^{(0)}(\bar{a}_\Lambda(h), h)] = \text{t}-\lim f_\Lambda[\mathcal{H}_\Lambda(h)] \equiv f_\infty(h), \tag{2.119}$$

it follows that the limit function $f_\infty(h)$ is concave in $h \in R$. Another difficulty arises from the fact that even though $f_\Lambda[\mathcal{H}_\Lambda^{(0)}(a, h)]$ depends real-analytically on the parameters a and h for all finite $|\Lambda|$, the approximating free energy density $f_\Lambda[\mathcal{H}_\Lambda^{(0)}(\bar{a}_\Lambda(h), h)]$ may have singularities in h which are brought into $\bar{a}_\Lambda(h)$ by the absolute minimum condition (2.117). That is why in the general case we have to assume that its right derivative with respect to the auxiliary field h exists and is finite for all $h \in I$. Note that in most applications of the AHM the free energy density of the approximating system is symple enough to admit explicit analysis of the above mentioned problems.

By taking the right derivative of the identity (2.118), and by taking into account that due to the extremum condition (2.117),

$$\left. \frac{\partial}{\partial a} f_\Lambda[\mathcal{H}_\Lambda^{(0)}(a, h)] \right|_{a=\bar{a}_\Lambda(h)} = 0, \tag{2.120}$$

one obtains

$$\Delta'_\Lambda(h + 0) = \langle Q \rangle_{\mathcal{H}_\Lambda(h)} - \langle Q \rangle_{\mathcal{H}_\Lambda^{(0)}(\bar{a}_\Lambda(h), h)}. \tag{2.121}$$

From the concavity of $f_\Lambda[\mathcal{H}_\Lambda(h)]$ in h it follows that

$$\Delta''_\Lambda(h + 0) \leq - \left. \frac{\partial^2}{\partial h^2} f_\Lambda[\mathcal{H}_\Lambda^{(0)}(a, h)] \right|_{a=\bar{a}_\Lambda(h)} -$$

$$\left. \frac{\partial^2}{\partial a \partial h} f_\Lambda[\mathcal{H}_\Lambda^{(0)}(a, h)] \right|_{a=\bar{a}_\Lambda(h)} \bar{a}'_\Lambda(h + 0). \tag{2.122}$$

Now, if there exists an interval $I \subset R$, such that $0 \in I$ and the right-hand side of the above inequality is bounded from above uniformly in $|\Lambda|$ and $h \in I$ by a positive number $D(I)$, then Eq. (2.115) of the lemma implies

$$\text{t}-\lim |\Delta'_\Lambda(0)| = \text{t}-\lim |\langle Q \rangle_{\mathcal{H}_\Lambda(0)} - \langle Q \rangle_{\mathcal{H}_\Lambda^{(0)}(\bar{a}_\Lambda(0), 0)}| = 0. \tag{2.123}$$

Usually, in the case of a system with spontaneous symmetry breaking in the low-temperature thermodynamic phase, and when the operator Q corresponds to an order parameter coupled to the field h, the critical value of which is $h = 0$, an upper bound of the above type can be found for $I_\delta = \{h \in R : |h| \geq \delta\}$, where $\delta > 0$. Then one can establish equivalence of the initial and approximating systems in the sense of quasiaverages (2.111) as well.

Chapter 3

Exactly Solved Models

In this chapter we consider two special classes of classical and quantum models which admit exact solution for the thermodynamic functions in any space dimensionsionality d. Models solvable exactly in two dimensions only, such as the ones considered in [Baxter (1982)], are out of the scope of our present interest. Our primary goal is to use these models in the remainder of the book for testing various predictions about quantum effects on critical phenomena and finite-size scaling.

The first class of exactly solved models comprises various generalizations of the classical spherical model, see Section 3.1. They can be obtained from a variety of basic models in solid state physics in the so-called spherical approximation. The Hamiltonians of these systems represent quadratic forms of continuous local dynamical variables, subject to one (or a finite number of) global constraint(s) with spherical symmetry. They can be exactly diagonalized by a canonical transformation even in the case of a finite region Λ under different boundary conditions. This feature of the spherical models allows one to study rigorously finite-size effects in the presence of external fields, long-range interactions, as well as the influence of boundary conditions on surface and layer critical properties. The relationship between the spherical model and the $n \to \infty$ limit of the classical n-vector model is considered briefly in Section 3.2, and the quantization of the classical spherical model is discussed in Section 3.3. It is shown that quantum effects in bulk systems can be fruitfully interpreted as a finite-size effect in an additional "quantum dimension" of the system. That is why the detailed study of the critical behaviour of both bulk and spatially finite quantum systems is postponed to the chapters following the exposition of finite-size scaling theory. In section 3.4 we considered in some detail a quantum system which undergoes a structural phase transition. The model is exactly solvable by the approximating Hamiltonian method presented in Chapter 2. As we have already mentioned, the rigorous study of finite-size effects in such quantum systems is possible only in the framework of the system defined by the approximating Hamiltonian.

3.1   Classical spherical models

In classical n-component spin models, the local spin variables $\mathbf{S_r}$, $\mathbf{r} \in \Lambda$, are considered as vectors of a fixed length, usually set equal to unity, $|\mathbf{S_r}| = 1$. This physically natural condition makes the calculation of the partition function a difficult, even unsolvable in higher dimensions, task. A radical way to simplify the problem is to completely relax the fixed-length condition and consider the spin components $S_\mathbf{r}^\alpha$, $\alpha = 1, \ldots, n$, as independent continuous variables which take values in the entire real axis R. Such a simplification leads to exactly solvable in any space dimensionality problem, provided the Hamiltonian $\mathcal{H}$ is a strictly positive quadratic form of the variables $S_\mathbf{r}^\alpha$, $\alpha = 1, \ldots, n$, $\mathbf{r} \in \Lambda$. However, this is not the case even for translation-invariant nearest-neighbor spin-spin interactions. The existence of the partition function can be ensured, at least partially, by adding to $\beta \mathcal{H}$ the positive definite quadratic form $s \sum_\mathbf{r} \mathbf{S_r^2}$, with some parameter $s > 0$. Then the integral

$$Z_\Lambda^{G\cdot}(\beta \mathbf{J}_\Lambda^{(\tau)}, \mathbf{h}_\Lambda, s) = \int_{R^{n|\Lambda|}} \prod_{\mathbf{r} \in \Lambda} \left(\prod_{\alpha=1}^{n} \mathrm{d}S_\mathbf{r}^\alpha \right)$$

$$\times \exp \left[\frac{1}{2} \sum_{\mathbf{r},\mathbf{r}' \in \Lambda} \beta J^{(\tau)}(\mathbf{r}, \mathbf{r}') \mathbf{S_r} \cdot \mathbf{S_{r'}} + \beta \sum_{\mathbf{r} \in \Lambda} \mathbf{H_r} \cdot \mathbf{S_r} - s \sum_{\mathbf{r} \in \Lambda} \mathbf{S_r^2} \right], \qquad (3.1)$$

whenever it exists, defines the partition function of the lattice *Gaussian model*, see also the discussion on the Gaussian approximation in Section 1.6.2. Here β is the inverse temperature, $\mathbf{J}_\Lambda^{(\tau)} = \{ J^{(\tau)}(\mathbf{r}, \mathbf{r}'), \mathbf{r}, \mathbf{r}' \in \Lambda \}$ is the spin-spin interaction matrix with elements $J^{(\tau)}(\mathbf{r}, \mathbf{r}')$ defining the (isotropic) exchange interaction potential between the spins at sites $\mathbf{r}$ and $\mathbf{r}'$; the dependence on the boundary conditions is denoted by the superscript (τ). The n-component vector $\mathbf{H_r}$ is the external magnetic field at site $\mathbf{r}$, and $\mathbf{h}_\Lambda = \{ \beta \mathbf{H_r}, \mathbf{r} \in \Lambda \}$.

Obviously, the Gaussian model is defined only for sufficiently high temperatures, such that $\beta\lambda_{\max}^{(\tau)} < \partial$, where $\lambda_{\max}^{(\tau)}$ is the largest eigenvalue of the matrix $\mathbf{J}_\Lambda^{(\tau)}$. One can overcome this serious defect by replacing in the n-vector model the set of local constraints $|\mathbf{S_r}|^2 = 1$, $\mathbf{r} \in \Lambda$, by one global constraint $\sum_{\mathbf{r}} |\mathbf{S_r}|^2 = |\Lambda|$, where the summation is taken over all the $|\Lambda|$ spins in the system. This procedure leads to the so called *spherical model*, the two main versions of which, see [Berlin and Kac (1952)] and [Lewis and Wannier (1952)], are considered below.

The spherical model is one of the few statistical mechanical models which have been exactly solved in any space dimensionality d, and exhibit a non-trivial (not of the mean field type) critical behavior for $d_l < d < d_u$, where d_l and d_u are the lower and upper critical dimensions, respectively.* Furthermore, the model has been solved exactly in the presence of an external field, for systems in confined geometry with different boundary conditions, in the cases of short-range and long-range power-law interactions [Joyce (1966)], [Joyce (1972)]. In addition, many different modifications of the spherical model have been studied: with random interactions [Kosterlitz et. al. (1976)], [Jagannathan and Rudnick (1989)], random magnetic fields [Hornreich and Schuster (1982)], [Vojta (1993)], dipole-dipole interactions [Toupin and Lax (1957)], [Danchev (1990)]. A close relationship between the critical behavior of the spherical model and that of the ideal Bose gas has been established and studied in detail [Gough and Pulé (1993)]. The spherical model has been applied also to disordered electronic systems with localized states [Vojta and Schreiber (1994)], fluctuation induced (Casimir) forces [Danchev (1998)], surface critical phenomena [Barber (1974)], [Barber et. al (1974)], [Danchev et. al (1997a)], [Danchev et. al (1997b)], phase separation [Patrick (1993)], [Brankov and Tonchev (1994)], [Garanin (1996)], etc. It has been widely used as a test case for the finite-size scaling hypotheses [Barber and Fisher (1973)], [Singh and Pathria (1985a)], [Rudnick (1990)], [Brankov and Tonchev (1992)]. It should be emphasized that the above list of cited works is by far not exhaustive; it simply illustrates the diversity of problems where the spherical approximation has been used.

Reviews on the model have been written by [Joyce (1972)] and, more recently, by [Khorunzhy et. al. (1992)] for spherical models with disorder;

---

*For all $O(n)$ models with short-range interactions the borderline dimensions are $d_l = 2$ and $d_u = 4$.

in [Rudnick (1990)] and [Brankov and Tonchev (1992)] the focus is set on finite-size effects.

3.1.1   *The spherical and mean spherical models*

The spherical model of a ferromagnet was first proposed by M. Kac in 1947 and later solved exactly in [Montroll (1949)], and [Berlin and Kac (1952)]. We shall consider the spherical model on the regular d-dimensional lattice Z^d. Initially, the system is defined in a finite region $\Lambda \in Z^d$ of the block geometry $L_1 \times L_2 \times \cdots \times L_d$, where the edges L_ν, $\nu = 1, \ldots, d$, are measured in units of the lattice spacings. With each lattice site $\mathbf{r} \in \Lambda$ one associates a real-valued scalar[†] variable (spin) $S_\mathbf{r}$. The allowed configurations $S_\Lambda = \{S_\mathbf{r}, \mathbf{r} \in \Lambda\}$ are confined to the surface of the $|\Lambda|$-dimensional sphere

$$\sum_{\mathbf{r} \in \Lambda} S_\mathbf{r}^2 = |\Lambda|, \tag{3.2}$$

where $|\Lambda|$ is the total number of spins in the system. The Hamiltonian is given by

$$\mathcal{H}_\Lambda(\{S_\mathbf{r}, \mathbf{r} \in \Lambda\} | \mathbf{J}_\Lambda^{(\tau)}, H_\Lambda) = -\frac{1}{2} \sum_{\mathbf{r}, \mathbf{r}' \in \Lambda} S_\mathbf{r} J^{(\tau)}(\mathbf{r}, \mathbf{r}') S_{\mathbf{r}'} - \sum_{\mathbf{r} \in \Lambda} H_\mathbf{r} S_\mathbf{r}, \tag{3.3}$$

where $H_\mathbf{r}$ is the scalar external field at site $\mathbf{r} \in \Lambda$, and $H_\Lambda = \{H_\mathbf{r}, \mathbf{r} \in \Lambda\}$. From Eqs. (3.2) and (3.3) it is clear that the spherical model can be formally obtained from the Ising model by relaxing the "strong" local constraints $\{S_\mathbf{r}^2 = 1, \mathbf{r} \in \Lambda\}$ and imposing instead the "weak" global constraint (3.2). The difference between these two models is clearly revealed by the following geometrical consideration. Each spin configuration $S_\Lambda = \{S_\mathbf{r}, \mathbf{r} \in \Lambda\}$ is represented by a point in the $|\Lambda|$-dimensional configuration space. Let us choose an orthogonal frame in that space, consisting of $|\Lambda|$ axes, one for each spin variable $S_\mathbf{r}$. Then, the allowed spin configurations of the spherical model are represented by the points on a hypersphere of radius $|\Lambda|^{1/2}$, while those for the Ising model are given by the $2^{|\Lambda|}$ vertices of the inscribed hypercube. As it is well known, the two models have essentially different critical behavior.

[†]The critical behavior of the resulting spherical model is independent of the dimensionality n of the original spin space, see Section 3.2.

The partition function of the spherical model can be written as

$$Z^s_\Lambda(\beta \mathbf{J}^{(\tau)}_\Lambda, h_\Lambda) = \int_{\mathbf{R}^{|\Lambda|}} \left(\prod_{\mathbf{r} \in \Lambda} \mathrm{d}S_{\mathbf{r}} \right) \delta \left(|\Lambda| - \sum_{\mathbf{r} \in \Lambda} S^2_{\mathbf{r}} \right)$$
$$\times \exp\left[-\beta \mathcal{H}_\Lambda(S_\Lambda | \mathbf{J}^{(\tau)}_\Lambda, H_\Lambda) \right], \tag{3.4}$$

where $h_\Lambda = \{\beta H_{\mathbf{r}}, \mathbf{r} \in \Lambda\}$ is the dimensionless external field configuration. This partition function can be thought of as corresponding to an assemble which is canonical with respect to the energy and microcanonical with respect to the quantity $\sum_{\mathbf{r}} S^2_{\mathbf{r}}$. By using the integral representation of the delta function,

$$\delta\left(|\Lambda| - \sum_{\mathbf{r} \in \Lambda} S^2_{\mathbf{r}} \right) = \frac{1}{2\pi i} \int_{a-i\infty}^{a+i\infty} \mathrm{d}s \, \exp\left[s\left(|\Lambda| - \sum_{\mathbf{r} \in \Lambda} S^2_{\mathbf{r}} \right) \right],$$

one obtains

$$Z^s_\Lambda(\beta \mathbf{J}^{(\tau)}_\Lambda, h_\Lambda) = \frac{1}{2\pi i} \int_{a-i\infty}^{a+i\infty} \mathrm{d}s \, e^{s|\Lambda|} Z^G_\Lambda(\beta \mathbf{J}^{(\tau)}_\Lambda, h_\Lambda, s). \tag{3.5}$$

Here $Z^G_\Lambda(\beta \mathbf{J}^{(\tau)}_\Lambda, h_\Lambda, s)$ is the partition function (3.1) of the scalar Gaussian model with a complex parameter s; the real part $a > 0$ of s is chosen so that the integral in (3.1) is finite.

One can consider also the ensemble which is canonical with respect to both $\mathcal{H}$ and $\sum_{\mathbf{r}} S^2_{\mathbf{r}}$. The partition function for this "grand canonical ensemble" is just $Z^G_\Lambda(\beta \mathbf{J}^{(\tau)}_\Lambda, h_\Lambda, s)$, where s is now a real variable (called *spherical field*) conjugate to $\sum_{\mathbf{r}} S^2_{\mathbf{r}}$, rather than a free parameter. The value of s is determined from the condition that the spherical constraint be satisfied in the mean,

$$\left\langle \sum_{\mathbf{r} \in \Lambda} S^2_{\mathbf{r}} \right\rangle = |\Lambda|, \tag{3.6}$$

or, equivalently,

$$-\frac{\partial}{\partial s} \ln Z^G_\Lambda(\beta \mathbf{J}^{(\tau)}_\Lambda, h_\Lambda, s) = |\Lambda|. \tag{3.7}$$

Note that in Eq. (3.6) the average value is taken with the Gibbs probability distribution for the temperature-dependent "Hamiltonian"

$$\mathcal{H}_\Lambda^{G\cdot}(S_\Lambda\,|\,\mathbf{J}_\Lambda^{(\tau)}, H_\Lambda, s) = \mathcal{H}_\Lambda(S_\Lambda\,|\,\mathbf{J}_\Lambda^{(\tau)}, H_\Lambda) + \beta^{-1}s\sum_{\mathbf{r}\in\Lambda} S_\mathbf{r}^2. \qquad (3.8)$$

The model defined by Eqs. (3.1) and (3.8), with

$$s = \bar{s}_\Lambda(\beta\mathbf{J}_\Lambda^{(\tau)}, h_\Lambda)$$

being the solution of Eq. (3.7), is the so-called *mean spherical model*. It has been introduced in [Lewis and Wannier (1952)].

From statistical point of view, the spherical model is defined by the joint probability distribution of the spin variables given by the measure

$$d\mu_\Lambda^{s\cdot}(S_\Lambda\,|\,\beta\mathbf{J}_\Lambda^{(\tau)}, h_\Lambda) = \frac{\exp\left[-\beta\mathcal{H}_\Lambda(S_\Lambda\,|\,\mathbf{J}_\Lambda^{(\tau)}, H_\Lambda)\right]}{Z_\Lambda^{s\cdot}(\beta\mathbf{J}_\Lambda^{(\tau)}, h_\Lambda)}$$

$$\times\,\delta\left(|\Lambda| - \sum_{\mathbf{r}\in\Lambda} S_\mathbf{r}^2\right)\prod_{\mathbf{r}\in\Lambda} dS_\mathbf{r}, \qquad (3.9)$$

where dS is the Lebesgue measure on R. The corresponding probability distribution for the mean spherical model is the Gaussian one,

$$d\mu_\Lambda^{m.s\cdot}(S_\Lambda\,|\,\beta\mathbf{J}_\Lambda^{(\tau)}, h_\Lambda) = \frac{\exp\left[-\beta\mathcal{H}_\Lambda^{G\cdot}(S_\Lambda\,|\,\mathbf{J}_\Lambda^{(\tau)}, H_\Lambda, \bar{s}_\Lambda)\right]}{Z_\Lambda^{G\cdot}(\beta\mathbf{J}_\Lambda^{(\tau)}, h_\Lambda, \bar{s}_\Lambda)}\prod_{\mathbf{r}\in\Lambda} dS_\mathbf{r}, \qquad (3.10)$$

with $\bar{s}_\Lambda$ satisfying Eq. (3.7). The two models are, in general, statistically inequivalent [Lewis and Wannier (1953)], [Yan and Wannier (1965)], [Brankov and Danchev (1987)].

Although the spherical model was originally introduced as an exactly solvable approximation to the Ising model, the closer similarities are between the spherical model and the $O(n)$, $n \geq 2$, models. For example, it has been shown that the isotropic Heisenberg model with an exchange interaction $J(\mathbf{r}, \mathbf{r}')$ cannot have ferromagnetic or antiferromagnetic ordering if the spherical model with the interaction $|J(\mathbf{r}, \mathbf{r}')|$ does not exhibit phase transition [Mermin and Wagner (1966)], [Joyce (1969)]. Furthermore, Stanley has established [Stanley (1968)], [Stanley (1969a)], [Stanley (1969b)], a precise correspondence between the translation-invariant spherical model and the $n \to \infty$ limit of the translation-invariant isotropic $O(n)$ model (see Section 3.2 for some details). It is interesting to note that the critical temperature of the three-dimensional spherical model with nearest-neighbor interactions underestimates by only 5% the critical temperature of the corresponding Heisenberg model, evaluated by high-temperature expansion techiques [Nagaev (1988)].

3.1.2 *Critical behavior of the spherical models*

As it could be expected, the spherical and mean spherical models have the same thermodynamic properties. That is why we present here the derivation of the corresponding expressions in the simpler case of the mean spherical model.

The Gibbs free energy density of the finite-size mean spherical model is defined by the Legendre transformation

$$f_\Lambda^{m.s.}(\beta \mathbf{J}_\Lambda^{(\tau)}, h_\Lambda) = \sup_{s>0} \left[f_\Lambda^{G.}(\beta \mathbf{J}_\Lambda^{(\tau)}, h_\Lambda, s) - \beta^{-1} s \right], \qquad (3.11)$$

where

$$f_\Lambda^{G.}(\beta \mathbf{J}_\Lambda^{(\tau)}, h_\Lambda, s) = -(\beta|\Lambda|)^{-1} \ln Z_\Lambda^{G.}(\beta \mathbf{J}_\Lambda^{(\tau)}, h_\Lambda, s) \qquad (3.12)$$

is the Gibbs free energy density of the Gaussian model with parameter s. The supremum in the right-hand side of Eq. (3.11) is attained at the *unique* solution of the mean spherical constraint (3.7).

In the remainder of this section we will assume periodic boundary conditions imposed on the system in Λ and, for simplicity of notation, the superscript (τ) will be omitted. We will focus on the case of translation-invariant and isotropic ferromagnetic interactions, when $J(\mathbf{r}, \mathbf{r}') = J(|\mathbf{r} - \mathbf{r}'|) > 0$. The calculation of the Gaussian partition function $Z_\Lambda^{G.}(\beta \mathbf{J}_\Lambda, h_\Lambda, s)$ is then straightforward: one readily performs the integration in Eq. (3.1) by first diagonalizing the quadratic form $\beta \mathcal{H}_\Lambda^{G.}(S_\Lambda | \mathbf{J}_\Lambda, H_\Lambda, s)$, see Eqs. (3.3) and (3.8), via a Fourier transformation of the spin variables $\{S_\mathbf{r}, \mathbf{r} \in \Lambda\}$ to a set of "normal coordinates". Denoting the Fourier transform of $J(\mathbf{r})$ by $\hat{J}(\mathbf{k})$, and setting

$$s = \tilde{K}(\tilde{\phi} + 1)/2, \qquad \tilde{K} = \beta \hat{J}(\mathbf{0}), \qquad (3.13)$$

one obtains for the free energy density of the Gaussian model in a homogeneous magnetic field $\{H_{\mathbf{r}} = H, \mathbf{r} \in \Lambda\}$, $\beta H = h$,

$$\beta f_{\Lambda}^{G\cdot}(\beta \mathbf{J}_{\Lambda}, h, \tilde{K}(\tilde{\phi}+1)/2) =$$

$$\frac{1}{2}\ln\frac{\tilde{K}}{2\pi} - \frac{h^2}{2\tilde{K}\tilde{\phi}} + \frac{1}{2|\Lambda|}\sum_{\mathbf{k}}\ln\left[1+\tilde{\phi}-\hat{J}(\mathbf{k})/\hat{J}(\mathbf{0})\right]. \qquad (3.14)$$

Here the vector $\mathbf{k} = \{k_1, \ldots, k_d\}$ takes values in the first Brillouin zone $\mathcal{B}_{\Lambda}$, which for L_{ν} odd integers is defined by

$$k_{\nu} = 2\pi n_{\nu}/L_{\nu}, \quad -(L_{\nu}-1)/2 \le n_{\nu} \le (L_{\nu}-1)/2, \quad \nu = 1, \ldots, d. \qquad (3.15)$$

The supremum in the right-hand side of Eq. (3.11) leads to the following equation for $\tilde{\phi}$:

$$\frac{h^2}{\tilde{K}\tilde{\phi}^2} + \frac{1}{|\Lambda|}\sum_{\mathbf{k}}\frac{1}{1+\tilde{\phi}-\hat{J}(\mathbf{k})/\hat{J}(\mathbf{0})} = \tilde{K}. \qquad (3.16)$$

In a similar way one obtains the net spin-spin correlation function, see Eq. (1.70),

$$G_{\Lambda}(\mathbf{r}, \beta \mathbf{J}_{\Lambda}, h, s) = \frac{1}{\tilde{K}|\Lambda|}\sum_{\mathbf{k}}\frac{\exp(i\mathbf{k}\cdot\mathbf{r})}{\tilde{\phi}-\hat{J}(\mathbf{k})/\hat{J}(\mathbf{0})+1}. \qquad (3.17)$$

Obviously, the thermodynamic properties and correlation functions depend directly on the Fourier transform of the interaction potential. That is why it is convenient to distinguish among the different interactions by fixing the function $\hat{J}(\mathbf{k})$. In the case of nearest-neighbor interactions one has

$$\hat{J}^{\mathrm{n.n.}}(\mathbf{k}) = 2J\sum_{\nu=1}^{d}\cos k_{\nu}. \qquad (3.18)$$

We will consider also long-range interaction potentials $J(r)$ which decay at large separations $r \gg 1$ as

$$J(r) \sim r^{-(d+\sigma)}, \qquad (\sigma > 0) \qquad r \to \infty. \qquad (3.19)$$

The parameter σ here must be strictly positive in order $\hat{J}(\mathbf{0})$ to be finite. It turns out that the critical properties of the model, as well as the leading-order finite-size effects, are determined by the long-wavelength (small $k = |\mathbf{k}|$) asymptotic form of the Fourier transform $\hat{J}(\mathbf{k})$. In the case of nearest-neighbor interactions $\hat{J}(\mathbf{k})$ is an analytic function of $\mathbf{k}$ at $k = 0$. More generally, for short-range interactions with $\sigma > 2$ in Eq. (3.19), due to the reflection symmetry and the isotropy of the interaction potential, the small-k expansion of its Fourier transform will have the leading-order form $\hat{J}(\mathbf{k}) \simeq \hat{J}(\mathbf{0})(1 - \rho_2 k^2)$. In the case of long-range potentials of the type (3.19) with $0 < \sigma < 2$, the Fourier transform behaves as

$$\hat{J}(\mathbf{k}) \simeq \hat{J}(\mathbf{0})(1 - \rho_\sigma k^\sigma). \tag{3.20}$$

Thus, short-range interactions correspond formally to $\sigma = 2$, $\hat{J}(\mathbf{0}) = 2dJ$, and $\rho_2 = (2d)^{-1}$ in Eq. (3.20). Note that the borderline case of power-law decay with exponent $\sigma = 2$ in Eq. (3.19) leads to logarithmic factors in the Fourier transform which are not allowed for in the present simple analysis. For ferromagnetic interactions one has $\hat{J}(\mathbf{0}) > 0$ and $\rho_\sigma > 0$. The price one pays for fixing the form of the Fourier transformed potential is that in the case of long-range interactions one actually considers effective potentials $J_\Lambda(r)$ which take into account interactions with repeated images of the system, i.e.

$$J_\Lambda(\mathbf{r}) = \sum_{\mathbf{m} \in Z^d} J\left(\left[\sum_{\nu=1}^{d} |r_\nu - m_\nu L_\nu|^2 \right]^{1/2} \right), \tag{3.21}$$

where $\mathbf{m} = \{m_1, \ldots, m_d\}$. Obviously, the above potential is a periodic function of $\mathbf{r}$ with period L_ν in the ν-th coordinate, $\nu = 1, \ldots, d$.

For interactions of the type (3.20), the free energy density of the mean spherical model reads

$$\beta f_\Lambda^{m.s.}(K, h) = \frac{1}{2} \sup_{\phi > 0} \left\{ -\frac{h^2}{K\phi} + \frac{1}{|\Lambda|} \sum_{\mathbf{k} \in \mathcal{B}_\Lambda} \ln(\phi + |\mathbf{k}|^\sigma) - K\phi \right\}$$
$$+ \frac{1}{2} \left[\ln \frac{K}{2\pi} - \frac{1}{\rho_\sigma} K \right], \tag{3.22}$$

where $K = \rho_\sigma \tilde{K}$ and $\phi = \tilde{\phi}/\rho_\sigma$. The supremum in the right-hand side of Eq. (3.22) is attained at the solution of the mean spherical constraint

$$\frac{1}{|\Lambda|} \sum_{\mathbf{k} \in \mathcal{B}_\Lambda} \frac{1}{\phi + |\mathbf{k}|^\sigma} + \frac{h^2}{K\phi^2} = K. \tag{3.23}$$

Taking the thermodynamic limit in Eq. (3.22) at fixed value of ϕ, we obtain the following expression for the bulk free energy density

$$\beta f_\infty(K, h) = \sup_{\phi > 0} \frac{1}{2}\left\{ U_{d,\sigma}(\phi) - K\phi - \frac{h^2}{K\phi} \right\} + \cdots, \qquad (3.24)$$

where we have introduced the function

$$U_{d,\sigma}(\phi) := (2\pi)^{-d} \int_{-\pi}^{\pi} dk_1 \cdots \int_{-\pi}^{\pi} dk_d \ln(\phi + |\mathbf{k}|^\sigma). \qquad (3.25)$$

The analysis of the above variational problem shows that when: (a) $d \le \sigma = d_l$; (b) $d > \sigma$ and $h \ne 0$, or $K < K_c$, where the critical coupling is (see below)

$$K_c = W_{d,\sigma}(0) < \infty, \qquad \text{if} \quad d > \sigma, \qquad (3.26)$$

then the solution $\phi = \phi_\infty(K, h)$ obeys the thermodynamic limit form of the mean spherical constraint (3.23), namely

$$W_{d,\sigma}(\phi) + \frac{h^2}{K\phi^2} = K, \qquad (3.27)$$

with

$$W_{d,\sigma}(\phi) := (2\pi)^{-d} \int_{-\pi}^{\pi} dk_1 \cdots \int_{-\pi}^{\pi} dk_d \frac{1}{\phi + |\mathbf{k}|^\sigma}. \qquad (3.28)$$

When $d > \sigma$ and $h = 0$, $K \ge K_c$, the solution of the variational problem Eq. (3.24) is given by the endpoint $\phi = 0$.

Let us mention some properties of the function $W_{d,\sigma}(\phi)$ that explain the results stated above. Firstly, $W_{d,\sigma}(\phi)$ is a strictly decreasing function of $\phi \ge 0$, attaining its maximum $W_{d,\sigma}(0)$ at $\phi = 0$ if $d > \sigma$; it diverges to plus infinity when $\phi \to 0^+$ if $d \le \sigma$. Clearly, $W_{d,\sigma}(\phi)$ is a real-analytic function of $\phi > 0$. Note that $\phi > 0$ is a necessary condition for the existence of the finite-size partition function. Finally, since the left-hand side of (3.27) is a monotonically decreasing function of $\phi > 0$, it is readily seen that the solution $\phi_\infty(K, h)$, if it exists, is unique for any fixed K and h. The physical meaning of this solution becomes immediately clear if one takes into account that, according to (3.24), the magnetization per spin in the thermodynamic limit equals

$$m_\infty(K, h) := -\frac{\partial}{\partial h} \beta f_\infty(K, h) = \frac{h}{K\phi_\infty(K, h)}. \qquad (3.29)$$

Therefore, the mean spherical constraint (3.27) can be rewritten in the form of *equation of state for the magnetization*,

$$(1 - m_\infty^2)K = W_{d,\sigma}\left(\frac{h}{Km_\infty}\right). \tag{3.30}$$

If $h \neq 0$, from Eqs. (3.27) and (3.30) one obtains $\phi_\infty(K,h) > 0$, hence, ϕ_∞ is a real-analytic function of h. The invariance of the free energy density (3.24) with respect to the change of sign of the magnetic field, $h \leftrightarrow -h$, shows that ϕ_∞ is actually an even function of h. Eq. (3.29) implies then that the magnetization $m_\infty(K,h)$ is, as expected, an odd function of h. When $h \to 0$ the equation of state admits two possibilities: *(i)* $m_\infty(K,h) \to m_0(K) \neq 0$, where $m_0(K)$ is the spontaneous magnetization density for the given K, and *(ii)* $m_\infty(K,h) \to 0$.

(i) When $m_\infty(K,h) \to m_0(K)$ as $h \to 0$, the equation of state becomes

$$m_0^2(K) = 1 - W_{d,\sigma}(0)/K.$$

Therefore, such a behavior may exist only for $K > W_{d,\sigma}(0)$, which identifies the critical coupling $K_c = W_{d,\sigma}(0)$. The critical temperature is nonzero when $W_{d,\sigma}(0)$ is finite, i.e. when $d > \sigma$.

(ii) When $m_\infty(K,h) \to 0$ as $h \to 0$, the equation of state becomes

$$K = W_{d,\sigma}(\phi).$$

This equation has a positive solution only if $K < W_{d,\sigma}(0) = K_c$. This is always the case when $W_{d,\sigma}(0) = \infty$, i.e. when $d \leq \sigma$.

From the existence of a unique solution $\phi(K,h)$, and the results described in *(i)* and *(ii)*, we conclude that

(a) For any $d > \sigma$ and $K > K_c$ there exists a spontaneous magnetization

$$m_0(K) = (1 - K_c/K)^{1/2}, \tag{3.31}$$

hence, the critical exponent for the order parameter is $\beta = 1/2$ independently of $d > \sigma$.

(b) For $K < K_c$ and $d > \sigma$, as well as for any K and $d \leq \sigma$, one obtains that $m_\infty(K,h) \to 0$ as $h \to 0$, i.e. there is no spontaneous magnetization.

From the above it follows that the lower critical dimension for the spherical model is $d_l = \sigma$. In the case of $\sigma = 2$ one recovers the well-known result $d_l = 2$ for $O(n)$ systems with $n \geq 2$ and short-range interactions.

To proceed further we need the asymptotic behavior of the function $W_{d,\sigma}(\phi)$ as $\phi \to 0^+$. It has been shown [Joyce (1972)] that

$$W_{d,\sigma}(\phi) - W_{d,\sigma}(0) \simeq$$
$$\simeq \begin{cases} -D_{d,\sigma}\, \phi^{(d-\sigma/\sigma}} & \text{if} \quad \sigma < d < 2\sigma \\ -D_{2\sigma,\sigma}\, \phi \ln(1/\phi) & \text{if} \quad d = 2\sigma \\ -|W'_{d,\sigma}(0)|\, \phi + D_{d,\sigma}\, \phi^{(d-\sigma)/\sigma} & \text{if} \quad 2\sigma < d < 3\sigma \end{cases} \qquad (3.32)$$

where, for integer m,

$$D_{d,\sigma} = \begin{cases} 2\pi[(4\pi)^{d/2}\sigma\Gamma(d/2)|\sin(\pi d/\sigma)|]^{-1}, & \text{if } \sigma < d \neq m\sigma \\ 2[(4\pi)^{\sigma}\sigma\Gamma(\sigma)]^{-1}, & \text{if } d = 2\sigma\,. \end{cases} \qquad (3.33)$$

Now we continue the study of the bulk critical behavior. By substituting (3.32) in Eq. (3.27) at $h = 0$, we easily obtain that as $t = K_c/K - 1 \to 0^+$

$$\phi_\infty(K,0) \simeq \begin{cases} (K_c/D_{d,\sigma})^{\sigma/(d-\sigma)}t^{\sigma/(d-\sigma)}, & \text{if } \sigma < d < 2\sigma \\ (K_c/D_{2\sigma,\sigma})\, t/(-\ln t), & \text{if } d = 2\sigma \\ (K_c/|W'_{d,\sigma}(0)|)\, t, & \text{if } 2\sigma < d. \end{cases} \qquad (3.34)$$

Similarly, when $K = K_c$ and $h \to 0$ one obtains:

$$\phi_\infty(K_c,h) \simeq \begin{cases} (K_c D_{d,\sigma})^{-\sigma/(d+\sigma)}|h|^{2\sigma/(d+\sigma)}, & \text{if } \sigma < d < 2\sigma \\ (2K_c D_{2\sigma,\sigma}/3)^{-1/3}|h|^{2/3}/(-\ln|h|)^{1/3}, & \text{if } d = 2\sigma \\ (K_c/|W'_{d,\sigma}(0)|)^{-1/3}|h|^{2/3}, & \text{if } 2\sigma < d. \end{cases}$$
$$(3.35)$$

When $K > K_c$ and $h \to 0$, by using the expansion (3.32) and Eq. (3.27), one obtains the following asymptotic behavior of the bulk magnetization per spin

$$m_\infty^2(K,h) \simeq m_0^2(K)$$
$$+K^{-1} \begin{cases} D_{d,\sigma}\phi_\infty^{(d-\sigma)/\sigma}(K,h), & \text{if } \sigma < d < 2\sigma \\ -D_{2\sigma,\sigma}\phi_\infty(K,h)\ln\phi_\infty(K,h), & \text{if } d = 2\sigma \\ |W'_{d,\sigma}(0)|\phi_\infty(K,h), & \text{if } 2\sigma < d < 3\sigma. \end{cases} \qquad (3.36)$$

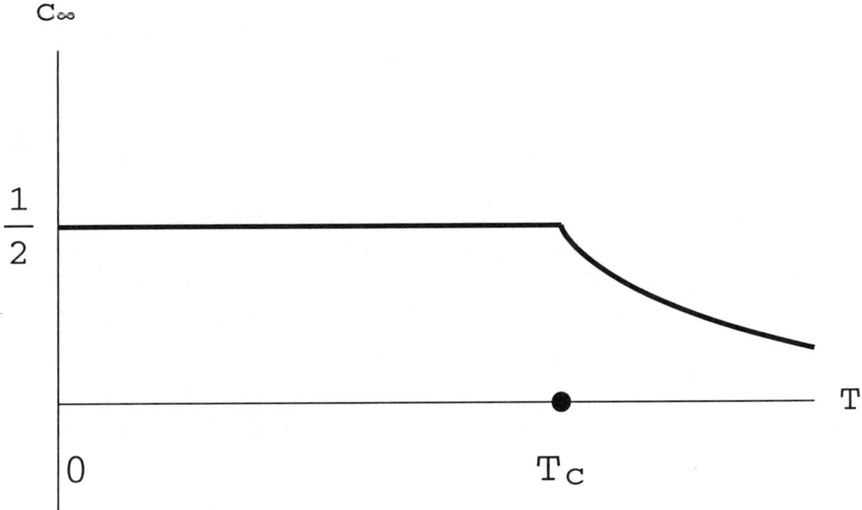

Fig. 3.1   The typical behavior of the specific heat for $3 < d < 4$.

By differentiating twice the free energy density (3.24) with respect to the temperature, one obtains the zero-field specific heat per spin,

$$c_\infty(K,0) := -K^2 \frac{\partial^2}{\partial K^2}[\beta f_\infty(K,0)] = \frac{1}{2} + \frac{1}{2}K^2\frac{\partial \phi_\infty}{\partial K} \qquad (3.37)$$

Since ϕ_∞ decreases when K increases, see Eq. (3.27), and $\phi_\infty \equiv 0$ for $K > K_c$, then the zero-field specific heat per spin keeps its maximum value $c_\infty(K,0) = 1/2$ for all $K > K_c$, see Fig. 3.1. Obviously, the famous Dulong-Petit law of classical thermodynamics holds for all $T \le T_c$, and the third law of thermodynamics (stating that the specific heat at constant volume must vanish as $T \to 0^+$) is violated. We note that a quantum version of the spherical model resolves these discrepancies, see [Nieuwenhuizen (1995)]. Since c_∞ has a cusp-type singularity at the critical temperature, according to definition (1.69) and Eq. (3.34) one concludes that

$$\alpha_s = \begin{cases} -(2\sigma - d)/(d - \sigma), & \text{if } \sigma < d < 2\sigma \\ 0, & \text{if } d > 2\sigma. \end{cases} \qquad (3.38)$$

The above critical exponent is non-positive for all space dimensions $d > \sigma$, and has a logarithmic correction at the upper critical dimension $d = 2\sigma$.

Note that in the case of the mean spherical model, due to the field dependence of the spherical field $\phi_\infty(K, h)$, one should distinguish beteen two kinds of susceptibilities. The "fluctuation part" of the susceptibility

$$k_B T\, \chi_\infty(K, h) := -\frac{\partial^2}{\partial h^2}[\beta f_\infty(K, h)]\Big|_{\phi_\infty(K,h)=const} = \frac{1}{K\phi_\infty} \qquad (3.39)$$

measures the fluctuations of the magnetization in the thermodynamic limit (more precisely, the variance of a properly normalized block-spin in the limit Gibbs state, see Chapter 9) and satisfies the fluctuation-dissipation relationship (1.60). On the other hand, by differentiating twice the free energy density (3.24) with respect to the magnetic field, taking into account the implicit dependence on h through the solution $\phi_\infty(K, h)$ of the mean spherical constraint (3.27), one obtains the *total* magnetic susceptibility per spin,

$$k_B T\, \chi_\infty^{\mathrm{tot}}(K, h) := -\frac{\mathrm{d}^2}{\mathrm{d}h^2}[\beta f_\infty(K, h)] = \frac{1}{K\phi_\infty}\left[1 - \frac{h}{\phi_\infty}\frac{\mathrm{d}\phi_\infty}{\mathrm{d}h}\right]. \qquad (3.40)$$

Note that these two quantities coincide in the zero-field case when there is no spontaneous magnetization in the system: either $d \le \sigma$, or $d > \sigma$ and $K \le K_c$. Their critical behavior at $h = 0$ and $K \to K_c^-$ follows from Eqs. (3.34) and (3.35). Thus one obtains the critical exponents

$$\gamma = \begin{cases} \sigma/(d - \sigma), & \text{if } \sigma < d < 2\sigma \\ 1, & \text{if } d > 2\sigma, \end{cases} \qquad (3.41)$$

and

$$\delta = \begin{cases} (d + \sigma)/(d - \sigma), & \text{if } \sigma < d < 2\sigma \\ 3, & \text{if } d > 2\sigma. \end{cases} \qquad (3.42)$$

Again there are logarithmic corrections at $d = 2\sigma$. Furthermore, it is clear that for $\sigma < d < 2\sigma$ one has the result $\chi_\infty(K, 0) \equiv \infty$ for all $K > K_c$. This is a common feature for all $O(n)$ models below the critical temperature and in the absence of a symmetry-breaking field. In particular, it implies that the critical exponent γ' is undefined. Above the upper critical dimension, $d > 2\sigma$, one obtains from Eqs. (3.34), (3.35) and (3.40) the mean field result $\gamma = \gamma' = 1$.

The scaling function for the equation of state of the spherical model can be explicitly determined. From (3.32) and (3.27) one obtains that near the critical point

$$h = K_c m \begin{cases} (K_c/D_{d,\sigma})^{\sigma/(d-\sigma)}|m|^{2\sigma/(d-\sigma)}(1 + t/m_\infty^2)^{\sigma/(d-\sigma)}, & \sigma < d < 2\sigma \\ (K_c/D_{d,\sigma})|m|^2(1 + t/m_\infty^2), & d > 2\sigma. \end{cases}$$

Hence, the equation of state has indeed the scaling form suggested in [Griffiths (1967)],

$$h \simeq m|m|^{\delta-1}h_s(t/|m|^{1/\beta}), \qquad h_s(x) = (1 + x)^\gamma, \tag{3.43}$$

where the scaling function h_s is given apart from a constant factor.

The singular behavior of the free energy density (3.24) follows from the asymptotic solutions (3.34) and (3.35), and the expansion of $U_{d,\sigma}(\phi)$ as $\phi \to 0^+$,

$$U_{d,\sigma}(\phi) - U_{d,\sigma}(0) - K_c\phi \simeq$$
$$\begin{cases} -(\sigma/d)\, D_{d,\sigma}\, \phi^{d/\sigma}, & \text{if } \sigma < d < 2\sigma \\ -\frac{1}{2}D_{2\sigma,\sigma}\, \phi^2 \ln(1/\phi), & \text{if } d = 2\sigma \\ -\frac{1}{2}|W'_{d,\sigma}(0)|\, \phi^2 + (\sigma/d)D_{d,\sigma}\phi^{d/\sigma}, & \text{if } 2\sigma < d < 3\sigma, \end{cases} \tag{3.44}$$

which can be obtained by term-by-term integration of Eq. (3.32). Thus, when $t \to 0^+$ at $h = 0$, the leading asymptotic form of the *singular part* of the free energy density (3.24) is

$$\beta f_{\infty,\text{sing}}(K, 0) \simeq$$
$$\begin{cases} \frac{1}{2}K_c t\phi_\infty - \frac{\sigma}{2d}D_{d,\sigma}\phi_\infty^{d/\sigma} \sim t^{d/(d-\sigma)}, & \text{if } \sigma < d < 2\sigma \\ \frac{1}{2}K_c t\phi_\infty - \frac{1}{4}D_{2\sigma,\sigma}\phi_\infty^2 \ln(1/\phi_\infty) \sim \frac{t^2}{-\ln t}, & \text{if } d = 2\sigma \\ \frac{\sigma}{2d}D_{d,\sigma}\phi_\infty^{d/\sigma} \sim t^{d/\sigma}, & \text{if } 2\sigma < d, \end{cases} \tag{3.45}$$

Note that in the case $d > 2\sigma$ we have omitted the regular in t term

$$\frac{1}{2}K_c t\, \phi_\infty - \frac{1}{4}|W'_{d,\sigma}(0)|\, \phi_\infty^2 \sim t^2, \qquad (d > 2\sigma). \tag{3.46}$$

From Eqs. (3.24), (3.35) and (3.44) one obtains the following behavior of the bulk free energy density when $h \to 0$ at $K = K_c$:

$$\beta f_{\infty,\mathrm{sing}}(K_c, h) \simeq$$
$$\begin{cases} -\frac{\sigma}{2d} D_{d,\sigma} \phi_\infty^{d/\sigma} - \frac{h^2}{2K_c \phi_\infty} \sim |h|^{2d/(d+\sigma)}, & \text{if } \sigma < d < 2\sigma \\[2mm] -\frac{1}{4} D_{2\sigma,\sigma} \phi_\infty^2 \ln(1/\phi_\infty) - \frac{h^2}{2K_c \phi_\infty} \sim \frac{|h|^{4/3}}{(-\ln|h|)^{1/3}}, & \text{if } d = 2\sigma \\[2mm] -\frac{1}{4} |W_{d,\sigma}(0)| \phi_\infty^2 - \frac{h^2}{2K_c \phi_\infty} \sim |h|^{4/3}, & \text{if } 2\sigma < d. \end{cases} \quad (3.47)$$

Finally, we consider the critical behavior of the bulk correlation function of the mean spherical model,

$$G_\infty(\mathbf{r}; K, h) = \frac{1}{K(2\pi)^d} \int_{-\pi}^{\pi} \cdots \int_{-\pi}^{\pi} \mathrm{d}^d\mathbf{k} \, \frac{\exp(i\mathbf{k} \cdot \mathbf{r})}{\phi_\infty(K, h) + |\mathbf{k}|^\sigma}. \quad (3.48)$$

At large distances $r = |\mathbf{r}|$, due to the rapid oscillations of the exponential function in the integrand, the limits of integration can be extended to infinity, which yields

$$\begin{aligned} G_\infty(\mathbf{r}; K, h) &\simeq \frac{1}{K(2\pi)^d} \int_{R^d} \mathrm{d}^d\mathbf{k} \, \frac{\exp(i\mathbf{k} \cdot \mathbf{r})}{\phi_\infty(K, h) + |\mathbf{k}|^\sigma} \\[2mm] &= \frac{r^{\sigma-d}}{K(2\pi)^d} \int_0^{\infty} \mathrm{d}x \, \frac{x^{d/2} J_{(d-2)/2}(x)}{x^\sigma + [r/\xi_\infty(K, h)]^\sigma}, \end{aligned} \quad (3.49)$$

where $J_\nu(\cdot)$ is the Bessel function of order ν and

$$\xi_\infty(K, h) = \phi_\infty^{-1/\sigma}(K, h) \quad (3.50)$$

is, by definition, the bulk characteristic length. From Eq.(3.49), taken at the critical point where $\lambda_\infty(K_c, 0) = \infty$, we obtain the critical exponent of the correlation function

$$\eta = 2 - \sigma, \qquad \text{if } d > \sigma. \quad (3.51)$$

From Eq. (3.50) at $h = 0$ and $K \to K_c, K > K_c$, and Eq. (3.34) we derive the critical exponent of the correlation length

$$\nu = \begin{cases} 1/(d-\sigma), & \text{if } \sigma < d < 2\sigma \\ 1/\sigma, & \text{if } d > 2\sigma. \end{cases} \quad (3.52)$$

The critical exponents of the (mean) spherical model satisfy all scaling laws, see Section 1.5.4. One should remember, however, that the hyperscaling relation (below the upper critical dimensionality) is satisfied by the exponent of the specific heat α_s, rather than by the exponent α.

From Eq. (3.49) one can obtain explicitly the scaling function for the pair correlation function [Joyce (1966)]

$$G_\infty(\mathbf{r}; K, h) = \frac{1}{K} \frac{1}{r^{d-2+\eta}} X(r/\xi_\infty),$$

namely,

$$X(y) = \int_0^\infty dx \, \frac{x^{d/2} J_{(d-2)/2}(x)}{y^\sigma + x^\sigma}.$$

When $K \geq K_c$ and $h \to 0$, as we already know $\phi_\infty \to 0$, so that

$$G_\infty(\mathbf{r}; K, 0) \simeq \frac{1}{K} \frac{1}{r^{d-\sigma}} X(0).$$

In the opposite case, when $K < K_c$, the solution for ϕ_∞ tends to a finite value as $h \to 0$ and, therefore, $\xi_\infty(K, 0) < \infty$. Then, for $r/\xi_\infty(K, 0) >> 1$ one has [Theumann (1970)] $X(y) \simeq D y^{-2\sigma}$, hence,

$$G_\infty(\mathbf{r}; K, 0) \simeq \frac{D}{K} \frac{\xi_\infty^{2\sigma}}{r^{d+\sigma}}.$$

This result should be compared with the exponential decay which holds in the case of finite-range interactions. Then, for $\sigma = 2$ one has

$$X(y) \sim y^{d/2-1} K_{d/2-1}(y).$$

When $y \to \infty$, this function behaves as $y^{(d-3)/2} \exp(-y)$, which implies exponential decay of correlations.

3.1.3 *Equivalent-neighbors spherical models*

The models with equivalent-neighbors interaction, an example of which is the Husimi–Temperley model considered in Section 2.5, are known also as mean field, or infinitely coordinated models [Silver et. al. (1972)], [Botet et. al. (1982)]. They constitute the simplest class of exactly solvable statistical mechanical systems that exhibit a second order phase transition with respect to the temperature, and a first order phase transition with respect to the external magnetic field. Within the framework of these models one readily obtains in an rigorous way the well-known mean field type critical behavior. A general description of the class is given in Section 6.4.1. Here we will confine our consideration to the definition and thermodynamic properties of the equivalent-neighbors spherical models; some of their non-trivial statistical properties will be studied in Section 9.3 in relation with the probabilistic view on finite-size scaling.

Consider a collection of $N = |\Lambda|$ continuous scalar variables (spins) $S_i \in R$, associated with the sites $\mathbf{r}_i$, $i = 1, \ldots, N$, of a finite d-dimensional lattice Λ. Under the crucial assumption that every two spins in the system interact with equal strength, the distance between particles and dimensionality of the space embedding Λ become irrelevant and, obviously, can be arbitrarily defined. For the sake of convenience, we will keep the notation $S_\Lambda = \{S_i, i = 1, \ldots, N\}$ for the spin configurations. To maintain extensivity of the total energy, the coupling constant has to be chosen inversely proportional to the number of spins N. Thus, the Hamiltonian of the system in a homogeneous external field is given by (for the sake of simplicity, the self-interaction terms are not excluded)

$$\beta \mathcal{H}_\Lambda(\{S_\Lambda\}|J,h) = -\frac{1}{2}\frac{K}{N}\left(\sum_{i=1}^{N} S_i\right)^2 - h\sum_{i=1}^{N} S_i. \tag{3.53}$$

It is instructive to consider [Brankov (1990b)] the Hamiltonian (3.53) as the $\sigma \to 0^+$ limit of Hamiltonian (3.3) for a system in the cubic domain $\Lambda = L^d$ with *periodic boundary conditions* and spin-spin interaction potential of the type, c.f. Eq. (3.19),

$$J^{(\sigma)}(r) = \sigma J_\sigma(r), \qquad J_\sigma(r) \sim J\, r^{-(d+\sigma)} \quad \text{as} \quad r \to \infty, \tag{3.54}$$

with a constant $J > 0$. Then, for the effective potential, see Eq. (3.21),

$$J_\Lambda^{(\sigma)}(\mathbf{r}) = \sum_{\mathbf{n}\in Z^d} J^{(\sigma)}(|\mathbf{r} - L\mathbf{n}|), \tag{3.55}$$

which takes into account interactions with repeated images of the system, one can obtain the asymptotic form

$$\begin{aligned}
J_\Lambda^{(\sigma)}(\mathbf{r}) &\simeq \sigma \sum_{\mathbf{n}\in Z^d, |\mathbf{n}|\leq M} J^{(\sigma)}(|\mathbf{r} - L\mathbf{n}|) + \frac{\sigma J}{L^{d+\sigma}} \sum_{\mathbf{n}\in Z^d, |\mathbf{n}|>M} |\mathbf{n} - \mathbf{r}/L|^{-d-\sigma} \\
&= \mathcal{A}_d(1)\, JL^{-d} + O(\sigma \ln L), \tag{3.56}
\end{aligned}$$

where

$$\mathcal{A}_d(r) = \frac{2\pi^{d/2}}{\Gamma(d/2)} r^{d-1} \tag{3.57}$$

is the surface area of a hypersphere of radius r in R^d, and the constant $M > 0$ has to be chosen sufficiently large. The last equality follows from the known asymptotic behaviour of the Epstein zeta function, see [Glasser and Zucker (1980)],

$$\sum_{\mathbf{n}\in Z^d, \mathbf{n}\neq\mathbf{g}} |\mathbf{n} - \mathbf{g}|^{-d-\sigma} = \mathcal{A}_d(1)/\sigma + c_d + O(\sigma), \qquad \sigma \to 0^+, \qquad (3.58)$$

valid for any vector $\mathbf{g}$, with some constant c_d.

By taking the limit $\sigma \to 0^+$ in the Fourier transform of the effective interaction potential (3.55),

$$\hat{J}_\Lambda^{(\sigma)}(\mathbf{k}) = \sum_{\mathbf{r}\in Z^d} J_\Lambda^{(\sigma)}(\mathbf{r})\, e^{-i\,\mathbf{r}\cdot\mathbf{k}}, \qquad (3.59)$$

where $\mathbf{k}$ takes values in the first Brillouin zone $\mathcal{B}_\Lambda$, see Eq. (3.15), one obtains

$$\lim_{\sigma\to 0^+} \hat{J}_\Lambda^{(\sigma)}(\mathbf{k}) = \mathcal{A}_d(1) J\, \delta_{\mathbf{k},\mathbf{0}}. \qquad (3.60)$$

The above Fourier transform of the equivalent-neighbors interaction formally corresponds to Eq. (3.20), if the validity of the latter is assumed over the whole Brillouin zone, together with $\rho_\sigma \to 1$ as $\sigma \to 0^+$.

Turning back to the definition of the spherical models with equivalent-neighbors interaction, we will make a helpful generalization of the spherical constraint (3.2) by introducing a free parameter $\zeta > 0$ in it,

$$\sum_{i=1}^{N} S_i^2 = N\zeta, \qquad (3.61)$$

so that the allowed spin configurations are confined now to a N-dimensional hypersphere of radius $(\zeta N)^{1/2}$.

The partition function of the model

$$Z_N^s(K, h, \zeta) = \int_{R^N} \exp\left[\frac{K}{2N} \left(\sum_{i=1}^{N} x_i\right)^2 + h \sum_{i=1}^{N} x_i \right]$$

$$\times \delta\left(\sum_{i=1}^{N} x_i^2 - \zeta N \right) dx_1 \ldots dx_N, \qquad (3.62)$$

can be readily evaluated by making an orthogonal change of variables $(x_1,\ldots,x_N) \to (y_1,\ldots,y_N)$, such that

$$\sum_{i=1}^{N} x_i^2 = \sum_{i=1}^{N} y_i^2, \qquad \sum_{i=1}^{N} x_i = N^{1/2} y_1, \qquad (3.63)$$

and integrating over $y_2,\ldots,y_N$. Then, by changing the variable in the remaining integral, $y_1 \to N^{1/2}x$, one obtains the following exact result

$$Z_N^s(K,h,\zeta) = \frac{1}{2} A_{N-1}(N^{1/2}) \int_{-\zeta^{1/2}}^{\zeta^{1/2}} dx (\zeta - x^2)^{-3/2} \exp[N g(x|K,h,\zeta)],$$
$$(3.64)$$

where $A_{N-1}(N^{1/2})$, see Eq. (3.57), is the surface area of a sphere of radius $N^{1/2}$ in the $(N-1)$-dimensional space, and

$$g(x|K,h,\zeta) = \frac{1}{2} K x^2 + hx + \frac{1}{2}\ln(\zeta - x^2). \qquad (3.65)$$

It is easy to see that if $K > \zeta^{-1}$, then $g(x|K,0,\zeta)$ has two global maxima at the points

$$x = \pm m_0(K,\zeta), \qquad m_0(K,\zeta) = (\zeta - K^{-1})^{1/2}, \qquad (3.66)$$

where, as it will be shown below, $m_0(K,\zeta)$ is the spontaneous magnetization of the model, c.f. Eq. (3.31). If $K < \zeta^{-1}$ then $g(x|K,0,\zeta)$ has only one global maximum at $x = 0$. If $K = \zeta^{-1}$, there is still one global maximum at $x = 0$, but now the lowest-order non-vanishing derivative at the maximum is the forth, rather than the second. Obviously, the model has a line of critical points, $K = \zeta^{-1}$, if $\zeta \in (0,\infty)$ is considered as a parameter. When $h \neq 0$, the maximum of the function $g(x|K,h,\zeta)$ is unique. By standard evaluation of the Laplace-type integral in Eq. (3.64), one obtains in the limit $N \to \infty$ the exact expression for the free energy density of the mean field spherical model,

$$\beta f_\infty^s(K,h,\zeta) = -\frac{1}{2}\ln(2\pi e) - \sup_{|x|<\zeta^{1/2}} \frac{1}{2}[K x^2 + 2hx + \ln(\zeta - x^2)]. \qquad (3.67)$$

The supremum in the right-hand side of the above equation is attained at the solution $x = x_\infty(K,h,\zeta)$ of the stationary-point equation

$$K x + h - \frac{x}{\zeta - x^2} = 0. \qquad (3.68)$$

In order to define the corresponding mean spherical model, we first introduce the equivalent-neighbors Gaussian model. Its partition function

$$Z_N^{G\cdot}(K,h,s) =$$

$$\int_{R^N} \exp\left[\frac{K}{2N}\left(\sum_{i=1}^N x_i\right)^2 + h\sum_{i=1}^N x_i - s\sum_{i=1}^N x_i^2\right] dx_1\cdots dx_N \quad (3.69)$$

is finite for all $s > K/2$. The condition for thermodynamic equivalence of the spherical and mean spherical models leads to the mean spherical constraint, c.f. Eq. (3.7),

$$-\frac{\partial}{\partial s}\ln Z_N^{G\cdot}(K,h,s) = \zeta N. \quad (3.70)$$

By using an orthogonal change of variables with the properties (3.63), and directly evaluating the resulting Gaussian integrals, one easily derives the exact expression for the free energy density of the equivalent-neighbors Gaussian model,

$$\beta f_N^{G\cdot}(K,h,K(\phi+1)/2) = \frac{1}{2}\left[\ln\left(\frac{K}{2\pi}\right) + \frac{\ln\phi}{N} + \frac{N-1}{N}\ln(1+\phi) - \frac{h^2}{K\phi}\right],$$
$$(3.71)$$

where $\phi = 2s/K - 1$. The free energy density of the mean spherical model is defined by the Legendre transformation, c.f. Eq. (3.11),

$$\beta f_N^{m.s.}(K,h,\zeta) = \sup_{\phi>0}\left[\beta f_N^{G\cdot}(K,h,K(\phi+1)/2) - \zeta K(\phi+1)/2\right]. \quad (3.72)$$

The mean spherical constraint (3.70) has the explicit form

$$\frac{1}{1+\phi}\left(1-\frac{1}{N}\right) + \frac{1}{N\phi} + \frac{h^2}{K\phi^2} = K\zeta. \quad (3.73)$$

By taking the thermodynamic limit $N \to \infty$ in expression (3.72), one obtains the bulk free energy density of the equivalent-neighbors mean spherical model

$$\beta f_\infty^{m.s.}(K,h,\zeta) = \frac{1}{2}\ln\left(\frac{K}{2\pi}\right) - \frac{1}{2}\zeta K$$

$$+ \frac{1}{2}\sup_{\phi>0}\left[-\frac{h^2}{K\phi} - \zeta K\phi + \ln(1+\phi)\right]. \quad (3.74)$$

The supremum over ϕ is attained at the solution $\phi = \phi_\infty(K, h, \zeta)$ of the bulk mean spherical constraint

$$\frac{1}{1+\phi} + \frac{h^2}{K\phi^2} - \zeta K = 0. \tag{3.75}$$

One can readily check the thermodynamic equivalence of the equivalent-neighbors spherical and mean spherical models. Indeed, let us set $\phi = h/(Kx)$ and write

$$\left[-\frac{h^2}{K\phi} - \zeta K\phi + \ln(1+\phi) \right]_{\phi = h/(Kx)} =$$
$$-2hx - \frac{h}{x}(\zeta - x^2) + \ln\left(1 + \frac{h}{Kx}\right). \tag{3.76}$$

On the other hand, the substitution $\phi = h/(Kx)$ transforms the mean spherical constraint (3.75) into a form equivalent, for $h/x > 0$ and $x^2 < \zeta$, to the spherical constraint (3.68). Hence, with the use of the equality

$$1 + h/Kx = [K(\zeta - x^2)]^{-1},$$

from Eqs. (3.74) and (3.76) it follows that $f_\infty^{m.s.}(K, h, \zeta) \equiv f_\infty^{s.}(K, h, \zeta)$.

Let us turn back to expressions (3.74) and (3.75) for the mean field mean spherical model. By comparing Eqs. (3.27) and (3.75), one concludes that the Watson type integral (3.28) yields in the limit $\sigma \to 0^+$

$$\lim_{\sigma \to 0} W_{d,\sigma}(\phi) = 1/(1+\phi).$$

Since any fixed space dimension $d > 0$, arbitrarily ascribed to the system, is larger than the $\sigma \to 0^+$ limit of the upper critical dimension $d_u = 2\sigma$, it follows that the critical exponents of the equivalent-neighbors spherical models take their classical values $\alpha = 0$, $\beta = 1/2$, $\gamma = 1$, $\delta = 3$. The critical temperature is determined from the equation, compare with Eq.(3.26), $\zeta K_c = W_{d,0}(0) \equiv 1$, i.e. $K_c = 1/\zeta$. The above results can be derived directly from the explicit expressions for the free energy densities (3.67) and/or (3.74). From Eq. (3.74) it follows that the bulk magnetization density is $m_\infty = h/(K\phi_\infty)$, where ϕ_∞ is the solution of the mean spherical constraint (3.75). The latter can be rewritten as an equation of state for the magnetization, c.f. Eq. (3.30),

$$(\zeta - m_\infty)K = 1 + \frac{h}{Km_\infty}. \tag{3.77}$$

Hence one obtains that the spontaneous magnetization per spin $m_0(K, \zeta)$ is given by the second equality in Eq. (3.66).

3.2 The classical n-vector model and its $n \to \infty$ limit

Here we consider classical lattice n-vector spin models, see Section 1.3.3, which describe a collection of n-component classical spins $\mathbf{S_r}$ at each site $\mathbf{r} \in \Lambda \subset Z^d$ of a d-dimensional cubic lattice. In the most natural class of such models the length of the spin vectors is fixed. We assume also short-range (not necessarily translational invariant) ferromagnetic spin-spin interaction. In the presence of an inhomogeneous n-component magnetic field $\mathbf{H_r}$, the Hamiltonian of the system is

$$H_\Lambda^{n-v} = -\frac{1}{2} \sum_{\mathbf{r}, \mathbf{r}' \in \Lambda} J(\mathbf{r}, \mathbf{r}') \mathbf{S_r} \cdot \mathbf{S_{r'}} - \sum_{\mathbf{r} \in \Lambda} \mathbf{H_r} \cdot \mathbf{S_r}, \qquad (3.78)$$

where $\mathbf{S_r} \in R^n$, $|\mathbf{S_r}|^2 = n$, and the external field is assumed to be a $O(n)$-symmetric vector $\mathbf{H_r} = H_\mathbf{r}(1, 1, \ldots, 1)$. The following conditions are imposed on the parameters of the above Hamiltonian:

$$\sum_{\mathbf{r}, \mathbf{r}' \in Z^d} J(\mathbf{r}, \mathbf{r}') < C < \infty, \qquad H_\mathbf{r} < \infty. \qquad (3.79)$$

We emphasize that the restriction on the length of the spins plays the role of the quartic term in the Ginzburg-Landau model, see Section 1.6.1, which makes the calculations extremely difficult. By relaxing this restriction one obtains the Gaussian model, see Section 3.1.1, which in the case of a translation-invariant interaction can be solved exactly by using Fourier transformation. As mentioned in Section 3.1.2, if the set of conditions $|\mathbf{S_r}|^2 = n$, $\mathbf{r} \in \Lambda$ is replaced by one global restriction on the mean square length of all the spins in Λ, one obtains the exactly solved spherical model.

The present consideration is based on the idea that drastic simplifications may occur when the $O(n)$ symmetry group of the model becomes very large. The arguments are the same as in the mean field theory: for n large, the squared length of the n-component spin vector, $\mathbf{S}^2 = \sum_{\alpha=1}^{n} (S^\alpha)^2$, is a sum of a large number of random variables, and, by the central limit theorem, its probability distribution should tend to the normal distribution. Therefore, the $O(n)$-invariant quantities should have small fluctuations due to "self-averaging". For example, in the limit $n \to \infty$ one may replace the quartic random variable $\mathbf{S}^4$ by a quadratic one $(\mathbf{S}^2)^2 \sim \langle \mathbf{S}^2 \rangle \mathbf{S}^2$, in concordance with the heuristic principle of linearization used in self-consistent mean field theories, as well as in the AHM, see Chapter 2.

The first systematic study of the above idea is due to H. E. Stanley, see [Stanley (1968)] and [Stanley (1969a)], who argued that the free energy density of the classical n-vector model converges precisely to that of the spherical model when the spin dimensionality n tends to infinity. For a detailed review of the subsequent efforts in this direction we refer the reader to [Khorunzhy et. al. (1992)]. It has been established that, in addition to the free energy density, all the correlation functions of the n-vector spin model converge as $n \to \infty$ to the corresponding correlation functions of the spherical model. There exists different techniques of proving that fact, which can be applied to many related problems. Bellow we sketch the method proposed in [Shcherbina (1988)]. Important advantages of this method are that in the case of translationally invariant interactions it does not require periodic boundary conditions, and it is quite useful in the study of disordered models as well, see [Khorunzhy et. al. (1992)].

The spherical model can be regarded as a weak limit of the n-vector model defined by Hamiltonian (3.78) in the sense that the limit Gibbs states (see Chapter 9) of the latter weakly converge as $n \to \infty$ to those of the sperical model described by the Hamiltonian

$$H_\Lambda^s = -\frac{1}{2} \sum_{\mathbf{r},\mathbf{r}' \in \Lambda} J(\mathbf{r},\mathbf{r}') S_\mathbf{r} S_{\mathbf{r}'} - \sum_{r \in \Lambda} H_\mathbf{r} S_\mathbf{r} + \frac{1}{2} \sum_{r \in \Lambda} \lambda(\mathbf{r}) \{ S_\mathbf{r}^2 - 1 \}, \quad (3.80)$$

Here $S_\mathbf{r} \in R$ are scalar spins, and the inhomogeneous spherical field $\lambda(\mathbf{r})$ has to satisfy a macroscopic number of constraints,

$$\langle S_\mathbf{r} \rangle_{H_\Lambda^s} = 1, \qquad \mathbf{r} \in \Lambda. \tag{3.81}$$

The main mathematical tool, used to prove that under conditions (3.81)

$$\lim_{V \to \infty} \lim_{n \to \infty} |f_{n\Lambda}[H_\Lambda^{n-v}] - f_\Lambda[H_\Lambda^s]| = 0, \tag{3.82}$$

is provided by the Bogolyubov inequalities (2.1). Note that here $f_{n\Lambda}[H_\Lambda^{n-v}]$ is the free energy density *per spin and per spin component* of the $O(n)$ vector model (3.78), and $f_\Lambda[H_\Lambda^s]$ is the free energy per spin of the scalar model (3.80). Referring the reader to the original paper [Shcherbina (1988)] (see also [Khorunzhy et. al. (1992)]) for the somewhat lengthy details of the proof, we point out only that the main idea is close in spirit to the Approximating Hamiltonian method considered in Chapter 2.

One can easily see that the different configuration spaces of the n-vector model (3.78) and the spherical model (3.80) make a direct comparison of the corresponding free energy densities rather difficult. The key solution to the problem is to introduce an auxiliary Hamiltonian Γ_Λ^n for a system of n-component spins $\mathbf{s_r} \in R^n$ which under some special condition leads to the same free energy density as the model Hamiltonian (3.78). On the other hand, it is defined in the same configuration space as the n-vector version of the spherical model with Hamiltonian

$$
\begin{aligned}
H_\Lambda^{n-s} \;=\; & -\frac{1}{2} \sum_{\mathbf{r},\mathbf{r'} \in \Lambda} J(\mathbf{r},\mathbf{r'})\mathbf{s_r} \cdot \mathbf{s_{r'}} - \sum_{\mathbf{r} \in \Lambda} \mathbf{h_r} \cdot \mathbf{s_r} \\
& + \frac{1}{2} \sum_{\mathbf{r} \in \Lambda} \lambda(\mathbf{r})\{|\mathbf{s_r}|^2 - n\}, \qquad \mathbf{s_r} \in R^n.
\end{aligned}
\tag{3.83}
$$

Let us note that the free energy density (per particle and per component) of the n-vector spherical model $f_{n\Lambda}[H_\Lambda^{n-s}]$ coincides with the free energy density $f_\Lambda[H_\Lambda^s]$ in the limit $n \to \infty$, since the partition function of the former factorizes into the product of n identical scalar spherical model partition functions. Now, for the Hamiltonians H_Λ^{n-s} and Γ_Λ^n it is possible to use the Bogolyubov inequalities and to prove the estimates

$$
-K\frac{\ln n}{n} \leq f_{n\Lambda}[\Gamma_\Lambda^n] - f_{n\Lambda}[H_\Lambda^{n-s}] \leq O\left(\frac{1}{n}\right),
\tag{3.84}
$$

where K is a n-independent constant. The importance of the above result is that it clarifies the status of the spherical model in the thermodynamic limit (3.82), as well as in the case of finite systems.

Generally speaking, the model described by the Hamiltonoan (3.80) is different from the mean spherical model studied in Section 3.1. It reduces to the latter only when the solution $\{\lambda(\mathbf{r}), \mathbf{r} \in \Lambda\}$ of the *set of constraints* (3.81) is spatially uniform, i.e., when $\lambda(\mathbf{r}) = \lambda = $ const. This is the case of a system with translationlly invariant interaction under periodic boundary conditions [Shcherbina (1988)]. Then the set of equations (3.81) can be replaced by a single mean spherical constraint (3.2).

The study of the correlation functions needs additional mathematical constructions in the form of infinitesimal symmetry-breaking terms in the Hamiltonians, compare with Eq. (1.70), as prescribed by the Bogolyubov quasiaverages below T_c. In this case the result depends on the chosen fields, due to the existence of various phases which are unstable under small perturbations. Let us note that the treatment of the non-translationally invariant case is rather sophisticated [Khorunzhy et. al. (1992)].

Formal expansions in powers of $1/n$ about the exact limit $n \to \infty$ have been developed independently in [Abe (72)], [Abe and Hikami (73)] and [Ma (1976)]. We note also that in [Kupiainen (1980)] the $1/n$-expansion has been considered for the n-component non-linear sigma model on a lattice of arbitrary dimension. It has been shown that the series expansions for the correlation functions and the free energy density are asymptotic for all temperatures above the critical temperature of the spherical model.

3.3 Quantum spherical models

The classical spherical models considered in Section 3.1 are one of the cornerstones in the field of exactly solvable models. They have been used for testing various hypotheses and approximations in classical critical phenomena. Since exact solvability is a rare event in statistical physics, possible quantum versions of the spherical models would yield a conspicuous possibility to investigate in an exact manner the interplay of quantum and classical fluctuations, in dependence on dimensionality d, external field h, and geometry of the system.

There exist different possible ways of quantizing the classical spherical model. The first one dates back to the work of G. Obermair (1972), in which a canonical quantization for a dynamical version of the model has been proposed. Recently, [Vojta (1996)] has renewed the interest in this quantization scheme in the context of quantum phase transitions at zero temperature. According to the quantum *ansatz* the classical continuous spin variables $S_{\mathbf{r}}$ are reinterpreted as coordinate operators $\mathcal{S}_{\mathbf{r}}$; then canonically conjugate momentum operators $\mathcal{P}_{\mathbf{r}}$ are defined, so that the commutation relations $[\mathcal{S}_{\mathbf{r}}, \mathcal{S}_{\mathbf{r}'}] = 0$, $[\mathcal{P}_{\mathbf{r}}, \mathcal{P}_{\mathbf{r}'}] = 0$, and $[\mathcal{S}_{\mathbf{r}}, \mathcal{P}_{\mathbf{r}'}] = i\delta_{\mathbf{r},\mathbf{r}'}$ are obeyed (at $\hbar = 1$). A new term containing the momentum operators has to be added to the initial Hamiltonian: the simplest choice is to add a kinetic energy term as the sum of the squared momentum operators. This is the so called *oscillator approach* to the problem. The appearance of coordinate and momentum operators in the Hamiltonian describes physics similar to that of the anharmonic crystal considered in Section 3.4, since the spherical constraint couples the oscillators. In order to clarify the properties of the model and its relations to other models, a functional-integral representation of the partition function based on the Trotter formula has been derived [Vojta (1996)]. It has been shown that the model is related to the quantum non-linear sigma model (QNLσM) in the limit of infinite-order parameter dimensionality (these models are in the same universality class). The QNLσM describes quantum rotors rather than Heisenberg-Dirac spins. Let us note that the functional-integral approach to the quantization of the spherical model opens various possibilities for the choice of the kinetic term in the effective action of the model, which in turn leads to different quantization schemes. For example, in the scheme of [Nieuwenhuizen (1995)] the effective action involves a first derivative with respect to the imaginary time, instead of the second order one which follows from the above commutation relations. This leads to a different physical behavior, closer to that of quantum Heizenberg-Dirac spins. In summary, the quantization scheme of the spherical model is not unique and depends on the physics one wants to consider. For details we refer the reader to a number of publications on the subject, in which the spherical approximation is used for studying magnetic systems with disorder [Vojta and Schreiber (1996)], [Kopec and Pirc (1997)], [Nieuwenhuizen and Ritort (1997)], quantum Lifshitz points [Dutta et. al. (1997)], finite-size effects [Chamati et. al. (1998)], [Chamati et. al. (1997)], multy-spin interactions and others [Nieuwenhuizen and Ritort (1997)], [Tu and Weichman (1994)]. Our choice here is made for the sake of simplicity.

The model we consider in this book is described by the Hamiltonian [Vojta (1996)]

$$\mathcal{H}_\Lambda = \frac{1}{2}g\sum_{\mathbf{r}}\mathcal{P}_{\mathbf{r}}^2 - \frac{1}{2}\sum_{\mathbf{r},\mathbf{r}'} J(\mathbf{r},\mathbf{r}')\mathcal{S}_{\mathbf{r}}\mathcal{S}_{\mathbf{r}'} - H\sum_{\mathbf{r}}\mathcal{S}_{\mathbf{r}} + \beta^{-1}s\sum_{\mathbf{r}}\mathcal{S}_{\mathbf{r}}^2, \quad (3.85)$$

where $\mathcal{S}_{\mathbf{r}}$ are quantum spin operators located at the sites $\mathbf{r} \in \Lambda$ of a finite hypercubic lattice Λ, the operators $\mathcal{P}_{\mathbf{r}}$ are the conjugate "momenta"; the coupling constants $J(\mathbf{r},\mathbf{r}') = J$ are between nearest neighbors only; the coupling constant g is a measure of the strength of quantum fluctuations, to be called below *quantum parameter* ; H is a uniform external magnetic field; finally, the spherical field s is introduced to insure the fulfilment of the mean spherical constraint (3.6).

The free energy density of the model in a finite region Λ is given by the

Legendre transformation, see Eq. (3.11),

$$\beta f_\Lambda(K, \lambda, h) = \sup_{s>0}\{-N^{-1}\ln Z_\Lambda^{G\cdot}(K, \beta g, h, s) - s\}, \qquad (3.86)$$

where $N = |\Lambda|$ is the total number of quantum spins, $\lambda = \sqrt{g/J}$, $K = \beta J$, $h = \beta H$ and $Z_\Lambda^{G\cdot}(K, \beta g, h, s)$ is the partition function of the Gaussian model with Hamiltonian (3.85). The latter Hamiltonian is of the type considered in Section 3.4 and describes a system of pseudo-harmonic oscillators. Therefore, $Z_\Lambda^{G\cdot}(K, \beta g, h, s)$ can be readily calculated by assuming periodic boudary conditions and performing a Fourier transformation of the Hamiltonian (3.85). The reduced free energy density takes the form

$$\beta f_\Lambda(K, \lambda, h) = -dK$$
$$+ \frac{K}{2}\sup_{\phi>0}\left\{\frac{2}{KN}\sum_{\mathbf{k}}\ln\left[2\sinh\left(\frac{\lambda K}{2}\Omega(\mathbf{k}, \phi)\right)\right] - \frac{h^2}{K^2\phi} - \phi\right\} \quad (3.87)$$

where

$$\Omega(\mathbf{k}, \phi) = \left[\phi + 2\sum_{\nu=1}^{d}(1 - \cos k_\nu)\right]^{1/2}. \qquad (3.88)$$

Here we have introduced the *shifted spherical field* $\phi = 2s/K - 2d$; the components of the vector $\mathbf{k}$, for L_ν odd integers, take values in the set (3.15). The supremum in (3.87) is attained at the solution of the mean spherical constraint Eq. (3.6), which explicitly reads

$$1 = \frac{\lambda}{2N}\sum_{\mathbf{k}}\frac{1}{\Omega(\mathbf{k}, \phi)}\coth\left[\frac{\lambda K}{2}\Omega(\mathbf{k}, \phi)\right] + \frac{h^2}{K^2\phi^2}. \qquad (3.89)$$

Eqs. (3.87) and (3.89) provide the basis for studying the critical behavior of the model under consideration. Note that in the classical limit $\lambda \to 0$ Eq. (3.89) reduces to the classical mean spherical constraint (3.16) with $\tilde{\phi} = \phi/(2d)$ and $\tilde{K} = 2dK$. The bulk critical behavior of the quantum mean spherical model will be studied in Chapter 10 in the context of finite-size scaling theory.

3.4 Reduced Q^4 model

There exist other types of models [Plakida and Tonchev (1986)], [Tonchev (1991)], [Verbeure and Zagrebnov (1992)], [Pisanova and Tonchev (1993)], [Chamati (1994)], [Pisanova and Tonchev (1995)], [Verbeure and Zagrebnov (1995)], [Momont et. al. (1997)], suitable for the joint description of classical and quantum fluctuations in an exact manner. If one considers $\mathcal{P}_\mathbf{r}$ and $\mathcal{S}_\mathbf{r}$ as canonically conjugated momentum and coordinate, respectively, of the $\mathbf{r}$-th atom with mass g^{-1} at each lattice point $\mathbf{r} \in \Lambda$, then the Hamiltonian Eq. (3.85) describes a system of quantum harmonic oscillators. One can see that in the absence of a spherical constraint such a lattice is unstable if $A = s/(2\beta) + dJ/2 < 0$, i.e., the parameter $A < 0$ defines the frequency of a mode which is unstable in the harmonic approximation. This suggests that an appropriate stabilization of the lattice may create a gap in the phonon spectrum. In the spirit of the self-consistent phonon approximation [Bruce and Cowley (1981)], [Aksenov et. al. (1989)], it is possible to add to the model Hamiltonian (3.85) the term

$$\frac{B}{4N}\left(\sum_\mathbf{r}\mathcal{S}_\mathbf{r}^2\right)^2, \qquad B > 0, \tag{3.90}$$

switching on an anharmonic interaction which stabilizes the lattice. Because of the factor $1/N$, the model under consideration turns out to be exactly solvable in the thermodynamic limit [Plakida and Tonchev (1986)]. This model is a quantum counterpart of the "soft" classical mean spherical model studied in Ref. [Shapiro and Rudnick (1986)] in the context of finite-size scaling theory. If B goes to zero, the quartic self-interaction term disappears and the model becomes a pure harmonic one. It can be seen that this limit is singular and that the parameter B, for d larger than the upper critical dimension, behaves as a dangerous irrelevant variable [Chamati (1994)], see Section 1.6.5.

3.4.1 *Model and Approximating Hamiltonians*

Here we consider the quantum version of the lattice model of structural phase transition proposed by [Schneider et. al. (1975)].The exact solution and the thermodynamic properties of the quantum version of the model have been studied in [Plakida and Tonchev (1986)], [Plakida and Tonchev

(1987)], and [Pisanova and Tonchev (1993)]. The Hamiltonian of the model reads

$$\mathcal{H} \;=\; 1/2 \sum_{\mathbf{r}}(P_{\mathbf{r}}^2/m - AQ_{\mathbf{r}}^2) + 1/4 \sum_{\mathbf{r},\mathbf{r}'}\varphi(\mathbf{r}-\mathbf{r}')(Q_{\mathbf{r}}-Q_{\mathbf{r}'})^2$$

$$+ \; (B/4N)\left(\sum_{\mathbf{r}} Q_{\mathbf{r}}^2\right)^2. \tag{3.91}$$

Here $Q_{\mathbf{r}}$ and $P_{\mathbf{r}}$ are the operators of displacement and momentum, respectively, of the particle of mass m at site $\mathbf{r}$ of a d-dimensional hypercubic lattice $\Lambda = L^d$ with periodic boundary conditions. The parameter $A = \nu_0^2 m > 0$ determines the frequency of a mode which is unstable in the harmonic approximation, and the parameter $B > 0$ introduces into the model an anharmonic interaction, which is inversely proportional to the particle number $N = L^d$. The harmonic force constants $\varphi(\mathbf{r}-\mathbf{r}')$, which are assumed to decrease at large distances $r = |\mathbf{r}-\mathbf{r}'|$ as $r^{-d-\sigma}$, describe a short-range ($\sigma = 2$) or a long-range ($0 < \sigma < 2$) interaction.

It is worthwhile pointing out the relation between model (3.91) and the standard non-reduced model, see Eq. (1.104) in Section 1.6.1, frequently used in the theory of structural phase transitions. The long-range anharmonic interaction in Eq. (3.91) can be obtained by modifying the usual single-particle anharmonic interaction $\sum_{\mathbf{r}} Q_{\mathbf{r}}^4$ according to the rule: $Q_{\mathbf{r}}^2 \to N^{-1}\sum_{\mathbf{r}} Q_{\mathbf{r}}^2$. Notice that, due to the periodic boundary conditions, the harmonic part is left unchanged by this modification.

It is important to note that the thermodynamic properties of the reduced model coincide with those of its usual (not reduced) version, if the latter is treated in the self-consistent phonon approximation. This fact attracts additional interest because it clarifies the status of the self-consistent phonon approximation. Note that in [Schneider et. al. (1975)] only the classical limit has been considered and the exact solvability of the model has been proven only in the para-phase region, excluding the phase transition point.

We shall study the critical behavior of the model with Hamiltonian (3.91) in the whole (T,λ)-plane, where T is the temperature and $\lambda = \hbar\nu_0/4E_0$ is a parameter which switches on quantum fluctuations; here $E_0 = A^2/4B$ is the barrier height of the double-well potential in Eq. (3.91) under uniform displacement of all particles: if $Q_{\mathbf{r}} \equiv x$, then $U(x) = -(A/2)x^2 + (B/4)x^4$. Note that λ is large, hence quantum effects are important, not only at large zero-point energy $\hbar\nu_0$ but also at small depth of the well E_0.

The model Hamiltonian (3.91) presents another nontrivial example of a system with repulsive separable interaction which can be treated by the AHM, see Section 2.3. The free energy density $f_N[\mathcal{H}]$ of the model can be estimated by using the Bogolyubov variational inequality (2.1),

$$N^{-1}\langle \mathcal{H} - \mathcal{H}_0 \rangle_{\mathcal{H}} \leq f_N[\mathcal{H}] - f_N[\mathcal{H}_0] \leq N^{-1}\langle \mathcal{H} - \mathcal{H}_0 \rangle_{\mathcal{H}_0}, \qquad (3.92)$$

where the approximating pseudo-harmonic Hamiltonian

$$\begin{aligned} \mathcal{H}_0 &= \sum_{\mathbf{r}} P_{\mathbf{r}}^2/2m + (1/2)\eta \sum_{\mathbf{r}} Q_{\mathbf{r}}^2 + (1/4)\sum_{\mathbf{r},\mathbf{r}'} \varphi(\mathbf{r}-\mathbf{r}')(Q_{\mathbf{r}} - Q_{\mathbf{r}'})^2 \\ &\quad - NC(\eta) \end{aligned} \qquad (3.93)$$

is chosen in such a way that the difference

$$\mathcal{H} - \mathcal{H}_0 = N\frac{B}{4}\left[\left(\frac{1}{N}\sum_{\mathbf{r}} Q_{\mathbf{r}}^2\right)^2 - 2\left(\frac{\eta+A}{B}\right)\frac{1}{N}\sum_{\mathbf{r}} Q_{\mathbf{r}}^2 + \frac{4}{B}C(\eta)\right]$$

be positive definite. This can be achieved by setting $C(\eta) = (\eta + A)^2/4B$, hence,

$$\mathcal{H} - \mathcal{H}_0 = N\frac{B}{4}\left[\frac{1}{N}\sum_{\mathbf{r}} Q_{\mathbf{r}}^2 - \left(\frac{\eta+A}{B}\right)\right]^2 \geq 0. \qquad (3.94)$$

By using the Fourier transformation to normal coordinates,

$$Q_{\mathbf{k}} = \left(\frac{m}{N}\right)^{1/2}\sum_{\mathbf{r}} Q_{\mathbf{r}}\, e^{-i\mathbf{k}\cdot\mathbf{r}}, \qquad P_{\mathbf{k}} = \frac{1}{(mN)^{1/2}}\sum_{\mathbf{r}} P_{\mathbf{r}}\, e^{-i\mathbf{k}\cdot\mathbf{r}}, \qquad (3.95)$$

we rewrite Hamiltonian (3.93) in the form

$$\mathcal{H}_0(\eta) = (1/2)\sum_{\mathbf{k}}(P_{\mathbf{k}}P_{-\mathbf{k}} + \omega_{\mathbf{k}}^2 Q_{\mathbf{k}}Q_{-\mathbf{k}}) - \frac{N}{4B}(A+\eta)^2. \qquad (3.96)$$

Here we have introduced the trial frequency

$$\omega_{\mathbf{k}}^2 = (1/m)(\eta + \varphi_0 - \varphi_{\mathbf{k}}), \qquad (3.97)$$

where $\varphi_{\mathbf{k}}$ is the Fourier transform of the interaction potential $\varphi(\mathbf{r})$. By using expression (3.96), the trial free energy density $f_N[\mathcal{H}_0]$ can easily be obtained:

$$f_N[\mathcal{H}_0] = \frac{A^2}{B}\left\{\frac{t}{N}\sum_{\mathbf{k}}\ln\left[2\sinh\left(\frac{\lambda\Omega_{\mathbf{k}}(\Delta)}{2t}\right)\right] - \frac{1}{4}(1+\Delta)^2\right\} \equiv \frac{A^2}{B}f_0(\Delta).$$

$$(3.98)$$

Here the summation index $\mathbf{k}$ runs over the Brillouin zone (3.15) and the following dimensionless quantities have been used

$$\begin{aligned}
\Delta &= \eta/A, & \Omega_{\mathbf{k}}^2 &= (m/A)\omega_{\mathbf{k}}^2 = \Delta + \tilde{\varphi}_0 - \tilde{\varphi}_{\mathbf{k}}, \\
\tilde{\varphi}_{\mathbf{k}} &= (1/A)\varphi_{\mathbf{k}}, & t &= k_B T(B/A^2).
\end{aligned}$$

$$(3.99)$$

In the further calculations we assume an isotropic spectrum of the Fourier transform of the interaction potential in the long-wavelength limit,

$$\tilde{\varphi}_0 - \tilde{\varphi}_{\mathbf{k}} \simeq sk^\sigma,$$

$$(3.100)$$

with $\sigma = 2$ for short-range, and $0 < \sigma < 2$ for long-range interaction potentials. The effects of spectrum anisotropy have been briefly discussed in [Plakida et. al. (1991)]. As it follows from inequalities (3.92) and (3.94),

$$0 \leq f - f_0(\Delta) \leq (B/NA^2)\langle\mathcal{H} - \mathcal{H}_0\rangle_{\mathcal{H}_0} \equiv k_N(\Delta),$$

$$(3.101)$$

where $f \equiv (B/A^2)f_N[\mathcal{H}]$, the best approximation is achieved if the value of the variational parameter Δ is determined by the condition[‡]

$$f_0(\Delta_0) = \max_{\Delta} f_0(\Delta).$$

$$(3.102)$$

Since the function $f_0(\Delta)$ is differentiable and $f_0''(\Delta) < 0$ for all real Δ, it has just one extremum – maximum. Thus, $\Delta = \Delta_0$ is defined by the equation $f_0'(\Delta) = 0$, which has the explicit form

$$1 + \Delta = \frac{1}{N}\sum_{\mathbf{k}}\frac{\lambda}{2\Omega_{\mathbf{k}}}\coth\left(\frac{\lambda\Omega_{\mathbf{k}}}{2t}\right) \equiv I_N(\Delta).$$

$$(3.103)$$

The inspection of the above equation shows that it is similar to Eq. (3.89) up to the linear in Δ term. This term appears to be essential only above the upper critical dimension. Thus, there is a clear mathematical analogy between the model defined in Section 3.3 and the model presented here. However, there is an important difference in the status of Eqs.

[‡]Note that for a finite-size system f, $f_0(\Delta)$ and Δ_0, are functions of N; for the sake of simplicity of notation we have omitted this dependence.

(3.89) and (3.103) in the context of finite-size systems. More specifically, in studying finite-size effects in models that can be solved by the AHM, one has to answer the question: is the finite-size description obtained for the approximating Hamiltonian (3.93) correct for the initial model (3.91)? Evidently, the answer would clarify the more general and elusive problem of the status of finite-size results obtained by approximate methods, e.g., by different self-consistent approximations, perturbative expansions, etc. A step in this direction has been presented in [Brankov (1990b)] on the example of the classical mean spherical with equivalent-neighbors interactions, see Section 3.1.3. An appropriate modification of the approximating Hamiltonian approach has been suggested there in order to reproduce some exact finite-size scaling results. Let us note that if we parallel that approach in the study of quantum systems, more sophisticated mathematical problems are to be solved, some of which remain as open problems up to now.

The self-consistency equation (3.103) has always a finite, non-vanishing in the thermodynamical limit solution for sufficiently large values of the quantum parameter, $\lambda > \lambda_c$. In fact, the right-hand side of Eq. (3.103) is an increasing function of t and a decreasing function of Δ, whereas the left hand side is an increasing function of Δ. Therefore, the existence of a solution of Eq.(3.103) at $T = 0$ implies the existence of a solution at $T > 0$. The parameter λ_c is found from the condition that the solution $\Delta(\lambda, T)$ vanishes at the point $\lambda = \lambda_c$, $T = 0$. In the thermodynamic limit one obtains from Eq. (3.103) the following equation for λ_c (in the remainder we set $s = 1$)

$$\lambda_c = 2 \left(1 - \frac{\sigma}{2d}\right) x_D^{\sigma/2}, \qquad d/\sigma > 1/2, \tag{3.104}$$

where $\mathcal{A}_d(1)$ has been defined in Eq. (3.57) and $x_D = 2\pi s[d/\mathcal{A}_d(1)]^{1/d}$ is the radius of the effective sphere replacing the Brillouin zone. Obviously, $\lambda_c \to 0^+$ as $d \to d_l^q \equiv \sigma/2$, where d_l^q is the quantum lower critical dimension.

Let us discuss now the thermodynamic equivalence of the systems described by the model Hamiltonian (3.91) and the approximating one (3.93).

The self-consistency equation (3.103) can be written identically as

$$1 + \Delta = \frac{B/m^2}{N} \sum_{\mathbf{k}} \langle Q_{\mathbf{k}} Q_{-\mathbf{k}} \rangle_{H_0}. \tag{3.105}$$

With the aid of this equation, the quantity $k_N(\Delta_0)$ in the right-hand side of inequality (3.101) can be represented as

$$k_N(\Delta_0) \sim \frac{1}{N^2} \left\langle \left[\sum_{\mathbf{k}} Q_{\mathbf{k}} Q_{-\mathbf{k}} - \sum_{\mathbf{k}} \langle Q_{\mathbf{k}} Q_{-\mathbf{k}} \rangle_{H_0} \right]^2 \right\rangle_{H_0}. \qquad (3.106)$$

Since the average values in the above expression are calculated by using the pseudoharmonic Hamiltonian (3.96), with the aid of the Wick theorem we obtain

$$k_N(\Delta_0) \sim \frac{1}{N^2} \sum_{\mathbf{k}} \left[\frac{\lambda}{2\Omega_{\mathbf{k}}(\Delta_0)} \coth \frac{\lambda \Omega_{\mathbf{k}}(\Delta_0)}{2t} \right]^2. \qquad (3.107)$$

Our aim now is to show that $k_N(\Delta_0)$ asymptotically vanishes as $N \to \infty$. It can be shown that a finite non-zero value of the solution Δ_0 of the self-consistency equation (3.103) corresponds to a vanishing function $k_N(\Delta_0)$ as $N \to \infty$. On the other hand, $k_N(\Delta_0)$ does not vanish in the region where $\Delta_0 = O(1/N)$. To understand this, let us note that Δ_0 plays the role of a gap in the spectrum of collective excitations, hence the fact that $\Delta_0 = O(1/N)$ may be a signal of a phase transition with the presence of strong fluctuations. A proper description of the phase transition can be given in terms of Bogolyubov's quasiaverages, see Section 2.5 and Chapter 9. To this end, one includes in the Hamiltonian interaction with an external homogeneous field h_e. Then the thermodynamic limit $N \to \infty$ is taken, and only after that the external field is eventually switched off.

Following the above procedure, we consider the Hamiltonians

$$\mathcal{H}(h) = \mathcal{H} - h_e \sum_{\mathbf{r}} Q_{\mathbf{r}}, \qquad \mathcal{H}_0(h) = \mathcal{H}_0 - h_e \sum_{\mathbf{r}} Q_{\mathbf{r}}. \qquad (3.108)$$

Obviously, the interaction with an external field does not change the difference $\mathcal{H} - \mathcal{H}_0$ in the Bogolyubov inequality; it redefines the approximating free energy density according to

$$f_0(h, \Delta) = f_0(\Delta) - h^2/(2\Delta), \qquad (3.109)$$

where $h = h_e(B/A^3)^{1/2}$. The difference between the model and approximating free energies is still bounded from above by $k_N(\Delta_0(h))$. In the thermodynamic limit $N \to \infty$ the self-consistency Eq. (3.103) takes the following form

$$1 + \Delta = \frac{\lambda \mathcal{A}_d(1)}{2(2\pi)^d} \int_0^{x_D} \frac{x^{d-1}}{\sqrt{\Delta - x^\sigma}} \coth\left(\frac{\lambda\sqrt{\Delta + x^\sigma}}{2t}\right) dx + \frac{h^2}{\Delta^2}. \quad (3.110)$$

The presence of the last term in the right-hand side of Eq. (3.110) assures the existence of a *finite non-zero solution* $\Delta_0(h)$ as long as h is kept finite. Now, it can be shown that $\lim_{N\to\infty} k_N(\Delta_0(h)) = 0$. Therefore, as it follows from the Bogolyubov inequality (3.101), the model (3.91) and approximating (3.93) systems are thermodynamically equivalent. After taking the thermodynamic limit $N \to \infty$, one can pass to $h \to 0$. Thus, the thermodynamic equivalence is understood in the sense that

$$\frac{B}{A^2} \lim_{N\to\infty} f_N[\mathcal{H}(h)] \equiv f(h,t) = \frac{B}{A^2} \max_{\Delta \in [0,\infty)} \lim_{N\to\infty} f_N[\mathcal{H}_0(h)] =$$

$$\frac{t\mathcal{A}_d(1)}{(2\pi)^d} \int_0^{x_D} \ln\left[2\sinh\frac{\lambda\sqrt{x^2 + \Delta_0}}{2t}\right] x^{d-1}\,dx - \frac{1}{4}(1 + \Delta_0)^2 + \frac{h^2}{2\Delta_0}. \quad (3.111)$$

Here Δ_0 is the unique solution of the self-consistency equation (3.110).

3.4.2 *Order parameter and critical exponents*

Let us consider the auxiliary Hamiltonians

$$\mathcal{H}(\{-\mathbf{k}_i\}) = \mathcal{H} - \sqrt{N} \sum_{\mathbf{k}_i} h_{-\mathbf{k}_i} Q_{\mathbf{k}_i},$$

$$\mathcal{H}_0(\{-\mathbf{k}_i\}) = \mathcal{H}_0 - \sqrt{N} \sum_{\mathbf{k}_i} h_{-\mathbf{k}_i} Q_{\mathbf{k}_i}, \quad (3.112)$$

which depend linearly on the finite set of real parameters (fields) $\{h_{-\mathbf{k}_i}, i = 1, \dots, p\}$. As it was shown above, we can prove the equality

$$f(\{h_{-\mathbf{k}_i}\}) = \max_\Delta f_0(\{h_{-\mathbf{k}_i}\}; \Delta), \quad (3.113)$$

where

$$f(\{h_{-\mathbf{k}_i}\}) \equiv (B/A^2) \lim_{N\to\infty} f_N[\mathcal{H}(\{-\mathbf{k}_i\})],$$

$$f_0(\{h_{-\mathbf{k}_i}\}; \Delta) \equiv (B/A^2) \lim_{N\to\infty} f_N[\mathcal{H}_0(\{-\mathbf{k}_i\})]. \quad (3.114)$$

Let us introduce the density of the Helmholtz free energy $a(\{x_{\mathbf{k}_i}\})$, where $\{x_{\mathbf{k}_i}\}$ is the set of parameters thermodynamically conjugate to $\{h_{-\mathbf{k}_i}\}$, by means of the Legendre transformation:

$$a(\{x_{\mathbf{k}_i}\}) = \max_{\{h_{-\mathbf{k}_i}\}} \left[f(\{h_{-\mathbf{k}_i}\}) + \sum_{\mathbf{k}_i} h_{-\mathbf{k}_i} x_{\mathbf{k}_i} \right]. \tag{3.115}$$

By substituting $f(\{h_{-\mathbf{k}_i}\})$ from Eq. (3.113) into Eq. (3.115), we obtain

$$
\begin{aligned}
a(\{x_{\mathbf{k}_i}\}) &= \max_{\{h_{-\mathbf{k}_i}\}} \left[\max_{\Delta} f_0(\{h_{-\mathbf{k}_i}\};\Delta) + \sum_{\mathbf{k}_i} h_{-\mathbf{k}_i} x_{\mathbf{k}_i} \right] \\
&= \max_{\Delta} \left[\max_{\{h_{-\mathbf{k}_i}\}} f_0(\{h_{-\mathbf{k}_i}\};\Delta) + \sum_{\mathbf{k}_i} h_{-\mathbf{k}_i} x_{\mathbf{k}_i} \right].
\end{aligned} \tag{3.116}
$$

We note that $f_0(\{h_{-\mathbf{k}_i}\};\Delta)$, being a differentiable function of the parameters $\{h_{-\mathbf{k}_i}\}$, attains its maximum with respect to the latter at the unique solution of the set of equations

$$-\frac{\partial f_0(\{h_{-\mathbf{k}_i}\};\Delta)}{\partial h_{-\mathbf{k}_i}} = x_{\mathbf{k}_i} \tag{3.117}$$

which yield

$$\sqrt{N} x_{\mathbf{k}_i} = \langle Q_{\mathbf{k}_i} \rangle_{\mathcal{H}_0} \quad \text{or, equivalently,} \quad x_{\mathbf{k}_i} = h_{-\mathbf{k}_i} \Omega_{\mathbf{k}_i}^{-2}. \tag{3.118}$$

Hence, as $h_{-\mathbf{k}_i} \to 0$, it is only for $\mathbf{k}_i = 0$ that we obtain $x_0 \equiv x \neq 0$ when $t < t_c$.

Equations (3.118) determine the order parameter in the model, since from Eq. (3.111) and the Griffits – Fisher lemma (see Section 2.2) we have the equalities ($h_0 \equiv h$)

$$
\begin{aligned}
x &= \lim_{h\to 0^+} \lim_{N\to\infty} \langle \frac{Q_0}{\sqrt{N}} \rangle_{\mathcal{H}(\{h_{-\mathbf{k}}\})} \equiv - \lim_{h\to 0^+} \frac{\partial f(h)}{\partial h} \\
&= - \lim_{h\to 0^+} \frac{\partial f_0(h;\Delta_0)}{\partial h} \equiv \lim_{h\to 0^+} \lim_{N\to\infty} \langle \frac{Q_0}{\sqrt{N}} \rangle_{\mathcal{H}_0(\{h_{-\mathbf{k}}\})}.
\end{aligned} \tag{3.119}
$$

A posteriori, by setting in Eq. (3.112) all $h_{-\mathbf{k}_i}$ with $\mathbf{k}_i \neq 0$ equal to zero, and by using the relationship $h = x\Delta$, which follows from Eqs. (3.99) and (3.118), we obtain from Eq. (3.116)

$$a(x) = \max_{\Delta}[f_0(x\Delta, \Delta) + x^2\Delta]$$

$$= \lim_{N\to\infty}\left\{\frac{1}{N}\sum_{\mathbf{k}}\ln\left[2\sinh\left(\frac{\lambda\Omega_{\mathbf{k}}(\Delta(x))}{2t}\right)\right]\right\}$$

$$+\frac{1}{2}\Delta(x)x^2 - \frac{1}{4}[\Delta(x) + 1]^2. \tag{3.120}$$

Here $\Delta = \Delta(x)$ solves the stationarity equation for $f_0(x\Delta, \Delta)$ with respect to Δ, which explicitly reads

$$1 + \Delta = y_t(\Delta) + x^2, \tag{3.121}$$

where

$$y_t(\Delta) := \int \frac{\lambda}{2\Omega_{\mathbf{k}}(\Delta)}\coth\frac{\lambda\Omega_{\mathbf{k}}(\Delta)}{2t}\mathrm{d}^d k. \tag{3.122}$$

In the above equation the integration is over the Brillouin zone.

The minimization of $a(x)$ with respect to the order parameter x yields the equation

$$\frac{da(x)}{dx} = \Delta(x)\,x = 0. \tag{3.123}$$

This is the self-consistency equation for the order parameter x. It has two types of solutions: (i) a trivial one, $x = 0$, which yields minimum of $a(x)$ at $t > t_c$; (ii) nontrivial solution $x \neq 0$, which yields minimum of $a(x)$ at $t < t_c$. The nontrivial solution can be obtained from the equation, see Eq. (3.122),

$$\Delta(x) \equiv -1 + x^2 + y_t(0) = 0. \tag{3.124}$$

Since the order parameter vanishes at the critical temperature $t = t_c(\lambda)$, we define t_c from the equation $y_{t_c}(0) = 1$. Hence, by using Eq. (3.122) for fixed $\lambda < \lambda_c$ and $t \sim t_c$, from we obtain the asymptotic behavior of the order parameter as the critical temperature is approached from below:

$$x^2 \simeq \lambda\left[\int \frac{\mathrm{d}k^d}{\sinh^2\frac{\lambda\Omega_{\mathbf{k}}(0)}{2t_c}}\right](t_c - t). \tag{3.125}$$

Therefore, the critical exponent of the order parameter always equals the mean field value $\beta = 1/2$.

Let us consider now the wavelength-dependent susceptibility $\chi(k,t)$ of the approximating system. At $k = 0$, when $t \to t_c$, it is given by

$$\chi(0,t) \sim \Delta_0^{-1}. \tag{3.126}$$

It is interesting to remark that the model provides an example of the situation in which

$$\chi(0,t) \neq \lim_{k \to 0} \chi(k,t), \tag{3.127}$$

i.e., the modulated susceptibility $\chi(k,t)$ is discontinuous at the point $k = 0$. This phenomenon has been considered in [Radosz (1988)] and [Radosz (1990)] in the classical limit, where it was called pseudo-Goldstone mode phenomenon, and in details including quantum effects in [Verbeure and Zagrebnov (1995c)]. It is related to the broken discrete symmetry $Q \to -Q$.

By setting $\Delta = 0$ and $h = 0$ in Eq. (3.110), the equation for the critical line $t = t_c(\lambda)$ takes the form

$$1 = \frac{\lambda \mathcal{A}_d(1)}{2(2\pi)^d} \int_0^{x_D} x^{d-1-\sigma/2} \coth\left(\frac{\lambda x^\sigma}{2t_c}\right) dx. \tag{3.128}$$

The phase diagram is sketched in Fig. 3.2. Of special interest are the classical multicritical point $(t = t_c, \lambda = 0)$ and the quantum multicritical point $t_c = 0$, $\lambda = \lambda_c(0)$. Two main possibilities of a phase transition can be considered: when the temperature t changes at fixed $\lambda = \overline{\lambda}$ (see the vertical arrows in Fig. 3.2); and when the parameter λ changes at fixed $t = \overline{t}$ (see the horizontal arrows). In real systems a phase transition driven by variation of λ can be observed by changing the external pressure, or the amount of doping. Evidently, by setting $t_c = 0$ in Eq. (3.128) one obtains $\lambda_c(0) = \lambda_c$, see Eq. (3.104). From Eq. (3.128) one obtains in the low-temperature region $t_c \to +0$,

$$t_c(\lambda) \simeq f^q(d,\sigma) \left[\lambda_c(0) - \lambda\right]^{\sigma/(2d-\sigma)}, \qquad d/\sigma > 1, \tag{3.129}$$

where $f^q(d,\sigma)$ is a known function of d and σ, see [Pisanova and Tonchev (1993)]. Hence the shift exponent ψ^q immediately follows[§]:

$$\psi^q = \sigma/(2d - \sigma), \qquad d/\sigma > 1. \tag{3.130}$$

[§]The subscripts q and cl mean quantum and classical, respectively.

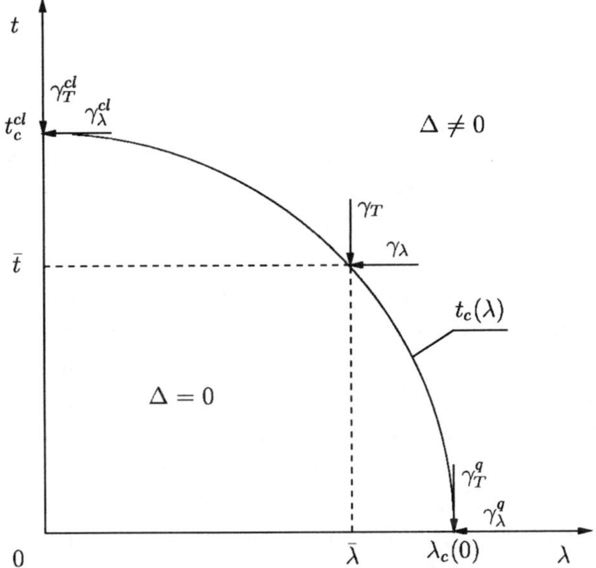

Fig. 3.2   Diagram for a phase transition with two parameters: temperature t, and parameter λ that switches on the quantum fluctuations.

Note that the critical line has an infinite slope at the quantum multicritical point $(0, \lambda_c(0))$,

$$\left.\frac{dt_c(\lambda)}{d\lambda}\right|_{\lambda=\lambda_c(0)} = \infty. \qquad (3.131)$$

Let us now consider the critical line near the classical multicritical point $(t_c^{cl}, 0)$. By setting $\lambda = 0$ in Eq. (3.128), we obtain in the classical limit

$$t_c^{cl} = \left(1 - \frac{\sigma}{d}\right) x_D^\sigma, \qquad d/\sigma > 1. \qquad (3.132)$$

Obviously, $t_c^{cl} \to 0^+$ as $d \to d_l^{cl} = \sigma$, where d_l^{cl} is the classical lower critical dimension.

When $t_c^{cl} - t \to +0$, the quantum parameter $\lambda_c(t)$ has the asymptotic behavior

$$\lambda_c(t) \simeq f^{cl}(d,\sigma)(t_c^{cl} - t)^{1/2}, \qquad d/\sigma > 1, \qquad (3.133)$$

hence the classical shift exponent is $\psi^{cl} = 1/2$. At the classical multicritical point $(t_c^{cl}, 0)$ one has

$$\frac{d\lambda(t_c)}{dt_c}\bigg|_{t_c = t_c^{cl}} = \infty. \tag{3.134}$$

Because of the infinite slope at the classical multicritical point, the critical behavior of the system with Hamiltonian (3.91) is different along the lines $t = t_c^{cl}$ and $\lambda = 0$. The same is true for the lines $t = 0$ and $\lambda = \lambda_c$ at the quantum multicritical point. Such type of critical behavior, with different critical exponents along two mutually perpendicular lines, is a common feature of quantum critical phenomena, see [Morf et. al. (1977)].

Finally, from the above consideration one can conclude that quantum fluctuations act unidirectionally with classical fluctuations in the tendency to suppress the critical temperature. This phenomenon has been observed in more general cases, beyond the reduced version of the Q^4 model. For example, a rigorous proof of the effect of quantum fluctuations has been given for the non-reduced model with the single-particle anharmonic interaction $\sum_r Q_r^4$ in [Verbeure and Zagrebnov (1995b)].

Let us consider the system with fixed quantum parameter $\lambda = \overline{\lambda}$. In the thermodynamic limit the asymptotic form of Eq. (3.103) near the critical line (as $\Delta \to +0$) can be written as

$$\Delta \simeq \frac{t - t_c(\overline{\lambda})}{t_c(\overline{\lambda})} + \frac{t_c(\overline{\lambda})}{(2\pi)^d}[I(\Delta) - I(0)], \tag{3.135}$$

where $I(\Delta) = \lim_{N \to \infty} I_N(\Delta)$. To obtain the small-$\Delta$ behavior of the difference $I(\Delta) - I(0)$, one can use the method proposed in [Lawrie (1978a)] and [Lawrie (1978b)] for studying another quantum critical model – the Ising model in transverse external field. The details of the calculations in the case under consideration can be found in [Pisanova and Tonchev (1993)]. Here we present only the final results. For the inverse susceptibility we obtain from Eq. (3.135)

$$\Delta \simeq [t - t_c(\overline{\lambda})]^{\gamma_T^{cl}}, \qquad 1 < d/\sigma < 2. \tag{3.136}$$

Here and below the subscripts T and λ denote that the critical exponents are with respect to the temperature and the quantum parameter, respectively, at the quantum (superscript q) or classical (superscript cl) multicritical point. For the critical exponent γ_T^{cl} in Eq. (3.136) we obtain in the classical limit $(\overline{\lambda} = 0)$, as well as in the part of the phase diagram where $0 < \overline{\lambda} <$

$\lambda_c(0)$ (see Fig. 3.2), the following values:

$$\gamma_T^{cl} = \gamma_T = \begin{cases} \sigma/(d-\sigma), & \text{if} \quad \sigma < d < 2\sigma \\ 1, & \text{if} \quad d \geq 2\sigma \end{cases} \qquad (3.137)$$

The logarithmic correction at the upper classical critical dimension $d_u^{cl} = 2\sigma$ has been neglected.

Let us note that Eqs. (3.135) and (3.136) are valid only when $0 \leq \overline{\lambda} < \lambda_c(0)$. For the inverse susceptibility at $\overline{\lambda} = \lambda_c(0)$ we obtain, instead of Eq. (3.136), $\Delta \simeq t^{\gamma_T^q}$, where

$$\gamma_T^q = \begin{cases} 2, & \text{if} \quad \sigma/2 < d < 3\sigma/2 \\ (2d-\sigma)/\sigma, & \text{if} \quad 3\sigma/2 \leq d < 2\sigma. \end{cases} \qquad (3.138)$$

The logarithmic correction at the upper quantum critical dimension $d_u^q = 3\sigma/2$ has been neglected again.

The critical exponent $\beta_T = \beta_T^{cl} = 1/2$ has been calculated in the previous subsection for all $0 \leq \overline{\lambda} < \lambda_c(0)$, $1 < d/\sigma$. Since no real order parameter exists above the transition point $t_c(\lambda)$, one can only formally define at $\overline{\lambda} = \lambda_c(0)$ an exponent playing the role of β_T. This critical exponent, called β_T^q, depends on the space dimensionality d,

$$\beta_T^q = (2d-\sigma)/(2\sigma), \qquad 1 < d/\sigma < 2, \qquad (3.139)$$

which reflects the fact that $(0, \lambda_c(0))$ is an end-point in the phase diagram. The reason to consider β_T^q is that it obeys scaling relations involving the other critical exponents.

Next we consider the system at fixed temperature $t = \overline{t}$. If one works through the analysis of [Pisanova and Tonchev (1993)], assuming that the phase transition is governed by the quantum parameter λ, then one obtains the following result for the exponent γ_λ^q

$$\gamma_\lambda^q = \begin{cases} 2\sigma/(2d-\sigma), & \text{if} \quad \sigma/2 < d < 3\sigma/2 \\ 1, & \text{if} \quad 3\sigma/2 \leq d < 2\sigma \end{cases} \qquad (3.140)$$

with logarithmic corrections at $d/\sigma = 3/2$. The critical exponent β_λ can also be calculated for all $0 \leq \overline{t} < t_c^{cl}$, and the result is

$$\beta_\lambda = \beta_\lambda^q = 1/2, \qquad d > \sigma/2. \qquad (3.141)$$

At $\overline{t} = t_c^{cl}$ one can formally define the critical exponent $\beta_\lambda^{cl} = 1 \ (d > \sigma)$.

Near the quantum multicritical point $(t_c = 0, \lambda = \lambda_c(0))$, the equation for the inverse susceptibility Δ can be written as

$$\Delta \simeq (\varepsilon/\bar{t})^{\gamma_\lambda}, \tag{3.142}$$

where $\varepsilon = 1 - \lambda/\lambda_c(\bar{t})$ and $\bar{t}$ take small nonzero values. In the above equation $\gamma_\lambda = \sigma/(d - \sigma)$ $(\sigma < d < 2\sigma)$ is the critical exponent along the line $t = \bar{t}$.

According to the crossover scaling theory [Fisher (1974)], we expect that sufficiently close to the quantum multicritical point Δ has the form

$$\Delta \simeq \varepsilon^{\gamma_\lambda^q} Z(\bar{t}/\varepsilon^{\phi_T}). \tag{3.143}$$

At large arguments the scaling function $Z(\cdot)$ has the asymptotic form

$$Z(z) \simeq Z_\infty z^{(\gamma_\lambda^q - \gamma_\lambda)/\phi_T}. \tag{3.144}$$

Eq. (3.142) can be rewritten identically in the form

$$\Delta \simeq \varepsilon^{\gamma_\lambda^q} (\bar{t}/\varepsilon^{\phi_T})^{(\gamma_\lambda^q - \gamma_\lambda)/\phi_T}, \tag{3.145}$$

where the exponent $\gamma_\lambda^q = 2\sigma/(2d - \sigma)$ and the crossover exponent $\phi_T = \sigma/(2d - \sigma) = \psi^q$ for $1 < d/\sigma < 3/2$, and $\gamma_\lambda^q = 1$, $\phi_T = (2\sigma - d)/\sigma \neq \psi^q$ for $3/2 < d/\sigma < 2$.

Returning to the self-consistency equation (3.110), we eliminate the variational parameter Δ with the aid of the relation $\Delta = h/x$, and obtain the equation of state,

$$h \simeq x|x|^{\delta-1} h_s(\tau|x|^{-1/\beta}). \tag{3.146}$$

Hence it follows that the field h is a homogeneous odd function of the order parameter x and the deviation from the critical point τ. The scaling function for the equation of state, $h_s(x) = (1 + x)^\gamma$, has the form which is well-known from the theory of the mean spherical model, see Eq. (3.43). The only difference is in the specific values of the critical exponents β, γ and δ. As we have already pointed out, these values depend on the location of the critical point in the phase diagram.

In the pure classical regime, $\lambda = 0$, $\tau = t/t_c^{cl} - 1 \to +0$, we have $\gamma = \gamma_T^{cl}$, $\beta = \beta_T^{cl}$, and $\delta = \delta_T^{cl}$, where the critical exponent δ_T^{cl} is given by

$$\delta_T^{cl} = \begin{cases} (d+\sigma)/(d-\sigma), & \text{if} \quad \sigma < d < 2\sigma \\ 3, & \text{if} \quad d \geq 2\sigma, \end{cases} \tag{3.147}$$

with logarithmic corrections at $d = 2\sigma$. Along the line $t = t_c^{cl}$ near the classical multicritical point $(\lambda \to +0, \tau \sim \lambda)$, we have respectively $\beta = \beta_\lambda^{cl} = 1$ and $\delta = \delta_\lambda^{cl} = \delta_T^{cl}$, for $\sigma < d$, and $\gamma = \gamma_\lambda^{cl}$, where the critical exponent γ_λ^{cl} is

$$\gamma_\lambda^{cl} = \begin{cases} 2\sigma/(d-\sigma), & \text{if} \quad \sigma < d < 2\sigma \\ 2, & \text{if} \quad \quad d \geq 2\sigma, \end{cases} \tag{3.148}$$

with logarithmic corrections at $d = 2\sigma$.

In the pure quantum regime, $t_c = 0$, $\tau = \lambda/\lambda_c - 1 \to 0$, in addition to the exponents β_λ^q and γ_λ^q obtained above, one has the critical exponent

$$\delta_\lambda^q = \begin{cases} (2d + 3\sigma)/(2d - \sigma), & \text{if} \quad \sigma/2 < d < 3\sigma/2 \\ 3, & \text{if} \quad \quad d \geq 3\sigma/2, \end{cases} \tag{3.149}$$

(with logarithmic correction at $d = 3\sigma/2$), which ensures the validity of Eq. (3.146). Along the line $\lambda = \lambda_c(0)$ near the quantum multicritical point $(t \to +0, \tau \sim t)$ one has $\delta_T^q = \delta_\lambda^q$, $\beta = \beta_T^q$, and $\gamma = \gamma_T^q$.

Thus we are able to study the critical properties of the model along the lines $\lambda = \overline{\lambda}$, where $\overline{\lambda} \in [0, \lambda_c]$, and $t = \overline{t}$, where $\overline{t} \in [0, t_c^{cl}]$. The neighborhoods of the classical $(t = t_c^{cl}, \lambda = 0)$ and quantum $(t = 0, \lambda = \lambda_c(0))$ multicritical points are of special interest, since in the remaining part of the phase diagram $\gamma_\lambda = \gamma_T = \gamma_T^{cl}$ and $\beta_\lambda = \beta_T = \beta_T^{cl}$, i.e., the critical exponents are independent of the way the points on the critical line $t_c(\lambda)$ are approached. For $t_c(\lambda) > 0$ the asymptotic critical behavior of the system is governed by classical fluctuations only, as it has been generally suggested by renormalization group arguments in the finite-temperature case, see e.g. [Hertz (1976)].

Along the line $\lambda = \overline{\lambda}$, where $\overline{\lambda} \in [0, \lambda_c(0))$ so that only the quantum multicritical point is excluded, the critical exponents obey *the classical to quantum crossover rule*, i.e. the $(d + \sigma/2)$-dimensional system is equivalent to the d-dimensional "purely quantum" $(t_c(\lambda) = 0)$ system.

We have found that in all cases the critical exponents obey the scaling relation $\gamma = \beta(\delta - 1)$. Let us note that the susceptibility at the critical point behaves as

$$\chi(k \to 0) \sim k^{-\sigma},$$

and the exponent η has the spherical model value $\eta = 2 - \sigma$. Now it is easy to verify that the hyperscaling relation $\nu d = 2\beta + \gamma$ holds in all cases below the corresponding classical $d_u^{cl} = 2\sigma$, or quantum $d_u^q = 3\sigma/2$, upper critical dimension. At the quantum multicritical point the critical exponents γ_T^q and β_T^q are related to γ_λ^q and β_λ^q through the shift exponent ψ^q: $\gamma_\lambda^q = \psi^q \gamma_T^q$, $\beta_\lambda^q = \psi^q \beta_T^q$, for $\sigma < d < 2\sigma$. At the classical multicritical point similar relations are valid: $\gamma_T^{cl} = \psi^{cl} \gamma_\lambda^{cl}$, $\beta_T^{cl} = \psi^{cl} \beta_\lambda^{cl}$.

Let us note that in the low-temperature region we have obtained the crossover exponent $\phi_T = \psi^q$, when $\sigma < d < 3\sigma/2$, while $\phi_T \neq \psi^q$ when $3\sigma/2 < d < 2\sigma$. The breakdown of the relation $\phi_T = \psi^q$ above the quantum borderline dimension $d_u^q = 3\sigma/2$ is not surprising. The same feature of the critical behavior has been established and discussed in [Lawrie (1978a)] for the transverse-field Ising model in the Hartree-Fock approximation. Note that the transverse-field Ising model in the Hartree-Fock approximation belongs to the same universality class as the exactly solved model with Hamiltonian (3.91).

3.4.3   *Some generalizations*

The single-particle anharmonic potential, usually used in models describing structural phase transitions, has the double-well form

$$U(x) = -Ax^2/2 + Bx^4/4,$$

with $A \geq 0$, $B > 0$. From both theoretical and experimental points of view, the asymptotic form $U(x^2) \sim x^4$ for large $|x|$ seems somewhat artificial. The double-well potential with a *Gaussian type anharmonicity*: $U(x^2) = ax^2 + b\exp(-cx^2)$, where $a, b, c > 0$, can be considered as more appropriate in some cases, see [Plakida (1981)]. For generalizations of the model (3.91) along this line see [Stamenković et. al. (1987)], [van Hemmen and Zagrebnov (1988)], [Verbeure and Zagrebnov (1992)], [Verbeure and Zagrebnov (1995)] and [Momont et. al. (1997)], where rigorous analysis of some quantum features of phase transitions has been given. Especially, it has been shown that the particular choice of the anharmonicity, provided it satisfies some general conditions, is unimportant for the critical properties of the model. It is a common feature of the whole class of models to possess exactly solvable modifications (belonging to the same universality class) which result under the substitution of Q_r^2 by its arithmetic mean over the N particles.

Some other problems are related to the n-component generalization of model (3.91),

$$\mathcal{H}^{(n)} = 1/2 \sum_{\mathbf{r},\alpha} [(P_{\mathbf{r}}^{\alpha})^2/m - A(Q_{\mathbf{r}}^{\alpha})^2] + 1/4 \sum_{\mathbf{r},\mathbf{r}',\alpha} \varphi(\mathbf{r} - \mathbf{r}')(Q_{\mathbf{r}}^{\alpha} - Q_{\mathbf{r}'}^{\alpha})^2$$

$$- (B_1/4N)\left[\sum_{\mathbf{r},\alpha}(Q_{\mathbf{r}}^{\alpha})^2\right]^2 + (B_2/4N)\sum_{\mathbf{r},\mathbf{r}',\alpha}(Q_{\mathbf{r}}^{\alpha})^2(Q_{\mathbf{r}'}^{\alpha})^2, \qquad (3.150)$$

where $\alpha = 1, 2, \ldots, n$ are the components of the displacement and $A > 0$, B_1, and B_2 are model constants. The exact solution for the model with Hamiltonian (3.150) has been obtained in the thermodynamic limit with the aid of the approximating Hamiltonian method, in [Plakida et. al. (1987)], [Plakida et. al. (1991)] for $n = 2$, and in [Chamati and Tonchev (1994)] for arbitrary n, only in the region of parameter space defined by the inequalities

$$nB_1 + B_2 > 0, \qquad B_2 > 0. \qquad (3.151)$$

The existence of an exact solution for a model with vector order parameter is very useful in considering the different measures of long-range order, i.e. the so-called "r-problem", see the next subsection.

Let us continue with some aspects of another interesting phenomenon: the fluctuation-induced first order phase transition. A large class of structural phase transitions in systems with cubic anisotropy can be described by the Hamiltonian

$$\mathcal{H}^{(2)} = 1/2 \sum_{\mathbf{r},\alpha} [(P_{\mathbf{r}}^{\alpha})^2/m - A(Q_{\mathbf{r}}^{\alpha})^2] + 1/4 \sum_{\mathbf{r},\mathbf{r}',\alpha} \varphi(\mathbf{r} - \mathbf{r}')(Q_{\mathbf{r}}^{\alpha} - Q_{\mathbf{r}'}^{\alpha})^2$$

$$- (B_1/4N)\sum_{\mathbf{r}}\left[\sum_{\alpha}(Q_{\mathbf{r}}^{\alpha})^2\right]^2 + (B_2/4N)\sum_{\mathbf{r},\alpha}(Q_{\mathbf{r}}^{\alpha})^4, \qquad (3.152)$$

where $\alpha = 1, 2$. The renormalization group analysis leads to the following conclusion: because of the cubic anysotropy, the phase transition turns out to be of the first order in the regions where the parameters B_1 and B_2 obey the inequalities, see [Rudnick (1978)],

$$B_1 + B_2 > 0 \qquad \text{and} \qquad 2B_1 + 3B_2 < 0, \qquad (3.153)$$

or

$$2B_1 + B_2 > 0 \qquad \text{and} \qquad B_1 < 0. \qquad (3.154)$$

According to a Landau-type analysis, in the range of parameters

$$B_1 + B_2 > 0, \qquad \text{and} \qquad 2B_1 + B_2 > 0 \qquad (3.155)$$

the phase transition should be of the second order.

For the system with Hamiltonian (3.152) the regions (3.153) and (3.154) are equivalent, as follows by rotation in the 2-dimensional space of single-site coordinates. The problem as to whether a similar equivalence holds for the exactly solvable model with Hamiltonian (3.150) has been studied in [Plakida et. al. (1987)], [Plakida et. al. (1991)] for $n = 2$. It turned out that regions (3.153) and (3.154) are not equivalent in this case. The exact solution can be obtained by the approximating Hamiltonian method in the part of the stability region with respect to Landau's theory (3.155) defined by inequalities (3.151) at $n = 2$. Here the phase transition was found to be of the second order, with critical exponents equal to those of the one-component version. In the remaining part of the region of interest (3.155), namely,

$$B_1 + B_2 > 0, \qquad B_2 < 0, \qquad (3.156)$$

the analysis has shown that the phase transition is of the first order. However, in this region there is no proof that the variational solution is exact.

3.4.4   *Long-range order and the r-problem*

The n-vector long-range order parameter $\mathbf{m_e}$, coupled to an external field $h\,\mathbf{e}$, where $\mathbf{e}$ is a vector of unit length, in a system undergoing a first order phase transition at $T < T_c$ and $h = 0$, can be defined as

$$\mathbf{m_e} = \langle\langle \mathbf{L}_\Lambda \rangle\rangle_\mathbf{e} = \lim_{h \to 0^+} \lim_{|\Lambda| \to \infty} \langle \mathbf{L}_\Lambda \rangle_{\mathcal{H}_\Lambda(h)}. \qquad (3.157)$$

Here $\mathbf{L}_\Lambda = (L_\Lambda^1, \ldots, L_\Lambda^n)$ is a n-vector operator with components

$$L_\Lambda^\alpha = |\Lambda|^{-1} \sum_{i \in \Lambda} L_i^\alpha, \qquad (\alpha = 1, \ldots, n) \qquad (3.158)$$

where L_i^α are the components of another n-vector operator $\mathbf{L}_i = (L_i^1, \ldots, L_i^n)$ associated with each lattice site $i \in \Lambda$; the average value in the right-hand side of Eq. (3.157) is taken with the Hamiltonian

$$\mathcal{H}_\Lambda(h) = \mathcal{H}_\Lambda - h\,\mathbf{e} \cdot \mathbf{L}_\Lambda |\Lambda|. \qquad (3.159)$$

The external field term is chosen so as to break the symmetry of the Hamiltonian $\mathcal{H}_\Lambda$. The spontaneous symmetry breaking ($m_e \neq 0$) in the low-temperature phase may occur only in the thermodynamic limit $|\Lambda| \to \infty$.

The double-limit procedure in Eq. (3.157) is well-defined mathematically, but has a considerable disadvantage: it is inconvenient for studying *finite-size effects* on phase transitions, as well as for analyzing data obtained from computer simulations. The standard recipe to avoid this problem is to use the root-mean-square (rms) order parameter for one of the components of the order operator $\mathbf{L}_\Lambda$, say,

$$ m_{rms}^\alpha = \left[\langle (L_\Lambda^\alpha)^2 \rangle_{\mathcal{H}_\Lambda} \right]^{1/2}. \tag{3.160} $$

The common statement is that the two definitions (3.157) and (3.160) are in some sense equivalent [Manousakis (1991)]. However, in the case of an order parameter with $n > 1$, a sort of "rotational diffusion" of the vector $\mathbf{e}$ on the surface of the unit sphere takes place [Binder (1992)]. Thus, the value of the ratio $r = m_e^\alpha / m_{rms}^\alpha$ is *a priori* not clear.

The so called *r-problem* is a rather sophisticated one and has a long history. For the Heisenberg model Griffiths proved [Griffiths (1966)] that $m_e \geq m_{rms}$. (Here and below, the superscript labeling the component of the order operator will be omitted.) Later Dyson, Lieb and Simon refined this result as $m_e \geq \sqrt{3}\, m_e$ [Dyson et. al. (1978)]. The above proofs made full use of the unique fact that for Heisenberg ferromagnets the order parameter is a constant of motion. The interest in this subject for models with non-commuting with the Hamiltonian order operator, e.g., Heisenberg antiferromagnets, was renewed in [Kaplan et. al. (1989)]. These authors obtained $m_e \geq m_{rms}$ at $T = 0$. An extension of the refined bound, $m_e \geq \sqrt{3} m_{rms}$, to the case of quantum Heisenberg antiferromagnets has been obtained in [Koma and Tasaki (1993)]. All the above articles discuss mainly model system which are not exactly solved.

The exactly solvable equivalent-neighbors Lieb-Mattis model has been considered in [Kaplan et. al. (1990)]. The exact value $r = \sqrt{3}$ has been obtained for the ground state, $T = 0$, in [Kaplan et. al. (1990)], and for arbitrary temperatures T and general spin values in [Gochev and Tonchev (1992)]. Following [Koma and Tasaki (1993)], we present here some heuristic arguments which clarify the appearance of the factor $\sqrt{3}$ in the above estimates. Let us note that the physical quantity m_e is related to a pure phase (see Section 9.1) in which the unit vector $\mathbf{e}$ may point in arbitrary direction. The quantity m_{rms} measures the long-range order in mixed phases as well. The following decomposition seems to hold, see [Bogolyubov (1970)], [Bogolyubov Jr. (1972)],

$$\langle \mathbf{L}_\Lambda^2 \rangle = K^{-1} \int d\mathbf{e}\, \langle\langle \mathbf{L}_\Lambda^2 \rangle\rangle_\mathbf{e}, \qquad (3.161)$$

where the integration is over the unit sphere and $K = \int d\mathbf{e}$. If in each pure phase one has

$$\langle\langle \mathbf{L}_\Lambda \rangle\rangle_\mathbf{e} = m_e \mathbf{e}, \qquad (3.162)$$

and the *clustering property* ([Glimm and Jaffe (1981)])

$$\langle\langle \mathbf{L}_\Lambda^2 \rangle\rangle_\mathbf{e} \to \langle\langle \mathbf{L}_\Lambda \rangle\rangle_\mathbf{e} \langle\langle \mathbf{L}_\Lambda \rangle\rangle_\mathbf{e}, \qquad \text{as} \quad |\Lambda| \to \infty, \qquad (3.163)$$

holds. Then from Eqs. (3.161),(3.162) and (3.163) one obtains $\langle \mathbf{L}_\Lambda^2 \rangle_{\mathcal{H}_\Lambda} = m_e^2$. On the other hand, from the $SU(2)$ invariance of the Heisenberg antiferromagnet and definition (3.160) for $\alpha = 3$, one has

$$\langle \mathbf{L}_\Lambda^2 \rangle_{\mathcal{H}_\Lambda} = 3\langle (\mathbf{L}_\Lambda^{(3)})^2 \rangle_{\mathcal{H}_\Lambda} = 3m_{rms}^2. \qquad (3.164)$$

Tnus, it immediately follows that $m_e = \sqrt{3}\, m_{rms}$.

These arguments are quite general and natural, but they do not have the status of a proof. For example, in [Koma and Tasaki (1993)] a trivial consideration is presented, which gives a counterexample of the abive result. This demonstrates that the detailed properties of the model are crucial for the relation between the magnetization in the state with broken symmetry $\langle\langle \cdots \rangle\rangle_\mathbf{e}$ and the symmetric equilibrium state $\lim_{|\Lambda|\to\infty}\langle \cdots \rangle_{\mathcal{H}_\Lambda}$.

Although all efforts in this area have been focused on magnetic systems, the r-problem is not of less interest for other physical systems exhibiting symmetry breaking, e.g., for the n-component generalization of model (3.91). The peculiarity of this model in the context of the r-problem comes from the presence of quantum and classical critical fluctuations, which depend on the temperature T and space dimensionality d, in conjunction with its exact solvability in the thermodynamic limit. The case of the d-dimentional, n-component model (3.150) has been considered in [Chamati and Tonchev (1994)] where the result $r = \sqrt{n}$ has been obtained. This result is independent of the type of fluctuations (quantum and/or classical) and gives a rare example of exact calculation of the ratio m_e/m_{rms} in a system with developed fluctuations.

Chapter 4

Finite-Size Scaling at Criticality

The most general and fundamental laws of modern theory of phase transitions can be formulated as scaling properties. They express asymptotically homogeneous dependence of the thermodynamic functions on the relevant thermodynamic parameters near a phase transition point.

The cornerstone of our current understanding of the way in which the thermodynamic singularities are modified by considering the system finite in some (or all) dimensions is the *finite-size scaling theory*. This theory describes how long-scale, universal features of collective phenomena associated with the onset of large fluctuations (classical or quantum) near a critical point manifest themselves in finite-size samples, as well as in numerical computer simulations. It provides the basis for interpretations of finite-size effects in experimental data for real systems. The theory gives the most reliable tool for extrapolation to the thermodynamic limit of numerical results of Monte Carlo and transfer matrix calculations. It reveals the intimate mechanism of how the critical singularities build up and, in particular, it can be fruitfully used in explaining the low-temperature behavior of quantum critical systems. The development of the finite-size scaling theory is interrelated with other theoretical studies like conformal invariance, surface and interface critical phenomena.

4.1   Finite-size systems and critical phenomena

We shall *call a finite-size system any system which has finite size in at least one space dimension.* We emphasize that singularities in the thermodynamic functions at a critical point may occur only in the (bulk) thermodynamic

125

limit, see Section 1.3, taken in at least d_l dimensions (d_l is the lower critical dimension of the system). If the system is fully finite, no such singularities exist and, strictly speaking, no phase transitions occur. These facts naturally give rise to the following questions: Why the bulk theory turns out adequate to the experimental evidence for finite macroscopic objects? How the singularities appear in the infinite system when no phase transitions occur in any fully finite one?

The answer to the first question lies in the fact that a macroscopic body is close in some sense to the idealized thermodynamic limit.

From a mathematical point of view, the answer to the second question becomes evident if one recalls that the limit function of a sequence of analytic functions needs not be analytic. Therefore, what happens is that in the thermodynamic limit some intesive thermodynamic functions of the system develop singularities which are attributes of phase transitions. So, in a finite system one expects an appreciable rounding of the critical point singularities. Let us consider the same event from a more physical point of view. To be more specific, let us focus on systems with short-range interactions. In the latter case, the only mechanism by which a spin in the bulk of a finite system can "learn" about the existence of boundaries is the step-wise propagation of "information" from a boundary, through the correlations built up by consequtive pairwise interactions. As we have already mentioned, the bulk correlation length $\xi_\infty(T)$ measures the distance over which the spins are significantly correlated. Assume that all relevant thermodinamic parameters (if any), except the temperature, are fixed at their critical values. Then, as the temperature T approaches the critical temperature T_c, the correlation length $\xi_\infty(T)$ diverges, see Section 1.5. When $\xi_\infty(T)$ attains a magnitude of the order of the characteristic size L of the finite system, then the boundary spins at the opposite sides of the system become well-correlated, and ordering cannot continue to build up further in the restricted dimensions. It is certainly reasonable to expect that the rounding of the phase transition is controlled by the criterion

$$\xi_\infty(T) \sim L. \tag{4.1}$$

Usually, the general geometry considered in the theory is of the type $L^{d-d'} \times \infty^{d'}$, i.e., when the system is infinite in d' dimensions and finite in the remaining $d - d'$ ones; the case $d' = 0$ corresponds to a fully finite system, $d' = 1$ to a system with the geometry of a cylinder, $d' = (d-1)$ – to a slab (or film) geometry. In Monte Carlo simulations one uses systems with $d' = 0$, having the shape of a d-dimensional hypercube, or parallelepiped; for transfer matrix calculations and phenomenological renormalization the cylindrical geometry is used. The geometry $L \times \infty^{d-1}$ is especially interesting from theoretical point of view, because for sufficient large d $(d \geq d_l + 1)$ the *finite-size system* may exhibit a phase transition with critical exponents of the $(d-1)$-dimensional system. Thus, in such a system one expects crossover effects from a d-dimensional to a $(d-1)$-dimensional type behavior to take place.

It is clear that as far as one deals with finite systems, one cannot avoid the question of the boundary conditions applied at the surfaces in the finite dimensions.

Let us illustrate some of the problems that appear in finite systems close to the bulk critical point by the example of "free" (zero-Dirichlet) boundary conditions, which will be denoted by the superscript (1); we refer the reader to Section 7.1 for the precise definition and notation of the boundary conditions used in this book. For such a d-dimensional system with a finite size L one can expand, on general grounds, the free energy density as

$$f_L^{(1)}(T) = f(T) + \frac{1}{L}f_{\rm s}(T) + \frac{1}{L^2}f_{\rm e}(T) + \cdots + \frac{1}{L^d}f_{\rm c}(T) + O\left(e^{-L/\xi(T)}\right),$$
$$(4.2)$$

where $f(T) \equiv f_\infty(T)$ is the bulk free energy density, $f_{\rm s}(T)$ is the contribution from the surfaces (the total surface free energy density), $f_{\rm e}(T)$, etc, and $f_{\rm c}(T)$, stem from the edges, etc, and the corners, respectively. For Ising type systems one may say that the above expansion is well established. If the geometry of the system lacks some of the above boundary objects, then the corresponding contribution in Eq. (4.2) is set identically to zero. Thus, for periodic boundary conditions (no surfaces, edges and corners) one has

$$f_L^{(p)}(T) = f(T) + O\left(e^{-L/\xi(T)}\right). \qquad (4.3)$$

This rapid exponential approach to the thermodynamic limit is one of the reasons why periodic boundary conditions are preferred in Monte Carlo simulations.

One observes that the left-hand side in the above expansions is a regular function of T, whereas the right-hand side is not: the bulk density $f(T)$, for example, is singular at the critical point T_c. Therefore, such expansions can hold only *away* from T_c. Obviously, for $T \simeq T_c$ one needs an alternative, finite-size scaling formulation.

Let us note that for Ising type models expansions (4.2) and (4.3) hold also below T_c. For systems with long-range interactions such expansions are, generally speaking, *not* valid. For example, in the case of interactions which at large distances r decay asymptotically as $r^{-(d+\sigma)}$, with $0 < \sigma < 2$, the finite-size corrections under periodic boundary conditions fall down as $(L/\xi_\infty)^{-(d+\sigma)}$ above T_c, see [Brankov and Danchev (1991)]. The above expansions break down even for short-range interactions in $O(n)$ models with $n \geq 2$, due to the fact that in such models $\xi_\infty(T) \equiv \infty$ for $T < T_c$.

4.2 Phenomenological finite-size scaling

The basic ideas of the phenomenological finite-size scaling theory at criticality have been suggested in [Fisher (1971)], and [Fisher and Barber (1972)]. In our concise exposition we follow the above cited original papers, as well as the review [Barber (1983)]. According to the phenomenological theory, rounding and shifting of the anomalies in the thermodynamic functions set in when the bulk correlation length ξ_∞ becomes comparable to the characteristic linear size L of the system. More specifically, it is predicted that finite-size effects are controlled by the ratio L/ξ_∞. Here we present some fundamental notions and facts of that theory.

Let us start with a system having the geometry $L^{d-d'} \times \infty^{d'}$, where $d' > d_l$. Then the *finite-size* system exhibits a phase transition at a temperature $T = T_{c,L}^{(\tau)}$, and the corresponding infinite system at $T = T_{c,\infty}^{(\tau)} \equiv T_c$, where the superscript τ denotes the boundary conditions imposed on the system. The so-called *fractional shift*, characterizing the shift of the critical temperature of the finite-size system is defined as

$$\varepsilon_L^{(\tau)} = \left(T_c - T_{c,L}^{(\tau)}\right) / T_c \simeq b^{(\tau)} L^{-\lambda}, \qquad (4.4)$$

where the expected asymptotic behavior for $L >> 1$ is characterized by the shift exponent λ. Let us now consider an arbitrary intensive quantity P (such as the specific heat c or susceptibility χ per spin) which in the bulk (infinite) system possesses a critical-point divergence of the type

$$P_\infty(T) \simeq At^{-\rho}, \qquad t \to 0^+, \qquad (4.5)$$

where $\rho = \rho(d)$ is the d-dimensional critical exponent and $t = (T - T_c)/T_c$.

On approaching the finite-size critical temperature from above at fixed L we should have

$$P_L^{(\tau)}(T) \simeq \dot{A}_L^{(\tau)} \dot{t}^{-\dot{\rho}}, \qquad \dot{t} \to 0^+, \tag{4.6}$$

where

$$\dot{t} = \left(T - T_{c,L}^{(\tau)}\right)/T_c = \varepsilon_L^{(\tau)} + t \tag{4.7}$$

and $\dot{\rho} = \rho(d')$ (in general, $\rho \neq \dot{\rho}$). Let $T_{*,L}^{(\tau)}$ denote the temperature at which the considered finite-size property $Q_L^{(\tau)}(T)$ first shows significant (of the relative order of unity) deviation from its bulk limit $P_\infty(T)$. Then we could define the *fractional rounding*

$$\delta_L^{(\tau)} = \left(T_{*,L}^{(\tau)} - T_c\right)/T_c \simeq c^{(\tau)} L^{-\theta}, \qquad L \gg 1. \tag{4.8}$$

Note that the "rounding" measures the region of crossover from bulk d-dimensional to d'-dimensional critical behavior.

The basic assertions of *phenomenological finite-size scaling* are:

(i) The only relevant variable on which the properties of the finite-size system depend in the neighborhood of T_c is $L/\xi_\infty(T) \sim L t^\nu$.

(ii) The rounding occurs when $\xi_\infty(T) \simeq L$.

The explicit formal expression of item *(i)* is

$$P_L^{(\tau)}(T) \simeq L^q \tilde{X}^{(\tau)}\left(L/\xi_\infty(T)\right), \tag{4.9}$$

or, equivalently,

$$P_L^{(\tau)}(T) \simeq L^q X^{(\tau)}(tL^{1/\nu}). \tag{4.10}$$

Here $X^{(\tau)}(x)$ is the *finite-size scaling function* describing the critical behavior of the quantity P.

It has been considered, see [Fisher (1971)], [Barber (1983)], that a more general formulation of this hypothesis is given by the equation

$$P_L^{(\tau)}(T) \simeq L^q X^{(\tau)}(\dot{t}L^{1/\nu}). \tag{4.11}$$

Apart from the allowed shift of T_c to $T_{c,L}$, Eqs. (4.10) and (4.11) are equivalent if ξ_∞ diverges algebraically with exponent $\nu \geq 1/\lambda$. If this is not the case, we will use Eq. (4.10).

It is easy to see that assumption *(ii)* leads directly to the conclusion that $\theta = 1/\nu$. Indeed, by using $\xi_\infty(T) \simeq A_\xi^\pm |t|^{-\nu}$ and Eq. (4.8) we obtain

$$L \simeq \xi_\infty(T_{*,L}^{(\tau)}) \simeq A_\xi^\pm \left| T_{*,L}^{(\tau)}/T_c - 1 \right|^{-\nu} \simeq A_\xi^\pm \left(c^{(\tau)} \right)^{-\nu} L^{\theta\nu}, \qquad (4.12)$$

hence, $c^{(\tau)} = (A_\xi^\pm)^{1/\nu}$, and $\theta = 1/\nu$, where one has to take A_ξ^+ or A_ξ^- depending on whether $T_{*,L}^{(\tau)}$ is larger or less than T_c. An interesting consequence from the above result is that $c^{(\tau)}$ is actually τ-independent. Let us now determine the exponent q. To this aim we let $L \to \infty$ in Eq. (4.10) and compare it with Eq. (4.5). In order to reproduce the proper bulk singularity in t, we have to require

$$X^{(\tau)}(x) \simeq X_\infty x^{-\rho}, \qquad x \to \infty, \qquad (4.13)$$

while to cancel the residual L-dependence we need $q = \rho/\nu$. Then, by comparing with Eq. (4.5) we identify $X_\infty = A$. On the other hand, to reproduce the asymptotic behavior (4.6) with respect to $\dot{t}$, we must have in addition

$$X^{(\tau)}(x) \simeq X_0^{(\tau)} x^{-\dot{\rho}} \quad \text{as} \quad x \to 0^+. \qquad (4.14)$$

By comparing with Eq. (4.6) we conclude that

$$\dot{A}_L^{(\tau)} = X_0^{(\tau)} L^{(\rho-\dot{\rho})/\nu}. \qquad (4.15)$$

The last prediction can be tested directly, since the determination of the exponents ρ, $\dot{\rho}$ and ν is independent of the amplitude $X_0^{(\tau)}$.

Summarizing the above results, we rewrite Eq. (4.10) in the form

$$P_L^{(\tau)}(T) \simeq L^{\rho/\nu} X^{(\tau)}(\dot{t}L^{1/\nu}), \qquad (4.16)$$

which is the appropriate finite-size scaling hypothesis for *any* intensive thermodynamic quantity with *algebraic* singularity at the critical temperature.

To allow for a logarithmic bulk singularity,

$$P_\infty(T) \simeq -A \ln t, \qquad t \to 0^+, \qquad (4.17)$$

which occurs, for example, in the specific heat capacity of the two-dimensional Ising model, one has to extend the hypothesis (4.10) to the form ([Fisher (1971)])

$$P_L^{(\tau)}(T) - P_L^{(\tau)}(T_0) \simeq X^{(\tau)}(\dot{t}L^{1/\nu}) - X^{(\tau)}(\dot{t}_0 L^{1/\nu}), \qquad (4.18)$$

where $t_0 > 0$ is some fixed noncritical value of t, say $t_0 = 1$, and T_0 is the corresponding temperature. The term $P_L^{(\tau)}(T_0)$ has only a weak L-dependence and, being of the order $O(1)$, can be omitted. The analysis can be carried out much along the same lines as in the algebraic case. By considering the limit $L \to \infty$ at fixed $T \neq T_{c,L}^{(\tau)}$, one recovers the bulk behavior (4.17) if

$$X^{(\tau)}(x) \simeq -A \ln x \qquad \text{as} \quad x \to +\infty. \tag{4.19}$$

An important result about the asymptotic behavior of $P_L^{(\tau)}(T)$ *at* the finite-size critical point follows by taking the limit $t \to 0$ in Eq. (4.18) at fixed size L. Assuming that $X^{(\tau)}(0) = O(1)$, one obtains in the above limit

$$P_L^{(\tau)}(T_{c,L}^{(\tau)}) \simeq (A/\nu) \ln L + O(1). \tag{4.20}$$

This prediction allows, at least in principle, the determination of the bulk critical exponent ν from data for small samples of varying size, obtained experimentally, or by Monte Carlo simulations.

Let us turn now to the general case of geometry $L^{d-d'} \times \infty^{d'}$ with $d' \leq d_l$. Under the last condition in the finite-size system there is no phase transition of its own, therefore, $\dot{\rho} \equiv 0$. The hypotheses stated above still hold, provided we set $\dot{\rho} = 0$ and replace $T_{c,L}^{(\tau)}$ by appropriately defined *pseudocritical* temperature $T_{m,L}^{(\tau)}$. The latter can be defined, for example, as the temperature at which the specific heat capacity reaches its maximum. Then, in the case of an algebraic bulk singularity of the type (4.5) we obtain

$$P_L^{(\tau)}(T_{m,L}^{(\tau)}) \simeq X_0^{(\tau)} L^{\rho/\nu}, \qquad L >> 1. \tag{4.21}$$

This asymptotic behavior, as well as Eq. (4.20) when appropriate, is exploited in one of the basic methods for evaluation of bulk critical exponents from finite-size data.

Finally, we emphasize that the use of the shifted temperature variable t in the above finite-size scaling hypotheses allows for any L-dependence of the shift $\varepsilon_L^{(\tau)}$, i.e. the shift exponent λ remains arbitrary. The assertion that the only criterion determining the finite-size scaling effects in the critical region is $\xi_\infty(T) \simeq L$ leads to the equalities

$$\lambda = \theta = 1/\nu. \tag{4.22}$$

This result follows from the renormalization group derivation of finite-size scaling, see [Barber (1983)]. It has also been confirmed to first order in the

ε-expansion ($\varepsilon = 4 - d$) for n-component systems with a film geometry and Dirichlet boundary conditions, see [Bray and Moore (1978)], where it has been found that

$$\lambda = 2 - \frac{n+2}{n+8}\varepsilon + O(\varepsilon^2) = 1/\nu + O(\varepsilon^2). \tag{4.23}$$

Except in some special cases (ideal Bose gas and spherical model with a film geometry and free boundaries, when $\lambda = 1$ in *all* dimensions), this relation seems to be generally valid. Before turning to the formulation of the simplified finite-size scaling hypothesis, in which $\lambda = \theta = 1/\nu$ is essentially used, we should mention that the relationship $\lambda = 1/\nu$ is *not* a necessary condition for finite-size scaling to hold in general. Let us now accept that $\lambda = 1/\nu$. Then, the reformulation of the hypothesis (4.16) in terms of t, instead of $\dot{t}$, reads

$$P_L^{(\tau)}(T) \simeq L^{\rho/\nu} X^{(\tau)}(tL^{1/\nu}). \tag{4.24}$$

The asymptotic analysis for large L and fixed $t \neq 0$ yields, as in the previous case, see Eq. (4.13).

In order to describe the behavior of a finite-size system with phase transition at $T_{c,L}^{(\tau)}$ we have to require that

$$X^{(\tau)}(x) \simeq X_0^{(\tau)}(x - x_c^{(\tau)})^{-\dot{\rho}}, \qquad x \to x_c^{(\tau)} + 0, \tag{4.25}$$

where the critical value $x_c^{(\tau)}$ of the scaling variable is explicitly given by

$$x_c^{(\tau)} = L^{1/\nu}\left(T_{c,L}^{(\tau)} - T_c\right)/T_c = -L^{1/\nu}\varepsilon_L^{(\tau)}. \tag{4.26}$$

Note that the scaling function $X^{(\tau)}(x)$ is supposed to diverge at $x = x_c^{(\tau)}$ rather than at $x = 0$. Therefore, Eq. (4.4) implies that now $\lambda = 1/\nu$ is a necessary condition for $x_c^{(\tau)}$ to be finite and nonvanishing as $L \to \infty$; then $x_c^{(\tau)} = -b^{(\tau)}$. The asymptotic behavior of the amplitude $\dot{A}_L^{(\tau)}$ is still given by Eq. (4.15).

One can ask the question: Why should the linear size L of the system be compared to the bulk correlation length ξ_∞ ? This may seem an artificial hypothesis of the theory, since ξ_∞ does not characterize the finite system. In Section 4.5 we make a clear distinction between the different assumptions underlying the conventional formulation of critical finite-size scaling. To this end we use the notion of asymptotic homogeneity of the thermodynamic functions, much alike the well known general formulation of bulk critical scaling, see [Stanley (1971)], [Fisher (1983)], [Huang (1987)].

4.3 Privman-Fisher hypothesis for the free energy

As mentioned in Section 1.5.4, it is often convenient to split the bulk free energy density $f(T, h)$ into a singular part $f_{\text{sing}}(T, h)$ and analytical (non-singular) "background" term $f_{\text{ns}}(T, h)$, see Eq. (1.81). The free energy density $f_L(T, h)$ of a system in region Λ of finite size L can also be written as a sum of "singular" and nonsingular parts,

$$f_L(T, h) = f_{L,\text{sing}}(T, h) + f_{L,\text{ns}}(T, h).$$

Here the "singular" part $f_{L,\text{sing}}(T, h)$ is actually a real-analytic function of its variables, *by definition* such that its thermodynamic limit reproduces the singular part of the bulk free energy:

$$\lim_{L \to \infty} f_{L,\text{sing}}(T, h) = f_{\text{sing}}(T, h). \tag{4.27}$$

Obviously, then one also has $\lim_{L \to \infty} f_{L,\text{ns}}(T, h) = f_{\text{ns}}(T, h)$.

Privman and Fisher suggested [Privman and Fisher (1984)] that the singular part of the reduced free energy density of the finite system with $d < d_u$ can be written as ($t = T/T_c - 1$)

$$\beta f_{L,\text{sing}}(T, h) \simeq L^{-d} Y \left(a t L^{1/\nu}, b h L^{\Delta/\nu} \right). \tag{4.28}$$

The finite-size scaling function Y is supposed to be *universal*, but dependent on the boundary conditions and the geometry of the system.

In this section, following the ideas of Privman and Fisher, we present a phenomenological and physically transparent derivation of Eq. (4.28).

Firstly, we recall that according to the original finite-size scaling hypothesis all the lengths diverging at the critical point should scale with the bulk correlation length ξ_∞. This assertion provides a natural generalization of the bulk scaling relation (1.82) to the form

$$\beta f_{L,\text{sing}}(T, h) \simeq A_1 |t|^{2-\alpha} \tilde{W}^{\pm} \left(A_2 h |t|^{-\Delta}, L/\xi_\infty \right) \tag{4.29}$$

where $\tilde{W}^{\pm}$ are universal functions. Even though the argument L/ξ itself does not entail any nonuniversal factor, the additional metric factor $A_\xi^{\pm}$ does enter into the above finite-size scaling form through the bulk critical behavior of $\xi_\infty(T)$, see Eq. (1.72). However, if one inserts A_1, given by Eq. (1.94), into Eq. (4.29), and makes use of the hyperscaling relation (1.96), then one obtains the Privman-Fisher hypothesis (4.28). By noting that the finite system undergoes no phase transition at $t = h = 0$, one concludes that the scaling function is analytic at the origin, hence, no distinct "$\pm$" functions are needed.

Before proceeding further, let us briefly consider the "nonsingular" term $f_{L,\mathrm{ns}}(T,h)$. Due to the $h \leftrightarrow -h$ symmetry, the first h-dependent term in the expansion of this function around $T = T_c$, $h = 0$, is

$$\frac{1}{2} f_{L,\mathrm{ns}}^{(2)}(T_c,0) h^2 \equiv \frac{1}{2} f_{L,\mathrm{ns}}^{(2)}(T_c,0) \left(h L^{\Delta/\nu} \right)^2 L^{-2\Delta/\nu},$$

where $f_{L,\mathrm{ns}}^{(2)}(T_c,0)$ is the second partial derivative of $f_{L,\mathrm{ns}}(T,h)$ with respect to h at $T = T_c$, $h = 0$. Therefore, in the "core" region of main interest, where $tL^{1/\nu} = O(1)$ and $hL^{\Delta/\nu} = O(1)$, the leading h-dependent term of $f_{L,\mathrm{ns}}$ is of the order of $L^{-2\Delta/\nu} = L^{-2d\Delta/(2-\alpha)} \ll L^{-d}$. Here we have taken into account that

$$\frac{2\Delta}{2-\alpha} = \frac{2\Delta}{2\beta+\gamma} = \frac{2\Delta}{2\Delta-\gamma} > 1.$$

Therefore, in the remainder we can ignore the dependence of the background term on the magnetic field variable h. Then one obtains for the

- magnetization

$$m_L(T,h) = -\frac{\partial \beta f}{\partial h} \simeq b L^{-\beta/\nu} Y^{(1)}\left(a t L^{1/\nu}, b h L^{\Delta/\nu} \right); \qquad (4.30)$$

- susceptibility

$$k_B T \chi_L(T,h) = \frac{\partial m}{\partial h} \simeq b^2 L^{\gamma/\nu} Y^{(2)}\left(a t L^{1/\nu}, b h L^{\Delta/\nu} \right); \qquad (4.31)$$

- nonlinear susceptibility (fourth partial derivative with respect to h)

$$k_B T \chi_L^{(4)}(T,h) = -\frac{\partial^4 \beta f}{\partial h^4} \simeq b^4 L^{(\gamma+2\Delta)/\nu} Y^{(4)}\left(a t L^{1/\nu}, b h L^{\Delta/\nu} \right);$$

$$(4.32)$$

- singular part of the specific heat

$$c_{L,\mathrm{sing}}(T,h) \simeq -k_B a^2 L^{\alpha/\nu} \left. \frac{\partial^2}{\partial x_1^2} Y(x_1, b h L^{\Delta/\nu}) \right|_{x_1 = a t L^{1/\nu}}. \qquad (4.33)$$

In the above expressions $Y^{(n)}$ denotes the n-th derivative of $Y(x_1, x_2)$ with respect to x_2. Note that thereby one obtains new universal critical-point ratios for the finite-size quantities evaluated at the bulk critical point. In particular, the coefficients of proportionality U_j in asymptotic relations such as $f_{L,\text{sing}}(0,0) \simeq U_0 L^{-d}$ for the singular part of the free energy* and $\chi_L^{(4)}(0,0)/\chi_L^2(0,0) \simeq U_1 L^d$, etc., should be universal. Quantities like U_1 can be straightforward evaluated from Monte Carlo or transfer matrix finite-size data.

Finally, we mention that in a finite-size geometry of the form $L_1 \times \cdots \times L_{d-d'} \times \infty^{d'}$, with different edge lengths $\mathbf{L} = \{L_1, \cdots, L_{d-d'}\}$ in the finite dimensions, the scaling hypothesis for the free energy density is naturally generalized to

$$f_{\mathbf{L},\text{sing}}(T,h) \simeq L_0^{-d}\tilde{Y}(atL_0^{1/\nu}, bhL_0^{\Delta/\nu}, \{L_i/L_0\}),$$

where L_0 is a diverging in the thermodynamic limit length, and the shape ratios $l_j = L_j/L_0$, $j = 1, \ldots, d - d'$, are supposed to approach bounded nonzero constants as $L_0 \to \infty$. Note that quantities like $f_{\mathbf{L},\text{sing}}(0,0)L_0^d$ are again universal, although dependent on the ratios $\{l_j\}$.

If $d' > d_l$, then the finite-size system has a critical point of its own with a singularity of the scaling function $Y(x_1, x_2)$ at some $x_1 = x_{1,c}$ and $x_2 = 0$, as we have already discussed. For systems with $d' = 0, 1$, one can define a pseudocritical temperature $T_{m,L}$, e.g., by the location of the specific heat maximum. To obtain it, we solve for x_1 the equation $\partial^3 Y(x_1, 0)/\partial x_1^3 = 0$. Its solution $x_1 = x_{1,c}$ is obviously universal and yields for the fractional shift $\varepsilon_L = -a^{-1}x_{1,c}L^{-1/\nu}$.

In fact, the hypotheses stated above represent a rather strong form of finite-size scaling which incorporates the hyperscaling conjecture and various properties which are known from the renormalization group approach, see Section 1.6, to hold for $d < d_u$. These properties include the equality of critical exponents defined for $t < 0$ and $t > 0$, hyperscaling exponent relations and the so-called two-scale factor universality (hyperuniversality) relations among the critical point amplitudes.

A review on the results available for the universal critical point ratios is given in [Privman et. al. (1991)].

---

*This amplitude is commonly known as the *Casimir amplitude* and is usually denoted by Δ. We do not use this notation currently in order to avoid confusion with the critical exponent Δ.

4.4 Definitions of correlation length

As we have seen, a central notion of the phenomenological finite-size scaling theory is the bulk correlation length ξ_∞. Its definition deserves a special attention, especially in the case of long-range interactions. Why should the linear size of a finite system be compared with the correlation length of the corresponding bulk system? For example, in [Singh and Pathria (1986)] and [Singh and Pathria (1987)] the spin-spin correlation function for the spherical model of geometry $L^{d-d'} \times \infty^{d'}$ ($2 < d < 4$, $d' \leq 2$), with nearest-neighbor interactions and periodic boundary conditions, has been studied in detail. One of the key findings is that the ratio L/ξ_L, where ξ_L is the correlation length in the actual finite-size system, rather than L/ξ_∞, emerges as the natural scaling variable. However, the subsidiary scaling hypothesis $\xi_L \simeq LX_\xi(L/\xi_\infty)$ reconciles the role of the variable L/ξ_L with the original hypothesis of phenomenological finite-size scaling about the role of L/ξ_∞. The same conclusion has been drawn in the case of power-law long-range interactions [Brankov and Danchev (1991)]. These facts illustrate the importance of defining a finite-size correlation length ξ_L too.

4.4.1 *Bulk correlation length*

The standard definitions of the bulk correlation length are based on the asymptotic behavior of the pair correlation function at large separations between the local observables. For example, in the case of a translation invariant bulk spin system, the correlation function (1.70) for the spin variables $S_\mathbf{r}$ and $S_{\mathbf{r}'}$, localized at the space points $\mathbf{r}$ and $\mathbf{r}'$, respectively, takes the form

$$G_\infty(\mathbf{r} - \mathbf{r}'; T) := \langle S_\mathbf{r} S_{\mathbf{r}'} \rangle - \langle S_\mathbf{r} \rangle \langle S_{\mathbf{r}'} \rangle. \tag{4.34}$$

The two most commonly used definitions of the bulk correlation length ξ_∞ are unambiguous in the case of exponential decay of the bulk pair correlation function $G_\infty(\mathbf{R}; T)$ with the distance $R = |\mathbf{R}|$,

$$G_\infty(\mathbf{R}; T) \sim \exp[-R/\xi_\infty(T)], \quad R \to \infty. \tag{4.35}$$

Hence, one defines

$$\xi_\infty^{(1)}(T) = - \lim_{R \to \infty} [R/\ln G_\infty(\mathbf{R}; T)]. \tag{4.36}$$

Alternatively, one may consider the second moment of the bulk pair correlation function and define the effective correlation radius,

$$\xi_\infty^{(2)}(T) = \left[\sum_{\mathbf{R}} R^2 G_\infty(\mathbf{R}; T) / \sum_{\mathbf{R}} G_\infty(\mathbf{R}; T) \right]^{1/2} \qquad (4.37)$$

If the asymptotic form (4.35) holds, then the two definitions (4.36) and (4.37) are equivalent in the sense that $\xi_\infty^{(1)}(T) \sim \xi_\infty^{(2)}(T) \sim t^{-\nu}$ as $t \to 0^+$.

Suppose that, eventually in the presence of long-range interactions, the function $G_\infty(\mathbf{R}; T)$ decays as $R^{-d+\sigma}$ when $R \to \infty$, for $T > T_c$. Then, if $\sigma < 2$, definitions (4.36) and (4.37) yield $\xi_\infty^{(1)}(T) = \xi_\infty^{(2)}(T) = \infty$ for all $T > T_c$. This is one of the difficulties we overcome by following the approach developed in [Brankov and Tonchev (1992)]. Its advantage is that the hypotheses are formulated in a way which treats on equal grounds the cases of short- and long-range interactions, as well as the standard and modified (above the upper critical dimensionality) finite-size scaling. Our approach exploits the idea of [Brézin (1982)] that when the exponential form (4.35) does not apply, there still exists some bulk *characteristic length* $\lambda_\infty(T)$ which: (i) determines the length-scale of variations of the correlation function and (ii) diverges at the critical point. We give an explicit definition of such a length and show its role in the formulation of finite-size scaling.

4.4.2 *Finite-size correlation length*

Obviously, in the case of a lattice system confined to a fully finite region $\Lambda \subset Z^d$, one can always define a finite-size correlation length $\xi_L^{(2)}(T)$ through the second moment of the finite-size pair correlation function

$$\xi_L^{(2)}(T) = \left[\sum_{\mathbf{R} \in \Lambda} R^2 G_L(\mathbf{R}; T) / \sum_{\mathbf{R} \in \Lambda} G_L(\mathbf{R}; T) \right]^{1/2} \qquad (4.38)$$

Even this definition involves some subtleties which have been discussed in the literature, see [Brézin (1982)], [Shapiro and Rudnick (1986)]. For example, since under periodic boundary conditions $G_L(\mathbf{R}; T)$ has the periodicity of the lattice, it has been proposed [Shapiro and Rudnick (1986)] that $R^2 = \sum_{\alpha=1}^d R_\alpha^2$ should be replaced by a length function which has the same periodicity but reduces to R^2 for $L \gg R$, e.g.,

$$\frac{L^2}{2\pi^2}\left[d - \sum_{\alpha=1}^{d} \cos\left(\frac{2\pi R_\alpha}{L}\right) \right].$$

Fortunately, each of these definitions becomes equivalent to the effective correlation range introduced [Fisher and Burford (1967)] in the large-L limit at a fixed $T > T_c$. On the other hand, each of them fails in the case of a long-range interaction when the second moment (4.38) diverges with $L \to \infty$, and the notion of a bulk correlation length needs a generalization.

4.5   Basic hypotheses

Let us start by realizing that a finite system has three characteristic length scales: a linear size L, a length $\lambda_L(T)$ which determines the scale of variation of the correlations, and a microscopic length a (e.g., a is the lattice spacing).

Our first hypothesis consists of three assumptions. The first one defines the characteristic length $\lambda_L(T)$ from the large-distance asymptotic behavior of the pair correlation function $G_L(\mathbf{R}, T)$. The second assumption is the homogeneity of any thermodynamic quantity of the finite-size system, such that its bulk limit is singular at $T = T_c$, as a function of the two dimensionless ratios $\lambda_L(T)/a$ and L/a. The third assumption is the existence of finite thermodynamic limits for the characteristic length and the pair correlation function.

The second hypothesis concerns the relationship between the finite-size length $\lambda_L(T)$ and its bulk limit $\lambda_\infty(T)$; the corresponding homogeneity assumption involves the dimensionless ratios $\lambda_\infty(T)/a$ and L/a.

4.5.1   *Hypothesis A*

Assumption A.1. When $L/a \gg 1$, $R/a \gg 1$ and $T \to T_c$, the leading-order asymptotic form of the finite-size pair correlation function is

$$G_L(\mathbf{R}; T) \simeq A(T) R^{-d+2-\eta} X_G(\mathbf{R}/\lambda_L, L/\lambda_L), \tag{4.39}$$

where $A(T)$ is a regular at $T = T_c$, temperature-dependent amplitude, $\lambda_L = \lambda_L(T)$ is some finite-size characteristic length, and $X_G(\cdot, \cdot)$ is a universal function.

We emphasize that the length $\lambda_L(T)$ is defined by the asymptotic form (4.39). It may, or may not coincide with the finite-size correlation length $\xi_L = \xi_L(T)$ defined, for example, by Eq. (4.38). The assumption (4.39) is equivalent to the assumption of ordinary homogeneity of the pair correlation function $G(\mathbf{R}/a, \lambda_L/a, L/a)$ with respect to the dimensionless ratios $\mathbf{R}/a$, λ_L/a and L/a. Similarly, we adopt

Assumption A.2. Any thermodynamic quantity $P_L(T)$ of the finite system, such that the corresponding bulk quantity $P_\infty(T)$ is singular at the critical point, is an ordinary homogeneous function of the dimensionless ratios λ_L/a and L/a near the bulk critical point:

$$P_L(T) = X_P(\lambda_L/a, L/a) \simeq (\lambda_L/a)^\theta \tilde{X}(L/\lambda_L) = (L/a)^\theta \bar{X}(L/\lambda_L). \quad (4.40)$$

Turning back to the assumption (4.39), it is natural to make

Assumption A.3. At any fixed $T \neq T_c$, there exists a finite bulk limit for the finite-size characteristic length,

$$\lim_{L \to \infty} \lambda_L(T) = \lambda_\infty(T), \quad (4.41)$$

and for the pair correlation function,

$$\lim_{L \to \infty} G_L(\mathbf{R}; T) = G_\infty(\mathbf{R}; T) \simeq A(T) R^{-d+2-\eta} X_G(\mathbf{R}/\lambda_\infty, \infty). \quad (4.42)$$

In view of the relationship between the bulk susceptibility per spin $\chi_\infty(T)$ and the bulk pair correlation function (4.42),

$$\chi_\infty(T) = (k_B T)^{-1} \sum_{\mathbf{R} \in \mathbb{Z}^d} G_\infty(\mathbf{R}; T), \quad (4.43)$$

assuming that $\lambda_\infty(T) \gg a$ and $\eta < 2$, we obtain

$$\chi_\infty(T) \sim \lambda_\infty^{2-\eta}(T). \quad (4.44)$$

Since $\chi_\infty(T) \sim t^{-\gamma}$ when $t \to 0^+$, the bulk quantity $\lambda_\infty(T)$ diverges as

$$\lambda_\infty(T) \sim t^{-\nu}, \qquad t \to 0^+, \quad (4.45)$$

where $\nu = \gamma/(2-\eta)$ is the critical exponent usually related to the correlation length singularity.

We need now a relationship between $\lambda_L(T)$ and the bulk, diverging at T_c, characteristic length $\lambda_\infty(T)$, which plays the role of a natural temperature variable. Most consistent with that role is the following hypothesis.

4.5.2   *Hypothesis B*

For a large system, $L \gg a$, near the bulk critical point, $\lambda_\infty(T)/a \gg 1$, the leading-order asymptotic form of the finite-size characteristic length,

$$\lambda_L(T)/a \simeq X_\lambda(\lambda_\infty(T)/a, L/a), \tag{4.46}$$

is a homogeneous function of the dimensionless ratios $\lambda_\infty(T)/a$ and L/a.

However, there are two possibilities for the homogeneity of the function $X_\lambda(\cdot,\cdot)$, namely, ordinary and generalized homogeneity. The consequences of each of the two cases will be considered separately.

Case (i). In the case of an ordinary homogeneity, one has

$$X_\lambda(x,y) = x^\theta \tilde{X}_\lambda(y/x) = y^\theta \bar{X}_\lambda(y/x). \tag{4.47}$$

Taking into account the existence of the thermodynamic limit, see Eq. (4.41), we conclude that $\theta = 1$ in Eq. (4.47) and our second hypothesis amounts to the relationship

$$\lambda_L(T) \simeq \lambda_\infty(T)\tilde{X}_\lambda(L/\lambda_\infty) = L\bar{X}_\lambda(L/\lambda_\infty). \tag{4.48}$$

From assumptions (4.39) and (4.48) the usual finite-size scaling hypothesis for the pair correlation function follows, with $\lambda_\infty(T)$ playing the role of the correlation length:

$$G_L(\mathbf{R};T) \simeq A(T)R^{-d+2-\eta}Y_G(\mathbf{R}/\lambda_\infty, L/\lambda_\infty). \tag{4.49}$$

Here the scaling function $Y_G(\cdot,\cdot)$ can be expressed in terms of the functions $X_G(\cdot,\cdot)$, defined in Eq. (4.39), and $\tilde{X}_\lambda(\cdot)$, defined in Eq. (4.40).

Similarly, from Eqs. (4.40) and (4.48), for the thermodynamic quantity $P_L(T)$ we obtain the standard finite-size scaling hypothesis:

$$P_L(T) \simeq (\lambda_\infty/a)^\theta \tilde{X}_\lambda(L/\lambda_\infty) = (L/a)^\theta \bar{X}_\lambda(L/\lambda_\infty), \tag{4.50}$$

where $\theta = \rho/\nu$ and ρ is the critical exponent of the bulk quantity $P_\infty(T)$, see Eq. (4.5).

Case (ii). In the case of a generalized homogeneity, one has

$$X_\lambda(x,y) = x^\theta \tilde{X}_\lambda(y^\omega/x) = y^{\theta\omega} \bar{X}_\lambda(y^\omega/x). \tag{4.51}$$

For the finite-size characteristic length λ_L we have to set again $\theta = 1$, due to the limit (4.41). Thus, for $\omega \neq 1$ we obtain from (4.46) *the modified finite-size scaling relationship for the characteristic length*

$$\lambda_L(T) \simeq \lambda_\infty(T)\tilde{X}_\lambda\left((L/a)^\omega a/\lambda_\infty\right) = a(L/a)^\omega \bar{X}_\lambda\left((L/a)^\omega a/\lambda_\infty\right). \quad (4.52)$$

On the one hand, this may be regarded as *breakdown of critical finite-size scaling*, since the microscopic length a remains (when $\omega \neq 1$) in the argument of the scaling function. On the other hand, equation (4.52) may be interpreted as appearance of another diverging bulk length, see also Chapter 6, namely,

$$l_\infty(T) = \left[\lambda_\infty(T)/a\right]^{1/\omega} a, \quad (4.53)$$

which controls finite-size effects through the ratio $L/l_\infty(T)$; we call it *bulk coherence length*. By defining the *finite-size coherence length*,

$$l_L(T) = \left[\lambda_L(T)/a\right]^{1/\omega} a, \quad (4.54)$$

we can rewrite Eq. (4.52) in an equivalent finite-size scaling form, compare with Eq. (4.48),

$$l_L(T) \simeq l_\infty(T)\tilde{X}_l(L/l_\infty) = L\bar{X}_l(L/l_\infty). \quad (4.55)$$

From the assumptions (4.39) and (4.52), using the definition (4.53), we obtain *the modified finite-size scaling prediction for the pair correlation function:*

$$
\begin{aligned}
G_L(\mathbf{R};T) &\simeq A(T)R^{-d+2-\eta}X_G\left(\frac{\mathbf{R}}{\lambda_\infty\tilde{X}_\lambda(L^\omega/l_\infty^\omega)}, \frac{L}{\lambda_\infty\tilde{X}_\lambda(L^\omega/l_\infty^\omega)}\right) \\
&:= A(T)R^{-d+2-\eta}Z_G(\mathbf{R}/\lambda_\infty, L/\lambda_\infty, L/l_\infty).
\end{aligned}
\quad (4.56)$$

Note that from Eq. (4.52), in the limit $L \to \infty$ at fixed $T \neq T_c$, it follows that $\tilde{X}_\lambda(\infty) = 1$, and the scaling function $Z_G(x,y,z)$ in the right-hand side of Eq. (4.56) has the limit

$$\lim_{z\to\infty} Z_G(x,y,z) - X_G(x,y). \quad (4.57)$$

Similarly, from equations (4.40) and (4.52) we obtain *the modified finite-size scaling prediction for the thermodynamic quantity $P_L(T)$,*

$$P_L(T) \simeq (L/a)^\theta \bar{X}_P\left(\frac{L}{\lambda_\infty\tilde{X}_\lambda(L^\omega/l_\infty^\omega)}\right) := (L/a)^\theta Z_P(L/\lambda_\infty, L/l_\infty).$$

$$(4.58)$$

Here again, since $\tilde{X}_\lambda(\infty) = 1$,

$$\lim_{y \to \infty} Z_P(x, y) = \bar{X}_P(x). \tag{4.59}$$

The appearance of the two dimensionless ratios L/λ_∞ and L/l_∞ in the arguments of the finite-size scaling function implies the existence of two asymptotically distinct neighborhoods of the critical temperature. When $\omega > 1$ in the outer neighborhood is defined by $t \sim L^{-1/\nu}$; then

$$L/\lambda_\infty(T) = O(1), \quad L/l_\infty(T) = O(L^{(\omega-1)/\omega}) \to \infty, \tag{4.60}$$

when $L \to \infty$. In this asymptotic regime Eq. (4.52) yields $\lambda_L(T) \simeq \lambda_\infty(T)$ and, due to (4.57) and (4.59), Eqs. (4.56) and (4.58) reduce to their standard finite-size scaling form (4.49) and (4.50), respectively.

The closer neighborhood of the critical point is reached at $t \sim L^{-\omega/\nu}$, when

$$L/\lambda_\infty(T) = O(L^{1-\omega}) \to 0, \quad L/l_\infty(T) = O(1), \tag{4.61}$$

as $L \to \infty$. Then, if $Z_P(x, y) \sim x^p \tilde{Z}_P(y)$, as $x \to 0$, we obtain

$$P_L(T) \simeq (L/a)^{\theta^*} \tilde{Z}_P(L/\lambda_\infty), \tag{4.62}$$

where

$$\theta^* = \theta - (\omega - 1)p \tag{4.63}$$

is called *the anomalous dimension* of the thermodynamic quantity Q, see also Chapter 6.

4.6   Extension to several scaling fields

Here we show how the above finite-size scaling hypotheses can be extended to take into account the dependence on several scaling variables (fields), see Section 1.6.5, denoted by $t, h, \zeta_1, \zeta_2, \ldots$, where t is the temperature variable and, for example, h is the magnetic field variable. It will be assumed that the critical point corresponds to vanishing values at all the relevant fields, see Section 1.6.4. Let the general scaling form of the singular part of the bulk free energy density be [Huang (1987)]

$$f_{sing}(t, h, \zeta_1, \zeta_2, \ldots) \simeq |t|^{2-\alpha} X_f^\pm(ht^{-\Delta}, \zeta_1 t^{-\Delta_1}, \zeta_2 t^{-\Delta_2}, \ldots), \tag{4.64}$$

where the superscript $\pm$ refers to $\pm t > 0$. The exponents Δ, Δ_1, Δ_2,..., are called *crossover exponents*, since they control the relative importance of the scaling fields as the critical point is approached. Obviously, Eq. (4.64) is asymptotically equivalent to a generalized homogeneity property of the function in its left-hand side, compare with Eq. (4.51),

$$f_{\text{sing}}(t, h, \zeta_1, \zeta_2, \ldots) =$$
$$|t|^{p/a_t} f_{\text{sing}}(\text{sign } t, ht^{-a_h/a_t}, \zeta_1 t^{-a_1/a_t}, \zeta_2 t^{-a_2/a_t}, \ldots), \quad (4.65)$$

with exponents

$$p/a_t = 2 - \alpha, \ a_h/a_t = \Delta, \ a_1/a_t = \Delta_1, \ a_2/a_t = \Delta_2, \ \ldots \quad (4.66)$$

Finite-size scaling can be incorporated in the above scheme by introducing a new scaling field, namely, L^{-1} which also vanishes at the bulk critical point. Thus, for the finite-size free energy density we may write

$$f_{L,\text{sing}}(t, h, \zeta_1, \zeta_2, \ldots) = g(t, h, \zeta_1, \zeta_2, \ldots, L^{-1}). \quad (4.67)$$

Now we show that the asymptotic homogeneity properties of the function g in the right-hand side of the above equation are uniquely determined by the thermodynamic scaling (4.65) and the finite-size scaling hypothesis about the scaled temperature variable. Indeed, by taking the limit $L \to \infty$ in Eq. (4.67) and comparing the result with Eq. (4.64), we obtain

$$g(t, h, \zeta_1, \zeta_2, \ldots, 0) \simeq |t|^{2-\alpha} g(\text{sign } t, ht^{-\Delta}, \zeta_1 t^{-\Delta_1}, \zeta_2 t^{-\Delta_2}, \ldots, 0). \quad (4.68)$$

On the other hand, if all the scaling fields, except the temperature field t, are set to their critical values, from *Hypothesis B, Case (i)*, it follows that $L/\lambda_\infty \simeq Lt^\nu$ and, therefore,

$$g(t, 0, 0, 0, \ldots, L^{-1}) \simeq |t|^{2-\alpha} g(1, 0, 0, 0, \ldots, L^{-1}t^{-\nu}). \quad (4.69)$$

This equation defines the crossover exponent Δ_L for the finite-size scaling field L^{-1},

$$\Delta_L \equiv a_L/a_t = \nu, \quad (4.70)$$

which turns out to be relevant to criticality, since $\nu > 0$.

Thus, from Eqs. (4.67) - (4.69) we conclude that in the general case

$$f_{L,\text{sing}}(t, h, \zeta_1, \zeta_2, \ldots) \simeq$$
$$|t|^{2-\alpha} g(\text{sign } t, ht^{-\Delta}, \zeta_1 t^{-\Delta_1}, \zeta_2 t^{-\Delta_2}, \ldots, L^{-1}t^{-\nu}), \quad (4.71)$$

which defines all the independent exponents of the generalized homogeneity property of the function in the right-hand side of Eq. (4.67).

Often it is more convenient to use the so-called L-scaled form of finite-size scaling, which is given in terms of variables linear in the scaling fields. To obtain it, we start with the generalized homogeneity property,

$$g(\lambda^{a_t} t, \lambda^{a_h} h, \lambda^{a_1}\zeta_1, \lambda^{a_2}\zeta_2, \ldots, \lambda^{a_L}L^{-1}) = \lambda^p g(t, h, \zeta_1, \zeta_2, \ldots, L^{-1}). \quad (4.72)$$

By setting here $\lambda = L^{1/a_L}$ and taking into account that due to Eqs. (4.66) and (4.70)

$$p/a_L = (2-\alpha)/\nu, \quad a_t/a_L = 1/\nu, \quad a_h/a_L = \Delta/\nu,$$
$$a_1/a_L = \Delta_1/\nu, \quad a_2/a_L = \Delta_2/\nu, \quad \ldots, \quad (4.73)$$

we obtain

$$f_{L,\text{sing}}(t, h, \zeta_1, \zeta_2, \ldots) \simeq$$
$$L^{-(2-\alpha)/\nu}g(tL^{1/\nu}, hL^{\Delta/\nu}, \zeta_1 L^{\Delta_1/\nu}, \zeta_2 L^{\Delta_2/\nu}, \ldots, 1). \quad (4.74)$$

Note that the familiar normalization factor L^{-d} for the free energy density of a fully finite system follows from the hyperscaling relation $d\nu = 2 - \alpha$.

In a similar way one can extend *Case (ii)* of *Hypothesis B* to allow for several scaling fields.

4.7   Testing finite-size scaling by the spherical models

Traditionally, basic ideas and methods in the theory of finite-size scaling are tested by the example of the lattice mean spherical model, or the classical $O(n)$ model in the $n \to \infty$ limit, see Section 3.1, which are the simplest model exactly solved at any space dimensionality, various finite-size geometries and boundary conditions.

4.7.1 *General finite-size expressions*

It is important to emphasize that the thermodynamical properties, as well as the leading-order finite-size effects, depend on the interaction potential $J_\Lambda(\mathbf{r})$ only through its long-wavelength asymptotic form. As it is mentioned in Section 3.1.2, the Fourier transform $\hat{J}(\mathbf{k})$ of a power-law interaction of the type (3.19) behaves as

$$\hat{J}(\mathbf{k}) \simeq \hat{J}(0) \times \begin{cases} (1 - \rho_\sigma k^\sigma), & \text{if } 0 < \sigma < 2, \\ (1 - \rho_2 k^2), & \text{if } \sigma \geq 2, \end{cases} \qquad (4.75)$$

when $k = |\mathbf{k}| \to 0$. In this chapter we shall consider the case of $\sigma = 2$, with $\hat{J}(0) = 2d$ and $\rho_2 = (2d)^{-1}$, which is related *ab initio* to short-range interactions. The reason to confine ourselves to the asymptotic form (4.75) with $\sigma = 2$, rather than to the exact Fourier transform (3.18) of the nearest-neighbors interaction, is twofold: *(a)* this form is most popular in field-theoretical models, and *(b)* the corresponding results demonstrate the essential role of the long-wavelength asymptotic behavior.

In studying finite systems of increasing size L_ν, $\nu = 1, \ldots, d$, we fix the function $\hat{J}(\mathbf{q})$ for $\mathbf{k} \in (-\pi, \pi]^d$ and account for the size of the system by choosing the appropriate discrete set of vectors $\mathbf{k} = \{k_1, \ldots, k_d\}$ in the set $\mathcal{B}_\Lambda$ defined by Eq. (3.15) for L_ν odd integers.

The basic expressions for our further analysis are Eqs. (3.22) and (3.23), which are specified here to the case of short-range interaction. Thus, the free energy density of the finite-size mean spherical model is defined by

$$\beta f_\Lambda^{m.s.}(K, h) = \frac{1}{2} \sup_{\phi > 0} \left\{ \frac{1}{|\Lambda|} \sum_{\mathbf{k} \in \mathcal{B}_\Lambda} \ln(\phi + |\mathbf{k}|^2) - K\phi - \frac{h^2}{K\phi} \right\}$$
$$+ \frac{1}{2} \ln \frac{K}{2\pi} - dK. \qquad (4.76)$$

The supremum in the right-hand side of the above equation is attained at the solution $\phi = \phi_\Lambda(K, h)$ of the mean spherical constraint

$$\frac{1}{|\Lambda|} \sum_{\mathbf{k} \in \mathcal{B}_\Lambda} \frac{1}{\phi + |\mathbf{k}|^2} + \frac{h^2}{K\phi^2} = K. \qquad (4.77)$$

The explicit expression for the pair correlation function is

$$G_\Lambda(\mathbf{r}; K, h) = \frac{1}{K|\Lambda|} \sum_{\mathbf{k} \in \mathcal{B}_\Lambda} \frac{\exp(i\,\mathbf{k} \cdot \mathbf{r})}{\phi_\Lambda(K, h) + |\mathbf{k}|^2}. \qquad (4.78)$$

It is in place here to mention that the thermodynamic limit leads to great simplifications in the analytical expressions. In the case of a finite system the derivation of an analytical result is a much more complicated task, since the L-dependence enters via the size and shape of the region Λ, being sensitive to the boundary conditions as well. Nevertheless, we shall demonstrate that it is possible to analyze rigorously the asymptotic behavior at large L of the finite d-fold sums which enter into the above equations.

4.7.2 *Leading finite-size behavior*

To test the hypotheses of finite-size scaling at criticality, one has to analyze first the asymptotic behavior of the solution $\phi_\Lambda(K, h)$ of the mean spherical constraint Eq. (4.77) when both the temperature and the magnetic field approach their critical values simultaneously with $L_\nu \to \infty$, $\nu = 1, \ldots, d$. Then one has to evaluate the asymptotic form of the sums in the expressions for the free energy density, Eq. (4.76), and the pair correlation function, Eq. (4.78), taken at $\phi = \phi_\Lambda(K, h)$. One of the frequently used methods of analysis is based on the application of the Poisson summation formula

$$\sum_{n=a}^{b} f(n) = \sum_{k=-\infty}^{\infty} \int_a^b dn \, e^{i2\pi kn} f(n) + \frac{1}{2}[f(a) + f(b)] . \tag{4.79}$$

Note that the term with $k = 0$ in the right-hand side of the above equation gives the integral approximation to the sum in the left-hand side. The generalization of Eq. (4.79) to d-fold sums allows one to explicitly separate out the bulk contribution. The analysis simplifies considerably, if one first uses some elementary identities to factorise the d-fold sums. For example, with the aid of

$$\frac{1}{1+z} = \int_0^\infty dx \, e^{-zx} e^{-x}, \qquad \text{Re } z > -1, \tag{4.80}$$

one obtains

$$\begin{aligned}
\Sigma_\Lambda^{d,2}(\mathbf{R}; \phi) &\equiv \frac{1}{|\Lambda|} \sum_{\mathbf{k} \in \mathcal{B}_\Lambda} \frac{\exp(i\mathbf{k} \cdot \mathbf{R})}{\phi + |\mathbf{k}|^2} \\
&= \frac{1}{\phi} \int_0^\infty dx \, e^{-x} \prod_{j=1}^{d} Q_{L_j}(x/\phi, R_j),
\end{aligned} \tag{4.81}$$

where

$$Q_L(z, R) = \frac{1}{L} \sum_{n=-(L-1)/2}^{(L-1)/2} \exp\left[-\left(\frac{2\pi n}{L}\right)^2 z + i\left(\frac{2\pi n}{L}\right) R\right] . \tag{4.82}$$

Now the Poisson summation formula (4.79) can be applied independently to each of the sums (4.82) in the right-hand side of Eq. (4.81). Thus, the d-dimensional problem is reduced to an effectively one-dimensional problem. Moreover, the general case of a system with finite as well as infinite dimensions, $\Lambda = L^{d-d'} \times \infty^{d'}$, can be obtained by setting $L_\nu = L$ for $\nu = 1, \ldots, d - d'$ and taking the limit $L_\nu \to \infty$ for $\nu = d - d' + 1, \ldots, d$.

The case of the free energy density can be treated similarly, see [Singh and Pathria (1985a)], if instead of (4.80) one uses

$$\ln(1 + z) = \int_0^\infty dx\, (1 - e^{-zx}) \frac{e^{-x}}{x}, \qquad \text{Re } z > -1. \tag{4.83}$$

The implementation of the identity (4.83) with

$$z = |\mathbf{k}|^2/\phi = \left(\frac{2\pi}{\phi^{1/2}}\right)^2 \sum_{j=1}^d \left(\frac{n_j}{L_j}\right)^2, \tag{4.84}$$

yields the following factorized expressions for the sum, see Eq. (4.76),

$$U_\Lambda^{d,2}(\phi) \equiv \frac{1}{|\Lambda|} \sum_{\mathbf{k} \in \mathcal{B}_\Lambda} \ln(\phi + |\mathbf{k}|^2)$$

$$= \ln \phi + \int_0^\infty dx\, \frac{e^{-x}}{x} \left\{ 1 - \prod_{j=1}^d Q_{L_j}(x\phi^{-1}, 0) \right\}. \tag{4.85}$$

Thus the initial problem has been reduced to the evaluation of the asymptotic behavior of standard sums of the type (4.82). First we outline the asymptotic analysis of these sums. The application of the Poisson summation formula to the sum (4.82) yields

$$Q_L(z, R) \simeq \sum_{k=-\infty}^\infty \frac{1}{L} \int_{-L/2}^{L/2} dn \exp\left[-\left(\frac{2\pi n}{L}\right)^2 z + i\, 2\pi \left(\frac{R}{L} + k\right) n \right]$$

$$= \frac{1}{\sqrt{4\pi z}} \sum_{k=-\infty}^\infty \exp\left[-\frac{(kL + R)^2}{4z} \right] \Phi\left(\pi z^{1/2}; \frac{|kL + R|}{z^{1/2}} \right), \tag{4.86}$$

where we have ignored terms of order $O(L^{-1} \exp(-\pi^2 z))$ and introduced the function

$$\Phi(u; v) := \frac{2}{\sqrt{\pi}\, v} e^{v^2/4} \int_0^{uv} dt\, e^{-(t/v)^2} \cos t. \tag{4.87}$$

Note that $\Phi(u;v)$ is an even function of v, and $\Phi(u;0) = \mathrm{erf}(u)$, where $\mathrm{erf}(\cdot)$ is the error function. For all the terms in the sum over k in the right-hand side of Eq. (4.86), the upper limit of the integral in Eq. (4.87) tends to infinity uniformly with respect to $z = x\phi^{-1}$ in the following cases: (a) when the summation index $k \neq 0$ and $L \to \infty$; (b) when the summation index $k = 0$ and $R \to \infty$: $u|v| = \pi|kL + R| \to \infty$. Taking into account the integral

$$\int_0^\infty \mathrm{d}t\, \mathrm{e}^{-(t/v)^2} \cos t = \frac{1}{2}\pi^{1/2} v\, \mathrm{e}^{-v^2/4}, \qquad v > 0, \qquad (4.88)$$

we conclude that the function $\Phi(u;v)$ in the right-hand side of Eq. (4.86) tends to unity. Therefore, the leading asymptotic form of $Q_L(z,R)$ as $L \to \infty$ and $R \to \infty$, $|R| < L$, is given for all $z > 0$ by the expression

$$Q_L(z,R) \simeq \frac{1}{\sqrt{4\pi z}} \sum_{k=-\infty}^{\infty} \exp\left[-\frac{(kL+R)^2}{4z}\right]. \qquad (4.89)$$

However, if $R = 0$ and $L \to \infty$, then

$$Q_L(z,0) \simeq \frac{1}{\sqrt{4\pi z}} \left\{ \mathrm{erf}(\pi z^{1/2}) + \sum_{k=-\infty}^{\infty}{}' \exp\left[-\frac{(kL)^2}{4z}\right] \right\}, \qquad (4.90)$$

where the primed summation means that the term with zero summation index has been omitted. In the limit $L \to \infty$ one obtains

$$
\begin{aligned}
Q_\infty(z,R) &\simeq (4\pi z)^{-1/2} \exp(-R^2/4z), \qquad R \gg 1, \\
Q_\infty(z,0) &= (4\pi z)^{-1/2} \mathrm{erf}(\pi z^{1/2}).
\end{aligned}
\qquad (4.91)
$$

From Eq. (4.89) with $z = x\phi^{-1}$ and Eq. (4.81), one obtains that when $L \to \infty$ and $R \to \infty$

$$\Sigma_\Lambda^{d,2}(\mathbf{R};\phi) \simeq \frac{\phi^{d/2-1}}{(4\pi)^{d/2}} \int_0^\infty \mathrm{d}x\, x^{-d/2} \mathrm{e}^{-x}$$

$$\times \sum_{\mathbf{k}(d)} \exp\left[-\frac{\phi}{4x} \sum_{j=1}^{d}(k_j L_j + R_j)^2\right], \qquad (4.92)$$

where the summation runs over all vectors $\mathbf{k} = \{k_1, \ldots, k_d\}$ with d integer components, $\mathbf{k} \in Z^d$. In the thermodynamic limit $L_j \to \infty$, $j = 1, \ldots, d$, only the term $\mathbf{k} = 0$ survives and one obtains

$$
\begin{aligned}
\Sigma_{\infty}^{d,2}(\mathbf{R};\phi) &\simeq \frac{\phi^{d/2-1}}{(4\pi)^{d/2}} \int_0^{\infty} dx\, x^{-d/2} \exp\left(-x - \frac{\phi|\mathbf{R}|^2}{4x}\right) \\
&= \frac{R^{2-d}}{(2\pi)^d} \int_0^{\infty} dx\, \frac{x^{d/2}\, J_{d/2-1}(x)}{x^2 + R^2\phi} \\
&= \frac{R^{2-d}}{(2\pi)^d}(R\phi^{1/2})^{\frac{d-2}{2}} K_{d/2-1}(R\phi^{1/2}).
\end{aligned} \tag{4.93}
$$

Hence, the result for the bulk correlation function (3.49) at $\sigma = 2$ follows, together with the identification of the bulk correlation length

$$
\xi_{\infty}(K,h) = \phi_{\infty}^{-1/2}(K,h). \tag{4.94}
$$

Let us turn now to the case $|\mathbf{R}| = 0$, which is of relevance to the mean spherical constraint Eq. (4.77). Note that, in view of Eq. (4.90), the product $\prod_{j=1}^{d} Q_{L_j}(z,0)$ contains sums of terms of the form

$$
(4\pi z)^{-d/2} \left[\mathrm{erf}\left(\pi z^{1/2}\right)\right]^m \exp\left\{-\sum_{j=1}^{d-m} \frac{(k_j L_j)^2}{4z}\right\}, \tag{4.95}
$$

with $1 \leq m \leq d-1$ and $k_j \neq 0$, $j = 1,\ldots,d-m$. In such terms the error function $\mathrm{erf}(\pi z^{1/2})$ can be replaced by unity, since the exponential function in the right-hand side of Eq. (4.95) cuts off the contribution from values of $z^{1/2} \ll L_j$. Therefore, by using Eq. (4.89) with $z = x\phi^{-1}$ one obtains

$$
\begin{aligned}
\prod_{j=1}^{d} &Q_{L_j}(x\phi^{-1},0) \simeq \\
&\frac{\phi^{d/2}}{(4\pi x)^{d/2}} \left\{\left[\mathrm{erf}\left(\frac{\pi x^{1/2}}{\phi^{1/2}}\right)\right]^d + \sum_{\mathbf{k}(d)}{}' \exp\left[-\frac{\phi}{4x}\sum_{j=1}^{d} L_j^2 k_j^2\right]\right\}.
\end{aligned} \tag{4.96}
$$

Hence, the leading asymptotic behavior of $\Sigma_{\Lambda}^{d,2}(0;\phi)$ is given by the expression

$$
\begin{aligned}
\Sigma_{\Lambda}^{d,2}(0;\phi) &\simeq W_{d,2}(\phi) \\
&+ \frac{\phi^{d/2-1}}{(4\pi)^{d/2}} \int_0^{\infty} dx\, x^{-d/2} \sum_{\mathbf{k}(d)}{}' \exp\left[-x - \frac{\phi}{4x}\sum_{j=1}^{d} L_j^2 k_j^2\right].
\end{aligned} \tag{4.97}
$$

To obtain the bulk term in the above equation,

$$W_{d,2}(\phi) := (2\pi)^{-d} \int_{-\pi}^{\pi} \cdots \int_{-\pi}^{\pi} d^d k \, \frac{1}{\phi + |\mathbf{k}|^2}, \tag{4.98}$$

we have used the identity

$$\frac{\phi^{d/2-1}}{(4\pi)^{d/2}} \int_0^{\infty} dx \, x^{-d/2} \exp(-x) \left[\mathrm{erf}\left(\frac{\pi x^{1/2}}{\phi^{1/2}}\right) \right]^d \equiv W_{d,2}(\phi). \tag{4.99}$$

Expression (4.97) will be studied in greater detail in the next section. The leading finite-size behavior of the sum (4.85) follows in a similar way,

$$U_\Lambda^{d,2}(\phi) \simeq U_{d,2}(\phi)$$

$$- \frac{\phi^{d/2}}{(4\pi)^{d/2}} \int_0^{\infty} \frac{dx}{x^{d/2+1}} {\sum_{\mathbf{k}(d)}}' \exp\left[-x - \frac{\phi}{4x} \sum_{j=1}^{d} L_j^2 k_j^2 \right], \tag{4.100}$$

where the function $U_{d,2}(\phi)$, defined by

$$U_{d,2}(\phi) := (2\pi)^{-d} \int_{-\pi}^{\pi} \cdots \int_{-\pi}^{\pi} d^d k \, \ln(\phi + |\mathbf{k}|^2), \tag{4.101}$$

can be written as

$$U_{d,2}(\phi) \equiv \ln \phi$$

$$+ \int_0^{\infty} \frac{dx}{x} e^{-x} \left\{ 1 - \frac{\phi^{d/2}}{(4\pi x)^{d/2}} \left[\mathrm{erf}\left(\frac{\pi x^{1/2}}{\phi^{1/2}}\right) \right]^d \right\}. \tag{4.102}$$

Equations (4.92), (4.97) and (4.100) constitute the basis of our further finite-size scaling analysis.

4.7.3　*Finite-size scaling for the correlation function*

One of the basic hypotheses of finite-size scaling at criticality, see *Assumption A.1*, is that the leading asymptotic form of the pair correlation function for a system of finite size L, as $L/a \to \infty$, $R/a \to \infty$ and $T \to T_c$, is given by

$$G_L(\mathbf{R}; T) \simeq A(T) R^{-d+2-\eta} X_G(\mathbf{R}/\xi_L; L/\xi_L). \tag{4.103}$$

From Eqs. (4.78) and (4.92) it follows that the spin-spin correlation function takes indeed the asymptotic form (4.103), generalized to account for the different linear sizes of Λ, namely, $L_j, j = 1, \ldots, d$. We can identify now the finite-size correlation length,

$$\xi_\Lambda(K, h) = \phi_\Lambda^{-1/2}(K, h), \tag{4.104}$$

the non-universal, regular at $T = T_c$ amplitude,

$$A(T) = \frac{k_B T}{\rho_2 \hat{J}(0)}, \tag{4.105}$$

the critical exponent $\eta = 0$, and the universal finite-size scaling function,

$$X_G^{d,2}(\mathbf{r}; y_1, \ldots, y_d) =$$
$$\frac{r^{-d+2}}{(4\pi)^{d/2}} \int_0^\infty \frac{dx}{x^{d/2}} \sum_{\mathbf{k}(d)} \exp\left[-x - \frac{1}{4x} \sum_{j=1}^d (y_j k_j + r_j)^2\right], \tag{4.106}$$

where $\mathbf{r} = \mathbf{R}/\xi_\Lambda$, $y_j = L_j/\xi_\Lambda$ $(j = 1, \ldots, d)$.

In the case when Λ is a cube of side L, Eq. (4.106) simplifies to:

$$X_G^{d,2}(\mathbf{r}; y) = \frac{r^{-d+2}}{(4\pi)^{d/2}} \int_0^\infty \frac{dx}{x^{d/2}} \sum_{\mathbf{k}(d)} \exp\left[-x - \frac{(y\mathbf{k} + \mathbf{r})^2}{4x}\right], \tag{4.107}$$

with $y = L/\xi_\Lambda$.

The case of a cylindrical domain $\Lambda = L^{d-d'} \times \infty^{d'}$, with $d - d' > 0$, can be obtained from Eq. (4.106) by setting $L_j = L$ for $j = 1, \ldots, d - d'$, and taking the limit $L_j \to \infty$ for all $j = d - d' + 1, \ldots, d$. The resulting universal finite-size scaling function is

$$X_G^{d,d',2}(\mathbf{r}; y) =$$
$$\frac{r^{-d+2}}{(4\pi)^{d/2}} \int_0^\infty \frac{dx}{x^{d/2}} \sum_{\mathbf{k}(d-d')} \exp\left[-x - \frac{(y\mathbf{k} + \mathbf{r}_\perp)^2 + \mathbf{r}_\parallel^2}{4x}\right], \tag{4.108}$$

where $\mathbf{r}_\perp = \mathbf{R}_\perp/\xi_L$, $\mathbf{r}_\parallel = \mathbf{R}_\parallel/\xi_L$, and $\mathbf{R}_\perp = \{R_1, \ldots, R_{d-d'}\}$, $\mathbf{R}_\parallel = \{R_{d-d'+1}, \ldots, R_d\}$.

Thus *Assumption A.1* has been verified in the case of short-range interaction. It turns out that the finite-size scaling function of the pair correlation function depends on the space dimensionality d, and on the shape of the domain Λ, in particular, on the number of infinite dimensions d'.

Finally, by using the fact that for all $\phi > 0$

$$\lim_{\Lambda \to \mathbf{Z}^d} \Sigma_\Lambda^{d,2}(0; \phi) = W_{d,2}(\phi), \tag{4.109}$$

one can easily prove that at fixed K and h, such that $(K, h) \notin [K_c, \infty) \times \{0\}$, there exists the finite bulk limit

$$\lim_{\Lambda \to \mathbf{Z}^d} \xi_\Lambda(K, h) = \xi_\infty(K, h). \tag{4.110}$$

Similarly, by using the existence of the bulk limits (4.93) and (4.110), one can prove Eq. (4.42) for the pair correlation function, which completes the verification of *Assumption A.3*. The remaining *Assumption A.2* of *Hypothesis A* will be checked bellow.

4.7.4 *Finite-size scaling for the equation of state*

In the general case of a domain $\Lambda = L^{d-d'} \times \infty^{d'}$, with $0 \le d' < d$, Eq. (4.97) takes the form

$$\Sigma_\Lambda^{d,2}(0; \phi) \simeq W_{d,2}(\phi) + \frac{1}{L^{d-2}} F_{d,d',2}(L\phi^{1/2}), \tag{4.111}$$

where we have introduced the universal finite-size scaling function

$$F_{d,d',2}(y) := \frac{y^{d-2}}{(4\pi)^{d/2}} \int_0^\infty \frac{\mathrm{d}x}{x^{d/2}} \sum_{\mathbf{k}(d-d')}{}' \exp\left[-x - \frac{y^2}{4x}|\mathbf{k}|^2\right]. \tag{4.112}$$

Although the asymptotic representation (4.111) seems rather complicated in comparison with the initial expression (4.81), it has the important advantage that the size-dependence does not enter into the summation limits. Eqs. (4.111), (4.112) turn out to be useful in the study of various models. For example, the function $F_{d,d',2}(\cdot)$ enters into the expressions for the shift of the critical temperature and the renormalization of the interaction constant in the study of finite-size effects in the φ^4-theory [Brézin and Zinn-Justin (1985)]. The same function is used in the expressions for the renormalized parameters in the dynamical finite-size scaling theory, see [Niel and Zinn-Justen (1987)], [Goldschmidt (1987)], [Diehl (1987)].

Let us now demonstrate the usefulness of the above asymptotic expression for the finite-size scaling analysis of the mean spherical model. By using the relationship (4.104) between the finite-size correlation length ξ_L

and the spherical field ϕ, and by intoducing the scaled temperature, x_1, and magnetic field, x_2, variables

$$x_1 = (K_c - K)L^{d-2}, \qquad x_2 = K^{-1/2}hL^{(d+2)/2}, \tag{4.113}$$

the mean spherical constraint (4.77) can be written as an equation for the finite-size correlation length ξ_L:

$$L^{d-2}[W_{d,2}(\xi_L^{-2}) - K_c] + F_{d,d',2}(L/\xi_L) + x_1 + x_2^2\,(\xi_L/L)^4 = 0. \tag{4.114}$$

Note that the mean spherical constraint in the thermodynamic limit, see Eq. (3.27) at $\sigma = 2$, can also be written as an equation for the bulk correlation length ξ_∞, see Eq. (4.94), as follows:

$$L^{d-2}[W_{d,2}(\xi_\infty^{-2}) - K_c] + x_1 + x_2^2\,(\xi_\infty/L)^4 = 0. \tag{4.115}$$

Only in the neighborhood of the critical point, when both $\xi_L \gg 1$ and $\xi_\infty \gg 1$, so that the asymptotic expansion (3.32) of $W_{d,2}(\phi)$ as $\phi \to 0^+$ at $2 < d < 4$ can be used, do these equations take the finite-size scaling form

$$-D_{d,2}(L/\xi_L)^{d-2} + F_{d,d',2}(L/\xi_L) + x_1 + x_2^2\,(\xi_L/L)^4 = 0, \tag{4.116}$$

and

$$-D_{d,2}(L/\xi_\infty)^{d-2} + x_1 + x_2^2\,(\xi_\infty/L)^4 = 0, \tag{4.117}$$

respectively. Here the constant $D_{d,2}$, defined by Eq. (3.33), reads explicitly

$$D_{d,2} = \pi\left[(4\pi)^{d/2}\,\Gamma(d/2)\,|\sin(\pi d/2)|\right]^{-1}, \qquad 2 < d < 4. \tag{4.118}$$

Of course, at fixed K and h Eq. (4.117) becomes independent of L.

Alternatively, by using the relationship between the finite-size magnetization per spin and the finite-size spherical field,

$$m_\Lambda(K,h) := -\frac{\partial}{\partial h}\beta f_\Lambda(K,h) = \frac{h}{K\phi_\Lambda(K,h)}, \tag{4.119}$$

the mean spherical constraint (4.77) can be written as equation of state for the finite-size magnetization m_Λ. Its bulk counterpart is given by Eq. (3.30) at $\sigma = 2$.

Let us analyze the above results in the light of the finite-size scaling hypotheses stated in Section 4.5. Apparently, Eq. (4.117) can be used to express x_1 in terms of ξ_∞/L and x_2: $x_1 = X_1^{d,2}(\xi_\infty/L, x_2)$, which after substitution in Eq. (4.116) yields the relationship

$$\xi_L(K, h) \simeq L \, X_\xi^{d,d',2}(\xi_\infty/L, x_2), \tag{4.120}$$

which verifies *Hypothesis B* in *Case (i)*, and extends it to account for the field dependence. Moreover, the conditions of validity of that hypothesis become transparent. The consideration of whether *Case (ii)* of *Hypothesis B* takes place under certain conditions will be left for Chapter 6.

The behavior of the scaling function $F_{d,d',2}(y)$ as $y \to 0$ is not evident from the representation given by Eq. (4.112). Obviously, it depends on the dimensions d and d'. To clarify that question, below we derive explicit representations of $F_{d,d',2}(y)$ suitable for analysis of the limit $L/\xi_L \to 0$.

In the remainder of this section we confine ourselves to the interval $2 < d < 4$, and the simplest geometry $\Lambda = L^d$.

Let us explain, first on heuristic level, the contents of our finite-size scaling analysis of the sum

$$\Sigma_\Lambda^{d,2}(\phi) = \frac{1}{L^d\phi} + \frac{1}{L^d} \sum_{\mathbf{k}\in\mathcal{B}_\Lambda}{}' \frac{1}{\phi + |\mathbf{k}|^2}$$

$$= \frac{1}{L^{d-2}} \left[\frac{1}{L^2\phi} + \sum_{n_1=-(L-1)/2}^{(L-1)/2} \cdots \sum_{n_d=-(L-1)/2}^{(L-1)/2}{}' \frac{1}{\phi L^2 + (2\pi|\mathbf{n}|)^2} \right]. \tag{4.121}$$

The last expression suggests that, apart from the factor L^{-d+2}, the sum $\Sigma_\Lambda^{d,2}(\phi)$ depends basically on the scaled variable ϕL^2. The only residual L-dependence comes from the limits of summation, which cannot be extended to infinity due to the divergence of the sum at $d > 2$. However, if $2 < d < 4$, one can split up the sum in Eq. (4.121) into two terms:

$$\frac{1}{L^d} \sum_{\mathbf{k}\in\mathcal{B}_\Lambda}{}' \frac{1}{\phi + |\mathbf{k}|^2} = \frac{1}{L^d} \sum_{\mathbf{k}\in\mathcal{B}_\Lambda}{}' \frac{1}{|\mathbf{k}|^2} - \frac{1}{L^d} \sum_{\mathbf{k}\in\mathcal{B}_\Lambda}{}' \frac{\phi}{|\mathbf{k}|^2(\phi + |\mathbf{k}|^2)}. \tag{4.122}$$

The first term in the right-hand side of the above identity is independent of ϕ and approaches as $L \to \infty$ the integral $W_{d,2}(0) := K_c$. Therefore,

$$\frac{1}{L^d} \sum_{\mathbf{k}\in\mathcal{B}_\Lambda}{}' \frac{1}{|\mathbf{k}|^2} = K_c + \varepsilon_L, \qquad \varepsilon_L \to 0, \qquad \text{as} \qquad L \to \infty, \tag{4.123}$$

where the correction term ε_L is to be explicitly evaluated later. Note that the second term in the right-hand side of Eq. (4.122) converges when the summation over $\mathbf{k} \in \mathcal{B}_\Lambda$, $|\mathbf{k}| \neq 0$, is extended to infinity, i.e., to the set (compare with Eq.(3.15))

$$\{\mathbf{k} \; : \; k_j = 2\pi n_j/L, \; n_j \in Z, \; j = 1,\ldots,d \, ; \; |\mathbf{n}| \neq 0\} \, . \tag{4.124}$$

Therefore, in the leading-order approximation we can write

$$\frac{1}{L^d} \sum_{\mathbf{k} \in \mathcal{B}_\Lambda}{}' \frac{\phi}{|\mathbf{k}|^2(\phi + |\mathbf{k}|^2)} \; \simeq \; \frac{1}{L^{d-2}} \sum_{\mathbf{n}(d)}{}' \frac{\phi L^2}{(2\pi|\mathbf{n}|)^2(\phi L^2 + (2\pi|\mathbf{n}|)^2)}$$

$$\equiv \; \frac{1}{L^{d-2}} S_{d,0,2}(\phi L^2). \tag{4.125}$$

Thus, the leading-order asymptotic behavior of the sum in Eq. (4.121) as $\phi \to 0$, $L \to \infty$, so that $\phi L^2 = O(1)$, is given by

$$\Sigma_\Lambda^{d,2}(\phi) \simeq K_c + \frac{1}{L^{d-2}} \left[\frac{1}{\phi L^2} - S_{d,0,2}(\phi L^2) \right] + \varepsilon_L. \tag{4.126}$$

We show next that the essential approximations involved in the above argument have already been made in the derivation of Eq. (4.97). Indeed, the result (4.126), including the explicit expression for ε_L, will be proved to follow by identical transformations of the right-hand side of Eq. (4.97), provided one neglects terms of order $O(\phi)$. Note that the d-fold sum in the right-hand side of Eq. (4.112) at $d' = 0$ can be written as

$$\sum_{\mathbf{k}(d)}{}' \exp\left[-\frac{y^2|\mathbf{k}|^2}{4x} \right] = \Theta_3^d \left(\frac{y^2}{4x} \right) - 1, \tag{4.127}$$

where

$$\Theta_3(z) := \sum_{k=-\infty}^{\infty} e^{-zk^2}, \qquad (\mathrm{Re}\; z > 0) \tag{4.128}$$

is a particular case of the Jacobi θ_3 function. Therefore, by making use of the Jacobi identity,

$$\Theta_3(z) = \left(\frac{\pi}{z} \right)^{1/2} \Theta_3 \left(\frac{\pi^2}{z} \right), \qquad z > 0, \tag{4.129}$$

we can write $F_{d,0,2}(y)$ in the form

$$F_{d,0,2}(y) = y^{-2} + y^{-2} \int_0^\infty \mathrm{d}x \, e^{-x} \left[\Theta_3^d \left(\frac{4\pi^2 x}{y^2} \right) - 1 - \frac{y^d}{(4\pi x)^{d/2}} \right] . \tag{4.130}$$

Next, we represent the integral in the right-hand side of Eq. (4.130) as a sum of three terms. The first term is given by

$$y^{-2} \int_0^\infty dx \ (e^{-x} - 1) \left[\Theta_3^d \left(\frac{4\pi^2 x}{y^2} \right) - 1 \right] = -S_{d,0,2}(y^2), \qquad (4.131)$$

the second term is

$$-\frac{y^{d-2}}{(4\pi)^{d/2}} \int_0^\infty dx \ x^{-d/2} \left(e^{-x} - 1 \right) = y^{d-2} D_{d,2}, \qquad (4.132)$$

and the third one equals the constant

$$C_{d,2} = \frac{1}{4\pi^2} \int_0^\infty dx \left[\Theta_3^d(x) - 1 - \left(\frac{\pi}{x} \right)^{d/2} \right], \qquad (4.133)$$

provided $d > 2$. Note that the constant $D_{d,2}$ in Eq. (4.132) is the same as the one defined in Eq.(4.118) for $2 < d < 4$. The constant $C_{d,2}$ in Eq. (4.133) is a Madelung-type lattice constant which can be written in the form [Brankov and Tonchev (1988)]

$$C_{d,2} = \frac{1}{4\pi^2} \lim_{\delta \to 0} \left\{ \sum_{\mathbf{n} \in Z^d}{}' \frac{e^{-\delta|\mathbf{n}|^2}}{|\mathbf{n}|^2} - \int_{R^d} d^d n \frac{e^{-\delta|\mathbf{n}|^2}}{|\mathbf{n}|^2} \right\}. \qquad (4.134)$$

Collecting the above results, one finds

$$F_{d,0,2}(y) = D_{d,2} y^{d-2} + y^{-2} - S_{d,0,2}(y^2) + C_{d,2}. \qquad (4.135)$$

By substitution of the above expression into Eq. (4.111), taking into account the expansion (3.32) of the integral $W_{d,2}(\phi)$ at $2 < d < 4$, and neglecting in that expansion terms of the order $O(\phi) = O(yL^{-2})$, one obtains Eq. (4.126) with the identification

$$\varepsilon_L = L^{-d+2} C_{d,2}. \qquad (4.136)$$

Note that in terms of the scaled temperature, x_1, and magnetic field, x_2, variables, see Eq. (4.113), the mean spherical constraint (4.77) takes the universal form

$$\frac{1}{y^2} - \sum_{\mathbf{n}(d)}{}' \frac{y^2}{(2\pi|\mathbf{n}|)^2 [y^2 + (2\pi|\mathbf{n}|)^2]} + C_{d,2} + x_1 + \frac{x_2^2}{y^4} = 0, \qquad (4.137)$$

where $y = L\phi^{1/2}$. Therefore, when $(K, h) \to (K_c, 0)$ simultaneously with $L \to \infty$, in the way prescribed by the equations

$$K = K_c - \frac{x_1}{L^{d-2}}, \qquad h = \frac{K_c^{1/2} x_2}{L^{(d+2)/2}}, \qquad (4.138)$$

with fixed x_1 and x_2, then the leading-order finite-size scaling form of the finite-size correlation length, see Eq. (4.104), is given by $\xi_L(K, h) \simeq L/y_{d,0,2}(x_1, x_2)$, where $y = y_{d,0,2}(x_1, x_2)$ is the positive solution of Eq. (4.137). Note that at the critical point $x_1 = x_2 = 0$ one obtains

$$\xi_L(K_c, 0) = A_\xi^{d,0,2} L, \qquad (4.139)$$

where $A_\xi^{d,0,2} = 1/y_{d,0,2}(0, 0)$ is an universal amplitude.

4.7.5 *Finite-size scaling for the free energy density*

In the general case when Λ has the geometry $L^{d-d'} \times \infty^{d'}$, with $0 \leq d' < d$, Eq. (4.100) takes the form

$$U_\Lambda^{d,2}(\phi) \simeq U_{d,2}(\phi) - \frac{1}{L^d} \Psi_{d,d',2}(L\phi^{1/2}), \qquad (4.140)$$

where we have introduced the universal finite-size scaling function

$$\Psi_{d,d',2}(y) := \frac{y^d}{(4\pi)^{d/2}} \int_0^\infty \frac{dx}{x^{d/2+1}} \sum_{\mathbf{k}(d-d')}{}' \exp\left[-x - \frac{y^2}{4x}|\mathbf{k}|^2\right]. \qquad (4.141)$$

By using Eq. (3.44) to expand the bulk term $U_{d,2}(\phi)$ as $\phi \to 0^+$ at $2 < d < 4$, one can cast Eq. (4.140) in the form

$$U_\Lambda^{d,2}(\phi) \simeq U_{d,2}(\phi) + K_c\phi - \frac{1}{L^d}\left[\frac{2}{d}D_{d,2} L^d \phi^{d/2} + \Psi_{d,d',2}(L\phi^{1/2})\right]. \qquad (4.142)$$

Naturally, in accordance with the extremum problem (4.76), we have

$$-\frac{1}{L^2}\frac{\partial}{\partial\phi}\Psi_{d,d',2}(L\phi^{1/2}) = F_{d,d',2}(L\phi^{1/2}). \qquad (4.143)$$

The above relationship can be obtained directly, by changing the integration variable $x \to y^2 t$ in Eq. (4.141) and differentiating then with respect to ϕ.

By using the asymptotic form (4.142) of the sum (4.85) in the expression (4.76) for the free energy density of the mean spherical model, we obtain that in the neighborhood of the critical point ($\phi \ll 1$):

$$\beta f_\Lambda^{m.s.}(K,h) \simeq \frac{1}{2}\left[\ln\frac{K}{2\pi} - \frac{1}{\rho_2}K + U_{d,2}(0)\right] + \frac{1}{L^d}X_f^{d,d',2}(x_1,x_2). \quad (4.144)$$

Here $X_f^{d,d',2}(x_1,x_2)$ is the universal finite-size scaling function for the *singular part* of the free energy density, which is given by

$$X_f^{d,d',2}(x_1,x_2) = \frac{1}{2}\sup_{y>0}\left\{x_1 y^2 - \frac{x_2^2}{y^2} - \frac{2}{d}D_{d,2}y^d - \Psi_{d,d',2}(y)\right\}. \quad (4.145)$$

At the critical point $x_1 = x_2 = 0$ the singular part of the free energy density becomes

$$f_\Lambda^{m.s.}(K_c,0)/k_B T_c = A_f^{d,d',2}\,L^{-d}, \quad (4.146)$$

where $A_f^{d,d',2} = X_f^{d,d',2}(0,0)$ is a universal amplitude.

Since all the finite-size thermodynamic functions $P_L(K,h)$, which in the bulk limit $L \to \infty$ become singular at the critical point $K = K_c$, $h = 0$, follow by differentiation of Eq. (4.145) with respect to the thermodynamic parameters K and h, respectively x_1 and x_2, the above result for the free energy density is in full conformity with *Assumption A.2* . This completes the verification of *Hypothesis A*.

Note that despite of the mathematical similarity between Eq. (4.100) for the finite-size term $U_\Lambda^{d,2}(\phi)$ in the expression for the free energy density and Eq. (4.97) for the finite-size term $\Sigma_\Lambda^{d,2}(0;\phi)$ in the mean spherical constraint, the attempt to apply the above technique to the sum

$$U_\Lambda^{d,2}(\phi) = \frac{1}{L^d}\sum_{k\in B_\Lambda}\ln(\phi + |k|^2) =$$

$$\ln\phi + \frac{1}{L^d}\sum_{n_1=-(L-1)/2}^{(L-1)/2}\cdots\sum_{n_d=-(L-1)/2}^{(L-1)/2}{}'\ln\left[1 + \frac{(2\pi|\mathbf{n}|)^2}{\phi L^2}\right],$$

runs into serious difficulties. They are caused by the strong divergence of the sum as its limits are extended to infinity and the lack of identical representation, analogous to Eq. (4.122), which could make possible the separation of the dependence on the scaled variable ϕL^2 from the L-dependence in the summation limits. That is why, following [Brankov (1989)], we give an alternative representation of the scaling function (4.141), suitable for the study of its asymptotic behavior as $y = L\phi^{1/2} \to 0$. It is based on the identity

$$\exp(-x) = \frac{1}{\sqrt{\pi x}} \int_0^\infty du \, \cos(u) \exp\left(-\frac{u^2}{4x}\right). \qquad (4.147)$$

With the aid of (4.147) one can transform Eq. (4.141) to

$$\Psi_{d,d',2}(y) = \frac{2}{\pi^{(d+1)/2}} \Gamma\left(\frac{d+1}{2}\right) \sum_{\mathbf{k}(d-d')}{}' |\mathbf{k}|^{-d} u_{d,2}(y|\mathbf{k}|), \qquad (4.148)$$

where $u_{d,2}(z)$ can be expressed in terms of the McDonald function (second modified Bessel function) $K_\nu(\cdot)$ as

$$u_{d,2}(z) = \int_0^\infty dx \frac{\cos(zx)}{(1+x^2)^{(d+1)/2}} = \frac{\pi^{1/2}}{\Gamma\left(\frac{d+1}{2}\right)} \left(\frac{z}{2}\right)^{d/2} K_{d/2}(z). \qquad (4.149)$$

The stationarity condition for the supremum problem in Eq. (4.145), with $\Psi_{d,d',2}(y)$ given by Eq. (4.148), yields the following new finite-size scaling representation of the mean spherical constraint

$$-D_{d,2} y^{d-2} + \frac{1}{\pi^{(d+1)/2)}} \Gamma\left(\frac{d+1}{2}\right) \sum_{\mathbf{k}(d-d')}{}' |\mathbf{k}|^{-d+2} w_{d,2}(y|\mathbf{k}|) + x_1 + \frac{x_2^2}{y^4} = 0,$$

$$(4.150)$$

where

$$w_{d,2}(z) := -z^{-1} \frac{d}{dz} u_{d,2}(z) = \frac{\pi^{1/2}}{2\Gamma\left(\frac{d+1}{2}\right)} \left(\frac{z}{2}\right)^{d/2-1} K_{d/2-1}(z). \qquad (4.151)$$

The asymptotic behavior of the function $w_{d,2}(z)$ as $z \to 0$ and $z \to \infty$ easily follows from the asymptotic behavior of the function $K_{d/2-1}(z)$.

It is interesting to note that in the important case of infinite film geometry, $d = 3$ and $d' = 2$, we obtain the simple result

$$w_{3,2}(z) = \frac{\pi^{1/2}}{2} \left(\frac{z}{2}\right)^{1/2} K_{1/2}(z) = \frac{\pi}{4} e^{-z}. \qquad (4.152)$$

In this case the finite-size scaling form (4.150) of the mean spherical constraint simplifies to

$$-D_{3,2} \, y + \frac{1}{2\pi} \sum_{k=1}^\infty \frac{1}{k} e^{-ky} + x_1 + \frac{x_2^2}{y^4} = 0. \qquad (4.153)$$

With the help of the identity

$$\sum_{k=1}^{\infty} \frac{1}{k} e^{-ky} = -\ln(1 - e^{-y}), \tag{4.154}$$

Eq. (4.153) takes the form

$$-\frac{1}{2\pi} \ln[2\sinh(y/2)] + x_1 + \frac{x_2^2}{y^4} = 0. \tag{4.155}$$

Hence, at the critical point $x_1 = x_2 = 0$, one obtains the "golden value" solution [Pathria (1972)], [Singh and Pathria (1985a)]

$$y_{L,c} \equiv \Theta = 2\ln\frac{1+\sqrt{5}}{2} = -2\ln\frac{\sqrt{5}-1}{2} = 0.962424\ldots \tag{4.156}$$

Note that an exact and transparent study [Danchev (1998)] of the Casimir effect in the framework of the mean spherical model is related to Eq. (4.155). Moreover, due to the quantum-to-classical crossover rule, the above results play certain role in the theory of two-dimensional quantum systems, see [Chubukov et. al. (1994)]. That is why, some related problems and results will be considered in more detail in the context of quantum systems.

Chapter 5

Long-Range Interactions

The range of interparticle interaction is one of the microscopic characteristics which are essential for the bulk critical properties. It is well known that systems with long-range microscopic interactions can exhibit *nonextensive* behavior of the thermodynamical functions, see Section 1.3. In order to avoid such type of problems, which are usually considered to extend beyond the scope of the standard statistical mechanics (see, e.g., [Jund et. al. (1989)] and references therein), we will confine ourselves to the class of models with power-low interactions which decay at large distances r as $r^{-(d+\sigma)}$. Here d is the space dimensionality and σ is a parameter which obeys the thermodynamic stability condition $\sigma > 0$. The long-range character of the interparticle potential ($0 < \sigma < 2$) causes, in general, considerably more difficulties in the study of finite-size effects near criticality. As we have seen in Chapter 4, a basic quantity of the finite-size scaling theory is the bulk correlation length ξ_∞, which is unambiguously defined for short-range interactions; for the considered class of models the latter formally correspond to $\sigma = 2$, see Section 3.1.2. In the long-range case the analogue of the correlation length has to be defined in a more sophisticated way [Brankov and Danchev (1991)]. The finite-size scaling hypotheses, formulated in Section 4.5, are based on the notions of finite-size characteristic length λ_L and its bulk limit λ_∞, which give the possibility to treat the cases of short- ($\sigma = 2$) and long-range ($0 < \sigma < 2$) interactions on equal grounds. This is demonstrated in the present chapter by the example of the classical mean spherical model with space dimensionality d *between* the lower, $d_l = \sigma$, and upper, $d = 2\sigma$, critical dimensionalities. We begin with the presentation of adequate mathematical techniques.

161

5.1   Mathematical techniques

Evidently, as in the case of short-range interaction considered in Section 4.7, the evaluation of the finite-size expressions will be greatly simplified if one can reduce the d-dimensional sums to effectively one-dimensional. In the case of long-range interactions, the following analogue of the identity (4.80) has been found [Brankov and Tonchev (1988)]

$$\frac{1}{1+z^{\sigma}} = \int_0^{\infty} \mathrm{d}t\, F_{\sigma/2}(t)\exp(-z^2 t), \tag{5.1}$$

where the function $F_{\sigma/2}(t)$ for $0 < \sigma < 2$ is given by

$$F_{\sigma/2}(t) = \frac{\sin(\sigma\pi/2)}{\pi} \int_0^{\infty} \mathrm{d}y\, \frac{y^{\sigma/2}\exp(-ty)}{1 + 2y^{\sigma/2}\cos(\sigma\pi/2) + y^{\sigma}}. \tag{5.2}$$

The identity (5.1) can be established via contour integration. One can check that in the particular case $\sigma = 1$

$$F_{1/2}(t) = (\pi t)^{-1/2} - \exp(t)[1 - \mathrm{erf}(t^{1/2})]. \tag{5.3}$$

From Eqs. ((5.1) and (5.2) one obtains the integral representation

$$\frac{1}{1+z^{\sigma}} = \int_0^{\infty} \frac{Q_{\sigma}(t)}{t+z^2}\mathrm{d}\,t, \tag{5.4}$$

where

$$Q_{\sigma}(t) = \frac{1}{\pi}\frac{\sin(\pi\sigma/2)t^{\sigma/2}}{1 + 2\cos(\pi\sigma/2)t^{\sigma/2} + t^{\sigma}}. \tag{5.5}$$

Equation (5.4) shows how the mathematical difficulties that arise in the case of long-range interaction potentials can be avoided: by an additional integration the problem is effectively reduced to the corresponding short-range one [Brankov and Tonchev (1988)]. A further step in the development of the analytical techniques for studying finite-size properties is the recognition that the function $F_{\sigma/2}(\cdot)$ is related to the entire functions of the Mittag-Leffler type defined by the power series

$$E_{\alpha,\beta}(z) = \sum_{k=0}^{\infty} \frac{z^k}{\Gamma(\alpha k + \beta)} \qquad \alpha > 0. \tag{5.6}$$

Indeed, it has been found [Brankov (1989)] that

$$F_{\sigma/2}(t) = t^{\sigma/2-1} E_{\sigma/2,\sigma/2}(-t^{\sigma/2}). \tag{5.7}$$

Thus, Eqs. (5.2) and (5.7) yield an integral representation of the Mittag-Leffler function $E_{\sigma/2,\sigma/2}(\cdot)$ with $0 < \sigma < 2$. On the other hand, the combination of Eqs. (5.1) and (5.7) yields the useful identity

$$\frac{1}{1+z^\alpha} = \int_0^\infty dt\, e^{-zt}\, t^{\alpha-1} E_{\alpha,\alpha}(-t^\alpha). \tag{5.8}$$

Note that in some special cases $E_{\alpha,\beta}(\cdot)$ reduces to elementary functions, for example,

$$E_{1,1}(z) = e^z, \quad E_{2,1}(z) = \cosh z^{1/2}, \quad E_{2,2}(z) = \frac{\sinh z^{1/2}}{z^{1/2}}. \tag{5.9}$$

The mean spherical model with long-range isotropic interactions has been treated by this techniques in [Brankov (1989)] and [Brankov and Tonchev (1990)]. A slight extension of the above formalism has been used in the study [Brankov and Danchev (1991)] of finite-size effects in the pair correlation function.

In the case of quantum models the situation is somewhat more complicated due to the different form of the propagator. However, in the zero-temperature case a similar trick can be used with the aid of the identity [Chamati (1994)]

$$[1+z^\sigma]^{-1/2} = \int_0^\infty dt \exp(-z^2 t)\, t^{\sigma/4} G_{\sigma/2,\sigma/4}(-t^{\sigma/2}), \tag{5.10}$$

where

$$G_{\alpha,\beta}(z) = \pi^{1/2} \sum_{k=0}^\infty \frac{\Gamma(k+1/2)z^k}{\Gamma(\alpha k+\beta)k!}, \qquad \alpha,\beta > 0. \tag{5.11}$$

One can check that the series (5.11) yields in the particular cases of short-range interaction ($\sigma = 2$),

$$G_{1,1/2}(z) = \pi^{-1/2} \exp(z), \tag{5.12}$$

and, in the special case of long-range interaction with $\sigma = 1$,

$$G_{1/2,3/4}(-z) = \pi^{-1} z^{1/2} \exp(z^2/2) K_{1/4}(z^2/2), \tag{5.13}$$

where $K_{1/4}(\cdot)$ is the MacDonald function.

Let us note that the above generalizations of identity (4.80) can be obtained by regarding the latter as Laplace transformation of the exponential function. Then the identities (5.1) and (5.10) follow just by using the fact that $F_{\sigma/2}(t)$ and $t^{\sigma/4}G_{\sigma/2,\sigma/4}(-t^{\sigma/2})$ are, essentially, Laplace originals of the corresponding left-hand sides.

To analyze the finite-size expression for the free energy density of the mean spherical model with long-range isotropic interaction, see Eq. (3.22) in Section 3.1.2, we replace identity (4.83) by the more general one [Brankov (1989)]

$$\ln(1 + z^\alpha) = \alpha \int_0^\infty \mathrm{d}x\, x^{-1}\left(1 - \mathrm{e}^{-zx}\right) E_{\alpha,1}(-x^\alpha), \qquad (5.14)$$

which follows by integration of the Laplace transform [Bateman end Erdélyi (1955)]

$$\int_0^\infty \mathrm{d}x e^{-px} E_{\alpha,1}(-x^\alpha) = \frac{p^{\alpha-1}}{1 + p^\alpha} \qquad (\mathrm{Re}\, p > 0) \qquad (5.15)$$

over p from zero to z. It might be useful to note also the relationship

$$\frac{\mathrm{d}}{\mathrm{d}z} E_{\alpha,1}(z) = \frac{1}{\alpha} E_{\alpha,\alpha}(z), \qquad (5.16)$$

which follows from the power-series representation. The implementation of the identities (5.8) and (5.14) at $\alpha = \sigma/2$, with

$$z = |\mathbf{k}|^2 \phi^{-2/\sigma} = \left(\frac{2\pi}{\phi^{1/\sigma}}\right)^2 \sum_{j=1}^d \left(\frac{n_j}{L_j}\right)^2, \qquad (5.17)$$

yields the following factorized expressions for the sums under consideration:

$$\Sigma_\Lambda^{d,\sigma}(\mathbf{R}; \phi) \equiv \frac{1}{|\Lambda|} \sum_{\mathbf{k} \in \mathcal{B}_\Lambda} \frac{1}{\phi + |\mathbf{k}|^\sigma} \exp(\mathrm{i}\,\mathbf{k} \cdot \mathbf{R})$$

$$= \phi^{-1} \int_0^\infty \mathrm{d}x\, x^{\sigma/2-1} E_{\sigma/2,\sigma/2}(-x^{\sigma/2}) \prod_{j=1}^d Q_{L_j}(x\phi^{-2/\sigma}, R_j), \qquad (5.18)$$

and

$$U_\Lambda^{d,\sigma}(\phi) \equiv \frac{1}{|\Lambda|} \sum_{\mathbf{k}\in\mathcal{B}_\Lambda} \ln(\phi + |\mathbf{k}|^\sigma)$$

$$= \ln\phi + \frac{\sigma}{2} \int_0^\infty dx\ x^{-1} E_{\sigma/2,1}(-x^{\sigma/2}) \left\{ 1 - \prod_{j=1}^d Q_{L_j}(x\phi^{-2/\sigma},0) \right\}. \quad (5.19)$$

Thus the initial problem can be reduced to the evaluation of the asymptotic behavior of standard sums of the type (4.82), which has already been carried out in Section 4.7.2. All the specific features of the long-range isotropic interaction are contained now in the analytical properties of the Mittag-Leffler type functions (5.6). The special case of a mean spherical model with both short-range and long-range anisotropic dipole-dipole interactions has been treated in [Danchev (1990)].

We conclude this section with a summary of the asymptotic properties of the Mittag-Leffler type functions with $0 < \alpha < 2$ and $\beta \geq 0$.

Lemma [Bateman end Erdélyi (1955)]. Let $0 < \alpha < 2$ and γ be a real number satisfying the condition

$$\alpha\pi/2 < \gamma < \min\{\pi, \alpha\pi\}.$$

Then, for any integer $p \geq 1$, the following asymptotic expressions hold as $|z| \to \infty$:

(1) At $|\arg z| \leq \gamma$,

$$E_{\alpha,\beta}(z) = \frac{1}{\alpha} z^{(1-\beta)/\alpha} \exp(z^{1/\alpha}) - \sum_{k=1}^p \frac{z^{-k}}{\Gamma(\beta - \alpha k)} + O(|z|^{-p-1}). \quad (5.20)$$

(2) At $\gamma \leq |\arg z| \leq \pi$,

$$E_{\alpha,\beta}(z) = - \sum_{k=1}^p \frac{z^{-k}}{\Gamma(\beta - \alpha k)} + O(|z|^{-p-1}). \quad (5.21)$$

Notice that since $\Gamma(0) = \infty$, from Eq. (5.21) it follows that

$$E_{\alpha,\alpha}(-x^\alpha) \simeq -\frac{x^{-2\alpha}}{\Gamma(-\alpha)} \qquad \text{as} \qquad x \to \infty \qquad (\alpha \neq 1). \quad (5.22)$$

5.2 Finite-size scaling for the correlation function

To test the hypotheses of finite-size scaling at criticality on the mean spherical model with long-range interaction, one has to analyze first the large-distance asymptotic behavior of the pair correlation function and, from its expected form (4.39) given by *Assumption A.1* in Section 4.5, to identify the appropriate finite-size characteristic length λ_L. Following these lines, we insert the asymptotic form (4.89) of the single sums $Q_{L_j}(z, R_j)$ as $L_j \to \infty$ and $R_j \to \infty$, $j = 1, \ldots, d$, into the factorized expression (5.18) of the d-fold sum, to obtain

$$\Sigma_\Lambda^{d,\sigma}(\mathbf{R}; \phi) \simeq \frac{\phi^{d/\sigma - 1}}{(4\pi)^{d/2}} \int_0^\infty dx \, x^{-(d-\sigma+2)/2} E_{\sigma/2, \sigma/2}(-x^{\sigma/2})$$

$$\times \sum_{\mathbf{k}(d)} \exp\left[-\frac{\phi^{2/\sigma}}{4x} \sum_{j=1}^d (k_j L_j + R_j)^2 \right]. \qquad (5.23)$$

Here the summation runs over all vectors $\mathbf{k} = \{k_1, \ldots, k_d\}$ with d integer components, $\mathbf{k} \in Z^d$.

By substitution of Eq. (5.23) in the general expression (3.17) for the net spin-spin correlation function, we obtain that its leading-order asymptotic behavior for a system in a parallelepiped with lagre (compared to the lattice spacing a) edges $\mathbf{L} = \{L_1, \ldots, L_d\}$, at large distances $\mathbf{R}$, and close to the critical point, has the form

$$G_{\mathbf{L}}(\mathbf{R}; K, h) \simeq A(T) R^{-d+2-\eta} X_G^{d,\sigma}(\mathbf{R}/\lambda_{\mathbf{L}}; \mathbf{L}/\lambda_{\mathbf{L}}). \qquad (5.24)$$

Here $A(T) = K^{-1} = k_B T / \rho_\sigma \hat{J}(0)$ is a non-universal, regular at $T = T_c$ amplitude, the critical exponent is $\eta = 2 - \sigma$, the finite-size characteristic length is identified as

$$\lambda_{\mathbf{L}}(K, h) = \phi_{\mathbf{L}}^{-1/\sigma}(K, h), \qquad (5.25)$$

where $\phi_{\mathbf{L}}(K, h)$ is the solution of the finite-size mean spherical constraint, see the next section, and

$$X_G^{d,\sigma}(\mathbf{r}; \mathbf{y}) = \frac{r^{-d+\sigma}}{(4\pi)^{d/2}} \int_0^\infty dx \, x^{-(d-\sigma+2)/2} E_{\sigma/2, \sigma/2}(-x^{\sigma/2})$$

$$\times \sum_{\mathbf{k}(d)} \exp\left[-\frac{1}{4x} \sum_{j=1}^d (y_j k_j + r_j)^2 \right], \qquad (5.26)$$

is a universal finite-size scaling function of the arguments $\mathbf{r} = \mathbf{R}/\lambda_{\mathbf{L}}$ and $\mathbf{y} = \mathbf{L}/\lambda_{\mathbf{L}}$.

The case of a cylindrical region $\Lambda = L^{d-d'} \times \infty^{d'}$, with $d - d' > 0$ finite and d' infinite dimensions, can be obtained from Eq. (5.26) by setting $L_j = L$ for $j = 1, \ldots, d - d'$, (then $\lambda_{\mathbf{L}} := \lambda_L$) and taking the limit $L_j \to \infty$ for $j = d - d' + 1, \ldots, d$. The resulting universal finite-size scaling function is

$$X_G^{d,d',\sigma}(\mathbf{r}; y) = \frac{r^{-d+\sigma}}{(4\pi)^{d/2}} \int_0^\infty dx \; x^{-(d-\sigma+2)/2} E_{\sigma/2,\sigma/2}(-x^{\sigma/2})$$

$$\times \sum_{\mathbf{k}(d-d')} \exp\left[-\frac{(y\mathbf{k} + \mathbf{r}_\perp)^2 + \mathbf{r}_\parallel^2}{4x} \right], \qquad (5.27)$$

where $\mathbf{r}_\perp = \mathbf{R}_\perp/\lambda_L$, $\mathbf{r}_\parallel = \mathbf{R}_\parallel/\lambda_L$, with $\mathbf{R}_\perp = \{R_1, \ldots, R_{d-d'}\}$, $\mathbf{R}_\parallel = \{R_{d-d'+1}, \ldots, R_d\}$, and $y = L/\lambda_L$.

Finally, by taking the thermodynamic limit $L_j \to \infty$, $j = 1, \ldots, d$, in Eq. (5.26), we see that only the term with $\mathbf{k} = 0$ gives a nonvanishing contribution, hence the result for the bulk spin-spin correlation function is

$$G_\infty(\mathbf{R}; K, h) \simeq \frac{\phi_\infty^{d/\sigma-1}}{(4\pi)^{d/2}} \int_0^\infty dx \; x^{-(d-\sigma+2)/2} E_{\sigma/2,\sigma/2}(-x^{\sigma/2})$$

$$\times \exp\left(-\frac{\phi_\infty^{2/\sigma}|\mathbf{R}|^2}{4x} \right)$$

$$= \frac{R^{\sigma-d}}{(2\pi)^d} \int_0^\infty dx \; \frac{x^{d/2}\, J_{d/2-1}(x)}{x^\sigma + (R/\phi_\infty)^\sigma}. \qquad (5.28)$$

Here $\phi_\infty = \phi_\infty(K, h)$ is the solution of the bulk mean spherical constraint (3.27).

From Eq. (5.23) it follows that the finite-size spin-spin correlation function takes indeed the asymptotic form (4.39), generalized to account for the different linear sizes $L_j, j = 1, \ldots, d$ of Λ. Thus *Assumption A.1* is verified in the case of long-range isotropic interaction with exponent $\sigma \in (0, 2)$, when $d \in (\sigma, 2\sigma)$. The finite-size scaling function of the pair correlation function depends on the space dimensionality d, on the shape of the domain Λ, in particular, on the number of infinite dimensions d', see Eq. (5.27), as well as on the decay exponent of the potential σ. Note that the definition of the finite-size characteristic length (5.25) holds true in both the short-range (see Eq. (4.104)) and long-range cases.

To verify *Assumption A.3* we have to study the bulk limits of the finite-size characteristic lenght and the pair correlation function. The finite-size characteristic lenght λ_L is directly related by Eq. (5.25) to the solution ϕ_L of finite-size mean spherical constraint (3.23).

5.3   Finite-size scaling for the equation of state

As we have already seen, due to the relationship Eq. (4.119) between the spherical field ϕ_Λ and the magnetization m_Λ for a system in a finite region Λ, the mean spherical constraint (3.23) has the meaning of equation of state. On the other hand, it is more suitable for our present aim to interpret the equation of state as equation for the finite-size characteristic length ϕ_Λ.

We start by considering Λ in the form of a finite d-dimensional parallelepiped with edges L_j, $j = 1, \ldots, d$. The implementation of the identity (5.8) to the basic d-fold sum $\Sigma_\Lambda^{d,\sigma}(0; \phi)$ yields the result:

$$\Sigma_\Lambda^{d,\sigma}(\phi) \equiv \Sigma_\Lambda^{d,\sigma}(0; \phi) = \frac{1}{|\Lambda|} \sum_{\mathbf{k} \in \mathcal{B}_\Lambda} \frac{1}{\phi + |\mathbf{k}|^\sigma}$$

$$= \frac{1}{\phi} \int_0^\infty \mathrm{d}x \, x^{\sigma/2-1} E_{\frac{\sigma}{2}, \frac{\sigma}{2}}(-x^{\sigma/2}) \prod_{j=1}^{d} Q_{L_j}(x\phi^{-2/\sigma}, 0). \tag{5.29}$$

The asymptotic analysis of the right-hand side of above expression goes along the same lines as in the short-range case: by using Eq. (4.90) with $z = x\phi^{-2/\sigma}$ one obtains

$$\prod_{j=1}^{d} Q_{L_j}(x\phi^{-2/\sigma}, 0) \simeq$$

$$\frac{\phi^{d/\sigma}}{(4\pi x)^{d/2}} \left\{ \left[\mathrm{erf}\left(\frac{\pi x^{1/2}}{\phi^{1/\sigma}} \right) \right]^d + \sum_{\mathbf{k}(d)}' \exp\left[-\frac{\phi^{2/\sigma}}{4x} \sum_{j=1}^{d} L_j^2 k_j^2 \right] \right\}. \tag{5.30}$$

Therefore, the leading asymptotic behavior of $\Sigma_\Lambda^{d,\sigma}(\phi)$ is given by the expression (compare with Eq. (4.97))

$$\Sigma_\Lambda^{d,\sigma}(\phi) \simeq W_{d,\sigma}(\phi) + \frac{\phi^{d/\sigma-1}}{(4\pi)^{d/2}} \int_0^\infty \mathrm{d}x\, x^{-(d-\sigma+2)/2} E_{\frac{\sigma}{2},\frac{\sigma}{2}}(-x^{\sigma/2})$$
$$\times \sum_{\mathbf{k}(d)}{}' \exp\left[-\frac{\phi^{2/\sigma}}{4x}\sum_{j=1}^d L_j^2 k_j^2\right]. \qquad (5.31)$$

For the bulk term in the above equation we have used the identity

$$\frac{\phi^{d/\sigma-1}}{(4\pi)^{d/2}} \int_0^\infty \mathrm{d}x\, x^{-(d-\sigma+2)/2} E_{\frac{\sigma}{2},\frac{\sigma}{2}}(-x^{\sigma/2}) \left[\mathrm{erf}\left(\frac{\pi x^{1/2}}{\phi^{1/\sigma}}\right)\right]^d \equiv W_{d,\sigma}(\phi),$$
$$(5.32)$$

where $W_{d,\sigma}(\phi)$ is defined in Eq. (3.28); its asymptotic expansion as $\phi \to 0^+$ is given by Eq. (3.32).

Confining ourselves to the case of a region $\Lambda = L^{d-d'} \times \infty^{d'}$, it is convenient to write Eq. (5.31) in the form

$$\Sigma_\Lambda^{d,\sigma}(\phi) \simeq W_{d,\sigma}(\phi) + \frac{1}{L^{d-\sigma}} F_{d,d',\sigma}(L/\lambda_L), \qquad (5.33)$$

where the universal finite-size scaling function

$$F_{d,d',\sigma}(y) := \frac{y^{d-\sigma}}{(4\pi)^{d/2}} \int_0^\infty \mathrm{d}x\, x^{-(d-\sigma+2)/2} E_{\sigma/2,\sigma/2}(-x^{\sigma/2})$$
$$\times \sum_{\mathbf{k}(d-d')}{}' \exp\left[-\frac{y^2}{4x}|\mathbf{k}|^2\right] \qquad (5.34)$$

is introduced. In terms of the finite-size scaling variables x_1 and x_2, compare with Eq. (4.113),

$$x_1 = (K_c - K)L^{d-\sigma}, \qquad x_2 = K^{-1/2}hL^{(d+\sigma)/2}, \qquad (5.35)$$

the finite-size mean spherical constraint (3.23) can be written as equation for the finite-size characteristic length λ_L, see Eq. (5.25),

$$L^{d-\sigma}[W_{d,\sigma}(\lambda_L^{-\sigma}) - K_c] + F_{d,d',\sigma}(L/\lambda_L) + x_1 + x_2^2\,(\lambda_L/L)^{2\sigma} = 0. \qquad (5.36)$$

Further, by using the fact that for all $\phi > 0$

$$\lim_{\Lambda \to Z^d} \Sigma_\Lambda^{d,\sigma}(0;\phi) = W_{d,\sigma}(\phi), \qquad (5.37)$$

one can easily prove that at fixed K and h, such that $(K, h) \notin [K_c, \infty) \times \{0\}$, there exists the limit

$$\lim_{L \to \infty} \lambda_L(K, h) := \lambda_\infty(K, h) = \phi_\infty^{-1/\sigma}(K, h), \qquad (5.38)$$

where $\phi_\infty(K, h)$ is the solution of the bulk spherical constraint (3.27). The latter, with the aid of Eq. (5.38), can be written as equation for the bulk characteristic length λ_∞:

$$L^{d-\sigma}[W_{d,\sigma}(\lambda_\infty^{-\sigma}) - K_c] + x_1 + x_2^2 \, (\lambda_\infty/L)^{2\sigma} = 0. \qquad (5.39)$$

Let us note that equations (5.36) and (5.39) are completely analogous to the corresponding Eqs. (4.114) and (4.115) obtained in Section 4.7.4 for the case of short-range interactions. Again, only in the neighborhood of the critical point, when both $\lambda_L \gg 1$ and $\lambda_\infty \gg 1$, so that the asymptotic expansion Eq. (3.32) can be used, do these equations take the finite-size scaling form

$$-D_{d,\sigma}(L/\lambda_L)^{d-\sigma} + F_{d,d',\sigma}(L/\lambda_L) + x_1 + x_2^2 \, (\lambda_L/L)^{2\sigma} = 0, \qquad (5.40)$$

and

$$-D_{d,\sigma}(L/\lambda_\infty)^{d-\sigma} + x_1 + x_2^2 \, (\lambda_\infty/L)^{2\sigma} = 0, \qquad (5.41)$$

respectively. Note that the constant $D_{d,\sigma}$ is defined in Eq. (3.33); at fixed K and h Eq. (5.41) is actually independent of L.

Let us analyze the above results in the light of the finite-size scaling hypotheses stated in Chapter 4. Apparently, Eq. (5.41) can be used to express x_1 in terms of λ_∞/L and x_2, namely, $x_1 = X_1^{d,\sigma}(\lambda_\infty/L, x_2)$, which after substitution in Eq. (5.40) yields

$$\lambda_L(K, h) \simeq L \, X_\lambda^{d,d',\sigma}(\lambda_\infty/L, x_2). \qquad (5.42)$$

This relationship verifies *Hypothesis B* in *Case (i)* under the conditions stated above.

Finally, by using the existence of the bulk limit for the pair correlation function of the corresponding Gaussian model,

$$\lim_{\Lambda \to Z^d} G_\Lambda^G(\mathbf{R}; \phi) = G_\infty^G(\mathbf{R}; \phi), \qquad (5.43)$$

as well as the limit (5.38), one can prove Eq. (4.42) which completes the verification of *Assumption A.3.* The remaining *Assumption A.2* of *Hypoth esis A* will be checked in Section 5.5, where the asymptotic behavior of the finite-size free energy density is studied.

In the remainder of this section we derive explicit representations of the function $F_{d,d',\sigma}(y)$, suitable for analysis of the limit $y = L/\lambda_L \to 0$. Since the results essentially depend on the dimensions d and d', as well as on the decay exponent σ, we confine ourselves to the parameter region $\sigma < d < 2\sigma$, and two representative geometries: fully finite cubic domain L^d and infinitely long cylinder $L^{d-1} \times \infty$.

5.3.1 *Finite cube* $\Lambda = L^d$

As in the case of short-range interaction, see Section 4.7.4, we first consider the sum over $\mathbf{k} \in \mathcal{B}_\Lambda$ in Eq. (5.29) on heuristic level. After singling out the term with $\mathbf{k} = 0$, the remaining sum can be written as follows

$$\frac{1}{L^d} \sum_{\mathbf{k}\in\mathcal{B}_\Lambda}{}' \frac{1}{\phi + |\mathbf{k}|^\sigma} = \frac{1}{L^d} \sum_{\mathbf{k}\in\mathcal{B}_\Lambda}{}' \frac{1}{|\mathbf{k}|^\sigma} - \frac{1}{L^d} \sum_{\mathbf{k}\in\mathcal{B}_\Lambda}{}' \frac{\phi}{|\mathbf{k}|^\sigma(\phi + |\mathbf{k}|^\sigma)}. \tag{5.44}$$

The first term in the right-hand side of the above identity is independent of ϕ and approaches the integral $W_{d,\sigma}(0) = K_c$ as $L \to \infty$, see Eqs. (3.28) and (3.26). Therefore, we can write

$$\frac{1}{L^d} \sum_{\mathbf{k}\in\mathcal{B}_\Lambda}{}' \frac{1}{|\mathbf{k}|^\sigma} = K_c + \varepsilon_L, \qquad \varepsilon_L \to 0, \quad \text{as} \quad L \to \infty, \tag{5.45}$$

where the correction term ε_L is to be explicitly evaluated below. The second term in the right-hand side of Eq. (5.44) converges when the summation over $\mathbf{k} \in \mathcal{B}_\Lambda$, $|\mathbf{k}| \neq 0$, is extended to infinity, i.e. to the set (4.124). Therefore, in the leading order approximation we can write

$$\frac{1}{L^d} \sum_{\mathbf{k}\in\mathcal{B}_\Lambda}{}' \frac{\phi}{|\mathbf{k}|^\sigma(\phi + |\mathbf{k}|^\sigma)} \simeq \frac{1}{L^{d-\sigma}} \sum_{\mathbf{n}(d)}{}' \frac{\phi L^\sigma}{(2\pi|\mathbf{n}|)^\sigma(\phi L^\sigma + (2\pi|\mathbf{n}|)^\sigma)}$$

$$\equiv \frac{1}{L^{d-\sigma}} S_{d,0,\sigma}(\phi L^\sigma). \tag{5.46}$$

Thus, the leading-order asymptotic form of $\Sigma_\Lambda^{d,\sigma}(\phi)$ as $\phi \to 0$, $L \to \infty$, so that $\phi L^\sigma = O(1)$, is given by

$$\Sigma_\Lambda^{d,\sigma}(\phi) \simeq K_c + \frac{1}{L^{d-\sigma}}\left[\frac{1}{\phi L^\sigma} - S_{d,0,\sigma}(\phi L^\sigma)\right] + \varepsilon_L. \tag{5.47}$$

Next we show that the result (5.47), including the explicit expression for ε_L, follows from Eq. (5.33) at $d' = 0$ by identical transformations of the universal finite-size scaling function $F_{d,0,\sigma}$, provided one neglects terms of order $O(\phi)$. To this end we note that the $d - d'$-fold sum over $k_1, \ldots, k_{d-d'}$ in the right-hand side of Eq. (5.34) is the same as in the case of short-range interaction, and apply to it the Jacoby identity, as shown in Eqs. (4.127) - (4.129). Thus, the function $F_{d,0,\sigma}(y)$ can be written in the form (compare with Eq. (4.130))

$$F_{d,0,\sigma}(y) = y^{-\sigma} + y^{-\sigma}\int_0^\infty dx\, x^{\sigma/2-1} E_{\frac{\sigma}{2},\frac{\sigma}{2}}(-x^{\sigma/2})$$
$$\times \left[\Theta_3^d\left(\frac{4\pi^2 x}{y^2}\right) - 1 - \frac{y^d}{(4\pi x)^{d/2}}\right]. \tag{5.48}$$

Here we have taken into account the equality

$$\int_0^\infty dx\, x^{\alpha-1} E_{\alpha,\alpha}(-x^\alpha) = 1, \qquad \alpha > 0, \tag{5.49}$$

which follows by integration of Eq. (5.16) over z from zero to infinity. Next, by making use of the identity, see Eq. (5.8),

$$\int_0^\infty dx\, e^{-zx}\, x^{\alpha-1}\left[E_{\alpha,\alpha}(-x^\alpha) - \frac{1}{\Gamma(\alpha)}\right] = -\frac{1}{(1+z^\alpha)z^\alpha}, \tag{5.50}$$

we represent the integral in the right-hand side of Eq. (5.48) as a sum of three terms. The first term is given by

$$y^{-\sigma}\int_0^\infty dx\, x^{\sigma/2-1}\left[E_{\frac{\sigma}{2},\frac{\sigma}{2}}(-x^{\sigma/2}) - \frac{1}{\Gamma(\sigma/2)}\right]\left[\Theta_3^d\left(\frac{4\pi^2 x}{y^2}\right) - 1\right]$$
$$= -S_{d,0,\sigma}(y^\sigma), \tag{5.51}$$

the second term is

$$-\frac{y^{d-\sigma}}{(4\pi)^{d/2}}\int_0^\infty dx\, x^{-(d-\sigma+2)/2}\left[E_{\frac{\sigma}{2},\frac{\sigma}{2}}(-x^{\sigma/2}) - \frac{1}{\Gamma(\sigma/2)}\right] = y^{d-\sigma} D_{d,\sigma}, \tag{5.52}$$

and the third one equals the constant (provided $d > \sigma$)

$$\frac{1}{(2\pi)^\sigma \Gamma(\sigma/2)} \int_0^\infty dx\, x^{\sigma/2-1} \left[\Theta_3^d(x) - 1 - \left(\frac{\pi}{x}\right)^{d/2} \right] = C_{d,\sigma}. \qquad (5.53)$$

Note that the constant $D_{d,\sigma}$ in Eq. (5.52) is the same as the one defined in Eq. (3.33) for $\sigma < d < 2\sigma$. To see that, we first write

$$x^{-d/2} = \frac{1}{\Gamma(d/2)} \int_0^\infty dt\, t^{d/2-1} e^{-xt}, \qquad (5.54)$$

by using the identity (5.50) take the integral over x, and then the integral over t ($\sigma < d < 2\sigma$):

$$D_{d,\sigma} \equiv -\frac{1}{(4\pi)^{d/2}\Gamma(d/2)} \int_0^\infty dt\, t^{d/2-1} \int_0^\infty dx\, e^{-tx} x^{\sigma/2-1}$$

$$\times \left[E_{\frac{\sigma}{2},\frac{\sigma}{2}}(-x^{\sigma/2}) - \frac{1}{\Gamma(\sigma/2)} \right]$$

$$= \frac{1}{(4\pi)^{d/2}\Gamma(d/2)} \int_0^\infty dt\, \frac{t^{d/2-\sigma/2-1}}{1+t^{\sigma/2}} = \frac{2\pi}{(4\pi)^{d/2}\sigma\Gamma(d/2)|\sin(\pi d/\sigma)|}.$$

The constant $C_{d,\sigma}$ in Eq. (5.53) is a Madelung-type lattice constant, which can be written in the form [Brankov and Tonchev (1988)] (compare with Eq. (4.134))

$$C_{d,\sigma} = \frac{1}{(2\pi)^\sigma \Gamma(\sigma/2)} \lim_{\delta \to 0} \left\{ \sum_{\mathbf{n}\in Z^d}{}' \frac{\Gamma(\sigma/2, \delta|\mathbf{n}|^2)}{|\mathbf{n}|^\sigma} \right.$$

$$\left. - \int_{R^d} d^d\mathbf{n}\, \frac{\Gamma(\sigma/2, \delta|\mathbf{n}|^2)}{|\mathbf{n}|^\sigma} \right\}, \qquad (5.55)$$

where $\Gamma(\alpha, x)$ is the incomplete gamma function, $\Gamma(\alpha, 0) = \Gamma(\alpha)$.

The collection of the above results yields

$$F_{d,0,\sigma}(y) = D_{d,\sigma} y^{d-\sigma} + y^{-\sigma} - S_{d,0,\sigma}(y^\sigma) + C_{d,\sigma}. \qquad (5.56)$$

By substitution of the above expression into Eq. (5.33), taking into account the expansion (3.32) of the integral $W_{d,\sigma}(\phi)$, and neglecting in that expansion terms of the order $O(\phi) = O(yL^{-\sigma})$, we obtain Eq. (5.47) with

$$\varepsilon_L = L^{-d+\sigma} C_{d,\sigma}. \qquad (5.57)$$

Note that in terms of the finite-size scaling temperature, x_1, and magnetic field, x_2, variables defined in Eq. (5.35), the finite-size mean spherical constraint (3.23) takes the universal form

$$\frac{1}{y^\sigma} - \sum_{\mathbf{n}(d)}{}' \frac{y^\sigma}{(2\pi|\mathbf{n}|)^\sigma[y^\sigma + (2\pi|\mathbf{n}|)^\sigma]} + C_{d,\sigma} + x_1 + \frac{x_2^2}{y^{2\sigma}} = 0, \qquad (5.58)$$

where $y = L\phi^{1/\sigma}$. Therefore, when $(K,h) \to (K_c,0)$ simultaneously with $L \to \infty$, in the way prescribed by the equations

$$K = K_c - \frac{x_1}{L^{d-\sigma}}, \qquad h = \frac{K_c^{1/2} x_2}{L^{(d+\sigma)/2}}, \qquad (5.59)$$

with $x_1 = O(1)$ and $x_2 = O(1)$, then the leading-order asymptotic form of the finite-size characteristic length $\lambda_L(K,h)$, see Eq. (5.25), is given by

$$\lambda_L(K,h) \simeq L/y_{d,0,\sigma}(x_1, x_2), \qquad (5.60)$$

where $y = y_{d,0,\sigma}(x_1, x_2)$ is the positive solution of Eq. (5.58). Hence, at the critical point $x_1 = x_2 = 0$, we obtain

$$\lambda_\Lambda(K_c, 0) = A_\lambda^{d,0,\sigma} L, \qquad (5.61)$$

where $A_\lambda^{d,0,\sigma} = 1/y_{d,0,\sigma}(0,0)$ is an universal amplitude.

5.3.2 *Infinitely long cylinder* $\Lambda = L^{d-1} \times \infty$

This case is especially interesting, since the long-range nature of the interaction provides, even in one infinite space dimension, two types of qualitatively different behavior of a finite-syze system: a true critical behavior and pseudocritical one, depending of the magnitude of σ.

Let us start again with an heuristic consideration of the sum (5.29) in the mean spherical constraint. The case of $\Lambda = L^{d-1} \times \infty$ is obtained by setting $L_j = L$ for $j = 1, \ldots, d-1$, and taking the limit $L_d \to \infty$:

$$\begin{aligned}
\Sigma_\Lambda^{d,\sigma}(\phi) &= \frac{1}{L^{d-1}} W_{1,\sigma}(\phi) \\
&+ \frac{1}{L^{d-1}} \sum_{\mathbf{k} \in \mathcal{B}_\Lambda(d-1)}{}' \frac{1}{2\pi} \int_{-\pi}^{\pi} \frac{dk_d}{\phi + (|\mathbf{k}|^2 + k_d^2)^{\sigma/2}},
\end{aligned} \qquad (5.62)$$

where $\mathcal{B}_\Lambda(d-1)$ is the Brillouin zone for a $(d-1)$-dimensional system, and

$$W_{1,\sigma}(\phi) = \frac{1}{\pi} \int_0^\pi \frac{dk}{\phi + k^\sigma} \tag{5.63}$$

is the Watson-type integral (3.28) for a one-dimensional infinite system. Now one has to distinguish three qualitatively different regimes determined by the value of σ. Formally, they correspond to different types of asymptotic behavior of the integral (5.63) as $\phi \to 0$:

$$W_{1,\sigma}(\phi) \simeq \begin{cases} [\pi^\sigma(1-\sigma)]^{-1}, & \text{if} \quad 0 < \sigma < 1 \\ \pi^{-1} \ln(1 + \pi/\phi), & \text{if} \quad \sigma = 1 \\ D_{1,\sigma}\phi^{-(\sigma-1)/\sigma}, & \text{if} \quad 1 < \sigma < 2. \end{cases} \tag{5.64}$$

From a physical point of view, *the infinite one-dimensional system*, which has a finite size L in the remaining $d - 1$ dimensions, can be found in three qualitatively different situations depending on the value of σ: (i) If $0 < \sigma < 1$, then the system is above its lower critical dimension $d_l = \sigma$ and, therefore, it exhibits a true critical behavior. A crossover from 1-dimensional to d-dimensional critical behavior takes place when $L \to \infty$. (ii) In the borderline case of $\sigma = d = 1$, the system is at its lower critical dimension and may have only a zero-temperature critical point. (iii) When $1 < \sigma < 2$, the system is below its lower critical dimension and a (d-dimensional) critical behavior appears only in the thermodynamic limit $L \to \infty$.

Here we confine ourselves to the consideration of the simplest case $1 < \sigma < 2$, when the finite-L system has no critical point. In this case, by taking into account Eq. (5.64), and by changing the integration variable $k_d \to t/L$ in the right-hand side of Eq. (5.62), we cast the latter equation in the form

$$\Sigma_\Lambda^{d,\sigma}(0;\phi) \simeq \frac{1}{L^{d-\sigma}} \left[\frac{1}{(L^\sigma \phi)^{1-1/\sigma}} \right.$$
$$\left. + \sum_{n_1=-(L-1)/2}^{(L-1)/2} \cdots \sum_{n_{d-1}=-(L-1)/2}^{(L-1)/2}{}' \frac{1}{2\pi} \int_{-\pi L}^{\pi L} \frac{dt}{\phi L^\sigma + [(2\pi|\mathbf{n}|)^2 + t^2]^{\sigma/2}} \right]. \tag{5.65}$$

We see that the same intrinsic combination $L^\sigma \phi$ emerges, apart from the factor $L^{-d+\sigma}$ and the L-dependence in the limits of integration and summation. Since these limits cannot be extended to infinity, due to the divergence at $d > \sigma$, we make use of the identity

$$
\frac{1}{L^{d-1}} \sum_{\mathbf{k} \in \mathcal{B}_\Lambda(d-1)}{}' \frac{1}{2\pi} \int_{-\pi}^{\pi} \frac{\mathrm{d}k_d}{\phi + (|\mathbf{k}|^2 + k_d^2)^{\sigma/2}} =
$$

$$
\frac{1}{L^{d-1}} \sum_{\mathbf{k} \in \mathcal{B}_\Lambda(d-1)}{}' \frac{1}{\pi} \int_{0}^{\pi} \frac{\mathrm{d}k_d}{(|\mathbf{k}|^2 + k_d^2)^{\sigma/2}}
$$

$$
- \frac{1}{L^{d-1}} \sum_{\mathbf{k} \in \mathcal{B}_\Lambda(d-1)}{}' \frac{\phi}{\pi} \int_{0}^{\pi} \frac{\mathrm{d}k_d}{(|\mathbf{k}|^2 + k_d^2)^{\sigma/2}[\phi + (|\mathbf{k}|^2 + k_d^2)^{\sigma/2}]}. \tag{5.66}
$$

The first term in the right-hand side of the above identity is independent of ϕ and, as $L \to \infty$, approaches the integral $W_{d,\sigma}(0) = K_c$. Therefore, we can write

$$
\frac{1}{L^{d-1}} \cdot \sum_{\mathbf{k} \in \mathcal{B}_\Lambda(d-1)}{}' \frac{1}{\pi} \int_{0}^{\pi} \frac{\mathrm{d}k_d}{(|\mathbf{k}|^2 + k_d^2)^{\sigma/2}} = K_c + \varepsilon_L^{(1)}, \tag{5.67}
$$

with a correction term $\varepsilon_L^{(1)} \to 0$ as $L \to \infty$, which is to be explicitly evaluated below. The second term in the right-hand side of Eq. (5.66) converges when the summation over $\mathbf{k} \in \mathcal{B}_\Lambda(d-1)$, $|\mathbf{k}| \neq 0$, is extended to infinity. Therefore, in the leading order we have

$$
\frac{\phi}{L^{d-1}} \sum_{\mathbf{k} \in \mathcal{B}_\Lambda(d-1)}{}' \frac{1}{\pi} \int_{0}^{\pi} \frac{\mathrm{d}k_d}{(|\mathbf{k}|^2 + k_d^2)^{\sigma/2}[\phi + (|\mathbf{k}|^2 + k_d^2)^{\sigma/2}]} \simeq
$$

$$
\frac{\phi L^\sigma}{L^{d-\sigma}} \sum_{\mathbf{n}(d-1)}{}' \frac{1}{\pi} \int_{0}^{\infty} \frac{\mathrm{d}x}{[(2\pi|\mathbf{n}|)^2 + x^2]^{\sigma/2}\{\phi L^\sigma + [(2\pi|\mathbf{n}|)^2 + x^2]^{\sigma/2}]\}}
$$

$$
\equiv \frac{1}{L^{d-\sigma}} S_{d,1,\sigma}(\phi L^\sigma). \tag{5.68}
$$

In deriving the last expression we have changed the integration variable $k_d \to x L^{-1}$ and extended the upper limit πL in the integral over x to infinity, since $d < 2\sigma$. Thus, the leading-order asymptotic behavior of $\Sigma_\Lambda^{d,\sigma}(\phi)$ in the cylinder geometry, as $\phi \to 0$, $L \to \infty$, so that $\phi L^\sigma = O(1)$, is given by ($1 < \sigma < 2$):

$$
\Sigma_\Lambda^{d,\sigma}(\phi) \simeq K_c + \frac{D_{1,\sigma}}{L^{d-1}\phi^{1-1/\sigma}} - \frac{1}{L^{d-\sigma}} S_{d,1,\sigma}(\phi L^\sigma) + \varepsilon_L^{(1)}. \tag{5.69}
$$

Let us now turn back to Eq. (5.34) and apply to its right-hand side at $d' = 1$ the transformations based on the Jacobi identity, see Eqs. (4.127) -

(4.129). Then, for $\sigma > 1$ we obtain

$$F_{d,1,\sigma}(y) = D_{1,\sigma}y^{1-\sigma} + \frac{y^{1-\sigma}}{\sqrt{4\pi}}\int_0^\infty dx\, x^{\sigma/2-3/2}E_{\frac{\sigma}{2},\frac{\sigma}{2}}(-x^{\sigma/2})$$

$$\times\left[\Theta_3^{d-1}\left(\frac{4\pi^2 x}{y^2}\right) - 1 - \frac{y^{d-1}}{(4\pi x)^{(d-1)/2}}\right]. \tag{5.70}$$

Here we have used the equality

$$\frac{1}{\sqrt{4\pi}}\int_0^\infty dx\, x^{\sigma/2-3/2}E_{\frac{\sigma}{2},\frac{\sigma}{2}}(-x^{\sigma/2}) = D_{1,\sigma}\,, \tag{5.71}$$

which can be proved by taking the integral with the aid of Eq. (5.54) and the basic identity (5.8).

Next, by subtracting and adding the constant $1/\Gamma(\sigma/2)$ to the function $E_{\sigma/2,\sigma/2}(\cdot)$, we represent the integral in Eq. (5.70) as a sum of three terms. With the aid of Eq. (5.54) at $d = 1$, the first term can be written in a form suitable for taking the integral over x by applying the identity (5.50) with $z = t + (2\pi|\mathbf{n}|/y)^2$. In the remaining integral over t we change the variable $t \to x^2/y^2$ and obtain

$$\frac{y^{1-\sigma}}{\sqrt{4\pi}}\int_0^\infty dx\, x^{\sigma/2-3/2}\left[E_{\frac{\sigma}{2},\frac{\sigma}{2}}(-x^{\sigma/2}) - \frac{1}{\Gamma(\sigma/2)}\right]$$

$$\times\left[\Theta_3^{d-1}\left(\frac{4\pi^2 x}{y^2}\right) - 1\right] = -S_{d,1,\sigma}(y^\sigma). \tag{5.72}$$

The second term is the same as in Eq. (5.52), and the third term yields the constant

$$C_{d,1,\sigma} = \frac{\pi^{1/2-\sigma}}{2^\sigma\Gamma(\sigma/2)}\int_0^\infty dx\, x^{\sigma/2-3/2}\left[\Theta_3^{d-1}(x) - 1 - \left(\frac{\pi}{x}\right)^{(d-1)/2}\right], \tag{5.73}$$

valid for $1 < \sigma < d$.

Collecting the above results we obtain

$$F_{d,1,\sigma}(y) = D_{d,\sigma}y^{d-\sigma} + D_{1,\sigma}y^{1-\sigma} - S_{d,1,\sigma}(y^\sigma) + C_{d,1,\sigma} \tag{5.74}$$

By substitution of this expression into Eq. (5.33) at $d' = 1$, taking into account the small-argument expansion (3.32) of $W_{d,\sigma}(\phi)$, and neglecting in that expansion terms of the order $O(\phi) = O(yL^{-\sigma})$, we obtain Eq. (5.69) with

$$\varepsilon_L^{(1)} = L^{-d+\sigma}C_{d,1,\sigma}. \tag{5.75}$$

Finally, in terms of the scaled temperature and magnetic field variables defined in Eq. (5.35), the mean spherical constraint for the considered geometry of Λ takes the universal finite-size scaling form ($y = L\phi^{1/\sigma}$):

$$D_{1,\sigma}y^{1-\sigma} \; - \; \sum_{\mathbf{n}(d-1)}{}' \frac{1}{\pi} \int_0^\infty \frac{\mathrm{d}x\, y^\sigma}{[(2\pi|\mathbf{n}|)^2 + x^2]^{\sigma/2}\{y^\sigma + [(2\pi|\mathbf{n}|)^2 + x^2]^{\sigma/2}]\}}$$

$$+ \; C_{d,1,\sigma} + x_1 + \frac{x_2^2}{y^{2\sigma}} = 0, \qquad (1 < \sigma < 2). \tag{5.76}$$

In complete analogy with the short-range case, when the critical point is approached as $L \to \infty$ in accordance with Eq. (5.59), the leading-order asymptotic form of the finite-size characteristic length $\lambda_L(K, h)$, see Eq.(5.25), is given by

$$\lambda_L(K, h) \simeq L/y_{d,1,\sigma}(x_1, x_2), \tag{5.77}$$

where $y = y_{d,1,\sigma}(x_1, x_2)$ is the positive solution of Eq. (5.76). At the critical point $x_1 = x_2 = 0$ we obtain

$$\lambda_\Lambda(K_c, 0) = A_\lambda^{d,1,\sigma}\, L, \tag{5.78}$$

where $A_\lambda^{d,1,\sigma} = 1/y_{d,1,\sigma}(0, 0)$ is a universal amplitude.

5.4   Finite-size shift of the critical temperature

In this section, following [Chamati and Tonchev (1996)]), we study the shift from the bulk value T_c of the critical (when $d' > d_l$), or pseudocritical (when $d' < d_l$) temperature of the finite-size mean spherical model with the geometry $L^{d-d'} \times \infty^{d'}$ under periodic boundary conditions in the finite dimensions. Now we are in the position to consider in a unified way the cases of short- and long-range interactions, distinguished by the value of the decay exponent σ.

We remind the reader, see also Section 4.2, that if the d-dimensional bulk system exhibits singularities at a critical temperature T_c, then a phase transition occurs in the corresponding finite-size system only if the number of space dimensions d', in which the latter extends infinitely, exceeds the lower critical dimensionality d_l. In this case the value of the critical temperature $T_{c,L}$, at which the finite-size thermodynamic functions exhibit singularities, is shifted from the bulk value T_c. In the opposite case, when the number of infinite dimensions is less than the lower critical dimension, the singularities of the bulk thermodynamic functions are rounded and no phase transition occurs in the finite-size system. Nevertheless, when $d' < d_l$ one can define a pseudocritical temperature $T_{m,L}$, corresponding to the position of the smeared singularities of the finite-size thermodynamic functions, and study its shift with respect to the bulk value T_c. Actually, for the sake of convenience, we will study here the shift of the dimensionless critical (pseudocritical) coupling $K_{c,L} = J/k_B T_{c,L}$ (respectively, $K_{m,L} = J/k_B T_{m,L}$), rather than the shift of the corresponding temperatures.

Confining ourselves to the zero-field case ($h = 0$), we write the finite-size mean spherical constraint (3.23) in the form

$$\Sigma_\Lambda^{d,\sigma}(\phi) = K. \tag{5.79}$$

The asymptotic behavior of $\Sigma_\Lambda^{d,\sigma}(\phi)$, given by Eq. (5.33), contains the universal finite-size scaling function $F_{d,d',\sigma}(\cdot)$ which has been studied in detail at $d' = 0$ and $d' = 1$ in the previous section, see expressions (5.56) and (5.74), respectively. These expressions can be written in a general form, valid at arbitrary real value of $d' < \sigma$, as follows

$$F_{d,d',\sigma}(y) = D_{d',\sigma} y^{d'-\sigma} + D_{d,\sigma} y^{d-\sigma} - S_{d,d',\sigma}(y^\sigma) + C_{d,d',\sigma}. \tag{5.80}$$

Here the constant $D_{d,\sigma}$ is explicitly given by Eq. (3.33); at $d < \sigma$ it has the integral representation, compare with Eq. (5.71),

$$D_{d,\sigma} = \frac{1}{(4\pi)^{d/2}} \int_0^\infty dx\, x^{(\sigma-d)/2-1} E_{\sigma/2,\sigma/2}(-x^{\sigma/2}) \qquad (d < \sigma), \tag{5.81}$$

and at $\sigma < d < 2\sigma$ its integral representation is given by Eq. (5.52); the function $S_{d,d',\sigma}(\cdot)$ is a straightforward generalization of Eq. (5.72) to real values of d',

$$S_{d,d',\sigma}(y^\sigma) = -\frac{y^{d'-\sigma}}{(4\pi)^{d'/2}} \int_0^\infty dx\, x^{(\sigma-d')/2-1}$$
$$\times \left[E_{\frac{\sigma}{2},\frac{\sigma}{2}}(-x^{\sigma/2}) - \frac{1}{\Gamma(\sigma/2)} \right] \left[\Theta_3^{d-d'}\left(\frac{4\pi^2 x}{y^2} \right) - 1 \right]; \tag{5.82}$$

the constant $C_{d,d',\sigma}$ is given by the following generalized version of Eq.

(5.73),

$$C_{d,d',\sigma} = \frac{\pi^{d'/2}}{(2\pi)^\sigma \Gamma(\sigma/2)} \int_0^\infty dx\, x^{(\sigma-d')/2-1}$$

$$\times \left[\Theta_3^{d-d'}(x) - 1 - \left(\frac{\pi}{x}\right)^{\frac{1}{2}(d-d')} \right], \qquad (d' < \sigma) \tag{5.83}$$

valid at $d' < \sigma < d$. Note that for integer d' the constant $C_{d,d',\sigma}$ can be interpreted as a generalization of the Madelung type constant (5.55):

$$C_{d,d',\sigma} = \frac{\pi^{d'/2}}{(2\pi)^\sigma \Gamma(\sigma/2)} \lim_{\lambda \to 0^+} \left[\sideset{}{'}\sum_{\mathbf{k}(d-d')} \int_\lambda^\infty dx\, x^{(\sigma-d')/2-1} e^{-x|\mathbf{k}|^2} \right.$$

$$\left. - \pi^{(d-d')/2} \int_\lambda^\infty dx\, x^{(\sigma-d)/2-1} \right]$$

$$= \frac{\pi^{d'/2}}{(2\pi)^\sigma \Gamma(\sigma/2)} \lim_{\lambda \to 0^+} \left\{ \sideset{}{'}\sum_{\mathbf{k}(d-d')} \frac{\Gamma((\sigma-d')/2, \lambda|\mathbf{k}|^2)}{|\mathbf{k}|^{\sigma-d'}} \right.$$

$$\left. - \int_{R^{d-d'}} d^{d-d'}\mathbf{k}\, \frac{\Gamma((\sigma-d')/2, \lambda|\mathbf{k}|^2)}{|\mathbf{k}|^{\sigma-d'}} \right\}. \tag{5.84}$$

By using Eqs. (5.33) and (5.80), we cast Eq. (5.79) for the spherical field in the form ($y = L\phi^{1/\sigma}$)

$$W_{d,\sigma}(\phi) + L^{\sigma-d} \left[D_{d',\sigma} y^{d'-\sigma} + D_{d,\sigma} y^{d-\sigma} - S_{d,d',\sigma}(y^\sigma) + C_{d,d',\sigma} \right] = K. \tag{5.85}$$

To obtain the key results of this section, we replace the integral $W_{d,\sigma}(\phi)$ by its small argument asymptotic expansion (3.32) at $\sigma < d < 2\sigma$. The independent of ϕ terms in the left-hand side of Eq. (5.85) are identified with the finite-size pseudocritical coupling $K_{m,L}$ for $d' < \sigma$, or with the finite-size critical coupling $K_{c,L}$ for $d' > \sigma$. Thus, for the shift δK_L of the above finite-size quantities from the bulk critical coupling K_c we obtain (c.f. Eqs. (5.57) and (5.75))

$$\delta K_L = L^{\sigma-d} C_{d,d',\sigma} \qquad (d' \neq \sigma). \tag{5.86}$$

Therefore, the shift depends on the dimensionalities d, d', as well as on the decay exponent σ. Since the critical exponent ν for the characteristic length of the d-dimensional bulk system equals $1/(d-\sigma)$ for $\sigma < d < 2\sigma$, we see from Eq. (5.86) that the L-dependence of the finite-size shift of the critical coupling is $L^{-1/\nu}$, in accordance with the standard finite-size scaling conjecture, see Section 4.2.

Let us now consider the properties of the constant $C_{d,d',\sigma}$ as a function of d, d' and σ. The mathematical tool of our analysis is an identity due to [Singh and Pathria (1989)]. Following that work, we start from the m-dimensional version ($m = 1, 2, 3, \ldots$) of the Jacoby identity (4.129):

$$1 + \sum_{\mathbf{k}(m)}{}' \exp(-\lambda|\mathbf{k}|^2) = \left(\frac{\pi}{\lambda}\right)^{m/2} \left[1 + \sum_{\mathbf{k}(m)}{}' \exp(-\pi^2|\mathbf{k}|^2/\lambda)\right]. \tag{5.87}$$

Multiplying both sides of Eq. (5.87) by $\lambda^{\frac{1}{2}m-1-\nu}$, taking an indefinite integral over λ (assuming $\nu \neq 0, m/2$), the resulting equality can be written in the form

$$C(\nu|m) = -\frac{\pi^{m/2}}{\nu\lambda^\nu} + \pi^{\frac{1}{2}m-2\nu} \sum_{\mathbf{k}(m)}{}' \int_{\pi^2|\mathbf{k}|^2/\lambda}^{\infty} \mathrm{d}x \frac{\mathrm{e}^{-x}x^{\nu-1}}{|\mathbf{k}|^{2\nu}}$$
$$-\frac{\lambda^{\frac{1}{2}m-\nu}}{\frac{1}{2}m-\nu} + \sum_{\mathbf{k}(m)}{}' \int_{\lambda|\mathbf{k}|^2}^{\infty} \mathrm{d}x \frac{\mathrm{e}^{-x}x^{\frac{1}{2}m-\nu-1}}{|\mathbf{k}|^{m-2\nu}}. \tag{5.88}$$

Here $C(\nu|m)$ is an integration constant independent of the parameter λ, therefore, the right-hand side must be independent of it too. By setting $m = d - d'$ and $\nu = (d - \sigma)/2$ in Eq. (5.88), we see that if $d' < \sigma < d$, then the second and third terms in the right-hand side vanish in the limit $\lambda \to 0^+$, while the first and fourth terms yield, see Eq. (5.83),

$$C\left(\frac{d-\sigma}{2}\bigg| d-d'\right) = \int_0^{\infty} \mathrm{d}x\, x^{(\sigma-d')/2-1}$$
$$\times \left[\Theta_3^{d-d'}(x) - 1 - \left(\frac{\pi}{x}\right)^{(d-d')/2}\right]. \tag{5.89}$$

On the other hand, if $d' > \sigma$, then the first, third and fourth terms in the right-hand side of Eq. (5.88) vanish in the limit $\lambda \to \infty$, while the second term yields

$$C\left(\frac{d-\sigma}{2}\bigg| d-d'\right) = \frac{\pi^\sigma}{\pi^{(d+d')/2}}\Gamma\left(\frac{d-\sigma}{2}\right) \sum_{\mathbf{k}(d-d')}{}' |\mathbf{k}|^{-d+\sigma} \quad (d' > \sigma).$$
$$\tag{5.90}$$

Note that Eqs. (5.89) and (5.90) are valid in complementary regions. The sum in the right-hand side of the last equation can be expressed in terms of the Epstein zeta function, see [Glasser and Zucker (1980)],

$$\mathcal{Z}\left.\begin{vmatrix} 0 \\ 0 \end{vmatrix}\right.(d-d',d-\sigma) = \sideset{}{'}\sum_{\mathbf{k}(d-d')} \frac{1}{|\mathbf{k}|^{d-\sigma}}, \tag{5.91}$$

which can be regarded as generalized $(d-d')$-dimensional analog of the Riemann zeta function $\zeta(d-\sigma)$. In the case under consideration the Epstein zeta function has only a simple pole at $d' = \sigma$, hence, it can be analytically continued to $0 \le d' < \sigma$. Therefore, the shift constant $C_{d,d',\sigma}$ in Eq. (5.86), which at $d' < \sigma$ is given by

$$C_{d,d',\sigma} = \frac{\pi^{d'/2}}{(2\pi)^{\sigma}\Gamma(\sigma/2)} C\left(\left.\frac{d-\sigma}{2}\right| d-d'\right), \tag{5.92}$$

see Eqs. (5.83) and (5.89), in the complementary region $d' > \sigma$ can be represented as

$$C_{d,d',\sigma} = \frac{\Gamma((d-\sigma)/2)}{2^{\sigma}\pi^{d/2}\Gamma(\sigma/2)} \sideset{}{'}\sum_{\mathbf{k}(d-d')} |\mathbf{k}|^{-d+\sigma}. \tag{5.93}$$

Indeed, from Eqs. (5.86) and (5.93) we recover the result for δK_L obtained in [Brankov and Danchev (1991)].

For some particular values of the finite dimensions of the lattice, the sum (5.91) can be expressed as a product of simple sums such as Dirichlet series. This has been done for $d - d' = 1, 2, 4, 6$ and 8 in [Singh and Pathria (1989)]. The case of $d = 3$ has been investigated numerically in [Glasser and Zucker (1980)]. The constant $C(x|y)$ at some physically interesting values of x and y has been calculated in [Chamati and Tonchev (2000a)].

The behavior of $\delta K_L L^{d-\sigma}$ as a function of σ, and some numerical results in the most interesting cases are presented in table 5.1 and Figs. 5.1-5.4.

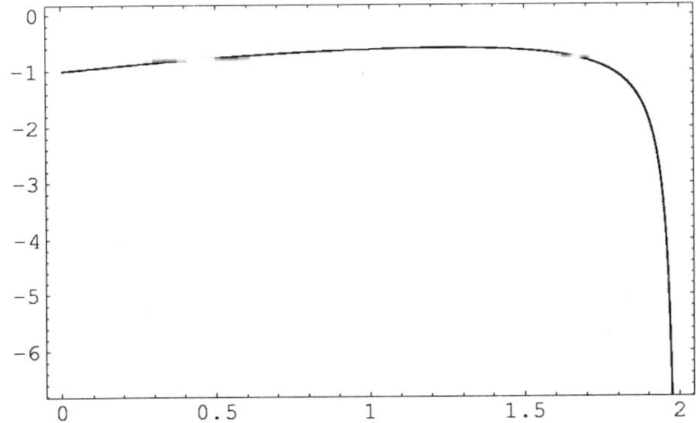

Fig. 5.1   The shift of the critical coupling as a function of σ at $d = 2$ and $d' = 0$.

$d = 3$	$\delta K_L L^{d-\sigma}$	
d'	$\sigma = 1$	$\sigma = 2$
0	-0.800387	-0.225785
1	$\mp\infty$	-0.310373
2	0.166667	$\mp\infty$
3	0	0
$d = 2$	$\delta K_L L^{d-\sigma}$	
d'	$\sigma = 1/2$	$\sigma = 1$
0	-0.766643	-0.620746
1	0.397469	$\mp\infty$
2	0	0

Table 5.1. The shift of the critical coupling $\delta K_L L^{d-\sigma} = C_{d,d'\sigma}$ at $d = 3$ for some particular values of d' and σ.

The result at $d = 3$, $d' = 0$ and $\sigma = 2$ has been obtained in [Brankov and Tonchev (1988)]. One can see that the shifted critical temperature $T_{c,L}$ is lower than the bulk critical temperature T_c at different values of d, d' and σ, which is the "normal case", see [Brézin (1983)], while the pseudocritical temperature $T_{m,L}$ is higher than T_c. Note that in the limit $d' \to \sigma$ the shift is infinite. This can be explained by the behavior of the Epstein zeta function near its pole at $d' = \sigma$ [Glasser and Zucker (1980)], where the shift is $\delta K_L \sim (d' - \sigma)^{-1}$.

The shift of the critical temperature and related finite-size effects have been studied by a number of authors, e.g., [Fisher and Barber (1972)],

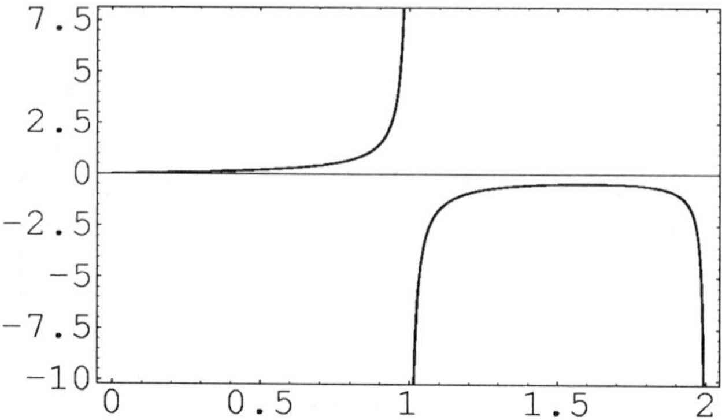

Fig. 5.2   The shift of the critical coupling as a function of σ for $d = 2$ and $d' = 1$.

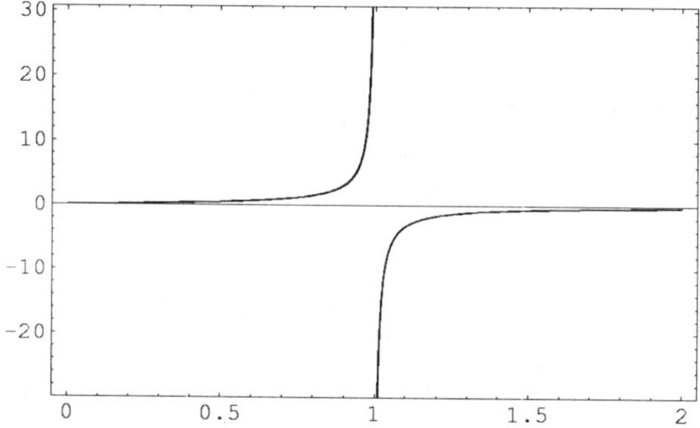

Fig. 5.3   The shift of the critical coupling as a function of σ at $d = 3$ and $d' = 1$.

[Barber (1983)], [Shapiro and Rudnick (1986)], [Allen and Pathria (1989)], [Brankov and Danchev (1991)]. Various techniques have been used to e-valuate the shift in the case of the spherical model: (i) In the case of a fully finite system with short-range interaction ($\sigma = 2$), an expression for the finite-size shift of the pseudocritical temperature has been derived in [Shapiro and Rudnick (1986)] by using an approach based on numerical approximations; (ii) in the case of a fully finite system with long-range interactions ($0 < \sigma \leq 2$), the shift has been obtained in [Brankov and

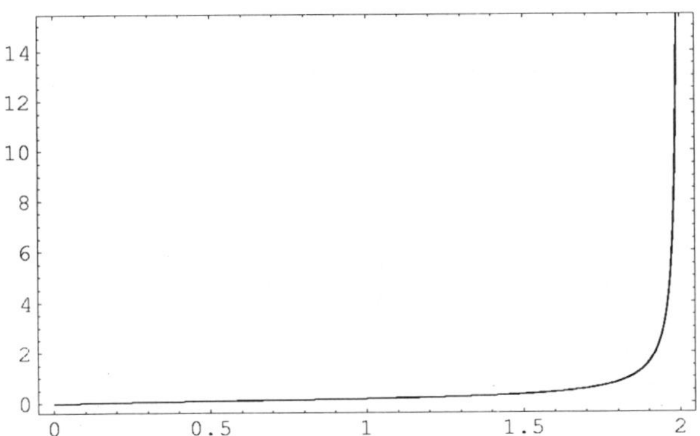

Fig. 5.4   The shift of the critical coupling as a function of σ at $d = 3$ and $d' = 2$.

Tonchev (1988)]. (*iii*) When the system exhibits a phase transition, the finite-size shift has been calculated in [Brézin (1983)] (for short-range interaction and $d' = d - 1$) and [Allen and Pathria (1989)] (for short-range interaction and $2 < d' < 4$); (*iv*) the case of long-range interaction and $\sigma < d' < 2\sigma$ has been considered in [Brankov and Danchev (1991)].

5.5   Finite-size scaling for the free energy density

To obtain the asymptotic form of the finite-size free energy density (3.22) of the mean spherical model in a d-dimensional parallelepiped with edges L_j, $j = 1, \ldots, d$, we make use of the factorized representation (5.19) of the basic d-fold sum $U_\Lambda^{d,\sigma}(\phi)$, which in combination with Eq. (5.30) yields

$$U_\Lambda^{d,\sigma}(\phi) \simeq U_{d,\sigma}(\phi) - \frac{\sigma\phi^{d/\sigma}}{2(4\pi)^{d/2}} \int_0^\infty dx \, x^{-d/2-1} E_{\frac{\sigma}{2},1}(-x^{\sigma/2})$$

$$\times \sum_{\mathbf{k}(d)}' \exp\left[-\frac{\phi^{2/\sigma}}{4x} \sum_{j=1}^d L_j^2 k_j^2\right]. \qquad (5.94)$$

Here we have used the fact that the integral $U_{d,\sigma}(\phi)$, defined in Eq. (3.25), can be written as

$$U_{d,\sigma}(\phi) \equiv \ln\phi$$

$$+ \frac{\sigma}{2} \int_0^\infty \frac{dx}{x} E_{\frac{\sigma}{2},1}(-x^{\sigma/2}) \left\{1 - \frac{\phi^{d/\sigma}}{(4\pi x)^{d/2}} \left[\mathrm{erf}\left(\frac{\pi x^{1/2}}{\phi^{1/\sigma}}\right)\right]^d\right\}. \qquad (5.95)$$

In the case when Λ has the geometry $L^{d-d'} \times \infty^{d'}$, with $0 \le d' < d$, Eq. (5.94) has the form

$$U_\Lambda^{d,\sigma}(\phi) \simeq U_{d,\sigma}(\phi) - \frac{1}{L^d}\Psi_{d,d',\sigma}(L\phi^{1/\sigma}), \qquad (5.96)$$

where $\Psi_{d,d',\sigma}(\cdot)$ is the universal finite-size scaling function

$$\Psi_{d,d',\sigma}(y) :=$$

$$\frac{\sigma y^d}{2(4\pi)^{d/2}} \int_0^\infty dx \, x^{-d/2-1} E_{\frac{\sigma}{2},1}(-x^{\sigma/2}) \sum_{\mathbf{k}(d-d')}' \exp\left[-\frac{y^2}{4x}|\mathbf{k}|^2\right]. \qquad (5.97)$$

By expanding the bulk term $U_{d,\sigma}(\phi)$ according to Eq. (3.44), we represent Eq. (5.96) in the form

$$U_\Lambda^{d,\sigma}(\phi) \simeq U_{d,\sigma}(0) + K_c\phi - \frac{1}{L^d}\left[\frac{\sigma}{d}D_{d,\sigma} \, L^d\phi^{d/\sigma} + \Psi_{d,d',\sigma}(L\phi^{1/\sigma})\right]. \qquad (5.98)$$

Note that from the extremum problem in the right-hand side of (3.22) we have

$$-\frac{1}{L^\sigma}\frac{\partial}{\partial\phi}\Psi_{d,d',\sigma}(L\phi^{1/\sigma}) = F_{d,d',\sigma}(L\phi^{1/\sigma}). \qquad (5.99)$$

The above relationship can be obtained directly, by changing the integration variable $x \to y^2 t$ in Eq. (5.97) and differentiating then with respect to ϕ by using Eq. (5.16).

After substitution of Eq. (5.98) into the definition of the free energy density (3.22), we obtain the following expression in the neighborhood of the critical point ($\phi \ll 1$):

$$\beta f_\Lambda^{m.s.}(K,h) \simeq \frac{1}{2}\left[\ln\frac{K}{2\pi} - \frac{1}{\rho_\sigma}K + U_{d,\sigma}(0)\right] + \frac{1}{L^d}X_f^{d,d',\sigma}(x_1,x_2). \quad (5.100)$$

Here $X_f^{d,d',\sigma}(\cdot,\cdot)$ is the universal finite-size scaling function for the *singular part* of the free energy density,

$$X_f^{d,d',\sigma}(x_1,x_2) = \frac{1}{2}\sup_{y>0}\left\{x_1 y^\sigma - \frac{x_2^2}{y^\sigma} - \frac{\sigma}{d}D_{d,\sigma}y^d - \Psi_{d,d',\sigma}(y)\right\}. \quad (5.101)$$

At the critical point $x_1 = x_2 = 0$ the singular part of the free energy density becomes ($\beta_c = k_B T_c$)

$$\beta_c f_\Lambda^{m.s.}(K_c,0) = A_f^{d,d',\sigma}\,L^{-d}, \quad (5.102)$$

where $A_f^{d,d',\sigma} = X_f^{d,d',\sigma}(0,0)$ is a universal amplitude.

Now we remark that all the finite-size thermodynamic functions $P_L(K,h)$, which in the bulk limit $L \to \infty$ become singular at the critical point $K = K_c$, $h = 0$, follow by differentiation of Eq. (5.101) with respect to the thermodynamic parameters K and h, respectively x_1 and x_2. Therefore the above result for the free energy density is in full conformity with *Assumption A.2* and this completes the verification of *Hypothesis A* in the case of long-range isotropic potential with decay exponent σ, provided $\sigma < d < 2\sigma$.

An alternative representation of the scaling function (5.97), suitable for the study of its asymptotic behavior as $y = L\phi^{1/\sigma} \to 0$, can be obtained in full analogy with the short-range case. It is based on the integral representation [Brankov (1989)]

$$E_{\alpha,1}(-x^\alpha) = \frac{1}{\sqrt{\pi x}}\int_0^\infty du\, E_{2\alpha,1}(-u^{2\alpha})\exp\left(-\frac{u^2}{4x}\right), \quad (5.103)$$

which can be obtained by means of term by term integration of the series representation of the integrand. With the aid of (5.103) one can transform Eq. (5.97) to

$$\Psi_{d,d',\sigma}(y) = \frac{\sigma}{\pi^{(d+1)/2}}\Gamma\left(\frac{d+1}{2}\right)\sum_{\mathbf{k}(d-d')}{}'|\mathbf{k}|^{-d}u_{d,\sigma}(y|\mathbf{k}|), \quad (5.104)$$

where the function $u_{d,\sigma}(\cdot)$ is defined as

$$u_{d,\sigma}(z) := \int_0^\infty dx (1+x^2)^{-(d+1)/2} E_{\sigma,1}(-z^\sigma x^\sigma). \qquad (5.105)$$

The following finite-size scaling representation of the mean spherical constraint can be obtained from the stationarity condition for the supremum problem in Eq. (5.101) with $\Psi_{d,d',\sigma}(y)$ given by Eq. (5.104):

$$-D_{d,\sigma} y^{d-\sigma} + \frac{1}{\pi^{(d+1)/2)}} \Gamma\left(\frac{d+1}{2}\right) \sum_{\mathbf{k}(d-d')}' |\mathbf{k}|^{-d+\sigma} w_{d,\sigma}(y|\mathbf{k}|)$$

$$+x_1 + \frac{x_2^2}{y^{2\sigma}} = 0, \qquad (5.106)$$

where

$$\begin{aligned} w_{d,\sigma}(z) &:= -z^{1-\sigma}\frac{d}{dz} u_{d,\sigma}(z) = \int_0^\infty dx \frac{x^\sigma}{(1+x^2)^{(d+1)/2}} E_{\sigma,\sigma}(-z^\sigma x^\sigma) \\ &= \frac{\sqrt{\pi}}{2^{d/2}\Gamma((d+1)/2)} \int_0^\infty dx \frac{x^{d/2}}{x^\sigma + z^\sigma} J_{d/2-1}(x). \qquad (5.107) \end{aligned}$$

The asymptotic behavior of the function $w_{d,\sigma}(z)$ as $z \to 0$ and $z \to \infty$ has been studied in [Brankov and Danchev (1991)]. Note that by using Eq. (5.4) one can obtain from Eq. (5.107) another representaion

$$w_{d,\sigma}(z) = \frac{\sqrt{\pi}}{2^{d/2}} \Gamma\left(\frac{d+1}{2}\right) z^{d-\sigma} \int_0^\infty dt \, Q_\sigma(t) K_{\frac{d}{2}-1}(z\sqrt{t}), \qquad (5.108)$$

which might be practically useful since the MacDonald function $K_\nu(\cdot)$ at half-integer ν can be expressed in terms of elementary functions.

5.6 Crossover to bulk asymptotic behavior

The finite-size scaling regime, studied in the previous sections, is defined by the condition $L/\lambda_L = O(1)$, which at $\sigma < d < 2\sigma$ implies that $O(x_1) \le O(1)$ and $O(x_2) \le O(1)$. However, the asymptotic validity of the expressions derived in this chapter is limited by the weaker condition $\lambda_\infty(K,h) \gg 1$, which means that K and h may be fixed close to the critical point. Then, from Eq. (5.35) it follows that as $L \to \infty$, one has $|x_1| \to \infty$, if $K \ne K_c$, and $|x_2| \to \infty$, if $h \ne 0$. Here we show that in these limits one obtains indeed a crossover from the finite-size scaling regime to the thermodynamic critical behavior. It is assumed that $\Lambda = L^{d-d'} \times \infty^{d'}$, with $d' = 0, 1$, $\sigma \in (d', 2)$ and $\sigma < d < 2\sigma$.

By a straightforward examination, one readily establishes the following properties of the term $S_{d,d',\sigma}(y^\sigma)$, defined by Eq. (5.51) at $d' = 0$, and Eq. (5.68) at $d' = 1$:

$$S_{d,d',\sigma}(0) = 0, \qquad \frac{\mathrm{d}}{\mathrm{d}z} S_{d,d',\sigma}(z) > 0, \qquad \text{for all} \quad z \geq 0, \tag{5.109}$$

and

$$S_{d,d',\sigma}(z) \simeq D_{d,\sigma} z^{d/\sigma - 1}, \qquad \text{as} \quad z \to \infty. \tag{5.110}$$

Consider now the case of fixed $K < K_c$, $|1 - K/K_c| \ll 1$, and $h = 0$, which implies that $x_1 \to \infty$, $x_2 = 0$. Then, from the corresponding equations of state, namely Eq. (5.58) at $d' = 0$, and Eq. (5.76) at $d' = 1$, it follows that $y_{d,d',\sigma}(x_1, 0) \to \infty$ as $x_1 \to \infty$. Therefore, by using the fact that the sum over $\mathbf{n} \in Z^{d-d'}$ in $S_{d,d',\sigma}(y^\sigma)$ approaches the corresponding integral, we find that independently of $d' = 0, 1$,

$$S_{d,d',\sigma}(y^\sigma) \simeq D_{d,\sigma} y^{d-\sigma}, \qquad \text{as} \quad y \to \infty. \tag{5.111}$$

In this case the equation of state takes the leading-order asymptotic form

$$-D_{d,\sigma} y^{d-\sigma} + x_1 = 0, \tag{5.112}$$

which implies that

$$y_{d,d',\sigma}(x_1, 0) \simeq (x_1/D_{d,\sigma})^{1/(d-\sigma)}, \qquad \text{as} \quad x_1 \to \infty. \tag{5.113}$$

The asymptotic behavior of the function $\Psi_{d,d',\sigma}(y)$, as $y \to \infty$ follows from its definition, Eq. (5.97) and the asymptotic expansion Eq. (5.21):

$$\Psi_{d,d',\sigma}(y) \sim y^{-\sigma}, \qquad \text{as} \quad y \to \infty. \tag{5.114}$$

Therefore, as $x_1 \to \infty$,

$$\begin{aligned}
X_f^{d,d',\sigma}(x_1, 0) &\simeq \frac{1}{2} \sup_{y>0} \left\{ x_1 y^\sigma - \frac{\sigma}{d} D_{d,\sigma} y^d \right\} \\
&= \frac{d-\sigma}{2d} (D_{d,\sigma})^{-\sigma/(d-\sigma)} x_1^{d/(d-\sigma)}.
\end{aligned} \tag{5.115}$$

Then from Eq. (5.100) it follows that

$$\beta f^{m.s.}_{\Lambda,\text{sing}}(K,h) \simeq \frac{d-\sigma}{2d}(D_{d,\sigma})^{-\sigma/(d-\sigma)}(K_c - K)^{d/(d-\sigma)}, \qquad (5.116)$$

in full conformity with the corresponding expansion (3.45) for the bulk free energy at $\sigma < d < 2\sigma$.

Let now $K = K_c$ and $|h| \ll 1$ be fixed, so that $x_1 = 0$ and $|x_2| \to \infty$. In this case the equation of state takes the leading-order asymptotic form

$$-D_{d,\sigma} y^{d-\sigma} + \frac{x_2^2}{y^{2\sigma}} = 0, \qquad (5.117)$$

which implies that as $|x_2| \to \infty$,

$$y_{d,d',\sigma}(0,x_2) \simeq \left(\frac{x_2^2}{D_{d,\sigma}}\right)^{1/(d+\sigma)}. \qquad (5.118)$$

Therefore, as $|x_2| \to \infty$,

$$\begin{aligned}
X_f^{d,d',\sigma}(0,x_2) &\simeq \frac{1}{2}\sup_{y>0}\left\{-\frac{x_2^2}{y^\sigma} - \frac{\sigma}{d}D_{d,\sigma}y^d\right\} \\
&= -\frac{d+\sigma}{2d}(D_{d,\sigma})^{\sigma/(d+\sigma)}x_2^{2d/(d+\sigma)}. \qquad (5.119)
\end{aligned}$$

Then from Eq. (5.100) it follows that

$$\beta_c f^{m.s.}_{\Lambda,\text{sing}}(K_c,h) \simeq -\frac{d+\sigma}{2d}(D_{d,\sigma})^{\sigma/(d+\sigma)}|h|^{2d/(d+\sigma)}, \qquad (5.120)$$

in full conformity with the corresponding expansion (3.47) for the bulk free energy density.

Chapter 6

Modified Finite-Size Scaling

The investigation of finite-size scaling above the upper critical dimension, where the mean field regime sets in, is of special interest for systems with long-range power-law interaction potential, since then $d_u = 2\sigma$ may become less then any physically attainable dimensionality $d = 1, 2, 3$, for small enough values of the decay exponent $\sigma \in (0, 2)$.

It has been found [Brézin (1982)], [Privman and Fisher (1983)], [Luck (1985)], [Brézin and Zinn-Justin (1985)], that the hypothesis about the fundamental role of the ratio L/ξ_∞, where L is the typical linear size of the system and ξ_∞ is the bulk correlation length, in the description of the finite-size effects near criticality fails at space dimensions d above the upper critical dimensionality d_u. Renormalization group analysis reveals that the violation of finite-size scaling, as well as the breakdown of hyperscaling, is a consequence of the appearance in the theory of a "dangerous irrelevant variable" at $d > d_u$, see Section 1.6.5.

6.1   Introduction

In the derivation of scaling laws one essentially uses the assumption of analytical dependence of the scaling functions on the irrelevant scaling variables, which is not the case at $d > d_u$. It has been suggested that if hyperscaling is not valid, then finite-size effects are controlled by another ratio, namely L/l_∞, where l_∞ is the so-called thermodynamic length [Binder et. al. (1985)], [Binder (1987a)]. For systems with fully finite block geometry this length diverges as

$$l_\infty \sim |t|^{-(2\beta+\gamma)/d}, \qquad t \to 0. \tag{6.1}$$

Obviously, when the hyperscaling relation $2\beta + \gamma = d\nu$ holds, then the thermodynamic length coincides with the bulk correlation length.

In this chapter the ideas suggested in [Binder et. al. (1985)] for systems with short-range interactions ($\sigma = 2$) and fully finite geometry ($d' = 0$) are extended to the case of arbitrary $\sigma \in (0, 2)$ and the geometry $\Lambda = L^{d-d'} \times \infty^{d'}$, $0 \le d' < \sigma$, with periodic boundary conditions in the finite dimensions [Brankov and Tonchev (1990)]. In Section 6.2 we show that above the upper critical dimension there naturally emerges a new, diverging at the critical point length $l_\infty(T)$, which we call *coherence length*, see Section 4.5.2. The bulk coherence length obeys Eq. (4.53), predicted in *Case (ii)* of *Hypothesis B*, with exponent $\omega > 1$. By establishing the validity of this equation for the mean spherical model we complete the verification of the hypotheses of finite-size scaling at criticality given in Chapter 4. In Section 6.3 the modified finite-size scaling is interpreted in the general framework of a dangerous irrelevant variable, which appears in $O(n)$ models at $d > d_u$ and leads to the modified finite-size scaling.

Note that a straightforward attempt to extend finite-size scaling to the limiting case $\sigma \to 0^+$ encounters the fact that the usual power-law singularity of the characteristic length, $\lambda_\infty \sim t^{-\nu}$ as $t \to 0^+$, breaks down due to the diverging exponent $\nu = 1/\sigma \to \infty$, see Eq. (3.52) for $d > 2\sigma$. It has been remarked in [Joyce (1966)] that the limit $\sigma \to 0$ at fixed d and the limit $d \to \infty$ in the case of nearest-neighbor interactions, both give rise to the same mean field type of behavior. However, no proof of this assertion has been given. As we have shown in Section 3.1.3, in case of a fully finite domain $\Lambda = L^d$ with periodic boundary conditions, by choosing the interaction potential proportional to the decay exponent σ, see Eq. (3.54), one can take the limit $\sigma \to 0$ and obtain the spherical model with equivalent-neighbors interactions in Λ [Brankov (1990b)]. This model is a member of the large family of equivalent-neighbors models studied in Section 6.4, which obey the same modified finite-size scaling, as the models with power-law interaction above the upper critical dimensionality.

The special formulation of finite size-scaling for models with equivalent-neighbors interaction, suggested in [Botet et. al. (1982)], [Botet and Jullien (1983)], on the basis of the concept of *coherence number*, is also included in our theory.

6.2 The mean spherical model above the upper critical dimension

In Chapter 5 we have analyzed the finite-size scaling behavior of the d-dimensional mean spherical model with a power-law interaction potential (3.19) in the case when the decay exponent σ obeys the inequalities $\sigma < d < 2\sigma$. Here we focus our attention on the case $d > 2\sigma$. Of course, Eq. (5.94) for the finite-size term $U_\Lambda^{d,\sigma}(\phi)$ in the general expression for the singular part of the free energy density, see Eqs. (3.22) and (5.19),

$$\beta f_{\Lambda,\mathrm{sing}}^{m.s.}(K,h) = \frac{1}{2}\sup_{\phi>0}\left\{U_\Lambda^{d,\sigma}(\phi) - K\phi - \frac{h^2}{K\phi}\right\}, \tag{6.2}$$

is valid irrespectively of the relative magnitude of the space dimensionality $d > 0$ and the exponent $\sigma \in (0,2)$. In the case of $\Lambda = L^{d-d'} \times \infty^{d'}$ with $0 \le d' < d$, the expression for $U_\Lambda^{d,\sigma}(\phi)$ has the form (5.96), where the universal scaling function $\Psi_{d,d',\sigma}(\cdot)$ has the general representation (5.104). The formal changes which take place above the upper critical dimension, $d > d_u = 2\sigma$, are related to the modified asymptotic expansion of the bulk term $U_{d,\sigma}(\phi)$, see Eq. (3.44), and, eventually, to the different asymptotic properties of the function $\Psi_{d,d',\sigma}(\cdot)$. In the neighborhood of the critical point, $\phi \ll 1$, by neglecting terms of the order of $\phi^{d/\sigma}$ at $d > 2\sigma$, we obtain

$$\beta f_{\Lambda,\mathrm{sing}}^{m.s.}(K,h) =$$
$$\frac{1}{2}\sup_{\phi>0}\left\{(K_c - K)\phi - \frac{1}{2}|W_{d,\sigma}'(0)|\phi^2 - \frac{h^2}{K\phi} - \frac{1}{L^d}\Psi_{d,d',\sigma}(L\phi^{1/\sigma})\right\}. \tag{6.3}$$

By taking into account relationship (5.99), we can write the stationarity condition of the supremum problem (6.3) in terms of the finite-size scaling variables x_1 and x_2, see Eq. (5.35), as follows

$$-|W_{d,\sigma}'(0)|(\phi L^\sigma)L^{d-2\sigma} + F_{d,d',\sigma}(L\phi^{1/\sigma}) + x_1 + \frac{x_2^2}{(L^\sigma\phi)^2} = 0. \tag{6.4}$$

The analysis of the above equation shows that its solution for $y = L\phi^{1/\sigma}$ tends to zero in the neighborhood of the critical point $x_1 = O(1)$, $x_2 = O(1)$, as $L \to \infty$. To find an explicit asymptotic solution we need to know the behavior of the function $F_{d,d',\sigma}(y)$ as $y \to 0$ for $d > 2\sigma$. The latter can be obtained by an extension of the procedure of identical transformations of the right-hand side of Eq. (5.34), used in Section 5.3 for $d \in (\sigma, 2\sigma)$ and $d' < \sigma$. Note that after the application of Eqs. (4.127) - (4.129) one may formally consider the number of infinite dimensions d' as a real number. Thus we obtain the following generalized version of Eqs. (5.48) and (5.70):

$$F_{d,d',\sigma}(y) = D_{d',\sigma} y^{d'-\sigma} + \frac{y^{d'-\sigma}}{(4\pi)^{d'/2}} \int_0^\infty dx \, x^{(\sigma-d')/2-1} E_{\frac{\sigma}{2},\frac{\sigma}{2}}(-x^{\sigma/2})$$
$$\times \left[\Theta_3^{d-d'}\left(\frac{4\pi^2 x}{y^2}\right) - 1 - \frac{y^{d-d'}}{(4\pi x)^{(d-d')/2}} \right], \quad (6.5)$$

where $D_{0,\sigma} = \lim_{d' \to 0} D_{d',\sigma} = 1$. To split up the integral into a sum of terms, one has to ensure their convergence at the lower limit by subtracting from the function $E_{\sigma/2,\sigma/2}(\cdot)$ several terms of its series representation (5.6), and then adding them in an appropriate way [Brankov and Tonchev (1990)]. One can easily verify that if I is a positive integer, such that $\sigma I < d < \sigma(I+1)$, then the number of terms to be subtracted is exactly I. Therefore, we write identically

$$F_{d,d',\sigma}(y) = D_{d',\sigma} y^{d'-\sigma}$$
$$+ \frac{y^{d'-\sigma}}{(4\pi)^{d'/2}} \int_0^\infty dx \, x^{(\sigma-d')/2-1} \left[E_{\frac{\sigma}{2},\frac{\sigma}{2}}(-x^{\sigma/2}) - \sum_{k=0}^{I-1} \frac{(-1)^k x^{\sigma k/2}}{\Gamma(\sigma(k+1)/2)} \right]$$
$$\times \left[\Theta_3^{d-d'}\left(\frac{4\pi^2 x}{y^2}\right) - 1 \right]$$
$$- \frac{y^{d-\sigma}}{(4\pi)^{d/2}} \int_0^\infty dx \, x^{(\sigma-d)/2-1} \left[E_{\frac{\sigma}{2},\frac{\sigma}{2}}(-x^{\sigma/2}) - \sum_{k=0}^{I-1} \frac{(-1)^k x^{\sigma k/2}}{\Gamma(\sigma(k+1)/2)} \right]$$
$$+ \sum_{k=0}^{I-1} \frac{(-1)^k}{\Gamma(\sigma(k+1)/2)} \frac{y^{d'-\sigma}}{(4\pi)^{d'/2}} \int_0^\infty dx \, x^{[\sigma(k+1)-d']/2-1}$$
$$\times \left[\Theta_3^{d-d'}\left(\frac{4\pi^2 x}{y^2}\right) - 1 - \frac{y^{d-d'}}{(4\pi x)^{(d-d')/2}} \right]. \quad (6.6)$$

The second term in the right-hand side of the above equation can be transformed with the aid of the identity

$$\int_0^\infty dx\, e^{-zx}\, x^{\alpha-1} \left[E_{\alpha,\alpha}(-x^\alpha) - \sum_{k=0}^{m-1} \frac{(-1)^k x^{\alpha k}}{\Gamma(\alpha(k+1))} \right] = \frac{(-1)^m}{(1+z^\alpha)z^{\alpha m}},$$

$$(6.7)$$

to one of the following forms, depending on the value of $d' = 0, 1$:

(i) For $d' = 0$ we obtain

$$(-1)^I \sum_{\mathbf{n}(d)}{}' \frac{y^{\sigma I}}{(2\pi|\mathbf{n}|)^{\sigma I}[y^\sigma + (2\pi|\mathbf{n}|)^\sigma]} \equiv -S^{(I)}_{d,0,\sigma}(y^\sigma). \qquad (6.8)$$

(ii) For $d' = 1$, by using equality (5.54), taking the integral over x and making the change of variable $t \to x^2/y^2$ in the remaining integral over t, we obtain

$$\frac{(-1)^I}{\pi} \int_0^\infty dx \sum_{\mathbf{n}(d-1)}{}' \frac{y^{\sigma I}}{[(2\pi|\mathbf{n}|)^2 + x^2]^{\sigma I/2}\{\phi L^\sigma + [(2\pi|\mathbf{n}|)^2 + x^2]^{\sigma/2}]\}}$$

$$\equiv -S^{(I)}_{d,1,\sigma}(y^\sigma). \quad (6.9)$$

Expressions (6.8) and (6.9) are obvious extensions of Eqs. (5.46) and (5.68) for $S_{d,d',\sigma}(\cdot)$, from space dimensionality $d \in (\sigma, 2\sigma)$ to $\sigma I < d < \sigma(I+1)$, where $I \geq 2$ is an integer.

The third term in the right-hand side of Eq. (6.6) is proportional to an integral which can also be taken with the aid of the identities (6.7) and (5.54); the result is $-(-1)^I D_{d,\sigma} y^{d-\sigma}$. The last term yields

$$\sum_{k=0}^{I-1} (-1)^k C^{(k)}_{d,d',\sigma}\, y^{\sigma k}, \qquad (6.10)$$

where the constants

$$C^{(k)}_{d,d',\sigma} \equiv \frac{\pi^{d'/2}}{(2\pi)^{(k+1)\sigma}\Gamma(\sigma(k+1)/2)} \int_0^\infty dx\, x^{[\sigma(k+1)-d']/2-1}$$

$$\times \left[\Theta_3^{d-d'}(x) - 1 - \left(\frac{\pi}{x}\right)^{(d-d')/2} \right] \qquad (6.11)$$

are well defined for $d' < \sigma(k+1) < d$.

We remark that when $d = \sigma I$, with integer $I \geq 2$, then the above argument has to be modified as described in [Brankov and Tonchev (1990)]. The appropriate modifications result in the replacement of $D_{d,\sigma}$ by another constant, depending also on d', and in the appearance of a logarithmic term of the order $O(y^{\sigma(I-1)} \ln y)$.

From the above analysis it follows that the leading asymptotic behavior of the universal finite-size scaling function for the equation of state, $F_{d,d',\sigma}(y)$, as $y \to 0$, for $d > 2\sigma$ and $d' < \sigma$, is given by

$$F_{d,d',\sigma}(y) \simeq D_{d',\sigma} y^{d'-\sigma} + C^{(0)}_{d,d',\sigma} \qquad (y \to 0) \qquad (6.12)$$

and the mean spherical constraint (6.4) takes the form

$$-|W'_{d,\sigma}(0)| y^\sigma L^{d-2\sigma} + D_{d',\sigma} y^{d'-\sigma} + x_1 + \frac{x_2^2}{y^{2\sigma}} + C^{(0)}_{d,d',\sigma} = 0, \qquad (6.13)$$

where, as usual, $y = L\phi^{1/\sigma}$. Note that one can define another finite-size scaling variable $\dot{x}_1 \equiv C^{(0)}_{d,d',\sigma} + x_1$, which allows for the finite-size shift of the critical temperature. Obviously, the solution of equation (6.13) has the form

$$y = y_{d,d',\sigma}(x_1, x_2, L^{d-2\sigma}) \qquad (6.14)$$

which violates the standard finite-size scaling.

On the other hand, by introducing the rescaled variables,

$$y^* = yL^{p_y}, \qquad x_1^* = x_1 L^{p_1}, \qquad x_2^* = x_2 L^{p_2}, \qquad (6.15)$$

where

$$p_y = \frac{d-2\sigma}{2\sigma - d'}, \qquad p_1 = -\frac{(\sigma - d')(d-2\sigma)}{2\sigma - d'}, \qquad p_2 = \frac{(\sigma + d')(d-2\sigma)}{2(2\sigma - d')}, \qquad (6.16)$$

we can cast the mean spherical constraint (6.13) in the scale-invariant (but not universal) form

$$-|W'_{d,\sigma}(0)|(y^*)^\sigma + D_{d',\sigma}(y^*)^{d'-\sigma} + x_1^* + \frac{(x_2^*)^2}{(y^*)^{2\sigma}} = 0. \qquad (6.17)$$

Denoting the solution of this equation by

$$y^* = y^*_{d,d',\sigma}(x_1^*, x_2^*), \qquad (6.18)$$

and taking into account that $y^* = L^{1+p_y}/\lambda_L$, we obtain the modified finite-size scaling behavior of the characteristic length

$$\lambda_L(K, h) \simeq L^{1+p_y}/y^*_{d,d',\sigma}(x_1^*, x_2^*). \qquad (6.19)$$

The remarkable and rather unexpected feature of the above result is that in the neighborhood of the critical point, defined by the conditions $O(x_1^*) \leq O(1)$ and $O(x_2^*) \leq O(1)$, the finite-size characteristic length λ_L grows *faster* than L as $L \to \infty$ when $d > d_u = 2\sigma$, since then $p_y > 0$ according to Eq. (6.16). This is in apparent contradiction with the assumption made in [Binder et. al. (1985)] that $p_y \leq 0$, which is based on the definition (4.37) of the effective correlation length $\xi_L^{(2)}$ as a second moment of the finite-size correlation function. Obviously, $\xi_L^{(2)}$ is always bounded by the maximal linear size of the system, while there are no such restrictions on the behavior of λ_L.

By taking into account the leading-order expansion (3.32) of $W_{d,\sigma}(\phi)$ at $d > 2\sigma$, we can write equation (5.39) for the bulk characteristic length λ_∞ in terms of the variables x_1^* and x_2^* as follows:

$$-|W'_{d,\sigma}(0)|(L^{1+p_y}/\lambda_\infty)^\sigma + x_1^* + (x_2^*)^2 \left(\frac{\lambda_\infty}{L^{1+p_y}}\right)^{2\sigma} = 0. \tag{6.20}$$

Hence, we can express

$$x_1^* = X_1^{d,\sigma}(\lambda_\infty/L^{1+p_y}, x_2^*). \tag{6.21}$$

The substitution of (6.21) into the right-hand side of Eq. (6.19) yields the following modified finite-size scaling relationship (4.52) for the characteristic length, see *Case (ii)* of *Hypothesis B*,

$$\lambda_L(K,h) \simeq L^{1+p_y} \, X_\lambda^{d,d'\sigma}(\lambda_\infty/L^{1+p_y}, x_2^*), \tag{6.22}$$

appropriately extended to account for the dependence on the scaled magnetic field variable. Therefore, we may conclude that, as $L \to \infty$ above the upper critical dimensionality, the finite-size characteristic length, see Eq. (4.46), $\lambda_L(K,h) \simeq X_\lambda(\lambda_\infty, h, L)$ asymptotically becomes a generalized homogeneous function. As predicted by Eq. (4.53), the latter property leads to the appearance of a new, diverging at the critical point bulk coherence length,

$$l_\infty(K,h) \sim \lambda_\infty^{1/\omega}(K,h), \qquad \omega = 1 + p_y = \frac{d - d'}{2\sigma - d'}, \tag{6.23}$$

which describes the finite-size effects through the ratio l_∞/L.

The fact that the dimensionality d', which characterizes the geometry of the finite-size system, appears in the above definition of the *bulk* coherence length needs some explanation. First of all, we have to emphasize that the finite-size scaling form of the pair correlation function, given by Eqs. (5.24) - (5.26), is valid for all $d > \sigma$ including the mean field regime $d > 2\sigma$. Therefore, the finite-size characteristic length $\lambda_L(K, h)$ is defined in terms of the spherical field $\phi_L(K, h)$ always by the relationship (5.25). As follows from the mean spherical constraint (6.13), at $d' < \sigma$ the leading-order asymptotic behavior of $\lambda_L(K, h)$ at the bulk critical point $x_1 = 0$, $x_2 = 0$ (as well as at the shifted critical point $\dot{x}_1 = 0$, $x_2 = 0$) is

$$\lambda_L(K_c, 0) \equiv L/y_L(K_c, 0) \simeq L^\omega \left[\frac{|W'_{d,\sigma}(0)|}{D_{d',\sigma}} \right]^{1/(2\sigma - d')} \qquad (L \to \infty),$$

$$(6.24)$$

where ω is given in Eq. (6.23). Therefore, the exponent ω describes the finite-L behavior of the characteristic length λ_L at the critical point. According to Eqs. (6.17) - (6.19) the same power-law dependence of λ_L on L persists in the close vicinity of the critical point defined by the conditions $x_1^* = O(1)$ and $x_2^* = O(1)$. On the other hand, the bulk limit $\lambda_\infty(K, h)$ of the characteristic length, see Eq. (4.41), is defined at fixed K and h *away from* the critical point. Of course, the bulk limit $\lambda_\infty(K, h)$ does not dependent on the finite-size geometry, and, particularly on d', since Eq. (6.20) is independent of L and d' at fixed values of K and h. However, from the same equation it follows that at fixed $x_1^* = O(1)$ and $x_2^* = O(1)$

$$\lambda_\infty \left(K_c - x_1^* L^{-\sigma\omega}, K^{1/2} x_2^* L^{-3\sigma\omega/2} \right) \sim L^\omega \qquad (L \to \infty). \qquad (6.25)$$

In the simple zero-field case, when $x_2^* = 0$ and $x_1^* = O(1)$ is positive, one obtains the explicit expression

$$\lambda_\infty(K_c - x_1^* L^{-\sigma\omega}, 0) \simeq \left(|W'_{d,\sigma}(0)|/x_1^* \right)^{1/\sigma} L^\omega = \left(|W'_{d,\sigma}(0)|/K_c \right)^{1/\sigma} t^{-1/\sigma}. \qquad (6.26)$$

Moreover, in the zero-field case the modified finite-size scaling relationship (6.22) for the characteristic length $\lambda_L(K, 0)$ of a fully finite system $(d' = 0)$ can be written explicitly in the form (4.55) of a standard finite-size relationship for the coherence length $l_L(K, 0)$:

$$l_L(K, 0) \simeq L \bar{X}_l(L/l_\infty(K, 0)), \qquad (6.27)$$

where

$$l_L(K,0) = |W'_{d,\sigma}(0)|^{-1/d}\lambda_L^{2\sigma/d}(K,0),$$
$$l_\infty(K,0) = |W'_{d,\sigma}(0)|^{-1/d}\lambda_\infty^{2\sigma/d}(K,0), \tag{6.28}$$

and the scaling function is

$$\bar{X}_l(x) = \frac{2}{x^{d/2} + (x^d + 4)^{1/2}}. \tag{6.29}$$

Thus we have shown that for d above the upper critical dimension $d_u = 2\sigma$, it is the "bulk" coherence length $l_\infty \sim \lambda_\infty^{1/\omega}$, see Eq. (6.23), which allows one to formulate the regime of modified finite-size scaling as $L/l_\infty(K,h) = O(1)$, in complete analogy with the description of the standard finite-size scaling regime for $d_l < d < d_u$ in terms of $L/\lambda_\infty(K,h) = O(1)$. Since the neighborhood of the critical point, where the modified finite-size scaling holds, depends on d', this dependence is inherited by the bulk coherence length. Hence, at the bulk critical point the latter diverges like $l_\infty(K,0) \sim t^{-1/\sigma\omega}$ as $t = K_c/K - 1 \to 0^+$. Thus, it is the interpretation of modified finite-size scaling in terms of the coherence length (4.54) that brings into the definition of its bulk limit (6.23) the dependence on the finite-size geometry.

Let us consider now the consequences of the second equation in (6.28). Since $\lambda_\infty(K,0) \sim t^{-1/\sigma}$ as $t \to 0^+$ at $d > 2\sigma$, one obtains that the bulk coherence length diverges like $l_\infty(K,0) \sim t^{-2/d}$, which coincides with the critical behavior of the thermodynamic length (6.1) with $\beta = 1/2$, see Eq. (3.31), and $\gamma = 1$, see Eq. (3.41).

Consider now Eq. (6.3) for the singular part of the finite-size free energy density. In view of Eq. (6.19), when $L \to \infty$ at fixed x_1^* and x_2^*, then $L\phi_L^{1/\sigma} = L/\lambda_L \to 0$. This justifies the following approximations for the universal finite-size scaling function of the free energy density, $\Psi_{d,d',\sigma}(L\phi_L^{1/\sigma})$, given by Eq. (5.104):

(i) At $d' = 0$,

$$\Psi_{d,0,\sigma}(y) \simeq \frac{\upsilon}{\pi^{(d+1)/2}}\Gamma\left(\frac{d+1}{2}\right)\frac{2\pi^{d/2}}{\Gamma(d/2)}\int_y^\infty dr\, r^{-1}u_{d,\sigma}(r)$$
$$\simeq -\sigma\ln y + \text{const}, \tag{6.30}$$

where we have taken into account that

$$u_{d,\sigma}(0) = \frac{\sqrt{\pi}\,\Gamma(d/2)}{2\Gamma((d+1)/2)}. \tag{6.31}$$

(ii) At $d' = 1 < \sigma$,

$$\Psi_{d,1,\sigma}(y) \simeq \frac{\sigma}{\pi^{(d+1)/2}} \Gamma\left(\frac{d+1}{2}\right) \left\{ u_{d,\sigma}(0) \sum_{\mathbf{k}(d-1)}' |\mathbf{k}|^{-d} - \right.$$

$$\left. -y \frac{2\pi^{(d-1)/2}}{\Gamma((d-1)/2)} \int_0^\infty dr\, r^{-2}[u_{d,\sigma}(0) - u_{d,\sigma}(r)] \right\}$$

$$= \frac{\sigma}{2} \frac{\Gamma(d/2)}{\pi^{d/2}} \sum_{\mathbf{k}(d-1)}' |\mathbf{k}|^{-d} - \sigma D_{1,\sigma}\, y, \qquad (6.32)$$

where we have taken into account the integral

$$\int_0^\infty dr\, r^{-2}[u_{d,\sigma}(0) - u_{d,\sigma}(r)] = \frac{\pi \Gamma((d-1)/2)}{2\Gamma((d+1)/2)}\, D_{1,\sigma}. \qquad (6.33)$$

(iii) For real values of d', such that $0 \le d' < \sigma$, there is a simple analytical expression which includes Eqs. (6.30) and (6.32) as special cases,

$$\Psi_{d,d',\sigma}(y) \simeq \text{const} - \frac{\sigma}{d'}\, (y^{d'} - 1)D_{d',\sigma}. \qquad (6.34)$$

Note that $\Psi_{d,0,\sigma}(y)$ follows in the limit $d' \to 0^+$.

From Eq. (6.34) we conclude that $\Psi_{d,d',\sigma}(y) = L^{-d'p_y}\Psi_{d,d',\sigma}(y^*)$, and, therefore, expression (6.3) for the singular part of the finite-size free energy density can be written in the modified form

$$\beta f^{m.s.}_{\Lambda,\text{sing}}(K,h) =$$

$$L^{-d^*} \frac{1}{2} \sup_{y^*>0} \left\{ x_1^*(y^*)^\sigma - \frac{1}{2}|W'_{d,\sigma}(0)|(y^*)^{2\sigma} - \frac{(x_2^*)^2}{(y^*)^\sigma} - \Psi_{d,d',\sigma}(y^*) \right\}. \qquad (6.35)$$

The exponent

$$d^* = d + d'p_y = \frac{2\sigma(d - d')}{2\sigma - d'} \qquad (6.36)$$

is related to Fisher's anomalous dimension [Fisher (1983)]. In view of Eq. (6.19), the normalization factor for the free energy density can be interpreted as a critical, finite-size characteristic volume ($d' = 0, 1$):

$$L^{d-d'} \lambda_L^{d'}(K_c, 0) \sim L^{d^*}. \qquad (6.37)$$

Note that all the three arguments given in [Binder et. al. (1985)] in favour of the hypothesis $d^* = d$ are based on the assumption that the system is fully finite, i.e., that $d' = 0$.

This completes the verification of the hypotheses of finite-size scaling at criticality in the two forms given in Section 4.5: standard, which holds for $d_l < d < d_u$, and modified, which holds for $d > d_u$. From Eq. (6.23) it is clear that for the mean spherical model above the upper critical dimension $(d > 2\sigma)$ one always has $\omega > 1$, which is the case considered in Section 4.5.2, see Eqs. (4.61) - (4.63).

6.3 Dangerous irrelevant variable

The results obtained in the previous section for the mean spherical model admit a simple interpretation in terms of finite-size scaling functions depending on a dangerous irrelevant variable u with exponent $p_u = d_u - d < 0$ at $d > d_u$. According to the renormalization group derivation, the standard finite-size scaling expressions for the correlation length and the singular part of the free energy density in the neighborhood of the critical point have the form [Binder et. al. (1985)]:

$$\lambda_L(K, h) \simeq L\lambda(tL^{p_t}, hL^{p_h}, uL^{p_u}), \tag{6.38}$$

$$\beta f_{L,\mathrm{sing}}(K, h) \simeq L^{-d} f(tL^{p_t}, hL^{p_h}, uL^{p_u}). \tag{6.39}$$

The irrelevant variable u is called "dangerous" [Fisher (1983)], if the finite-size scaling functions are singular as $u \to 0$, i.e., if

$$\lambda(x, y, z) \simeq z^{q_\lambda} X_\lambda(xz^{q_t}, yz^{q_h}), \qquad \text{as} \quad z \to 0, \tag{6.40}$$

$$f(x, y, z) \simeq z^{q_f} X_f(xz^{q_t}, yz^{q_h}), \qquad \text{as} \quad z \to 0, \tag{6.41}$$

with some *negative exponents* q_f and q_λ. We see that, apart from the additional divergence condition, the effect of the dangerous irrelevant variable is equivalent to a generalized homogeneity of the scaling functions with respect to that variable. Thus, Eqs. (6.40) and (6.41) imply the modified scaling behavior ($u = 1$) of the characteristic length,

$$\lambda_L(K, h) \simeq L^\omega X_\lambda(tL^{p_t^*}, hL^{p_h^*}), \tag{6.42}$$

and the singular part of the free energy density,

$$\beta f_{L,\mathrm{sing}}(K, h) \simeq L^{-d^*} X_f(tL^{p_t^*}, hL^{p_h^*}), \tag{6.43}$$

with effective exponents given by

$$\omega = 1 + p_u q_\lambda, \quad d^* = d - p_u q_f, \quad p_t^* = p_t + p_u q_t, \quad p_h^* = p_h + p_u q_h.$$
$$(6.44)$$

By comparing Eqs. (6.42), (6.43) and (6.44) with Eqs. (6.15), (6.19) and (6.35), one identifies

$$
\begin{aligned}
q_\lambda &= p_y/p_u &&= -\frac{1}{2\sigma - d'} \\[2mm]
q_f &= -d' p_y/p_u &&= \frac{d'}{2\sigma - d'} \\[2mm]
q_t &= p_1/p_u &&= \frac{\sigma - d'}{2\sigma - d'} \\[2mm]
q_h &= p_2/p_u &&= -\frac{\sigma + d'}{2(2\sigma - d')}.
\end{aligned}
$$

Strictly speaking, the variable u turns out to be dangerous only for the characteristic length, since $q_\lambda < 0$, but $q_f \geq 0$.

Next, by taking into account that the standard finite-size scaled temperature and field variables are given by Eq. (5.35) as $x_1 \sim t L^{p_t}$ with $p_t = d - \sigma$ and $x_2 \sim h L^{p_t}$ with $p_h = (d + \sigma)/2$, we obtain the modified finite-size scaling variables:

$$
\begin{aligned}
x_1^* &\sim t L^{p_t^*} &&\text{with} &&p_t^* &&= \frac{\sigma(d - d')}{2\sigma - d'} \\[2mm]
x_2^* &\sim h L^{p_h^*} &&\text{with} &&p_h^* &&= \frac{3\sigma(d - d')}{2(2\sigma - d')}.
\end{aligned}
\qquad (6.45)
$$

Remarkably, in the case of a fully finite geometry, $d' = 0$, one finds

$$x_1^* \sim t L^{d/2}, \qquad x_2^* \sim h L^{3d/4}, \qquad (6.46)$$

independently of the potential decay exponent $\sigma \in (0, 2]$.

Recently Chen and Domb have argued in [Chen and Dohm (1998a)], [Chen and Dohm (1998b)] that the correct results for both the continuum and lattice versions of the φ^4 model disagree with the modified finite-size scaling hypothesis (6.43). This issue has been a matter of debate in the literature, see [Luijten et. al. (1999)] and references therein. Our considerations support the asymptotic validity of (6.43) and we are in the position to shed some light on this controversy at least in the framework of the mean spherical model. For simplicity, we confine ourselves to the short-range case ($\sigma = 2$) and fully finite geometry ($d' = 0$). Then, from Eq. (6.13) at $x_2 = 0$ we obtain the finite-size zero-field susceptibility $\chi_L = (K\phi_L)^{-1}$ in the form equvalent to Eqs. (138) - (140) of Ref. [Chen and Dohm (1998b)]:

$$\chi_L \simeq L^2 2K_c^{-1} \left[\delta(x,z) + \sqrt{\delta^2(x,z) + 4z} \right]^{-1}. \tag{6.47}$$

Here

$$\delta(x,z) = x + C_{d,0,2}^{(0)} z, \quad x = t(L/\xi_0)^2, \quad z = (L/l_0)^{4-d},$$

and

$$\xi_0 = \left[|W_{d,2}'(0)|/K_c \right]^{1/2}, \quad l_0 = |W_{d,2}'(0)|^{1/(4-d)}.$$

Note that ξ_0 is the critical amplitude of the bulk correlation length above the upper critical dimension, $\xi_\infty(K,0) \simeq \xi_0 t^{-1/2}$, see Eqs. (3.50) and (3.34); the parameter l_0 has been introduce in formal analogy with the definition of the dangerous irrelevant variable in the lattice φ^4 model [Chen and Dohm (1998a)], [Chen and Dohm (1998b)]; $C_{d,0,2}^{(0)} = -a_2(d)$, where $a_2(d)$ is the constant used in the above references. The L-dependent shift in $\delta(x,z)$ is the very reason for the statement in [Chen and Dohm (1998a)], [Chen and Dohm (1998b)] that the finite-size scaling function of the susceptibility $P(x,z)$ cannot be written as a function of the single scaling variable

$$x/z^{1/2} = tL^{d/2}\xi_0^{-2}l_0^{(4-d)/2} = (L/l_\infty)^{d/2},$$

where l_∞ is the zero-field bulk coherence length defined in Eq. (6.28), taken at $\sigma = 2$. However, as we have emphasized already, the modified finite-size scaling holds in the temperature region $x_1^* \sim tL^{d/2} = O(1)$ in which the finite-size shift $C_{d,0,2}^{(0)} z^{1/2}$ of the new scaling variable $x/z^{1/2}$ is of the order of $(L/l_0)^{(4-d)/2}$ and, therefore, has to be neglected. Then, the finite-size behavior of the zero-field susceptibility which follows from Eq. (6.47) is

$$\chi_L \simeq L^{d/2}\xi_0^2 l_0^{(d-4)/2} \tilde{P}\left(tL^{d/2}\xi_0^{-2}l_0^{(4-d)/2} \right). \tag{6.48}$$

with $\tilde{P}(x) := 2\left(x + \sqrt{x^2 + 4}\right)^{-1}$. This result is in complete agreement with the conclusion drawn in [Luijten et. al. (1999)] about the validity of the modified finite-size scaling. Obviously, the above arguments hold also in the case of long-range interaction with $0 < \sigma < 2$.

The existence of the thermodynamic limit for the characteristic length $\lambda_L(K, h)$ at fixed $t > 0$ and h in the neighborhood of the critical point,

$$\lim_{L \to \infty} L^{\omega} X_{\lambda}(tL^{p_t^*}, hL^{p_h^*}) = \lambda_{\infty}(K, h) \sim t^{-\nu} Y_{\lambda}(h/t^{\Delta}), \qquad (6.49)$$

with $\Delta = \beta + \gamma$, implies that

$$\Delta = p_h^*/p_t^* = 3/2, \qquad \nu = \omega/p_t^* = 1/\sigma, \qquad (6.50)$$

in full conformity with the modified values of the critical exponents for $d > 2\sigma$.

Similarly, the existence of the thermodynamic limit for the free energy density $f_{L,\mathrm{sing}}(K, h)$ at fixed $t > 0$ and h in the neighborhood of the critical point,

$$\lim_{L \to \infty} L^{-d^*} X_f(tL^{p_t^*}, hL^{p_h^*}) = f_{\infty,\mathrm{sing}}(K, h)/k_B T \sim t^{2-\alpha} Y_f(h/t^{\Delta}), \quad (6.51)$$

implies that the critical exponent of the heat capacity takes the value $\alpha = 2 - d^*/p_t^* = 0$, which violates the hyperscaling relation $d\nu = 2 - \alpha$.

6.4 The limit of equivalent-neighbors interaction

In Section 3.1.3 we have seen that the equivalent-neighbors mean spherical model with Hamiltonian (3.53) can be considered as the extremely long-range limit ($\sigma \to 0^+$) of the mean spherical model with power-law interaction potential (3.54), defined in a finite region $\Lambda = L^d$ with fixed dimensionality d and periodic boundary conditions. Note that at $d' = 0$ we have $d^* = d$, and from Eq. (6.45) it follows that $p_t^* = d/2$ and $p_h^* = 3d/4$, independently of the value of $\sigma \in (0, d/2)$. Therefore, by formally taking the limit $\sigma \to 0^+$ in the modified finite-size scaling prediction (6.43), we obtain that the free energy density of the equivalen-neighbors mean spherical model in $\Lambda = L^d$ should have the asymptotic form

$$f_{L,\mathrm{sing}}(K, h)/k_B T \simeq L^{-d} X_f(tL^{d/2}, hL^{3d/4}), \qquad (6.52)$$

near the critical point $t = h = 0$. However, if the equivalent-neighbors model is defined *ad hoc*, then it has no intrinsic geometric structure and the notions of space dimensionality and length remain ambigously defined.

A finite-size scaling theory, appropriate for equivalent-neighbors models, has been suggested in [Botet et. al. (1982)] and [Botet and Jullien (1983)] on basis of the notion of a *coherence number*, rather than a correlation (or characteristic, coherence) length. It is conjectured that the coherence number $\mathcal{N}_{c,\infty}$ of a bulk d-dimensional system near criticality obeys the relation $\mathcal{N}_{c,\infty} \sim \lambda_\infty^{d_u}$. Hence, as the critical point is approached along the line $h = 0$,

$$\mathcal{N}_{c,\infty}(K,0) \sim \lambda_\infty^{2\sigma}(K,0) \sim t^{-2} \qquad (t = K_c/K - 1 \to 0^+), \qquad (6.53)$$

independently of σ for all $\sigma < d/2$; therefore, the limit $\sigma \to 0^+$ can be trivially taken. The main postulate of the theory is that finite-size scaling for such systems can be formulated in terms of the ratio

$$N/\mathcal{N}_{c,\infty}(K,0) \sim Nt^2 \qquad (t \to 0^+), \qquad (6.54)$$

where $N = L^d$ is the number of particles in Λ. From Eq. (6.23) it follows the above ratio can be equivalently rewritten in terms of

$$[L/l_\infty(K,0)]^d \sim L^d t^2 \qquad (t \to 0^+), \qquad (6.55)$$

since at $d' = 0$ one has $1/\omega = 2\sigma/d$.

Thus, at any fixed space dimensionality $d > 0$ the limit $\sigma \to 0^+$ brings the system in the mean field regime, $d > d_u$, in which the hyperscaling relation $d\nu = 2 - \alpha$ is violated and the finite-size scaling variables are modified according to Eq. (6.46).

In this section we present an independent derivation of modified finite-size scaling for a general family of classical spin models with equivalent-neighbors interaction, which is a slight extension of the one defined and studied in [Ellis and Newman (1978a)], [Ellis and Newman (1978b)]. Remarkably, it turns out [Brankov and Zagrebnov (1983)] that for these models the order parameter of the finite system obeys an *exact partial differential equation* equivalent to the Burgers equation [Whitham (1974)] known in hydrodynamics. This fact allows one to formally describe the appearence (in the thermodynamic limit) of a jump in the magnetization as a function of the magnetic field as development of a *shock wave* in a non-linear medium with vanishing diffusion, see also Section 8.5. Moreover, the symmetries of the exact Burgers equation can be used for the derivation of finite-size scaling laws near the critical point [Brankov (1990a)].

6.4.1 *The class of models*

Consider a finite collection of N spin variables $\{S_i \in R, i = 1, \ldots, N\}$ with the Hamiltonian (energy function), see Eq. (3.53) in Section 3.1.3,

$$\mathcal{H}_N(\{S_i\}_{i=1}^N) = -\frac{J}{2N} \sum_{i,j=1}^N S_i S_j - H \sum_{i=1}^N S_i, \qquad (6.56)$$

where $J > 0$ is the ferromagnetic interaction constant, $H \in R$ is an external magnetic field. Let the joint probability distribution of the set of spins be given by the measure μ_ρ on R^N of the form

$$\mu_\rho(\mathrm{d}x_1, \ldots, \mathrm{d}x_N | K, h) = \left[Z_N^{(\rho)}(K, h) \right]^{-1}$$

$$\times \exp\left[\frac{K}{2N} \left(\sum_{i=1}^N x_i \right)^2 + h \sum_{i=1}^N x_i \right] \mu_0(\mathrm{d}x_1, \ldots, \mathrm{d}x_N | \rho). \qquad (6.57)$$

Here $K = J/k_\mathrm{B}T$ and $h = H/k_\mathrm{B}T$ are the dimensionless thermodynamic parameters, μ_0 is usually chosen as a free measure of the product type,

$$\mu_0(\mathrm{d}x_1, \ldots, \mathrm{d}x_N | \rho) = \prod_{i=1}^N \rho(\mathrm{d}x_i), \qquad (6.58)$$

where ρ is a non-degenerate probability measure on R which satisfies the conditions

$$\int_R \mathrm{e}^{x^2} \rho(\mathrm{d}x) < \infty, \qquad \int_R \mathrm{e}^{sx} \rho(\mathrm{d}x) < \mathrm{e}^{s^2/2}, \qquad \forall s \in R \setminus \{0\}. \qquad (6.59)$$

Note that the normalization coefficient $Z_N^{(\rho)}$ in Eq. (6.57) is the partition function of the model. To include in our consideration the equivalent-neighbors spherical model, see Section 3.1.3, we allow for the non-product type free measure of the form

$$\mu_0^s(\mathrm{d}x_1, \ldots, \mathrm{d}x_N) \sim \delta\left(\sum_{i=1}^N x_i^2 - N \right) \prod_{i=1}^N \mathrm{d}x_i, \qquad (6.60)$$

where $\mathrm{d}x$ is the Lebesgue measure on R.

6.4.2 *Finite-size magnetization and the Burgers equation*

Let us formally introduce a "time" variable t and a "space" coordinate x by setting[*]

$$t = K - K_c, \qquad x = -h, \tag{6.61}$$

where $K_c > 0$ is a parameter (dimensionless critical coupling) to be determined below. Consider now the magnetization per spin of a finite system

$$m_N^{(\rho)}(t,x) := \int_{R^N} \left(N^{-1} \sum_{i=1}^{N} x_i \right) \mu_\rho(\mathrm{d}x_1, \dots, \mathrm{d}x_N | K_c + t, -x). \tag{6.62}$$

By taking into account expression (6.57) for the measure μ_ρ and differentiating Eq.(6.62) with respect to the variables t and x, we obtain that $m_N^{(\rho)}(t,x)$ obeys the Burgers equation

$$\frac{\partial}{\partial t} m + m \frac{\partial}{\partial x} m = \frac{1}{2N} \frac{\partial^2}{\partial x^2} m. \tag{6.63}$$

Note that the various models in the class under considerations differ only by the *initial condition*:

$$\Phi_N^{(\rho)}(x) := m_N^{(\rho)}(t = -K_c, x) = \left[Z_N^{(\rho)}(0, -x) \right]^{-1}$$

$$\times \int_{R^N} \left(N^{-1} \sum_{i=1}^{N} x_i \right) \exp \left(-x \sum_{i=1}^{N} x_i \right) \mu_0(\mathrm{d}x_1, \dots, \mathrm{d}x_N | \rho), \tag{6.64}$$

which depends on the choice of the measure μ_0.

Equation (6.63) describes the evolution of perturbations in a dispersive medium in which the velocity of propagation depends linearly on the amplitude and the perturbations disperse according to the diffusion law. In our case the diffusion coefficient is inversely proportional to the number of particles and vanishes in the thermodynamic limit. This implies that the properties of the magnetization as a function of the magnetic field are qualitatively different for finite and infinite systems.

It is instructive to begin with the explicit investigation of these properties in the simplest case of the Ising-type Husimi – Temperley model defined

---

[*]Note the opposite sign of t as compared to the definition of the dimensionless temperature variable elsewhere in the book.

in Section 2.5. The model is specified by the product measure (6.58) with

$$\rho(\mathrm{d}x) = \frac{1}{2}[\delta(x-1) + \delta(x+1)]\mathrm{d}x. \tag{6.65}$$

In this case $K_c = 1$ and the initial condition (6.64) becomes

$$\Phi_N(x) = -\tanh(x). \tag{6.66}$$

By making use of the Cole – Hopf substitution [Whitham (1974)], the Burgers equation (6.63) with initial condition (6.66) can be solved explicitly:

$$m_N(t,x) = \left\{\int_{-\infty}^{\infty} \mathrm{d}\eta \exp[-NG(\eta;t,x)]\right\}^{-1}$$
$$\times \int_{-\infty}^{\infty} \mathrm{d}\eta \frac{x-\eta}{1+t} \exp[-NG(\eta;t,x)], \tag{6.67}$$

where

$$\begin{aligned}
G(\eta;t,x) &= \int_{0}^{\infty} \mathrm{d}\eta' \, \Phi_N(\eta') \quad + \frac{(x-\eta)^2}{2(1+t)}\\
&= -\ln(\cosh\eta) + \frac{(x-\eta)^2}{2(1+t)}.
\end{aligned} \tag{6.68}$$

From the mathematical theory of Laplace integrals it follows that if $G(\eta;t,x)$ has a unique absolute minimum with respect to $\eta \in R$ at $\eta = \bar{\eta}(t,x)$, then the finite-size magnetization per spin has the bulk limit

$$\lim_{N\to\infty} m_N(t,x) = [x - \bar{\eta}(t,x)]/(1+t). \tag{6.69}$$

From the stationarity condition for the function given by Eq. (6.68), we obtain that $\bar{\eta}(t,x)$ obeys the well known mean-field equation for the magnetization,

$$\frac{\eta - x}{1+t} = \tanh\eta, \tag{6.70}$$

and, therefore,

$$m_N(t,x) \to -\tanh\bar{\eta}(t,x), \quad \text{as} \quad N \to \infty. \tag{6.71}$$

Actually, there are three different regimes of behaviour of the right-hand side of Eq. (6.67), depending on the range of t: *high-temperature regime*, when $t \in [-1,0)$, *bulk critical point*, at $t = 0$, and *low-temperature regime*, when $t \in (0,\infty)$. Here we will confine ourselves to the former two regimes which are relevant to critical finite-size scaling. The low-temperature regime will be considered in Chapter 8.

High-temperature regime. For $t \in [-1, 0]$ one readily finds that Eq. (6.70) has only one solution which provides the single global minimum of $G(\eta; t, x)$, since

$$\frac{\partial^2}{\partial \eta^2} G(\eta; t, x) = -\frac{1}{\cosh^2 \eta} + \frac{1}{1+t} \tag{6.72}$$

is positive for all $-1 \le t \le 0$. At $x = 0$ the global minimum of $G(\eta; t, 0)$ is reached at $\bar{\eta}(t, 0) = 0$.

Note that for all finite N the magnetization per spin is a smooth, infinitely differentiable function of x, as follows from the explicit expression (6.67). It is interesting to follow the "time" evolution of its slope at $x = 0$ which is proportional to the initial magnetic susceptibility per spin $\chi_N(t, 0)$. By differentiating Eq. (6.67) with respect to x, we obtain

$$\begin{aligned}
\frac{\partial}{\partial x} m_N(t, x) \bigg|_{x=0} &= \frac{1}{1+t} - \frac{N}{(1+t)^2} \left\{ \int_{-\infty}^{\infty} d\eta \exp[-NG(\eta; t, 0)] \right\}^{-1} \\
&\times \int_{-\infty}^{\infty} d\eta \, \eta^2 \exp[-NG(\eta; t, 0)].
\end{aligned} \tag{6.73}$$

The leading asymptotic form of the right-hand side as $N \to \infty$ can be evaluated by expanding $G(\eta; t, 0)$ around $\eta = 0$ up to the quadratic term, and evaluating the resulting Gaussian integrals. Thus we obtain

$$\frac{\partial}{\partial x} m_N(t, x) \bigg|_{x=0} \simeq -\frac{1}{|t|}. \tag{6.74}$$

Hence, the critical exponent of the magnetic susceptibility takes the mean field value $\gamma = 1$, see Table 1.1.

Bulk critical point. When $t = 0$, the global minimum is reached again at $\bar{\eta}(0, 0) = 0$, but now the first three derivatives of $G(\eta; 0, 0)$ vanish at $\eta = 0$, see Eq. (6.72), and the series expansion starts with the quartic term. Hence,

$$\frac{\partial}{\partial x} m_N(0, x) \bigg|_{x=0} \simeq -\sqrt{12} \, \frac{\Gamma(\frac{3}{4})}{\Gamma(\frac{1}{4})} \, N^{1/2}. \tag{6.75}$$

Obviously, the bulk singularity of the magnetic susceptibility at the critical point is rounded in the finite system to a peak of height $\sim N^{1/2}$. Note that the modified finite-size scaling form of the singular part of the free energy density for a finite system of $N = L^d$ sites above the upper critical dimension, see Eq. (6.52), can be written as

$$f_{N,\text{sing}}(K, h)/k_{\mathrm{B}}T \simeq N^{-1} X_f(tN^{1/2}, hN^{3/4}). \qquad (6.76)$$

Hence, by differentiating twice with respect to the field variable h, we obtain that the finite-size magnetic susceptibility *at the critical point* behaves as

$$k_{\mathrm{B}}T_c\, \chi_N(K_c, 0) \simeq N^{1/2} \left[-\frac{\partial^2}{\partial z^2} X_f(0, z) \right]_{z=0}, \qquad (6.77)$$

in full conformity with Eq. (6.75).

6.4.3 *Self-similar solutions and bulk scaling*

On the basis of the Burgers equation (6.63) with the initial condition (6.66), one can draw analogy between, on the one hand, the appearance of a bulk critical point at $K = K_c = 1$, followed by a first-order phase transition with respect to the field variable x at $K > K_c$, and, on the other hand, the development of a shock wave at times $t \geq 0$ [Brankov and Zagrebnov (1983)]. This topic is discussed furhter in Chapter 8 in the context of finite-size scaling at first-order transitions. Here we proceed by showing that the scaling law for the bulk magnetization follows from the existence of self-similar solutions of the Cauchy problem (6.63), (6.64) in the limit $N \to \infty$.

Thermodynamic scaling predicts, see Eq. (1.88) in Section 1.5.4, that in the neighborhood of the critical point $t = 0$, $h = 0$, the bulk magnetization per spin $m_\infty(t, x)$ is a generalized homogeneous function of the variables x and t of the form

$$m_\infty(t, x) \simeq |t|^\beta v_\pm(x|t|^{-\Delta}), \qquad (6.78)$$

where the two branches $v_\pm(\cdot)$ of the scaling function, corresponding to $t > 0$ and $t < 0$, may be different. As usual, β is the critical exponent of the order parameter and $\Delta = \beta + \gamma$, where γ is the critical exponent of the susceptibility. Since $m_\infty(t, x)$ obeys the equation,

$$\frac{\partial}{\partial t} m + m \frac{\partial}{\partial x} m = 0, \qquad (6.79)$$

then, if bulk scaling holds, expression (6.78) should be among the self-similar solutions of the above equation, at least in the neighborhood of the critical point. Let us consider this important issue in greater detail.

High-temperature regime. In the case when $t \in [-K_c, 0)$ we look for self-similar solutions of Eq. (6.79) of the form

$$m = (-t)^\beta v_+(y), \qquad y = x(-t)^{-\Delta}, \tag{6.80}$$

with arbitrary positive exponents β and Δ. By inserting (6.80) into (6.79) we obtain an ordinary differential equation for the unknown function $v_+(\cdot)$:

$$\left[(-t)^{\beta+1-\Delta}v_+ + \Delta y\right] v'_+ - \beta v_+ = 0. \tag{6.81}$$

Obviously, this equation defines a function of the single variable y only if $\Delta = \beta+1$. Since by definition $\Delta = \beta+\gamma$, it follows that self-similar solution of Eq. (6.79) exists only for the mean field value $\gamma = 1$. Then, integrating Eq. (6.81) we obtain that $v_+ = v_+(y)$ is defined as an implicit function of y by the equation

$$|v_+| = A|v_+ + y|^{\beta/(\beta+1)}. \tag{6.82}$$

Here $A > 0$ is an arbitrary integration constant and the exponent β may still take any positive value.

Let us now take into consideration that the magnetization (6.80) must obey the initial condition on the function v_+:

$$K_c^\beta v_+(xK_c^{-\beta-1}) = \Phi_\infty^{(\rho)}(x). \tag{6.83}$$

Obviously, the initial condition (6.66) for the Husimi – Temperley model does not satisfy in the limit $N \to \infty$ the general constraint imposed on $\Phi_\infty^{(\rho)}(x)$ by Eq. (6.82) at the bulk critical temperature $t = -K_c$:

$$|\Phi_\infty^{(\rho)}(x)| = A|K_c\Phi_\infty^{(\rho)}(x) + x|^{\beta/(\beta+1)}. \tag{6.84}$$

Neither does so, e.g., the initial condition for the equivalent-neighbors spherical model,

$$\Phi_N^{(s.)}(x) == -\frac{I_{N/2}(Nx)}{I_{N/2-1}(Nx)} = -\frac{2x}{1 + (4x^2 + 1)^{1/2}} + O(N^{-1}), \tag{6.85}$$

where I_n is the modified Bessel function of order n. Moreover, from Eq. (6.84) it follows that the function $\Phi_\infty^{(\rho)}(x)$ must diverge when $x \to \infty$, while actually $\Phi_\infty^{(\rho)}(x \to \pm\infty) \to \mp m_0(\infty)$, where $m_0(\infty)$ is the saturation magnetization per spin. Therefore, the models with equivalent-neighbors interaction considered here may not have globally self-similar magnetization of the form (6.80).

Consider now the local properties of the solutions of the Cauchy problem when $x \to 0$. Confining ourselves to the case of symmetric measures μ_0, which implies $\Phi_N^{(\rho)}(-x) = -\Phi_N^{(\rho)}(x)$, and assuming the analyticity of $\Phi_\infty^{(\rho)}(x)$ at the point $x = 0$, we consider the expansion

$$\Phi_\infty^{(\rho)}(x) = \frac{\partial}{\partial x}\Phi_\infty^{(\rho)}(0)\, x + \frac{1}{3!}\frac{\partial^3}{\partial x^3}\Phi_\infty^{(\rho)}(0)\, x^3 + \ldots \quad (x \to 0). \qquad (6.86)$$

Now we can satisfy the constraint (6.84) by setting

$$\beta = \frac{1}{2}, \quad K_c^{-1} = -\frac{\partial}{\partial x}\Phi_\infty^{(\rho)}(0) > 0, \quad A^{-3} = K_c^4 \frac{1}{3!}\left|\frac{\partial^3}{\partial x^3}\Phi_\infty^{(\rho)}(0)\right|. \qquad (6.87)$$

For the coefficients of higher powers of x one obtains a system of recursion relations.

Thus the condition that the self-similar solution (6.80) locally satisfies the initial condition (6.86) in the neighborhood of the point $x = 0$ implies the mean field value $\beta = 1/2$ of the magnetization critical exponent. Moreover, we have expressed the parameters K_c and A in terms of quantities depending on the specific model.

6.4.4   *Derivation of modified finite-size scaling*

Let us consider separately the consequences for the critical finite-size scaling which follow fron the differential equation for the order parameter of the finite system, see Eq. (6.63), and from the initial condition (6.64). Since the equivalent-neighbors models lack a natural geometric description, they can be defined in a space of arbitrary dimensiona. To avoid ambiguities related to the definition of a "linear size" of the system, we choose the number of particles N as a finite-size parameter. Let us then look for solutions of the Burgers equation (6.63) in the self-similar form

$$m = N^{-p}w(N^q t, N^r x) \qquad (6.88)$$

with arbitrary positive exponents p, q and r. By introducing the variables

$$w = N^p m, \qquad x_1 = N^q t, \qquad x_2 = N^r x \qquad (6.89)$$

we may recast the Burgers equation in the form

$$\frac{\partial}{\partial x_1} w + N^{-p-q+r} w \frac{\partial}{\partial x_2} w = \frac{1}{2} N^{-1-q+2r} \frac{\partial^2}{\partial x_2^2} w. \tag{6.90}$$

Since the function $w(x_1, x_2)$ must not depend explicitly on N, the existence of a self-similar solution requires the equalities $q = 1 - 2p$, $r = 1 - p$, where $0 < p < 1/2$ is an arbitrary parameter at present. Therefore, the most general finite-size scaling law compatible with the Burgers equation is

$$m_N(t, x) = N^{-p} w(N^{1-2p} t, N^{1-p} x) \qquad (0 < p < 1/2). \tag{6.91}$$

Here the function $w(x_1, x_2)$ is a solution of the size-independent Burgers equation

$$\frac{\partial}{\partial x_1} w + w \frac{\partial}{\partial x_2} w = \frac{1}{2} \frac{\partial^2}{\partial x_2^2} w. \tag{6.92}$$

Let us now consider the initial condition (6.64) which in the scaled variables (6.89) has the form

$$w\left(x_1 = -K_c N^{1-2p}, \ x_2\right) = N^p \Phi_N^{(\rho)}(N^{p-1} x_2). \tag{6.93}$$

To study the above expression we need the asymptotic form of $w(x_1, x_2)$ as $x_1 \to -\infty$. Let us note that at any fixed $t \neq 0$ the limit $N \to \infty$ in the magnetization (6.91) should lead to the thermodynamic scaling form (6.78), which may happen only if

$$w(x_1, x_2) \simeq |x_1|^\beta v_\pm(x_2 |x_1|^{-\beta-1}), \qquad x_1 \to \pm\infty \tag{6.94}$$

and, moreover, under the condition that p and β satisfy the equalities

$$1 - p = (1 - 2p)(\beta + 1), \qquad (1 - 2p)\beta = p. \tag{6.95}$$

Hence we obtain $p = \beta/(2\beta + 1)$ and, in view of $\beta = 1/2$, it follows that $p = 1/4$. Now we may readily check that the initial condition (6.93) at fixed x_2 and $N \to \infty$ reduces to the equality

$$(K_c)^\beta v_+(0) = \Phi_\infty^{(\rho)}(0) \tag{6.96}$$

which holds due to condition (6.83) at $x = 0$.

Here we have given an independent derivation of the critical finite-size scaling form for the magnetization,

$$m_N(t, x) = N^{-1/4} w(N^{1/2} t, N^{3/4} x), \tag{6.97}$$

which follows from Eq. (6.91) at $p = 1/4$. Note that the prediction of the modified finite-size scaling, which follows from Eq. (6.76),

$$m_N(K,h) \simeq N^{-1/4} \left[-\frac{\partial}{\partial z} X_f(tN^{1/2}, z) \right]_{z=hN^{3/4}}, \qquad (6.98)$$

is in full conformity with Eq. (6.97).

Generalizations. We have derived the general one-parameter family (6.91) of finite-size scaling laws compatible with the Burgers equation. Remarkably, the value $\beta = 1/2$ of the magnetization critical exponent, and hence the value $p = 1/4$, is determined solely by the power of the second term in the Taylor expansion (6.86) of the initial condition and does not depend on the details of the model, i.e. on the specific choice of the measure μ_0. If, instead of (6.86), the following expansion were valid when $x \to 0$,

$$\Phi_\infty^{(\rho)}(x) = \frac{\partial}{\partial x} \Phi_\infty^{(\rho)}(0)\, x + \frac{1}{(2k+1)!} \frac{\partial^{2k+1}}{\partial x^{2k+1}} \Phi_\infty^{(\rho)}(0)\, x^{2k+1} + \dots, \qquad (6.99)$$

we would obtain

$$\beta = \frac{1}{2k}, \qquad p = \frac{1}{2k(k+1)}. \qquad (6.100)$$

Thus, the role of the initial condition consists in the reduction of the one-parameter family of self-similar solutions to a single representative with $p = \beta/(2\beta + 1)$. This reduction takes place under the condition that the thermodynamic limit for the magnetization per spin exists. The explicit expression for the finite-size scaling function $w(x_1, x_2)$ can be easily obtained from the well known integral representation for the solution of the Cauchy problem for the Burgers equation. For any finite N the one-parameter family of solutions (6.91) of the Burgers equation analytically depends on the temperature and the magnetic field with the only exception of the point at infinity $t = \infty, x = 0$, which corresponds to the zero-temperature limit at vanishing field.

Chapter 7

Boundary Effects

The presence of physical boundaries contributes to the observable properties of a finite system due to effects of missing neighbors (in the case of free surfaces), or, in general, due to modified interparticle interactions close to and across the boundary. These modified interactions can be caused by, say, structural deffects and/or impurities near the boundary, different nature of the envirenment, etc. One can also model the effect of a boundary by intoducing *surface fields*, localized at it, which may have various physical nature. All these effects break the translation invariance and isotropy which are usually built in the formal Hamiltonian of the infinite system. As a result, the average values of a local dynamical observable, calculated in a Gibbs ensemble for the finite system, may differ near the surface and in the bulk of the system. In cases when the surface contributions are small compared to the extensive properties of the finite system, one can split off the thermodynamic functions into bulk terms, proportional to the volume, and surface terms, proportional to the area of the surface. Then one can define appropriate surface densities in the limit of an infinite system and study the *surface properties* much in the same way as in the case of bulk densities, see Sections 1.3 and 1.5.

It is well known that in systems with surfaces a variety of surface phase transitions can take place, depending on the boundary conditions, the enhancement of the surface couplings and on the (magnetic) fields applied at the boundaries (for a general review on the subject see, e.g., [Barber (1983)], [Diehl (1986)], [Dietrich (1990)]). If the surface couplings enhancement and the dimensionality are large enough, the surface orders at some temperature T_s larger than the bulk critical one, T_c, of the corresponding infinite system. The lowering of the temperature leads then to the so-called *extraordinary surface phase transition*, when the bulk orders at T_c in the presence of an already ordered surface. In the opposite case, when the surface enhancement is not sufficient to compensate the effect of missing neighbors at the surface (in the case of the so-called "free" boundary conditions), the *surface critical behavior* will be driven by the bulk: this is *the ordinary surface phase transition*. The borderline case between these two types is termed *special, or surface-bulk, surface phase transition*. To characterize the singular behavior of the different surface-introduced quantities, such as the surface magnetization, a variety of surface critical exponents has been defined for each class of surface phase transitions. It is well established fact now that for the ordinary phase transition there is only one independent surface critical exponent (in addition of the bulk ones). In the extraordinary surface universality class all the surface exponents can be expressed in terms of the bulk ones. The special phase transition, being a borderline case and hence a multicritical phase transition, is characterized by two new independent critical exponents. In this case one additional crossover exponent ϑ appears, which describes how in the phase diagram the line of surface phase transitions joins the line of extraordinary transitions.

For systems having the film geometry $L \times \infty^{d-1}$, with broken translation invariance in one space dymension, it makes sense to define layer densities and study the corresponding *layer critical behavior*, if such occurs. The above possibilities motivate the definition and evaluation of *local critical exponents*. Thus, for each of the L layers parallel to the boundaries of the film, one can define layer magnetization density m_l and, if the layer supports spontaneous long-range order, the corresponding critical exponent β_l ($l = 1, \ldots, L$). However, the notions of surface critical exponents and layer critical exponents are not yet sufficient for a complete description of the variety of related phenomena. For example, one can define two different local susceptibilities at the surface $l = 1$: one of them, χ_1, describing the response of the surface magnetization m_1 to a uniform field, and the other, $\chi_{1,1}$, describing the response of the surface magnetization to a local field H_1 acting only on that surface. Such local quantities are measurable by using experimental techniques like low-energy electron diffraction, nuclear magnetic resonance, Mössbauer spectroscopy, etc.

The outline of this chapter is as follows. In Section 7.1 we specify the boundary conditions for lattice spin systems with nearest neighbor interactions. The solution of the spectral problem for the one-dimensional discrete

Laplacian on a finite chain under different boundary conditions is presented in Section 7.1.2. In Section 7.1.3 these results are applied to the spectrum of the interaction Hamiltonian describing three-dimensional spherical models with nearest-neighbor interactions in a fully finite region Λ. In Section 7.2 we consider some new finite-size effects, which appear due to the presence of nonperiodec boundaries. In Section 7.3 the finite-size scaling behavior of the three-dimensional mean spherical model with a L-layer film geometry is studied in the presence of layer fields. A modified three-dimensional mean spherical model with the film geometry is introduced and studied in Section 7.4 under Neumann boundary conditions. The closing Section 7.5 contains a discussion of the results on the surface and layer critical behavior of the mean (standard and modified) spherical model.

7.1 Boundary conditions for spin systems

7.1.1 *General aspects*

In an equilibrium Gibbs ensemble, the formal expression for the Hamiltonian defines uniquely the *conditional probability distribution* on the configuration space of the system, confined to a finite space domain Λ, see Section 1.2. This probability distribution is conditioned by the fixed "environment" outside Λ. To be more specific, consider lattice spin systems with a finite single-spin space. The Gibbs distribution then defines the probabilty of finding a configuration $S_\Lambda = \{S(\mathbf{r}), \mathbf{r} \in \Lambda\}$ in the finite space of all the allowed spin configurations in the region Λ, provided the spin configuration in its infinite complement $\Lambda^c = Z^d \setminus \Lambda$ is fixed. Note that the fixation of the spin configuration in Λ^c is a formal expression for what is known in physics as *boundary conditions*. The above construction leads to a correct formulation of the problems concerning the probability distributions arising in the thermodynamic limit, and has a natural counterpart in the mathematical theory of random fields and processes. The boundary conditions can be enriched by allowing for modified interactions of the spins in Λ with spins in Λ^c, see, e.g., [Barber et. al (1974)], [Singh et. al. (1975)]. Let us note also that two different types of boundary conditions can be considered: macroscopic boundary conditions, when the average magnetization at the boundary is fixed, and microscopic ones, when the microscopic spin variables at the boundary are fixed. In some cases they lead to identical results, as shown in [Brézin et. al. (1993)]. Here we consider only microscopic boundry conditions.

Let us take $\Lambda \in Z^d$ to be the parallelepiped $\Lambda = \mathcal{L}_1 \times \cdots \times \mathcal{L}_d$, where $\times$ denotes the direct (Cartesian) product of the finite sets $\mathcal{L}_\nu = \{1, \ldots, L_\nu\}$. The boundary conditions, distinguished by the superscript (τ) for each pair of opposite faces of Λ, define the interaction of the spins in the region Λ with a specified configuration $S_{\Lambda^c} = \{S(\mathbf{r}), \mathbf{r} \in \Lambda^c\}$ in its complement $\Lambda^c = Z^d \setminus \Lambda$. Obviously, in the case of unmodified nearest-neighbor couplings it suffices to define the pairwise interaction energies across the boundaries. We shall explicitly consider the case of a three-dimensional film geometry $(d = 3)$ which results in the limit $L_2, L_3 \to \infty$ at finite values of $L_1 = L$. To define this limit, we start by considering the system in a fully finite parallelepiped with periodic boundary conditions along the space axes $\nu = 2, 3$. For the sake of simplicity, we confine ourselves to the case of scalar, real-valued spin variables.

Periodic boundary conditions $(\tau = p)$ are most often used when the boundary effects are undesirable and, therefore, have to be minimized. Such are the studies, both analytical and by computer simulations, of bulk critical phenomena and finite-size scaling in the absence of boundary effects. In general, periodic boundary conditions for a fully finite d-dimensional system mean that the spin configuration S_{Λ^c} of the "environment" is represented by repeated copies of $S_\Lambda = \{S(\mathbf{r}), \mathbf{r} \in \Lambda\}$, shifted along each coordinate axis ν $(\nu = 1, \ldots, d)$ by all the multiples of the edge size L_ν. Formally, this is expressed by the set of equalities

$$S(r_1 + m_1 L_1, r_2 + m_2 L_2, \ldots, r_d + m_d L_d) = S(r_1, r_2, \ldots, r_d), \qquad (7.1)$$

for all $(r_1, r_2, \ldots, r_d) \in \Lambda$ and all integers $m_1, m_2, \ldots, m_d \in Z^d$.

In the case of nearest neighbor interactions and three dimensional film geometry $L \times \infty^2$, it suffices to specify the spin configurations in the layers with coordinates $(r_1 = 0, r_2, r_3)$ and $(r_1 = L+1, r_2, r_3)$ for all $(r_2, r_3) \in Z^2$. Then, periodic boundary conditions imply

$$S(0, r_2, r_3) = S(L, r_2, r_3), \quad S(L+1, r_2, r_3) = S(1, r_2, r_3). \qquad (7.2)$$

Quite often it is physically more appropriate to consider the so called "free" boundary conditions which mean that the complement Λ^c of Λ is "empty", i.e., outside Λ there is nothing for the spins to interact with. It is convenient to assume that the spin system is defined on the whole lattice Z^d, but its configuration outside Λ is identically fixed to zero spin value. The above construction leads to the so called *zero Dirichlet boundary conditions* ($\tau = 1$). In the case of nearest neighbor interactions they are

$$S(0, r_2, r_3) = S(L + 1, r_2, r_3) = 0. \tag{7.3}$$

Note that non-zero Dirichlet boundary conditions are usually considered when the spin variables take values from a finite set $S(\mathbf{r}) \in \{-S_m, \ldots, S_m\}$. For example, the $(+, +)$ boundary conditions for Ising-type models mean that

$$S(0, r_2, r_3) = S(L + 1, r_2, r_3) = S_m. \tag{7.4}$$

The definition of $(-, -)$ boundary conditions is analogous. In such systems important role (e.g., for the creation of interfaces below T_c) play the so-called $(+, -)$ boundary conditions:

$$S(0, r_2, r_3) = S_m, \quad S(L + 1, r_2, r_3) = -S_m. \tag{7.5}$$

There are boundary conditions, alternative to the periodic ones, which also model the environment by means of identical copies of the original system. Thus, under the *Neumann boundary conditions* ($\tau = 0$) the spins surrounding Λ are set identically equal to their nearest neughbors inside the system:

$$S(0, r_2, r_3) = S(1, r_2, r_3), \quad S(L + 1, r_2, r_3) = S(L, r_2, r_3). \tag{7.6}$$

Of course, one may consider different types of boundary conditions at the opposite surfaces of Λ. For the combination of *Neumann – Dirichlet boundary conditions* ($\tau = c$) one has

$$S(0, r_2, r_3) = S(1, r_2, r_3), \quad S(L + 1, r_2, r_3) = 0. \tag{7.7}$$

The Newmann boundary conditions can be reasonably extended by assuming mere *proportionality*, and not identity, of the spin variables associated with each pair of sites which are nearest neighbors across the boundary. Thus, if one sets for some real ω

$$S(0, r_2, r_3) = (1 - \omega)S(1, r_2, r_3), \quad S(L + 1, r_2, r_3) = (1 - \omega)S(L, r_2, r_3), \tag{7.8}$$

one obtains Neumann boundary conditions at $\omega = 0$ and Dirichlet boundary conditions at $\omega = 1$. For $\omega \neq 0, 1$ equations (7.8) define the so-called *mixed boundary conditions* ($\tau = \omega$), which appear in Section 7.4.

Finally, we note that a convenient way of creating interfaces between domains of opposite magnetization in systems with continuous spin consists in imposing *antiperiodic boundary conditions* ($\tau = a$),

$$S(0, r_2, r_3) = -S(L, r_2, r_3), \qquad S(L+1, r_2, r_3) = -S(1, r_2, r_3). \qquad (7.9)$$

The terminology used here is justified by analogy with the continuum limit, see also Section 7.1.2. Note that the case of free surfaces in a system with the film geometry, considered in [Barber and Fisher (1973)], [Barber (1974)], [Barber et. al (1974)], [Singh et. al. (1975)], corresponds to (zero) Dirichlet boundary conditions ($\tau = 1$). In the literature on lattice spin systems the terms "free" and "fixed" boundary conditions have both been used instead of Dirichlet boundary conditions, see, e.g., [Barber (1983)], [Barber and Fisher (1973)], [Gelfand and Fisher (1988)], [Gelfand and Fisher (1990)], and for what we call Neumann boundary conditions the term "free" boundary conditions has also been used [Gelfand and Fisher (1988)], [Gelfand and Fisher (1990)]. To avoid misunderstanding, we will adhere to the corresponding classical terminology stemming from continuous models.

7.1.2 *Spectrum of the discrete Laplacian*

The one-dimensional discrete Laplacian $\Delta_L^{(\tau)}$ on a finite chain $\mathcal{L} = \{1, \ldots, L\}$ is defined by its action on any L-component vector $S_L = \{S(1), \ldots, S(L)\}$. At the internal points of $\mathcal{L}$ it acts as a second central difference operator:

$$(\Delta_L^{(\tau)} S)(r) = S(r-1) - 2S(r) + S(r+1), \qquad 2 \leq r \leq L-1. \qquad (7.10)$$

Obviously, one has to define the action of $\Delta_L^{(\tau)}$ at the end-points $\partial\mathcal{L} = \{1, L\}$, which means to specify the boundary conditions. Starting from the definitions given in Section 7.1.1, we consider the extended chain $\mathcal{L}^e = \{0, 1, \ldots, L, L+1\}$ and the corresponding vectors $S_{L+2} = \{S(0), S(1), \ldots, S(L), S(L+1)\}$; the extended Laplacian $\Delta_{L+2}^{(\tau)}$ is assumed to act at the sites $r = 1$ and $r = L$ according to Eq. (7.10). Clearly, it suffices to consider explicitly the cases of periodic, mixed, Neumann-Dirichlet, and antiperiodic boundary conditions; the cases of Dirichlet and Neumann boundary conditions follow from the mixed ones at $\omega = 1$ and $\omega = 0$, respectively. For the sake of convenience, below we consider S_L as a column-vector in a L-dimensional linear space and write down the matrix representation of $\Delta_L^{(\tau)}$ for different boundary conditions τ.

Under *periodic boundary conditions* $(\tau = p)$, see (7.2), one sets $S(0) = S(L)$ and $S(L+1) = S(1)$. Hence one obtains

$$(\Delta_L^{(p)} S_L)(1) = S(L) - 2S(1) + S(2),$$
$$(\Delta_L^{(p)} S_L)(L) = S(L-1) - 2S(L) + S(1), \tag{7.11}$$

which yields the matrix representation

$$\Delta_L^{(p)} = \begin{pmatrix} -2 & 1 & 0 & \ldots & 0 & 1 \\ 1 & -2 & 1 & \ldots & 0 & 0 \\ 0 & 1 & -2 & \ldots & 0 & 0 \\ \ldots & \ldots & \ldots & \ldots & \ldots & \ldots \\ 0 & 0 & 0 & \ldots & -2 & 1 \\ 1 & 0 & 0 & \ldots & 1 & -2 \end{pmatrix}. \tag{7.12}$$

Under *mixed boundary conditions* $(\tau = \omega)$, see (7.8), one sets $S(0) = (1-\omega)S(1)$ and $S(L+1) = (1-\omega)S(L)$. Then one obtains

$$(\Delta_L^{(\omega)} S_L)(1) = -(1+\omega)S(1) + S(2),$$
$$(\Delta_L^{(\omega)} S_L)(L) = S(L-1) - (1+\omega)S(L), \tag{7.13}$$

and the matrix representation of $\Delta_L^{(\omega)}$ is

$$\Delta_L^{(\omega)} = \begin{pmatrix} -(1+\omega) & 1 & 0 & \ldots & 0 & 0 \\ 1 & -2 & 1 & \ldots & 0 & 0 \\ 0 & 1 & -2 & \ldots & 0 & 0 \\ \ldots & \ldots & \ldots & \ldots & \ldots & \ldots \\ 0 & 0 & 0 & \ldots & -2 & 1 \\ 0 & 0 & 0 & \ldots & 1 & -(1+\omega) \end{pmatrix}. \tag{7.14}$$

Under *Neumann-Dirichlet boundary conditions* $(\tau = c)$, see (7.7), one sets $S(0) = S(1)$ and $S(L+1) = 0$, hence

$$(\Delta_L^{(c)} S_L)(1) = -S(1) + S(2), \qquad (\Delta_L^{(c)} S_L)(L) = S(L-1) - 2S(L). \tag{7.15}$$

The corresponding matrix representation is

$$
\Delta_L^{(c)} = \begin{pmatrix}
-1 & 1 & 0 & \cdots & 0 & 0 \\
1 & -2 & 1 & \cdots & 0 & 0 \\
0 & 1 & -2 & \cdots & 0 & 0 \\
\cdots & \cdots & \cdots & \cdots & \cdots & \cdots \\
0 & 0 & 0 & \cdots & -2 & 1 \\
0 & 0 & 0 & \cdots & 1 & -2
\end{pmatrix}.
\tag{7.16}
$$

Under *antiperiodic boundary conditions* $(\tau = a)$, see (7.9), one sets $S(0) = -S(L)$ and $S(L+1) = -S(1)$. Thus one obtains

$$
(\Delta_L^{(p)} S_L)(1) = -S(L) - 2S(1) + S(2),
$$
$$
(\Delta_L^{(p)} S_L)(L) = S(L-1) - 2S(L) - S(1),
\tag{7.17}
$$

which yields the matrix representation:

$$
\Delta_L^{(a)} = \begin{pmatrix}
-2 & 1 & 0 & \cdots & 0 & -1 \\
1 & -2 & 1 & \cdots & 0 & 0 \\
0 & 1 & -2 & \cdots & 0 & 0 \\
\cdots & \cdots & \cdots & \cdots & \cdots & \cdots \\
0 & 0 & 0 & \cdots & -2 & 1 \\
-1 & 0 & 0 & \cdots & 1 & -2
\end{pmatrix}.
\tag{7.18}
$$

Below we give a list of the complete sets of orthonormal eigenfunctions, $\{u_L^{(\tau)}(r,k),\ k = 1,\ldots,L\}$, of the one-dimensional discrete Laplacian under the boundary conditions used in this book:

$$
u_L^{(p)}(r,k) = L^{-1/2} \exp[-ir\varphi_L^{(p)}(k)];
\tag{7.19}
$$

$$
u_L^{(a)}(r,k) = L^{-1/2} \exp[-ir\varphi_L^{(a)}(k)];
\tag{7.20}
$$

$$
u_L^{(1)}(r,k) = [2/(L+1)]^{1/2} \sin r\varphi_L^{(1)}(k);
\tag{7.21}
$$

$$
u_L^{(0)}(r,k) = \begin{cases}
L^{-1/2} & \text{for} \quad k = 1 \\
(2/L)^{1/2} \cos(r - \tfrac{1}{2})\varphi_L^{(0)}(k) & \text{for} \quad k = 2,\ldots,L;
\end{cases}
\tag{7.22}
$$

$$
u_L^{(c)}(r,k) = 2(2L+1)^{-1/2} \cos(r - 1/2)\varphi_L^{(c)}(k).
\tag{7.23}
$$

For $|1 - \omega| \neq 1$,

$$u_L^{(\omega)}(r, k) = \sqrt{\frac{2}{L}}$$

$$\times \frac{\sin[r\varphi_L^{(\omega)}(k)] - (1 - \omega)\sin[(r - 1)\varphi_L^{(\omega)}(k)]}{\{(1 - \omega)^2 - 2(1 - \omega)\cos\varphi_L^{(\omega)}(k) + 1 + L^{-1}[1 - (1 - \omega)^2]\}^{1/2}}. \qquad (7.24)$$

The quantities $\varphi_L^{(\tau)}$, $k = 1, \ldots, L$, are defined as follows

$$\varphi_L^{(p)}(k) = 2\pi k/L, \qquad\qquad \varphi_L^{(1)}(k) = \pi k/(L + 1),$$
$$\varphi_L^{(0)}(k) = \pi(k - 1)/L, \qquad\qquad \varphi_L^{(c)}(k) = \pi(2k - 1)/(2L + 1)$$
$$\varphi_L^{(a)}(k) = \pi(2k + 1)/L.$$

Note that in the case of mixed boundary contitions with $\omega \neq 0, 1$ there is no closed form expression for $\varphi_L^{(\omega)}(k)$. Nevertheless, from [Singh et. al. (1975)] and [Zuk (1992)] one can obtain the following information. For given L and ω, where $|1 - \omega| \neq 1$, the numbers $\varphi_L^{(\omega)}(k)$, $k = 1, \ldots, L$ are the L roots of the equations

$$1 - \omega = \frac{\sin[\frac{1}{2}(L + 1)\varphi]}{\sin[\frac{1}{2}(L - 1)\varphi]} \qquad (7.25)$$

and

$$1 - \omega = \frac{\cos[\frac{1}{2}(L + 1)\varphi]}{\cos[\frac{1}{2}(L - 1)\varphi]} \qquad (7.26)$$

with $0 < \mathrm{Re}(\varphi) < \pi$ and $\mathrm{Im}(\varphi) > 0$. For concreteness and simplification of the notations below, without loss of generality in the final results, we assume L to be an odd integer. Then, it is easy to see that equation (7.25) possesses $(L - 1)/2$ solutions of the specified type, whereas equation (7.26) gives the remaining $(L + 1)/2$ solutions. From (7.25) and (7.26) one obtains that (for fixed L and k)

$$\frac{d\varphi}{d\omega} = \frac{1}{L}\frac{2\sin\psi}{(1 - \omega)^2 - 2(1 - \omega)\cos\varphi + 1 + L^{-1}[1 - (1 - \omega)^2]}. \qquad (7.27)$$

Further, if $|1 - \omega| < 1$ all the L roots are real. In that case there is only one root of equation (7.25) per interval $(2\pi k/(L - 1), 2\pi(k + 1)/(L - 1))$, $k = 0, \ldots, (L - 3)/2$. Similarly, equation (7.26) has only one root per interval $(\pi(2k - 1)/(L - 1), \pi(2k + 1)/(L - 1))$, $k = 1, \ldots, (L - 3)/2$, and one root in each of the intervals $(0, \pi/(L - 1))$ and $(\pi - \pi/(L - 1), \pi)$. Let us now consider the case $|1 - \omega| > 1$. Then, if $\omega < 0$, one again has only one root of equation (7.25) per interval $(2\pi k/(L - 1), 2\pi(k + 1)/(L - 1))$, $k = 1, \ldots, (L-3)/2$, and, similarly, one root of equation (7.26) per interval $(\pi(2k-1)/(L-1), \pi(2k+1)/(L-1))$, $k = 1, \ldots, (L-3)/2$, and one root in $(\pi - \pi/(L-1), \pi)$, i.e. altogether $L - 2$ real roots in the interval $(0, \pi)$. The remaining two roots are given by $\varphi_0^{\pm} = i\ln(1-\omega) \pm O\left((1 + |\omega|)^{-(L-1)}\right)$. In the case $\omega > 2$ one again has $L - 2$ real roots in the interval $(0, \pi)$ and the remaining two roots are $\varphi_0^{\pm} = \pi + i\log(\omega - 1) \pm O\left((\omega - 1)^{-(L-1)}\right)$. For the sake of completeness we mention that if $\omega = 2$, then one has the explicit result $u_L^{(2)}(r, k) = (2/L)^{1/2} \sin[(r - 1/2)\varphi_L^{(2)}(k)$, where $\varphi_L^{(2)}(k) = \pi k/L$, $k = 1, \ldots, L$.

The eigenvalues of $\Delta_L^{(\tau)}$, corresponding to the eigenvectors given by Eqs. (7.19) - (7.24), are

$$\lambda_L^{(\tau)}(k) = -2 + 2\cos\varphi_L^{(\tau)}(k), \qquad k = 1, \ldots, L. \qquad (7.28)$$

All the above results will be used in the remainder for explicit calculations of thermodynamic functions, local spin observables and probability distributions in the framework of the mean spherical model under different boundary conditions.

7.1.3 *Spectrum of the interaction Hamiltonian*

Consider a system of continuous scalar spins $S(\mathbf{r}) \in R$ associated with the sites of the infinite three dimensional lattice $\mathbf{r} = (r_1, r_2, r_3) \in Z^3$. In the case of nearest-neighbor ferromagnetic coupling, the spin-spin interaction part of the Hamiltonian (3.3), see Section 3.1.1, for the system of spins in a fully finite region $\Lambda \subset Z^3$, can be written in the form

$$\beta\mathcal{H}_{\Lambda,int}(S_\Lambda|S_{\Lambda^c}; K) = -\frac{1}{2}K \sum_{\mathbf{r},\mathbf{r}' \in \Lambda} Q_\Lambda^{(\tau)}(\mathbf{r} - \mathbf{r}')S(\mathbf{r})S(\mathbf{r}')$$

$$-K \sum_{\mathbf{r} \in \Lambda, \mathbf{r}' \in \Lambda^c} Q_\Lambda^{(\tau)}(\mathbf{r} - \mathbf{r}')S(\mathbf{r})S(\mathbf{r}'). \qquad (7.29)$$

Here $Q_\Lambda^{(\tau)}(\mathbf{r} - \mathbf{r}')$, with $\mathbf{r}, \mathbf{r}' \in Z^3$, is the adjacency matrix for the infinite cubic lattice: $Q_\Lambda^{(\tau)}(\mathbf{r} - \mathbf{r}') = 1$ if and only if $|\mathbf{r} - \mathbf{r}'| = 1$ and $Q_\Lambda^{(\tau)}(\mathbf{r} - \mathbf{r}') = 0$ otherwise. The first sum in the right-hand side of (7.29) describes the pairwise interaction between the spins in Λ, while the second sum is the boundary term which depends on the boundary conditions: it describes the interaction of the spins in the region Λ with a specified configuration $S_{\Lambda^c} = \{S(\mathbf{r}), \mathbf{r} \in \Lambda^c\}$ in the complement $\Lambda^c = Z^3 \setminus \Lambda$.

Let us take $\Lambda = \mathcal{L}_1 \times \mathcal{L}_2 \times \mathcal{L}_3$, where $\mathcal{L}_\nu = \{1, \ldots, L_\nu\}$ is a finite set. It is convenient to consider the configuration space $\Omega_\Lambda = R^{|\Lambda|}$ as an Euclidean vector space in which each configuration is represented by a column-vector S_Λ with components labeled according to the lexicographic order of the set $\{(r_1, r_2, r_3) \in \Lambda\}$. Let $S_\Lambda^\dagger$ be the corresponding transposed row-vector and let the dot $(\cdot)$ denote matrix multiplication. Then, for given boundary conditions $\tau = (\tau_1, \tau_2, \tau_3)$, specified for each pair of opposite faces of Λ by some τ_ν, see Section 7.1.1, the interaction Hamiltonian (7.29) takes the form

$$\beta \mathcal{H}_{\Lambda, int}^{(\tau)}(S_\Lambda | K) = -\frac{1}{2} K S_\Lambda^\dagger \cdot Q_\Lambda^{(\tau)} \cdot S_\Lambda. \tag{7.30}$$

Here the $|\Lambda| \times |\Lambda|$ interaction matrix $Q_\Lambda^{(\tau)}$ can be written as

$$Q_\Lambda^{(\tau)} = (\Delta_1^{(\tau_1)} + 2\,E_1) \times (\Delta_2^{(\tau_2)} + 2\,E_2) \times (\Delta_3^{(\tau_3)} + 2\,E_3), \tag{7.31}$$

where $\Delta_\nu^{(\tau_\nu)}$ is the one-dimentional discrete Laplacian defined on the finite chain $\mathcal{L}_\nu$ under boundary conditions τ_ν, and E_ν is the $L_\nu \times L_\nu$ unit matrix.

By using the results of Section 7.1.2, we can write down the eigenfunctions of the interaction matrix (7.31) in the form

$$u_\Lambda^{(\tau)}(\mathbf{r}, \mathbf{k}) = u_{L_1}^{(\tau_1)}(r_1, k_1)\, u_{L_2}^{(\tau_2)}(r_2, k_2)\, u_{L_3}^{(\tau_3)}(r_3, k_3), \qquad \mathbf{k} \in \Lambda, \tag{7.32}$$

and obtain the corresponding eigenvalues

$$\mu_\Lambda^{(\tau)}(\mathbf{k}) = 2 \sum_{\nu=1}^{3} \cos \varphi_{L_\nu}^{(\tau_\nu)}(k_\nu), \qquad \mathbf{k} \in \Lambda. \tag{7.33}$$

Note that the interaction Hamiltonian (7.29) has negative eigenvalues, which makes necessary the inclusion of a positive-definite quadratic form in the Gibbs exponent, to ensure the existence of the mean spherical partition function.

7.2 Finite-size scaling for systems with nonperiodic boundary conditions

Here, for the sake of simplicity, we focus our attention mainly on systems with zero Dirichlet boundary conditions, see Section 7.1, and flat boundaries. For such systems one expects that near T_c the nonsingular part $f^{(1)}_{L,\text{ns}}$ of the free energy density has an expansion similar to that given by Eq. (4.2), [Privman (1990a)],

$$
\begin{aligned}
f^{(1)}_{L,\text{ns}}(T) &= f_{\text{ns}}(T) + \frac{1}{L} f_{\text{s,ns}}(T) + \frac{1}{L^2} f_{\text{e,ns}}(T) \\
&\quad + \cdots + \frac{1}{L^d} f_{\text{c,ns}}(T) + O(L^{-(d+1)}),
\end{aligned}
\tag{7.34}
$$

where $f_{\text{ns}}(T)$ is the bulk contribution, $f_{\text{s,ns}}(T)$, $f_{\text{e,ns}}(T)$, $\ldots$, and $f_{\text{c,ns}}(T)$ are the nonsingular contributions from the surfaces, edges, etc, and corners (curvature) of the system boundaries, respectively. All the terms in the right-hand side of Eq. (7.34) are supposed to be regular at T_c. Let us assume for a while that the finite-size scaling form of the singular part of the free energy density is given by Eq. (4.28). Note that this term is proportional to L^{-d}, even for a system with continuously varying dimensionality d, provided $d_l < d < d_u$. On the other hand, one observes that for systems with corners the expansion of the nonsingular part of the free energy contains all the terms associated with the geometry of the boundary: these terms are proportional to integer powers of L^{-1}, and there is a term of the order of L^{-d} which is due to the contribution of the corners. It has been conjectured [Privman (1990a)], that when d passes through integer values D (say, $D = 2,3$), "resonant" poles develop in the finite-size scaling function Y and in the amplitude $f_{\text{c,ns}}$, which combine in such a way that a "resonant" logarithmic in L term appears. Since Y is universal, the resulting logarithmic term $uL^{-d}\ln L$ should have an *universal* amplitude, denoted here by u. The above statement can be cast formally as follows. Let $d \to D$, where D is an integer, and suppose that [Privman (1990a)]

$$
Y(x_1, x_2) = -\frac{u}{d - D} + \tilde{Y}(x_1, x_2) + O(d - D),
\tag{7.35}
$$

and

$$
f_{\text{c,ns}}(T) = \frac{u}{d - D} + \tilde{f}_{\text{c,ns}}(T) + O(d - D).
\tag{7.36}
$$

Hence, for the full free energy density at integer $d = D$ one obtains

$$
\begin{aligned}
f^{(1)}_L(T) &\simeq L^{-D}\tilde{Y}(atL^{1/\nu}, bhL^{\Delta/\nu}) + f_{\text{ns}}(T) + \frac{1}{L} f_{\text{s,ns}}(T) \\
&\quad + \frac{1}{L^2} f_{\text{e,ns}}(T) + \cdots + \frac{1}{L^D} \tilde{f}_{\text{c,ns}}(T) + u \lim_{d \to D} \frac{L^{-D} - L^{-d}}{d - D},
\end{aligned}
\tag{7.37}
$$

or, in a final form,

$$f_L^{(1)}(T) \simeq L^{-D}\tilde{Y}(atL^{1/\nu}, bhL^{\Delta/\nu}) + f_{\text{ns}}(T) + \frac{1}{L}f_{\text{s,ns}}(T)$$

$$+ \cdots + \frac{1}{L^D}\tilde{f}_{\text{c,ns}}(T) + \frac{u}{L^D}\ln L. \tag{7.38}$$

Here, for simplicity of notation, all the free energies are in units of $k_B T$. Note that both the scaling function Y and the coefficient u are universal. The resonant logarithmic in L contributions in the free energy density have been derived first for two-dimensional systems by conformal invariance methods, and then have been extended to the $d = 3$ case on the ground of the general arguments presented above. For a review of the results available for $d = 2$ one can consult, e.g. [Privman (1990a)]. For $d > 2$ the above predictions have been tested in the framework of a few exactly solved models, namely, in a Gaussian-type model [Gelfand and Fisher (1988)], [Gelfand and Fisher (1990)], in the constrained monomer-dimer model [Brankov and Priezzhev (1992)], and in the mean spherical model with "free" (Neumann) [Brankov and Danchev (1993b)] and "fixed" (zero Dirichlet) [Brankov and Danchev (1993a)] boundary conditions.

One can define surface, edge and corner finite-size free energy densities by *postulating* that expansion (4.2) formally holds in the critical region too. Let us consider, for simplicity, two identical three-dimensional fully finite systems in a region $\Lambda = L \times L_2 \times L_3$, one of which is periodic in all dimensions, and the other has Diriclet boundary conditions along the first dimension of size L, and is periodic in the remaining two dimensions. The thermodynamic functions of these two systems will be distinguished by the superscripts (p, p, p) and $(1, p, p)$, respectively. For finie $L, L_2, L_3 \gg 1$, up to higer order corrections we can write

$$f_L^{(p,p,p)}(T) \simeq f(T), \quad f_L^{(1,p,p)}(T) \simeq f(T) + \frac{1}{L}f_{\text{s}}(T), \tag{7.39}$$

where $f_{\text{s}}(T)$ is the total contribution from the surfaces. Obviously, due to the considered type of geometry, the term $f_{\text{s}}(T)$ is twice the surface free energy density (per $L_2 \times L_3$), $f_{\text{s}}(T) = 2f_{\text{surf}}(T)$. Then, by taking the limit $L_2, L_3 \to \infty$, one comes to the following definition of the *finite-size* surface free energy density

$$f_{L,\text{surf}}(T) \equiv \frac{L}{2}\left[f_L^{(1)}(T) - f_L^{(p)}(T)\right], \tag{7.40}$$

where $f_L^{(1)}(T)$ and $f_L^{(p)}(T)$ are the free energy densities of systems with the film geometry $L \times \infty^2$ under Dirichlet, respectively, periodic boundary conditions in the finite dimension. Of course, in the limit $L \to \infty$ one obtains the standard definition of the surface free energy density $f_{\text{surf}}(T)$. The finite-size scaling behavior of $f_{L,\text{surf}}(T)$ follows from the corresponding properties of the free energy densities $f_L^{(p)}(T)$ and $f_L^{(1)}(T)$, hence,

$$\beta f_{L,\text{surf, sing}}(T) \simeq L^{-(d-1)} Y_{\text{surf}}(atL^{1/\nu}), \tag{7.41}$$

where

$$Y_{\text{surf}}(x) = \frac{1}{2}\left[Y^{(1)}(x) - Y^{(p)}(x)\right]. \tag{7.42}$$

Here $Y^{(1)}$ and $Y^{(p)}$ are the finite-size scaling functions for the "singular part" of the free energy of the system under zero Dirichlet, respectively, periodic boundary conditions, see Eq. (4.28).

Similar to the finite-size surface free energy density, one can define a finite-size *interfacial free energy density* as the difference beteen the free energy densities of two samples of identical systems: one with, and the other without, an interface that separates two coexisting phases. Consider, e.g., an Ising system in a fully finite region $L \times L_2^{d-1}$ with periodic boundary conditions in all the $d-1$ dimensions except the first, where it has a linear extent L. By imposing either antiperiodic, or $(+,-)$ boundary conditions along the first dimension, one forces a "floating" interface of area L_2^{d-1} to emerge within the system. The appropriate reference systems, with no interface and no additional free energy contributions from the boundaries, are those with periodic boundery conditions in the first case, and $(+,+)$, or $(-,-)$, boundary conditions in the second one. Then the finite-size interfacial free energy density (per unit ares) is defined as

$$\sigma_L^{+,-}(T) = L\left[f_L^{+,-}(T) - f_L^{+,+}(T)\right], \tag{7.43}$$

where, for simplicity, we have taken the limit $L_2 \to \infty$. A similar difference can be defined for antiperiodic versus periodic boundary conditions,

$$\sigma_L^{(a)}(T) = L\left[f_L^{(a)}(T) - f_L^{(p)}(T)\right], \tag{7.44}$$

where $f_L^{(a)}(T)$ is the free energy density of an infinite film $L \times \infty^2$ with antiperiodic boundary conditions. Obviously, the above defined finite-size interfacial free energies depend on the choice of the nonperiodic boundary conditions, i.e., in general, $\sigma_L^{(a)}(T) \neq \sigma_L^{+,-}(T)$. However, they must have a unique "bulk" limit $\sigma(T)$ as $L \to \infty$, which has the well-known properties: $\sigma(T) \sim |t|^\mu$ when $t \to 0^-$, and $\sigma(T) \equiv 0$ when $T \geq T_c$. By analogy with the finite-size surface free energy, the corresponding finite-size scaling hypothesis for the interfacial finite-size free energy reads

$$\sigma_L^{(\tau)}(T) \simeq L^{-(d-1)}\Theta^{(\tau)}(atL^{1/\nu}). \tag{7.45}$$

Taking into account that when $L \to \infty$ at fixed $|t| \ll 1$ one should reproduce the bulk behavior of $\sigma(T)$, one immediately obtains that $\mu = (d-1)\nu$. Note the following important difference from the finite-size behavior of the surface free energy behavior: we have not required Eq. (7.45) to be valid only for the "singular part" of the finite-size interface free energy density. That is, we have assumed that *for "floating" periodic interfaces all "nonsingular" terms in the difference of the free energy densities cancel out.* Note further, that in order $\sigma(T)$ to exist we have tacitly assumed that

$$f_L^{(a)}(T) - f_L^{(p)}(T) = \frac{1}{L}\sigma(T) + o(L^{-1}) \tag{7.46}$$

when $T < T_c$. For Ising-like systems this is indeed the case, but the above is not true for $O(n)$ systems: then at $T < T_c$ one has

$$f_L^{(a)}(T) = f(T) + \frac{1}{2}\left(\frac{\pi}{L}\right)^2 \Upsilon(T) + o(L^{-2}), \tag{7.47}$$

where $\Upsilon(T)$ is the *helicity modulus*. In Eq. (7.47) it has been taken into account that the existence of extra degrees of freedom in systems with vector order parameter allows the phase change (induced by the antiperiodic boundary conditions) to take place smoothly, through a "Bloch wall". If the distance between the antiperiodic boundaries is L, then the relative angle $\delta\theta$ between neighboring spins in an $O(n)$, $n \geq 2$, model is of the order of $\delta\theta \sim \pi/L$. Then, the amount of the increase in energy due to this twist can be estimated as being proportional to $1 - \cos(\delta\theta) \sim \frac{1}{2}\pi^2/L^2$, wherefrom one obtains that expansion (7.47) for the antiperiodic free energy density should be valid with some $\Upsilon(T) = O(1)$. Hence, Eq. (7.47) can be used to define the helicity modulus Υ (per unit area) as [Fisher et. al. (1973)]

$$\Upsilon(T) = \lim_{L\to\infty} \frac{2L^2}{\pi^2}\left[f_L^{(a)}(T) - f_L^{(p)}(T)\right]. \tag{7.48}$$

For the definition of the helicity modulus in the $O(n)$ vector model under "twisted" boundary conditions, and its evaluation by using $\varepsilon = d - 2$, $\varepsilon = 4 - d$, and large-n expansions, see [Brézin et. al. (1993)].

The definition (7.48) can easely be extended to define a finite-size helicity modulus $\Upsilon_L(T)$ via [Danchev (1993)]

$$\Upsilon_L(T) = \frac{2L^2}{\pi^2}\left[f_L^{(a)}(T) - f_L^{(p)}(T)\right]. \tag{7.49}$$

The stardard finite-size scaling hypotesis for the singular part of $\Upsilon_L(T)$ would read

$$\Upsilon_{L,\text{sing}}(T) = L^{-(d-2)}\Phi(atl^{1/\nu}), \tag{7.50}$$

where Φ is a universal scaling function. In order to reproduce the L-independent bulk behavior of $\Upsilon(T)$ when $T \to T_c^-$ we have to require that

$$\Phi(x) = \Phi_{-\infty}|x|^{(d-2)\nu}, \qquad x \to -\infty, \tag{7.51}$$

hence we obtain the critical behavior $\Upsilon_{\text{sing}}(T) \sim |t|^v$, as $t \to 0^+$, with exponent v wich satisfies the Josephson relation $v = (d - 2)\nu$.

However, it has been argued [Privman (1990b)] that in full analogy with the case of the finite-size free energy of a system with corners, additional resonant logarithmic in L term might appear in dimensionality $d = 3$. Then, the leading nonsingular term and the singular part of $\Upsilon_L(T)$ are again of the same order in L^{-1}. So, Privman suggested that at $d = 3$

$$\Upsilon_L(T) = L^{-1}\left[\tilde{\Phi}(atL^{1/\nu}) + w\ln L\right] + \Phi_{\text{ns},2}(T)/L + o(L^{-1}), \tag{7.52}$$

where $\Phi_{\text{ns},2}(T)$ is supposed to be regular at $T = T_c$. The amplitude w and the scaling function $\tilde{\Phi}$ are universal. In [Danchev (1993)] the hypothesis (7.52) has been checked on the example of the mean spherical model. No logarithmic corrections of the type suggested above have been found. It turns out that $\Upsilon_L(T)$ behaves in a way similar to that of the interfacial free energy due to floating periodic interfaces in Ising-like systems. In the latter case the leading nonsingular contributions in the free energy densities used in the definition of the finite-size interfacial tension are supposed to fully cancel out. Furthermore, logarithmic corrections have been found in the finite-size behavior of the interfacial free energy [Privman (1990a)], [Gelfand and Fisher (1990)] *below* T_c [Gelfand and Fisher (1990)]. For example, it has been shown that in the capillary-wave d-dimensional Gaussian-type model the singular part (per $k_B T$) of the interfacial free energy density (when the interface is created by applying surface fields which fix its mean position) has the form [Gelfand and Fisher (1990)]

$$\sigma_L(T) = g_0(T) + g_1(T)/L + \cdots - 2^{d-1} \ln L/L^{d-1} + \cdots , \qquad (7.53)$$

at any fixed $T < T_c$ as $L \to \infty$. As shown above, there is a *universal* logarithmic term present, even though at any fixed T the functions g_k are *not* universal (for more details see Ref. [Gelfand and Fisher (1990)]). Similarly to the case of the interfacial free energy, logarithmic corrections (within the spherical model) have been found for the helicity modulus only *below* the critical temperature [Danchev (1993)], where a geometry of the type $L^{3-d'} \times \infty^{d'}, 0 \leq d' < 3$ has been considered. In the case of a film ($d' = 2$) and cylinder ($d' = 1$) with antiperiodic and/or periodic boundary conditions, no logarithmic corrections have been found. In the case of a fully finite geometry with different boundary conditions: in $d_p \geq 0$ space dimensions – periodic, in $d_a \geq 0$ – antiperiodic, in $d_0 \geq 0$ – Newmann, and in $d_1 \geq 0$ – Dirichlet, so that $d_p + d_a + d_0 + d_1 = 3$, a universal logarithmic correction term

$$-[(d_0 - d_1)/4\pi + 2^{d_a - 1}/\pi^2] \ln L/L \qquad (7.54)$$

has been obtained *only below* T_c, when $(T_c - T)/T_c \gg L^{-1} \ln L$. In the vicinity of T_c it has been found that the finite-size scaling form

$$\Upsilon_L(T) = L^{-(d-2)} \Phi(atL^{1/\nu}). \qquad (7.55)$$

holds (not only for "the singular part of" the helicity modulus).

7.3 Surface and layer critical exponents in the mean spherical model

The mean spherical model has been extensively studied with respect to both the finite-size scaling theory and (relatively less) the theory of surface phase transitions, for a review see, e.g.,[Barber (1983)], [Binder (1983)], [Privman (1990a)] and [Krech (1994)]. The finite-size scaling behavior of the mean spherical model in the presence of surface fields has been studied under periodic, antiperiodic, Dirichlet, Newmann, and Newmann-Dirichlet boundary conditions, see [Barber and Fisher (1973)], [Barber (1974)], [Barber et. al

(1974)], [Singh et. al. (1975)], [Danchev et. al (1997a)]. Nevertheless, the situation with its surface critical exponents, especially those for the special phase transition, is not completely clear.

The mean spherical model with the film geometry $L \times \infty^{d-1}$, under zero Dirichlet boundary conditions, in the presence of an external homogeneous magnetic field h and a surface field h_1 acting on the first layer, has been considered in [Barber (1974)]. For $d = 3$ it has been found that the singular part of the free energy density of the system is of the form

$$\beta f_{L,\text{sing}}(K, h, h_1) = L^{-3} X(a t L^{1/\nu}, b h L^{\Delta/\nu}, c h_1 L^{\Delta_1/\nu}). \qquad (7.56)$$

Here $\nu = 1$, $\Delta = 5/2$, $\Delta_1 = 1/2$ are the scaling exponents (at $d = 3$), t is the shifted reduced critical temperature $t = (T - T_c)/T_c + \varepsilon_L$, where ε_L is the shift obeying $\lim_{L\to\infty} \varepsilon_L = 0$ and a, b, c are some (non-universal) metric factors. The critical exponent $\gamma_{1,1}$ of the local surface susceptibility

$$\beta^{-1} \chi_{1,1}(K) = \frac{1}{2} \lim_{L\to\infty} \left[-L \partial^2 \beta f_{L,\text{sing}}(K, h, h_1)/\partial h_1^2 \right]_{h=h_1=0} \qquad (7.57)$$

is related to the surface exponent Δ_1 in Eq. (7.56) by the equality $\gamma_{1,1} = 2\Delta_1 - 2$.* For Dirichlet boundary conditions its value has been found to be $\gamma_{1,1}^0 = -1$, which means that $\chi_{1,1}(K)$ does not diverge at $T = T_c$ in this case. Actually, the functional dependence given by Eq. (7.56) is expected to hold for any system undergoing ordinary phase transition. The only independent new surface critical exponent in this case is $\Delta_1 = \Delta_1^0$.

As already mentioned, the infinite translational invariant spherical model is equivalent to the $n \to \infty$ limit of such a n-component system [Stanley (1968)],[Kac and Thompson (1971)], but the spherical model with free surfaces (or, more generally, without translation-invariant symmetry) is in fact *not* such a limit [Knops (1973)], see also Section 3.2. The last becomes apparent if one investigates surface phase transitions for an $O(n)$ model in the limit $n \to \infty$. In that case one obtains [Barber (1983)] $\Delta_1 = 1/(d-2)$ (i.e. $\Delta_1 = 1$ for $d = 3$) for ordinary and $\Delta_1 = 2/(d-2)$ for special phase transitions. An attempt to clarify the situation with the special and extraordinary phase transitions within the spherical model has been made in [Barber et. al (1974)] and in [Singh et. al. (1975)], where the spherical model with Dirichlet boundary conditions and an enhancement of the surface

---

*The general relationship, assuming the validity of hyperscaling, see Eq. (1.96), reads $\gamma_{1,1} = 2\Delta_1 - (d-1)\nu$.

coupling $K_s = K(1+w)$ (K is the bulk dimensionless coupling and $w > 0$), has been considered. It turns out that if, as usual, only one global mean spherical constraint is imposed, the model predicts quite unphysically the existence of an extraordinary phase transition [Barber et. al (1974)] when $w > 1/(2d-2)$ for $d \geq 3$ (in [Barber et. al (1974)] only integer dimensionalities have been considered). But, if an additional constraint is involved to ensure the proper behavior of the surface spins, one obtains [Singh et. al. (1975)], that for any $w > 0$ there is no other critical temperature except the bulk one, provided $d \leq 3$. This is in agreement with the results for the $O(n)$ model in the limit $n \to \infty$, for which the crossover exponent $\vartheta = (d-3)/(d-2)$.

In this section, following [Danchev et. al (1997a)], we consider the finite-size scaling behavior of the 3-dimensional mean spherical model on a hypercubic lattice system with the film geometry $L \times \infty^2$. At the surfaces of the film Neumann ($\tau = 0$) and Neumann-Dirichlet ($\tau = c$) boundary conditions are imposed. Surface fields h_1 and h_L are supposed to act at the surfaces bounding the system.

7.3.1 *The model*

Consider the three-dimensional mean spherical model with nearest-neighbor ferromagnetic interactions, see Section 3.1.2, in the fully finite region $\Lambda = \mathcal{L}_1 \times \mathcal{L}_2 \times \mathcal{L}_3$. Let $h_\Lambda = \{h(\mathbf{r}), \mathbf{r} \in \Lambda\}$ be a column-vector representing (in units of $k_B T$) the inhomogeneous magnetic field configuration, and let $h_\Lambda^\dagger$ be the transposed row-vector. Then, for given boundary conditions $\tau = (\tau_1, \tau_2, \tau_3)$, specified for each pair of opposite faces of Λ by some $\tau_i = p$ (periodic), 1 (Dirichlet), 0 (Neumann) or c (Neumann-Dirichlet), the matrix representation of the Hamiltonian is

$$\beta\mathcal{H}_\Lambda^{(\tau)}(S_\Lambda | K, h_\Lambda; s) = -\frac{1}{2} K S_\Lambda^\dagger \cdot Q_\Lambda^{(\tau)} \cdot S_\Lambda + s\, S_\Lambda^\dagger \cdot S_\Lambda - h_\Lambda^\dagger \cdot S_\Lambda. \quad (7.58)$$

Here the interaction matrix $Q_\Lambda^{(\tau)}$ has been defined in Eq. (7.31), and its eigenvalues are given by Eq. (7.33).

In order to ensure the existence of the partition function, all the eigenvalues $-\frac{1}{2} K \mu_\Lambda^{(\tau)}(\mathbf{k}) + s$, $\mathbf{k} \in \Lambda$, of the quadratic form in $\beta\mathcal{H}_\Lambda^{(\tau)}(S_\Lambda | K, h_\Lambda; s)$, see Eqs. (7.33) and (7.58), must be positive. Hence, the spherical field s

must satisfy the inequality

$$s > \frac{1}{2}K \max_{\mathbf{k}\in\Lambda} \mu_\Lambda^{(\tau)}(\mathbf{k}) = \frac{1}{2}K\mu_\Lambda^{(\tau)}(\mathbf{k}_0). \tag{7.59}$$

In view of this condition, it is convenient to introduce a shifted and rescaled spherical field $\phi > 0$ by setting $s = s(\phi)$, where

$$s(\phi) = \frac{1}{2}K[\phi + \mu_\Lambda^{(\tau)}(\mathbf{k}_0)]. \tag{7.60}$$

Note that under fully periodic boundary conditions, $\tau = (p,p,p)$, one has $\mathbf{k}_0 = (L_1, L_2, L_3)$, hence $\mu_\Lambda^{(p,p,p)}(\mathbf{k}_0) = 6$, and Eq. (7.60) reduces to $s = K(\phi+6)/2$. In the general case, the eigenvalues of Hamiltonian (7.58) can be written in the form

$$s(\phi) - \frac{1}{2}K\mu_\Lambda^{(\tau)}(\mathbf{k}) = \frac{1}{2}K[\phi + \omega_\Lambda^{(\tau)}(\mathbf{k})], \tag{7.61}$$

where $\omega_\Lambda^{(\tau)}(\mathbf{k}) = \mu_\Lambda^{(\tau)}(\mathbf{k}_0) - \mu_\Lambda^{(\tau)}(\mathbf{k})$ is the normalized excitation spectrum.
   The free energy density of a system in a finite region Λ is

$$\beta f_\Lambda^{(\tau)}(K, h_\Lambda) = \frac{1}{2}\left\{ \ln(K/2\pi) - K\mu_\Lambda^{(\tau)}(\mathbf{k}_0) \right.$$
$$\left. + \mathcal{U}_\Lambda^{(\tau)}(\phi_\Lambda^{(\tau)}) - P_\Lambda^{(\tau)}(K, h_\Lambda; \phi_\Lambda^{(\tau)}) - K\phi_\Lambda^{(\tau)} \right\}, \tag{7.62}$$

where $\phi_\Lambda^{(\tau)} = \phi_\Lambda^{(\tau)}(K, h_\Lambda)$ is the solution of the mean spherical constraint

$$|\Lambda|^{-1} \sum_{\mathbf{r}\in\Lambda} \langle S^2(\mathbf{r})\rangle_\Lambda^{(\tau)}(K, h_\Lambda; \phi) = 1, \tag{7.63}$$

and $\langle\cdots\rangle_\Lambda^{(\tau)}(K, h_\Lambda; \phi)$ denotes expectation value in the Gibbs ensemble with Hamiltonian (7.58). In Eq. (7.62) we have introduced the function

$$\mathcal{U}_\Lambda^{(\tau)}(\phi) = |\Lambda|^{-1} \sum_{\mathbf{k}\in\Lambda} \ln[\phi + \omega_\Lambda^{(\tau)}(\mathbf{k})], \tag{7.64}$$

which describes the contribution of the spin-spin interaction (to be called "interaction term"), and the function

$$P_\Lambda^{(\tau)}(K, h_\Lambda; \phi) = \frac{1}{K|\Lambda|} \sum_{\mathbf{r}\in\Lambda} \frac{|\hat{h}_\Lambda^{(\tau)}(\mathbf{k})|^2}{\phi + \omega_\Lambda^{(\tau)}(\mathbf{k})}, \tag{7.65}$$

which represents the "field term". Here $\hat{h}_\Lambda^{(\tau)}(\mathbf{k})$ denotes the projection of the magnetic field configuration h_Λ on the eigenfunction $\{\bar{u}_\Lambda^{(\tau)}(\mathbf{r},\mathbf{k}),\ \mathbf{r} \in \Lambda\}$:[†]

$$\hat{h}_\Lambda^{(\tau)}(\mathbf{k}) = \sum_{\mathbf{r} \in \Lambda} h(\mathbf{r})\bar{u}_\Lambda^{(\tau)}(\mathbf{r},\mathbf{k}). \tag{7.66}$$

The mean spherical constraint (7.63) can be written as

$$\frac{\mathrm{d}}{\mathrm{d}\phi}\mathcal{U}_\Lambda^{(\tau)}(\phi) - \frac{\partial}{\partial\phi}P_\Lambda^{(\tau)}(K,h_\Lambda;\phi) = K. \tag{7.67}$$

Now we set $\tau_1 = \tau \in \{1,0,c\}$, $\tau_2 = \tau_3 = p$, note that for these boundary conditions $\mathbf{k}_0 = \{1,L_2,L_3\}$, and take the limit $L_2,L_3 \to \infty$ in expression (7.64) at fixed $L_1 = L$:

$$\mathcal{U}_{L,3}^{(\tau)}(\phi) = \lim_{L_2,L_3 \to \infty} \mathcal{U}_\Lambda^{(\tau,p,p)}(\phi)$$

$$= \frac{1}{L}\sum_{k=1}^{L}\mathcal{V}_2\left(\phi + 2\cos\varphi_L^{(\tau)}(1) - 2\cos\varphi_L^{(\tau)}(k)\right), \tag{7.68}$$

where

$$\mathcal{V}_d(z) := \frac{1}{(2\pi)^d}\int_{-\pi}^{\pi}\mathrm{d}\theta_1 \cdots \int_{-\pi}^{\pi}\mathrm{d}\theta_d \ln\left[z + 2\sum_{\nu=1}^{d}(1 - \cos\theta_\nu)\right]. \tag{7.69}$$

Next we confine ourselves to the consideration of uniform in the space dimensions $\nu = 2,3$ magnetic fields, $h(\mathbf{r}) = h_{\mathrm{surf}}(r_1)$, $\mathbf{r} \in \Lambda$, and by taking the same limit in (7.65) we obtain

$$P_L^{(\tau)}(K,h_{\mathrm{surf}};\phi) := \lim_{L_2,L_3 \to \infty} P_\Lambda^{(\tau,p,p)}(K,h_\Lambda;\phi)$$

$$= \frac{1}{KL}\sum_{k=1}^{L}\frac{[\hat{h}_{\mathrm{surf}}^{(\tau)}(k)]^2}{\phi + 2\cos\varphi_L^{(\tau)}(1) - 2\cos\varphi_L^{(\tau)}(k)}, \tag{7.70}$$

where

$$\hat{h}_{\mathrm{surf}}^{(\tau)}(k) := \sum_{r=1}^{L} h_{\mathrm{surf}}(r)u_L^{(\tau)}(r,k), \qquad \tau \in \{1,0,c\}. \tag{7.71}$$

[†]By $\bar{u}$ we denote the complex conjugate of $u \in \mathbf{C}$

Thus, the mean spherical constraint (7.67) can be written in the form

$$\mathcal{W}_{L,3}^{(\tau)}(\phi) - \frac{\partial}{\partial \phi} P_L^{(\tau)}(K, h_{\text{surf}}; \phi) = K, \tag{7.72}$$

where

$$\mathcal{W}_{L,3}^{(\tau)}(\phi) := \frac{1}{L} \sum_{k=1}^{L} \mathcal{W}_2 \left(\phi + 2\cos\varphi_L^{(\tau)}(1) - 2\cos\varphi_L^{(\tau)}(k) \right) \tag{7.73}$$

and

$$\mathcal{W}_d(z) = \frac{1}{(2\pi)^d} \int_{-\pi}^{\pi} \cdots \int_{-\pi}^{\pi} \frac{\mathrm{d}\theta_1 \cdots \mathrm{d}\theta_d}{z + 2\sum_{\nu=1}^{d}(1 - \cos\theta_\nu)} \tag{7.74}$$

is the generalized Watson function (note that $\mathcal{W}_d(z) = \mathrm{d}\mathcal{V}_d(z)/\mathrm{d}z$).

Note that having evaluated $\mathcal{W}_{L,3}^{(\tau)}(\phi)$, the corresponding interaction term $\mathcal{U}_{L,3}^{(\tau)}(\phi_L^{(\tau)})$ in the singular (in the limit $L \to \infty$) part of the free energy density, see Eq. (7.62),

$$\beta f_{L,\text{sing}}^{(\tau)}(K, h_{\text{surf}}) = \frac{1}{2} \left\{ \mathcal{U}_{L,3}^{(\tau)}(\phi_L^{(\tau)}) - P_L^{(\tau)}(K, h_{\text{surf}}; \phi_L^{(\tau)}) - K\phi_L^{(\tau)} \right\}, \tag{7.75}$$

can be obtained by integration:

$$\mathcal{U}_{L,3}^{(\tau)}(\phi_L^{(\tau)}) = \mathcal{U}_{L,3}^{(\tau)}(\phi_0) + \int_{\phi_0}^{\phi_L^{(\tau)}} \mathrm{d}\phi \, \mathcal{W}_{L,3}^{(\tau)}(\phi). \tag{7.76}$$

Here $\phi_L^{(\tau)} = \phi_L^{(\tau)}(K, h_{\text{surf}})$ is the solution of equation (7.72), and $\phi_0 \geq 0$ is a suitably chosen constant.

Equations (7.70) - (7.76) provide the starting expressions for our further finite-size scaling analysis.

7.3.2 *Finite-size scaling under Neumann boundary conditions*

Let us study first the finite-size scaling behavior of the mean spherical constraint and the free energy density in the case of Neumann boundary conditions. We consider external fields h_1 and h_L which act at the surfaces bounding the system:

$$h_{\text{surf}}(r_1) = h_1 \delta_{r_1,1} + h_L \delta_{r_1,L}. \tag{7.77}$$

For the projection (7.71) of the boundary fields on the k-th eigenvector (7.22) we obtain

$$\hat{h}^{(0)}_{\text{surf}}(k) = \begin{cases} L^{-1/2}(h_1 + h_L) & \text{for} \quad k = 1 \\ (2/L)^{1/2}[h_1 - (-1)^k h_L]\cos \tfrac{1}{2}\varphi^{(0)}_L(k) & \text{for} \quad k = 2,\ldots,L. \end{cases} \tag{7.78}$$

From Eqs. (7.70) and (7.78), assuming L even, we find that the field term takes the form

$$P^{(0)}_L(K,h_1,h_L;\phi) =$$
$$\frac{(h_1 + h_L)^2}{KL^2}\left[\frac{1}{\phi} - \frac{L}{4} + \frac{1}{2} + \left(1 + \frac{\phi}{4}\right)\sum_{k=1}^{L/2-1}\left(1 + \frac{\phi}{2} - \cos\frac{2k\pi}{L}\right)^{-1}\right]$$
$$+\frac{(h_1 - h_L)^2}{KL^2}\left[-\frac{L}{4} + \left(1 + \frac{\phi}{4}\right)\sum_{k=1}^{L/2}\left(1 + \frac{\phi}{2} - \cos\frac{(2k-1)\pi}{L}\right)^{-1}\right]. \tag{7.79}$$

Since $\phi > 0$, we set $1 + \phi/2 = \cosh x$, and consider first the sum

$$S_1(x) = \sum_{k=1}^{L/2-1}\left(\cosh x - \cos\frac{2k\pi}{L}\right)^{-1}. \tag{7.80}$$

By making use of the identity [Gradshteyn and Ryzhik (1973)]

$$\sum_{k=1}^{n-1}\ln\left(2\cosh x - 2\cos\frac{k\pi}{n}\right) = \ln(\sinh nx) - \ln(\sinh x), \tag{7.81}$$

and differentiating it with respect to x, at $n = L/2$ one obtains

$$S_1(x) = \frac{L\coth(Lx/2)}{2\sinh x} - \frac{\cosh x}{\sinh^2 x}. \tag{7.82}$$

Similarly one considers the sum

$$S_2(x) = \sum_{k=1}^{L/2}\left(\cosh x - \cos\frac{(2k-1)\pi}{L}\right)^{-1}, \tag{7.83}$$

which can be obtained by differentiating the identity [Gradshteyn and Ryzhik (1973)]

$$\sum_{k=1}^{n} \ln\left(2\cosh x - 2\cos\frac{(2k-1)\pi}{2n}\right) = \ln(2\cosh nx), \tag{7.84}$$

with respect to x, and setting $n = L/2$:

$$S_2(x) = \frac{L\tanh(Lx/2)}{2\sinh x}. \tag{7.85}$$

From equations (7.79), (7.82) and (7.85), one obtains the exact expression

$$\begin{aligned}
P_L^{(0)}&(K, h_1, h_L; \phi) = \\
&\frac{(h_1 + h_L)^2}{2KL}\left[\phi^{-1/2}(1 + \phi/4)^{1/2}\coth(Lx/2) - 1/2\right] \\
&+\frac{(h_1 - h_L)^2}{2KL}\left[\phi^{-1/2}(1 + \phi/4)^{1/2}\tanh(Lx/2) - 1/2\right].
\end{aligned} \tag{7.86}$$

Hence, in the limit

$$\phi \to 0, \qquad L \to \infty, \qquad \text{so that} \quad y := \phi^{1/2}L = O(1), \tag{7.87}$$

by taking into account that $x = \phi^{1/2}[1 + O(\phi)]$, we obtain the asymptotic form of the field term

$$\begin{aligned}
P_L^{(0)}&(K, h_1, h_L; \phi) \simeq -\frac{1}{2KL}(h_1^2 + h_L^2) \\
&+\frac{1}{2Ky}\left[(h_1 + h_L)^2\coth\left(\frac{1}{2}y\right) + (h_1 - h_L)^2\tanh\left(\frac{1}{2}y\right)\right],
\end{aligned} \tag{7.88}$$

which holds up to corrections of order $O(\phi) = O(L^{-2})$.

Next, we evaluate the interaction term in the mean spherical constraint by using an improved version [Danchev et. al (1997a)] of the method developed in [Barber and Fisher (1973)]. Following the latter work we set

$$\mathcal{W}_2(z) := -(1/4\pi)\ln z + (5/4\pi)\ln 2 + Q_2(z), \tag{7.89}$$

where $Q_2(z)$, defined by the above equation, has the asymptotic behavior

$$Q_2(z) = O(z\ln z), \qquad dQ_2/dz = O(\ln z), \qquad z \to 0. \tag{7.90}$$

Then the interaction term in equation (7.72) takes the form

$$
\mathcal{W}_{L,3}^{(0)}(\phi) := \frac{1}{L}\sum_{k=1}^{L}\mathcal{W}_2\left(\phi + 2 - 2\cos\phi_L^{(0)}(k)\right)
$$
$$
= g_1^{(0)}(\phi) + g_2^{(0)}(\phi) + (5/4\pi)\ln 2, \tag{7.91}
$$

where

$$
g_1^{(0)}(\phi) = -\frac{1}{4\pi L}\sum_{k=0}^{L-1}\ln\left(2\cosh x - 2\cos\frac{\pi k}{L}\right) \tag{7.92}
$$

and

$$
g_2^{(0)}(\phi) = \frac{1}{L}\sum_{k=0}^{L-1}Q_2\left(\phi + 4\sin^2\frac{\pi k}{2L}\right). \tag{7.93}
$$

From (7.81) at $n = L$ we obtain

$$
g_1^{(0)}(\phi) = -\frac{1}{4\pi L}[\ln\phi + \ln(\sinh Lx) - \ln(\sinh x)]. \tag{7.94}
$$

In the limit (7.87) the above expression yields

$$
g_1^{(0)}(\phi) = \frac{\ln L}{4\pi L} - \frac{1}{4\pi L}\ln(y\sinh y) + O(L^{-3}). \tag{7.95}
$$

Since $Q_2(\phi + 4\sin^2(\pi k/2L))$ is periodic with respect to k, with period $2L$, the function $g_2^{(0)}(\phi)$ can be written as

$$
g_2^{(0)}(\phi) = \frac{1}{2L}\sum_{k=0}^{2L-1}Q_2\left(\phi + 4\sin^2\frac{\pi k}{2L}\right) + \frac{1}{2L}[Q_2(\phi) - Q_2(\phi + 4)]. \tag{7.96}
$$

By using the Poisson summation formula (4.79) for the sum in (7.96), the above expression takes the form

$$
g_2^{(0)}(\phi) = \frac{1}{\pi}\int_0^\pi Q_2(\phi + 4\sin^2\theta)\mathrm{d}\theta + \frac{1}{2L}[Q_2(\phi) - Q_2(\phi + 4)]
$$
$$
+ \frac{2}{\pi}\sum_{q=1}^{\infty}\int_0^\pi Q_2(\phi + 4\sin^2\theta)\cos(4Lq\theta)\,\mathrm{d}\theta. \tag{7.97}
$$

Integration by parts, with the aid of (7.90), yields the result that the last term in equation (7.97) is of order $O(L^{-2})$. Thus

$$
\begin{aligned}
g_2^{(0)}(\phi) = {}& \frac{1}{\pi} \int_0^\pi \mathcal{W}_2(\phi + 4\sin^2\theta)\, d\theta + \frac{1}{4\pi^2} \int_0^\pi \ln(\phi + 4\sin^2\theta)\, d\theta \\
& + \frac{1}{2L}[Q_2(\phi) - Q_2(\phi + 4)] - (5/4\pi)\ln 2 + O(L^{-2}) \\
= {}& \mathcal{W}_3(\phi) + (1/4\pi)\ln\left[1 + \phi/2 + \phi^{1/2}(1 + \phi/4)^{1/2}\right] \\
& + \frac{1}{2L}[Q_2(\phi) - Q_2(\phi + 4)] - (5/4\pi)\ln 2 + O(L^{-2}).
\end{aligned} \quad (7.98)
$$

When $\phi \to 0$, in view of (7.90) and the asymptotic expansion of $\mathcal{W}_3(\phi)$, see [Barber and Fisher (1973)], expression (7.98) simplifies to

$$
\begin{aligned}
g_2^{(0)}(\phi) = {}& K_c - (5/4\pi)\ln 2 \\
& - \frac{1}{2L}\left[\mathcal{W}_2(4) - (3/4\pi)\ln 2 + O(\phi\ln\phi)\right] + O(\phi) + O(L^{-2}),
\end{aligned} \quad (7.99)
$$

where $K_c = \mathcal{W}_3(0)$ is the bulk critical coupling.

Therefore, from equations (7.91), (7.95) and (7.99), in the limit (7.87) we obtain the following asymptotic form of the interaction term

$$
\begin{aligned}
\mathcal{W}_{L,3}^{(0)}(\phi) = {}& K_c + \frac{\ln L}{4\pi L} \\
& - \frac{1}{L}\left\{ \frac{1}{4\pi}\ln[L\phi^{1/2}\sinh(L\phi^{1/2})] - \frac{3\ln 2}{8\pi} + \frac{1}{2}\mathcal{W}_2(4) \right\} + O(L^{-2}).
\end{aligned} \quad (7.100)
$$

Hence, by taking the derivative of (7.88) with respect to ϕ and ignoring the $O(L^{-2})$ corrections, the mean spherical constraint (7.72) takes the finite-size scaling form

$$
\begin{aligned}
\frac{1}{4\pi}\ln(y\sinh y) &- \frac{(\eta_1 + \eta_L)^2}{8y^2}\left[\frac{2}{y}\coth\frac{y}{2} + \frac{1}{\sinh^2\frac{y}{2}}\right] \\
&- \frac{(\eta_1 - \eta_L)^2}{8y^2}\left[\frac{2}{y}\tanh\frac{y}{2} - \frac{1}{\cosh^2\frac{y}{2}}\right] = \tau.
\end{aligned} \quad (7.101)
$$

Here we have introduced the scaled spherical field $y = \phi^{1/2}L$, see Eq. (7.87), and scaled variables,

$$
\tau = (K_{c,L}^{(0)} - K)L, \qquad \eta_1 = K^{-1/2}h_1 L^{3/2}, \qquad \eta_L = K^{-1/2}h_L L^{3/2}.
$$

$$
(7.102)
$$

where $K_{c,L}^{(0)}$ is the shifted critical coupling,

$$K_{c,L}^{(0)} = K_c + \frac{1}{4\pi L}\left[\ln L + \frac{3}{2}\ln 2 - 2\pi \mathcal{W}_2(4)\right].$$ (7.103)

The free energy density for Neumannn boundary conditions can be found from equations (7.62), (7.88), and (7.100):

$$\beta f_L^{(0)}(K, h_1, h_L) = \frac{1}{2}\left\{\ln\frac{K}{2\pi} - 3K + \mathcal{U}_L^{(0)}(0) + \frac{1}{2KL}(h_1^2 + h_L^2)\right\}$$

$$-\frac{1}{2L^3}\left\{\frac{1}{2\pi}\int_0^{y_L^{(0)}} x\ln(\sinh x)\,dx + \frac{1}{4\pi}(y_L^{(0)})^2(\ln y_L^{(0)} - 1/2)\right.$$

$$\left. +\frac{(\eta_1 + \eta_L)^2}{2y_L^{(0)}}\coth\frac{1}{2}y_L^{(0)} + \frac{(\eta_1 - \eta_L)^2}{2y_L^{(0)}}\tanh\frac{1}{2}y_L^{(0)} - (y_L^{(0)})^2\tau\right\}$$

$$+O(L^{-4}). \quad (7.104)$$

Here $y_L^{(0)} = y_L^{(0)}(\tau, \eta_1, \eta_L)$ is the solution of the mean spherical constraint (7.101) in the neighborhood of the critical point defined by

$$\tau = O(1), \quad \eta_1 = O(1), \quad \eta_L = O(1). \quad (7.105)$$

Thus, for the three-dimensional L-layer mean spherical model Eq. (7.104) predicts that the finite-size scaling form of the singular (in the thermodynamic limit) part of the free energy density under Neumann boundary conditions is

$$f_{L,\text{sing}}^{(0)}(K, h_1, h_L) = L^{-3}X(atL^{1/\nu}, bh_1 L^{\Delta_1^{\text{sb}}/\nu}, bh_L L^{\Delta_1^{\text{sb}}/\nu}),$$ (7.106)

where $t := (K_{c,L}^{(0)} - K)$ is the shifted coupling, $\nu = 1$ and $\Delta_1^{\text{sb}} = 3/2$ are the scaling exponents at $d = 3$, and the metric factor is $b = K^{-1/2}$. According to the general theory of finite-size scaling, this is in consistence with the expected form of the singular part of the free energy for a finite system undergoing special phase transition in the case when $\vartheta = 0$ [Krech (1994)]. The corresponding scaling function X depends, of course, on the boundary conditions.

7.3.3   *Finite-size scaling under Neumann-Dirichlet boundary conditions*

In the case of Neumann-Dirichlet boundary conditions ($\tau = c$), the projection of the boundary fields (7.77) on the k-th eigenvector (7.23) is

$$\hat{h}^{(c)}_{\text{surf}}(k) = \frac{2}{\sqrt{2L+1}} \left[h_1 \sin L\varphi^{(c)}_L(k) + h_L \sin \varphi^{(c)}_L(k) \right]. \qquad (7.107)$$

The corresponding field term (7.70) in the free energy density can be written in the form

$$P^{(c)}_L(K, h_1, h_L; \phi) = \frac{1}{KL(2L+1)} \{ h_1^2[C_L(\phi,0) + C_L(\phi,1)] +$$

$$2h_1 h_L[C_L(\phi, L-1) - C_L(\phi, L+1)] + h_L^2[C_L(\phi,0) - C_L(\phi,2)] \}, \quad (7.108)$$

where

$$C_L(\phi, q) = \sum_{k=1}^{L} \frac{\cos[q\varphi^{(c)}_L(k)]}{\phi/2 + \cos \varphi^{(c)}_L(1) - \cos \varphi^{(c)}_L(k)}, \qquad (7.109)$$

with $\varphi^{(c)}_L(k) = \pi(2k-1)/(2L+1)$, see Eq. (7.1.2).

The summation in the right-hand side of (7.109) can be performed exactly by using the techniques suggested in [Patrick (1994)]. The resulting analytic expression depends on the value of $\phi/2 + \cos \varphi^{(c)}_L(1)$ compared to 1.

Case 1. When $\phi/2 + \cos \varphi^{(c)}_L(1) > 1$ we set

$$\phi/2 + \cos \varphi^{(c)}_L(1) = \cosh x \qquad (7.110)$$

and for any integer $q \in \{0, \ldots, 2L\}$ obtain

$$C_L(\phi, q) = \left(L + \frac{1}{2} \right) \frac{\sinh(L - q + \frac{1}{2})x}{\sinh x \cosh(L + \frac{1}{2})x} - \frac{(-1)^q}{2(1 + \cosh x)}. \qquad (7.111)$$

In this case the field term (7.108) becomes

$$P^{(c)}_L(K, h_1, h_L; \phi) = \frac{1}{KL \cosh[(L+1/2)x]} \left\{ h_1^2 \frac{\sinh(Lx) \cosh(x/2)}{\sinh x} \right.$$

$$+ 2h_1 h_L \cosh(x/2) + h_L^2 \cosh[(L - 1/2)x] \}. \qquad (7.112)$$

In the limit (7.87) equation (7.110) yields that for any fixed $y > \pi/2$

$$x = y_c L^{-1} + O(L^{-2}), \qquad y_c \equiv (y^2 - \pi^2/4)^{1/2}. \qquad (7.113)$$

Therefore, the asymptotic behavior of Eq. (7.112) is

$$P_L^{(c)}(K, h_1, h_L; \phi) = \frac{h_L^2}{KL}$$

$$+ \frac{1}{KL^3}[h_1^2 L^3 Y_1(y) + 2h_1 h_L L^2 Y_2(y) + h_L^2 LY_3(\varphi)][1 + O(L^{-1})], \quad (7.114)$$

where

$$Y_1(y) = \frac{\tanh y_c}{y_c}, \quad Y_2(y) = \frac{1}{\cosh y_c}, \quad Y_3(y) = -y_c \tanh y_c. \quad (7.115)$$

Hence, by differentiation with respect to ϕ, after ignoring the $O(L^{-2})$ corrections, we find that the field term in the mean spherical constraint (7.72) has the leading-order asymptotic form

$$-\frac{\partial}{\partial \phi} P_L^{(c)}(K, h_1, h_L; \phi) \simeq -\frac{1}{2KLy}[h_1^2 L^3 Y_1'(y) + 2h_1 h_L L^2 Y_2'(y) + h_L^2 LY_3'(y)],$$

$$(7.116)$$

where $Y_i'(y) = \mathrm{d}Y_i(y)/\mathrm{d}y$, $i = 1, 2, 3$.

The interaction term in the mean spherical constraint can be evaluated by using the same method as in Section 7.3.2. First, by making use of Eqs, (7.89) and (7.110), we write the interaction term in the form

$$\begin{aligned}
\mathcal{W}_{L,3}^{(c)}(\phi) &:= \frac{1}{L} \sum_{k=1}^{L} \mathcal{W}_2 \left(\phi + 2\cos \varphi_L^{(c)}(1) - 2\cos \varphi_L^{(c)}(k) \right) \\
&= g_1^{(c)}(\phi) + g_2^{(c)}(\phi) + (5/4\pi)\ln 2, \quad (7.117)
\end{aligned}$$

where

$$g_1^{(c)}(\phi) = -\frac{1}{4\pi L} \sum_{k=1}^{L} \ln \left(2\cosh x - 2\cos \varphi_L^{(c)}(k) \right), \quad (7.118)$$

and

$$g_2^{(c)}(\phi) = \frac{1}{L} \sum_{k=1}^{L} Q_2 \left(2\cosh x - 2\cos \varphi_L^{(c)}(k) \right). \quad (7.119)$$

By using the identity [Gradshteyn and Ryzhik (1973)]

$$\prod_{k=0}^{L-1} 2 \left(\cosh x - \cos \frac{\pi(2k+1)}{2L+1} \right) = \frac{\cosh(L + \frac{1}{2})x}{\cosh \frac{1}{2}x}, \quad (7.120)$$

we obtain

$$g_1^{(c)}(\phi) = -\frac{1}{4\pi L}\ln\left[\frac{\cosh(L+\frac{1}{2})x}{\cosh\frac{1}{2}x}\right]. \tag{7.121}$$

In the limit (7.87), the above expression has the asymptotic form

$$g_1^{(c)}(\phi) = -\frac{1}{4\pi L}\ln(\cosh y_c) + O(L^{-2}). \tag{7.122}$$

Since $Q_2\left(2\cosh x - 2\cos\varphi_L^{(c)}(k)\right)$ is periodic with respect to k, with period $2L+1$, we can write $g_2^{(c)}(\phi)$ in the form

$$g_2^{(c)}(\phi) = \frac{1}{2L}\sum_{k=0}^{2L}Q_2\left(2\cosh x - 2\cos\frac{(2k+1)\pi}{2L+1}\right) - \frac{1}{2L}Q_2(2\cosh x + 2). \tag{7.123}$$

By using the Poisson summation formula (4.79) for the sum in (7.123), we obtain

$$g_2^{(c)}(\phi) = \left(1+\frac{1}{2L}\right)\frac{1}{\pi}\int_0^\pi Q_2(2\cosh x - 2\cos 2\theta)\,d\theta$$
$$-\frac{1}{2L}Q_2(2\cosh x + 2)$$
$$+\left(1+\frac{1}{2L}\right)\sum_{q=1}^\infty\frac{2}{\pi}\int_0^\pi Q_2(2\cosh x - 2\cos 2\theta)\cos[2q(2L+1)\theta]\,d\theta. \tag{7.124}$$

Integration by parts, with the aid of (7.89), yields again that the last term in equation (7.124) is of order $O(L^{-2})$. Then, in the limit (7.87) one has

$$2\cosh x - 2\cos 2\theta = 4\sin^2\theta + y_c^2 L^{-2} + O(L^{-3}), \tag{7.125}$$

hence,

$$g_2^{(c)}(\phi) = K_c - \frac{5}{4\pi}\ln 2 + \frac{1}{2L}\left[K_c - \frac{\ln 2}{2\pi} - W_2(4)\right] + O(L^{-2}). \tag{7.126}$$

Therefore, by collecting (7.117), (7.122) and (7.126), one obtains the following asymptotic form of the interaction term

$$W_{L,3}^{(c)}(\phi) = K_{c,L}^{(c)} - \frac{1}{4\pi L}\ln(\cosh y_c) + O(L^{-2}), \tag{7.127}$$

where

$$K_{c,L}^{(c)} = K_c + \frac{1}{2L}\left[K_c - \mathcal{W}_2(4) - \frac{\ln 2}{2\pi}\right] \qquad (7.128)$$

is the shifted critical coupling.

Finally, by combining (7.116) and (7.127), after ignoring the $O(L^{-2})$ corrections, the mean spherical constraint (7.72) takes the finite-size scaling form

$$\frac{1}{4\pi}\ln(\cosh y_c) + \frac{1}{2y}[\eta_1^2 Y_1'(y) + 2\eta_1\eta_L Y_2'(y) + \eta_L^2 Y_3'(y)] = \tau, \qquad (7.129)$$

with scaled variables given by

$$\tau = (K_{c,L}^{(c)} - K)L, \qquad \eta_1 = K^{-1/2}h_1 L^{3/2}, \qquad \eta_L = K^{-1/2}h_L L^{1/2}. \qquad (7.130)$$

The free energy density for Neumann-Dirichlet boundary conditions in the limit (7.87) can be found from (7.62), (7.114) and (7.127). By choosing the integration constant $\phi_0 = 2 - 2\cos\varphi_L^c(1)$, we obtain

$$\beta f_L^{(c)}(K, h_1, h_2) = \frac{1}{2}\left\{\ln\frac{K}{2\pi} - 6K + \mathcal{U}_L^{(c)}(\phi_0) - \frac{h_L^2}{KL}\right\}$$

$$- \frac{1}{2L^3}\left\{\frac{1}{2\pi}\int_{\pi/2}^{y_L^{(c)}} x\ln[\cosh(x^2 - \pi^2/4)^{1/2}]\,\mathrm{d}x - [(y_L^{(c)})^2 - \pi^2/4]\tau\right.$$

$$\left. + \eta_1^2 Y_1(y_L^{(c)}) + 2\eta_1\eta_L Y_2(y_L^{(c)}) + \eta_L^2 Y_3(y_L^{(c)})\right\} + O(L^{-4}). \quad (7.131)$$

Here $y_L^{(c)} = y_L^{(c)}(\tau, \eta_1, \eta_L)$ is the solution of the mean spherical constraint (7.129) in the neighborhood of the critical point defined by Eq. (7.105).

Case 2. When $\phi/2 + \cos\varphi_L^{(c)}(1) < 1$ we set

$$\phi/2 + \cos\varphi_L^{(c)}(1) = \cos x \qquad (7.132)$$

and for any integer $q \in \{0, \ldots, 2L\}$ obtain

$$C_L(\phi, q) = \left(L + \frac{1}{2}\right)\frac{\sin[(L - q + \frac{1}{2})x]}{\sin x \cos[(L + \frac{1}{2})x]} - \frac{(-1)^q}{2(1 + \cos x)}. \qquad (7.133)$$

In this case the field term (7.108) in the free energy density is given by equation (7.112) with the hyperbolic functions replaced by the corresponding trigonometric ones. Tn the limit (7.87) Eq. (7.132) yields at any fixed

$$0 \leq y < \pi/2$$

$$x = \tilde{y}_c L^{-1} + O(L^{-2}), \quad \tilde{y}_c \equiv (\pi^2/4 - y^2)^{1/2} = \mathrm{i}\, y_c. \tag{7.134}$$

Therefore, the asymptotic behavior of the field term (7.108) is given by equation (7.114) with the functions

$$Y_1(y) = \frac{\tan \tilde{y}_c}{\tilde{y}_c}, \quad Y_2(y) = \frac{1}{\cos \tilde{y}_c}, \quad Y_3(y) = \tilde{y}_c \tan \tilde{y}_c. \tag{7.135}$$

Obviously, these functions are analytical continuation of the functions (7.115) to the domain $0 \leq y^2 \leq \pi^2/4$. Hence, by differentiation with respect to ϕ, the field term in the mean spherical constraint (7.72) takes the leading-order asymptotic form (7.116), with $Y_i(y), i = 1, 2, 3$, defined in (7.135).

The interaction term in the mean spherical constraint can be obtained along the same lines as in case 1. Thus, by using the identity [Gradshteyn and Ryzhik (1973)]

$$\prod_{k=0}^{L-1} 2 \left(\cos x - \cos \frac{\pi(2k+1)}{2L+1} \right) = \frac{\cos(L + \frac{1}{2})x}{\cos \frac{1}{2}x}, \tag{7.136}$$

we obtain in the limit (7.87)

$$g_1^{(c)}(\phi) = -\frac{1}{4\pi L} \ln(\cos \tilde{y}_c) + O(L^{-2}). \tag{7.137}$$

For $g_2^{(c)}(\phi)$ we obtain Eqs. (7.123) - (7.125) with $\cosh x$ replaced by $\cos x$. In the limit (7.87) the result (7.126) remains true. Therefore, from equations (7.117), (7.126) and (7.137) one finds that the interaction term takes the asymptotic form

$$\mathcal{W}_{L,3}^{(c)}(\phi) = K_{c,L}^{(c)} - \frac{1}{4\pi L} \ln(\cos \tilde{y}_c) + O(L^{-2}), \tag{7.138}$$

where $K_{c,L}^{(c)}$ is given by (7.128). Hence, by combining Eqs. (7.116) and (7.138), after ignoring the $O(L^{-2})$ corrections, the mean spherical constraint (7.72) takes the finite-size scaling form

$$\frac{1}{4\pi} \ln(\cos \tilde{y}_c) + \frac{1}{2y} [\eta_1^2 Y_1'(y) + 2\eta_1 \eta_L Y_2'(y) + \eta_L^2 Y_3'(y)] = \tau, \tag{7.139}$$

with scaled variables given by Eq. (7.130).

The free energy density can be obtained in the limit (7.87) from Eqs. (7.62), (7.114), (7.135), and the integral of Eq. (7.138) over y in the limits from $y_L^{(c)}$ to $\phi_0 L^2$ with the same choice of the integration constant $\phi_0 = 2 - 2\cos\varphi_L^c(1)$. Thus we obtain

$$\beta f_L^{(c)}(K, h_1, h_L) = \frac{1}{2}\left\{\ln\frac{K}{2\pi} - 6K + \mathcal{U}_L^{(c)}(\phi_0) - \frac{h_L^2}{KL}\right\}$$

$$-\frac{1}{2L^3}\left\{-\frac{1}{2\pi}\int_{y_L^{(c)}}^{\pi/2} x\ln[\cos(\pi^2/4 - x^2)^{1/2}]\,dx + [\pi^2/4 - (y_L^{(c)})^2]\tau\right.$$

$$\left.+\eta_1^2 Y_1(y_L^{(c)}) + 2\eta_1\eta_L Y_2(y_L^{(c)}) + \eta_L^2 Y_3(y_L^{(c)})\right\} + O(L^{-4}). \quad (7.140)$$

Here $y_L^{(c)} = y_L^{(c)}(\tau, \eta_1, \eta_L)$ is the solution of the mean spherical constraint (7.139) in the neighborhood of the critical point defined by (7.105). Obviously, equation (7.140) is an analytical continuation of equation (7.131) from the domain $y_L^{(c)} > \pi/2$ to the domain $0 \le y_L^{(c)} < \pi/2$.

Thus, for the three dimensional L-layer mean spherical model equations (7.131) and (7.140) predict that the finite-size scaling form of the singular part of the free energy density under Neumann-Dirichlet boundary conditions is

$$f_{L,\text{sing}}^{(c)}(K, h_1, h_L) \simeq L^{-3} X(atL^{1/\nu}, bh_1 L^{\Delta_1^{\text{sb}}/\nu}, bh_L L^{\Delta_1^{\text{o}}/\nu}). \quad (7.141)$$

Here $t := (K_{c,L}^{(c)} - K)$ is the shifted coupling, $\nu = 1$, $\Delta_1^{\text{sb}} = 3/2$, and $\Delta_1^{\text{o}} = 1/2$ are the scaling exponents at d=3, and the metric factor is $b = K^{-1/2}$.

7.3.4 *Layer field exponents under Neumann-Dirichlet boundary conditions*

In the light of the results of the previous section, one may naturally guess that the layer critical exponent $\Delta_1(l)$ for a magnetic field acting only on the l-th layer, $1 \le l \le L$, should interpolate between the two extreme values: $\Delta_1^{\text{sb}} = 3/2$ at the Neumann boundary ($l = 1$) and $\Delta_1^{\text{o}} = 1/2$ at the Dirichlet boundary ($l = L$). In this section we prove that this is indeed the case, provided one takes into account the following important fact: we establish that the effective exponent $\Delta_1(l)$ depends not on l itself, but on the exponent which scales l with the finite size L.

Consider a field of strength h_l which acts on the l-th layer, i.e., set

$h_{\text{surf}} = h_l \delta_{l,r}$. Its projection on the k-th eigenvector (7.23) is given by

$$\hat{h}_{\text{surf}}^{(c)}(k) = \frac{2h_l}{(2L+1)^{1/2}} \cos[(l - 1/2)\varphi_L^{(c)}(k)]. \tag{7.142}$$

Hence, the field term (7.70) takes the form

$$P_L^{(c)}(K, h_l; \phi) = \frac{h_l^2}{KL(2L+1)}[C_L(\phi, 0) + C_L(\phi, 2l - 1)], \tag{7.143}$$

where $C_L(\phi, q)$ has been defined in equation (7.109).

Again we consider separately the two cases depending on whether the value of $\phi/2 + \cos\varphi_L^{(c)}(1)$ is greater than or less than 1.

Case 1. When $\phi/2 + \cos\varphi_L^{(c)}(1) > 1$, we use the substitution (7.110) and with the aid of (7.111) obtain the following exact expression for the field term,

$$P_L^{(c)}(K, h_l; \phi) = \frac{h_l^2}{KL} \frac{\sinh[(L - l + 1)x]\cosh[(l - \frac{1}{2})x]}{\sinh x \cosh[(L + \frac{1}{2})x]}. \tag{7.144}$$

Assume now that the distance l from the Neumann boundary scales with L as

$$l = \rho L^\alpha, \qquad 0 \leq \alpha \leq 1. \tag{7.145}$$

If $\alpha = 1$ and $0 \leq \rho < 1$, with the aid of equation (7.113) we obtain in the limt (7.87) that the leading-order asymptotic behavior of (7.144) is

$$P_L^{(c)}(K, h_l; \phi) \simeq \frac{h_l^2}{K} \frac{\sinh[(1 - \rho)y_c]\cosh(\rho y_c)}{y_c \cosh y_c}, \tag{7.146}$$

with $y_c = (y^2 - \pi^2/4)^{1/2}$, see Eq. (7.113). This implies that on the macroscopic scale $l = \rho L$, with $0 \leq \rho < 1$, the finite-size scaled field variable is

$$\eta_l = K^{-1/2}h_l L^{3/2}, \tag{7.147}$$

i.e. $\Delta_1(l) = 3/2$. From expression (7.144) it is evident that the case when $0 \leq \alpha < 1$ in (7.145) is equivalent to setting $\rho = 0$ in (7.146).

The situation may change qualitatively only when l is asymptotically close to the Dirichlet boundary $(l = L)$. Indeed, let us assume

$$l = L - \rho L^\alpha, \qquad 0 \leq \alpha < 1. \tag{7.148}$$

Then, in the limit (7.87) one obtains for $0 < \alpha < 1$

$$P_L^{(c)}(K, h_l\,;\phi) = \frac{h_l^2}{K}\left(\rho L^{-(1-\alpha)} + L^{-1}\right)$$

$$-\frac{h_l^2}{K}L^{-2(1-\alpha)}\rho^2 y_c \tanh y_c + h_l^2 O(L^{-2+\alpha}). \qquad (7.149)$$

Since the first term in the right-hand side of the above equation is independent of y and must be attributed to the regular part of the free energy density, we conclude that the proper finite-size scaled field variable in this case is

$$\eta_l = K^{-1/2}h_l L^{1/2+\alpha}, \qquad (7.150)$$

i.e. there exists a continuous family of layer exponents $\Delta_1(l) = 1/2 + \alpha$ depending on the parameter α in (7.148). In particular, the layer critical exponent is the same for any fixed distance from the boundary.

For the sake of completeness, we give below the explicit finite-size scaling forms of the mean spherical constraint and the free energy density in this non-trivial case.

Since the interaction term is the same as in Section 7.3.3, by combining Eq. (7.127) and the derivative of (7.149) with respect to ϕ, we obtain to leading order the finite-size scaling form of the mean spherical constraint,

$$\frac{1}{4\pi}\ln(\cosh y_c) - \frac{\rho^2}{2}\eta_l^{\,2}[Y_1(y) + Y_2^2(y)] = \tau. \qquad (7.151)$$

Similarly, for the singular part of the free energy density we obtain

$$\beta f_{L,sing}^{(c)}(K, h_l) = -\frac{1}{2L^3}\left\{\frac{1}{2\pi}\int_{\pi/2}^{y_L^{(c)}} x \ln[\cosh(x^2 - \pi^2/4)^{1/2}]\,\mathrm{d}x\right.$$

$$\left. -(y_L^{(c)})^2\tau + \rho^2\eta_l^{\,2}Y_3(y)\right\} + O(L^{-3-\alpha}). \quad (7.152)$$

Here $Y_i(y)$, $i = 1, 2, 3$, are defined in (7.115), and τ is defined in (7.130); $y_L^{(c)} = y_L^{(c)}(\tau, \eta_l)$ is the solution of the mean spherical constraint (7.151) in the neighborhood of the critical point defined by $\tau = O(1)$ and $\eta_l = O(1)$. *Case 2.* When $\phi/2 + \cos\varphi_L^{(c)}(1) < 1$, we use the substitution (7.132) and with the aid of (7.133) obtain

$$P_L^{(c)}(K, h_l; \phi) = \frac{h_l^2}{KL}\frac{\sin[(L - l + 1)x]\cos[(l - \frac{1}{2})x]}{\sin x \cos[(L + \frac{1}{2})x]}. \qquad (7.153)$$

Under the assumption (7.145), with $\alpha = 1$, and $0 \leq \rho < 1$, we obtain in the limt (7.87) that the leading-order asymptotic behavior of the field term (7.153) is

$$P_L^{(c)}(K, h_l; \phi) \simeq \frac{h_l^2}{K} \frac{\sin[(1 - \rho)\tilde{y}_c]\cos(\rho\tilde{y}_c)}{\tilde{y}_c \cos \tilde{y}_c}, \tag{7.154}$$

with $\tilde{y}_c = (\pi^2/4 - y^2)^{1/2}$, see Eq. (7.134). This implies, as in case 1, that $\Delta_1(l) = 3/2$.

Under the assumption (7.148) with $0 < \alpha < 1$, we obtain in the limit (7.87) that

$$P_L^{(c)}(K, h_l; \phi) = \frac{h_l^2}{K}(\rho L^{-(1-\alpha)} + L^{-1})$$
$$+ \frac{h_l^2}{K} L^{-2(1-\alpha)}\rho^2 \tilde{y}_c \tan \tilde{y}_c + h_l^2 O(L^{-2+\alpha}). \tag{7.155}$$

Hence one easily derives the finite-size scaling forms of the mean spherical constraint,

$$\frac{1}{4\pi}\ln(\cos\tilde{y}_c) - \frac{\rho^2}{2}\eta_l^2[Y_1(y) + Y_2^2(y)] = \tau, \tag{7.156}$$

and the singular part of the free energy density,

$$\beta f_{L,sing}^{(c)}(K, h_l) = -\frac{1}{2L^3}\left\{-\frac{1}{2\pi}\int_{y_L^{(c)}}^{\pi/2} x\ln[\cos(\pi^2/4 - x^2)^{1/2}]\,dx\right.$$
$$\left. - (y_L^{(c)})^2\tau + \rho^2\eta_l^2 Y_3(y)\right\} + O(L^{-3-\alpha}). \tag{7.157}$$

Here $y_L^{(c)} = y_L^{(c)}(\tau, \eta_l)$ is the solution of the mean spherical constraint (7.156) in the neighborhood of the critical point defined by $\tau = O(1)$ and $\eta_l = O(1)$.

Thus, for the three-dimensional L-layer mean spherical model equations (7.152) and (7.157) predict that the finite-size scaling form of the singular part of the free energy density under assumption (7.148) is

$$f_{L,sing}^{(c)}(K, h_l) \simeq L^{-3}X(atL^{1/\nu}, bh_l L^{\Delta_1(l)/\nu}). \tag{7.158}$$

Here $\nu = 1$ is the bulk correlation length exponent, and $\Delta_1(l) = 1/2 + \alpha$ is a family of layer critical exponents depending on the parameter α in (7.148).

7.4 Surface critical exponents in a modified mean spherical model

The results obtained in Section 7.3.3 imply the well known exponent $\Delta_1^o = 1/2$ for the ordinary surface phase transition at a Dirichlet boundary, and the emergence of a new critical exponent $\Delta_1^{sb} = 3/2$ at the Neumann boundary. Hence, the value $\gamma_{1,1}^{sb} = 1$ of the critical exponent for the local surface susceptibility $\chi_{1,1}$ follows. The same value is known to hold also for the spherical model with enhanced surface couplings under Dirichlet boundary conditions [Barber et. al (1974)]. However, in the latter case the $d = 3$ model unphysically exhibits surface ordering, for a sufficiently large enhancement, at some temperature above the bulk critical one. This is no more the case when one improves the model by introducing a second spherical constraint on the spins at the boundaries [Singh et. al. (1975)], since the only critical point that remains for $d \leq 3$ is the bulk one. Then, the exponent $\gamma_{1,1}^o = -1$ for $d = 3$ corresponds to an ordinary phase transition [Barber (1974)], [Barber (1983)].

In this section, following [Danchev et. al (1997b)], we aim at answering the question whether and how the surface behavior of the mean spherical model with Neumann boundary conditions will change under an additional mean spherical constraint on the spins in the surface layers. To this end we consider the so-called *modified mean spherical model*, which contains an additional spherical field used to constrain the mean square value of the spins at the boundaries to take an arbitrary positive value ρ. We evaluate explicitly the critical behavior of the local surface susceptibility

$$\beta^{-1}\chi_{1,1}^{(\tau)}(T;\rho) = \lim_{L\to\infty}\left[-L\partial^2\beta f_L^{(\tau)}(K,h_1,h_L;\rho)/\partial h_1^2\right]_{h_1=h_L=0}, \quad (7.159)$$

where h_1 and h_L are local boundary fields, and the local layer susceptibility

$$\beta^{-1}\chi_{l,l}^{(\tau)}(K;\rho) = \lim_{L\to\infty}\left[-L\partial^2\beta f_L^{(\tau)}(K,h_l;\rho)/\partial h_l^2\right]_{h_l=0}, \quad (7.160)$$

where h_l is a local field acting on the l-th layer.

7.4.1 *The model*

Consider the three-dimensional modified mean spherical model with nearest-neighbor ferromagnetic interactions, under Neumann boundary conditions

$(\tau_1 = 0)$ at the surfaces $\partial\Lambda$,

$$\partial\Lambda = \{(1, r_2, r_3) \cup (L_1, r_2, r_3), \, r_2 = 1, \ldots, L_2, \, r_3 = 1, \ldots, L_3\};$$

and periodic boundary conditions along the axes $\nu = 2, 3$. The Hamiltonian of the system in Λ is

$$\beta\mathcal{H}_\Lambda^{(0,p,p)}(S_\Lambda | K, h_\Lambda; s, v) =$$
$$-\frac{1}{2} K S_\Lambda^\dagger \cdot Q_\Lambda^{(0,p,p)} \cdot S_\Lambda + s\, S_\Lambda^\dagger \cdot S_\Lambda + v\, S_{\partial\Lambda}^\dagger \cdot S_{\partial\Lambda} - h_\Lambda^\dagger \cdot S_\Lambda. \quad (7.161)$$

Here the interaction matrix $Q_\Lambda^{(\tau)}$ is given by Eq. (7.31), s and v are two spherical fields which are to be determined from the mean spherical constraints, see Eqs. (7.164) and (7.165) below; $S_{\partial\Lambda} = \{S(\mathbf{r}), \mathbf{r} \in \partial\Lambda\}$ is the column-vector of spins located at the nonperiodic boundaries, and $S_{\partial\Lambda}^\dagger$ is the corresponding row-vector, the transposed of $S_{\partial\Lambda}$.

The free-energy density of the modified mean shperical model is given by the Legendre transformation

$$\beta f_\Lambda^{(\tau)}(K, h_\Lambda; \rho) := \sup_{s,v} \left\{ -|\Lambda|^{-1} \ln Z_\Lambda^{(\tau)}(K, h_\Lambda; s, v) - s - \rho v |\partial\Lambda| / |\Lambda| \right\},$$
$$(7.162)$$

where $|\partial\Lambda| = 2L_2 L_3$ is the total number of spins at the nonperiodic boundaries, and

$$Z_\Lambda^{(\tau)}(K, h_\Lambda; s, v) = \int_{R^{|\Lambda|}} \exp\left[-\beta\mathcal{H}_\Lambda^{(\tau)}(S_\Lambda | K, h_\Lambda; s, v) \right] \prod_{\mathbf{r} \in \Lambda} dS(\mathbf{r}) \quad (7.163)$$

is the partition function. The supremum in Eq. (7.162) is attained at the solutions of the mean spherical constraints

$$\langle S_\Lambda^\dagger \cdot S_\Lambda \rangle = |\Lambda| \qquad (7.164)$$

and

$$\langle S_{\partial\Lambda}^\dagger \cdot S_{\partial\Lambda} \rangle = \rho |\partial\Lambda|, \qquad (7.165)$$

where $\langle \cdots \rangle$ denotes the expectation value corresponding to the Hamiltonian $\beta\mathcal{H}_\Lambda^{(\tau)}(S_\Lambda | K, h_\Lambda; s, v)$.

The calculations can be easily performed, by noticing that Hamiltonian (7.161) can be identically written as

$$\beta\mathcal{H}_\Lambda^{(\omega,p,p)}(S_\Lambda|K, h_\Lambda; s) =$$
$$-\frac{1}{2}KS_\Lambda^\dagger \cdot Q_\Lambda^{(\omega,p,p)} \cdot S_\Lambda + s\, S_\Lambda^\dagger \cdot S_\Lambda - h_\Lambda^\dagger \cdot S_\Lambda. \qquad (7.166)$$

Here the interaction matrix $Q_\Lambda^{(\omega,p,p)}$ is defined by Eq. (7.31) with $\tau_1 = \omega$, where $\omega = 2v/K$, $\tau_2 = \tau_3 = p$. Therefore, the modified mean spherical model (7.161) is equivalent with the standard mean spherical model (7.166) under mixed boundary conditions with a parameter ω depending on K, h_Λ, and ρ through the spherical constraints (7.164) and (7.165). Then, by direct evaluation of the integrals in the corresponding partition function, after taking the limit $L_2, L_3 \to \infty$ at a fixed $L_1 = L$, we obtain for the free energy, c.f. Eq. (7.75),

$$\beta f_L^{(0)}(K, h_1, h_L; \rho) = \frac{1}{2}\ln\frac{K}{2\pi} - 3K$$
$$+\frac{1}{2}\sup_{\phi,\omega}\left\{\frac{1}{L}\sum_{k=1}^{L}\mathcal{V}_2\left(\phi + 2 - 2\cos\varphi_L^{(\omega)}(k)\right)\right.$$
$$\left.-\frac{1}{KL}\sum_{k=1}^{L}\frac{\left|h_L^{(\omega)}(k)\right|^2}{\phi + 2[1 - \cos\varphi_L^{(\omega)}(k)]} - K\left(\phi + \frac{2}{L}\rho\omega\right)\right\}. \qquad (7.167)$$

Here the function $\mathcal{V}_2(z)$ is defined in Eq. (7.69); note that the numbers $-2 + 2\cos\varphi_L^{(\omega)}(k)$, $k = 1,\ldots,L$ are the eigenvalues of the one-dimensional Laplacian $\Delta_L^{(\omega)}$ under mixed boundary conditions, see Eq. (7.14); $h_L^{(\omega)}(k)$ is the projection of the surface magnetic field (7.77) on the k-th eigenvector $\{u_L^{(\omega)}(r, k), r = 1,\ldots,L\}$, see Eq. (7.24),

$$h_L^{(\omega)}(k) = h_1 u_L^{(\omega)}(1, k) + h_L u_L^{(\omega)}(L, k), \qquad (7.168)$$

and we have introduced the rescaled spherical fields

$$\phi = 2s/K - 6, \qquad \omega = 2v/K. \qquad (7.169)$$

From the requirement for existence of the partition function (7.163) one obtains the condition

$$\phi + 2\min_{k\in\{1,\ldots,L\}}[1 - \cos\varphi_L^{(0)}(k;\omega)] > 0. \qquad (7.170)$$

To determine the behavior of the spherical fields ϕ and ω one has to analyze equations (7.164) and (7.165). From Eqs. (7.162), (7.167), (7.169), and (7.27) one obtains explicitly the set of equations

$$K = \frac{1}{L} \sum_{k=1}^{L} \mathcal{W}_2 \left(\phi + 2 - 2 \cos \varphi_L^{(\omega)}(k) \right) \tag{7.171}$$

and

$$K\rho = \frac{1}{L} \sum_{k=1}^{L} \mathcal{W}_2 \left(\phi + 2 - 2 \cos \varphi_L^{(\omega)}(k) \right) 2 \sin^2 \varphi_L^{(\omega)}(k)$$
$$\times \left\{ (1 - \omega)^2 - 2(1 - \omega) \cos \varphi_L^{(\omega)}(k) + 1 + L^{-1}[1 - (1 - \omega)^2] \right\}^{-1}, \tag{7.172}$$

where the function $\mathcal{W}_2(z)$ has been defined in Eq. (7.74). Equations (7.171) and (7.172) determine the point at which the finite-size free energy density (7.167), which is analytical and strictly concave function of ϕ and ω in the domain given by inequality (7.170), reaches its global maximum. Clearly, in the thermodynamic limit the free energy density is independant of the surface spherical field ω. As it is well known, for all $K \geq K_c$ its supremum sticks to the endpoint $\phi_0 = 0$ of the allowed interval $\phi_0 > 0$, where the bulk free energy density is finite. When $K < K_c$, the supremum is attained at a point $\phi_0 = \phi_0(K) > 0$, which satisfies the limit form of equation (7.171) [Barber and Fisher (1973)],

$$K = \mathcal{W}_3(\phi_0), \tag{7.173}$$

where $\mathcal{W}_3(z)$ is the 3-dimensional Watson integral (7.74), and $\mathcal{W}_3(0) = K_c$. In general, the solutions ϕ and ω of equations (7.171) and (7.172) can be written in the form $\phi = \phi_0 + \Delta\phi$, $\omega = \omega_0 + \Delta\omega$, where $\Delta\phi$ and $\Delta\omega$ tend to zero when $L \to \infty$ and ϕ_0 and ω_0 are solutions of the corresponding equations where the limit $L \to \infty$ is taken.

The surface spherical field ω_0 satisfies the corresponding limit form of the spherical constraint (7.172) at fixed $\phi_0 = 0$ for $K \geq K_c$, and $\phi_0 = \phi_0(K)$ for $K < K_c$. When $|1 - \omega| < 1$, due to the properties of the roots $\varphi_L^{(\omega)}(k)$, $k = 1, \ldots L$, the sum in (7.172) converges as $L \to \infty$ to the corresponding well defined integral, which can be taken exactly. Performing this procedure, one obtains

$$K\rho = G_3(\phi_0, \omega_0), \tag{7.174}$$

where

$$G_d(\phi,\omega) = \frac{2}{\pi^d} \int_0^\pi d\theta_1 \cdots \int_0^\pi d\theta_{d-1} \left\{ \phi + 2\omega + 2\sum_{\nu=1}^{d-1}(1 - \cos\theta_\nu) \right.$$

$$\left. + \left[\phi + 2\sum_{\nu=1}^{d-1}(1 - \cos\theta_\nu)\right]^{1/2}\left[\phi + 4 + 2\sum_{\nu=1}^{d-1}(1 - \cos\theta_\nu)\right]^{1/2} \right\}^{-1} . \tag{7.175}$$

When $|1 - \omega| > 1$, one has to treat the contribution from the two complex roots separately. The contribution from the $L-2$ real roots leads again to a well defined integral that can be taken exactly. The final result is given by the same analytical expression as in the case $|1 - \omega| < 1$, i.e., Eq. (7.174) is actually valid for all ω_0 (the restrictions on ω_0 and ϕ_0 stemming from the condition (7.170) are stated below).

Let us denote by $G_3^+(\phi,\omega)$ the branch of the function $G_3(\phi,\omega)$ defined for $\omega \geq 0$ and by $G_3^-(\phi,\omega)$ the one for $\omega < 0$. Then, by means of identical transformations it is easy to show that

$$G_3^-(\phi,\omega) = (1-\omega)^{-2}G_3^+\left(\phi,\frac{|\omega|}{1-\omega}\right) - \frac{\omega(2-\omega)}{(1-\omega)^2}\mathcal{W}_2\left(\phi - \frac{\omega^2}{1-\omega}\right),$$

$$\tag{7.176}$$

and

$$G_3^+(\phi,\omega) = (1-\omega)^{-1}\left[2\mathcal{W}_3(\phi) - \frac{1}{6} + \frac{1}{6}\phi\mathcal{W}_3(\phi)\right]$$

$$- \frac{\omega}{(1-\omega)(2-\omega)}\frac{2}{\pi^2}\int_0^\pi d\theta_2 \int_0^\pi d\theta_3 \left\{ \phi + 2\sum_{\nu=2}^{3}(1 - \cos\theta_\nu) + \frac{\omega}{2-\omega} \right.$$

$$\left. \times \left[\phi + 2\sum_{\nu=2}^{3}(1 - \cos\theta_\nu)\right]^{1/2}\left[\phi + 4 + 2\sum_{\nu=2}^{3}(1 - \cos\theta_\nu)\right]^{1/2} \right\}^{-1} . \tag{7.177}$$

Note that in the limit $L \to \infty$ condition (7.170) for the existence of the partition function yields the allowed domain of values of the spherical fields,

$$\phi_0 \geq \begin{cases} 0, & \text{if } \omega_0 \geq 0 \\ \omega_0^2/(1-\omega_0), & \text{if } \omega_0 \leq 0. \end{cases} \tag{7.178}$$

As it is evident from equation (7.175), $G_3(\phi,\omega)$ is a monotonically decreasing function of ω which tends to zero from above as $\omega \to +\infty$. Due to inequalities (7.178), at $\phi = 0$ we have to consider it on the half-line $\omega \geq 0$, where it is bounded from above by its value at $\omega = 0$, see equation (7.177),

$$G_3(0,0) = 2K_c - 1/6 := K_c \rho_c. \tag{7.179}$$

On the other hand, if $\phi > 0$, the definition domain of $G_3(\phi,\omega)$ is restricted by (7.178) to the half-line $\omega \geq \omega_1(\phi)$, where

$$\omega_1(\phi) = -(\phi + \phi^2/4)^{1/2} - \phi/2. \tag{7.180}$$

From Eq. (7.176) and the known expansion of $\mathcal{W}_2(x)$ as $x \to 0^+$,

$$\mathcal{W}_2(x) = (4\pi)^{-1} \ln x^{-1} + O(1), \tag{7.181}$$

it follows that $G_3(\phi,\omega)$ diverges logarithmically to $+\infty$ as $\omega \to \omega_1(\phi) + 0$.

Before passing to the analysis of the solutions for the spherical fields we note that, instead of considering ω as a variable that has to be determined from equations (7.171) and (7.172), one can consider it as an additional free parameter. Then, from the definitions given in Section 7.1.1 it is clear that $\omega = 0$ yields the standard spherical model with Neumann boundary conditions, whereas $\omega = 1$ yields the same model under Dirichlet boundary conditions. When $0 < \omega < 1$ we have mixed (or "intermediate" [Gelfand and Fisher (1990)]) boundary conditions which interpolate between the above two extreme cases. Therefore, in this way one should reproduce the previously known results for the properties of the local susceptibilities. In addition, as we shall see later, by choosing ω to be a given function of the temperature, one can define an effective mean spherical model with $\gamma_{1,1} = 2$, which corresponds to the critical exponent for the surface-bulk phase transition within the $O(n)$ model in the limit $n \to \infty$.

7.4.2 *Local surface susceptibility at mixed boundary conditions*

Here we study the critical behavior of the local surface susceptibility at fixed value of ω, i.e., for the standard mean spherical model with layer geometry under mixed boundary conditions. From Eqs. (7.159) and (7.167) we obtain the expression

$$k_B T\, \chi_{1,1}^{(\omega)}(K) = \frac{1}{K} \lim_{L \to \infty} \sum_{k=1}^{L} \frac{|u_L^{(\omega)}(1,k)|^2}{\phi + 2[1 - \cos\varphi_L^{(\omega)}(k)]}. \tag{7.182}$$

Hence, by using the explicit form (7.24) of the eigenfunctions under mixed boundary conditions, we obtain

$$k_B T\, \chi_{1,1}^{(\omega)}(K) = \frac{2}{K} \lim_{L\to\infty} \frac{1}{L} \sum_{k=1}^{L} \frac{\sin^2 \varphi_L^{(\omega)}(k)}{\phi + 2[1 - \cos \varphi_L^{(\omega)}(k)]}$$

$$\times [1 - 2(1 - \omega)\cos \varphi_L^{(\omega)}(k) + (1 - \omega)^2 + [1 - (1 - \omega)^2]/L]^{-1}, \quad (7.183)$$

where the limit $L \to \infty$ is to be taken at fixed ω over the finite-size solutions for ϕ of Eq. (7.171). If $|1 - \omega| < 1$, from the properties of $\varphi_L^{(\omega)}(k)$ ($k = 1, \ldots, L$) described in Section 7.1.2, it follows that as $L \to \infty$ the sum in equation (7.183) tends to the well defined integral

$$k_B T\, \chi_{1,1}^{(\omega)}(K) = \frac{2}{K\pi} \int_0^\pi \frac{\sin^2 \varphi\, d\varphi}{[\phi_0 + 2(1 - \cos \varphi)][1 - 2(1 - \omega)\cos \varphi + (1 - \omega)^2]}, \quad (7.184)$$

where ϕ_0 is the solution of the limit form (7.173) of the bulk spherical constraint. The integral can be taken exactly [Gradshteyn and Ryzhik (1973)] with the result

$$k_B T\, \chi_{1,1}^{(\omega)}(K) = \frac{2}{K} \frac{1}{[\phi_0(4 + \phi_0)]^{1/2} + \phi_0 + 2\omega}. \quad (7.185)$$

When $|1 - \omega| > 1$ one has to take into account the contribution of the two complex roots which turns out to be of the same order as the contribution of all other roots. The contribution of the latter $L - 2$ roots is again given by the integral in the right-hand side of equation (7.184). Performing the calculations one obtains the same analytical expression for $\chi_{1,1}^{(\omega)}(K)$ as the one given by equation (7.185). Note that if ω satisfies inequalities (7.178) with ω_0 replaced by ω, then $\chi_{1,1}^{(\omega)}(K) \geq 0$, as it should be expected on general physical grounds.

From (7.185) and the well known behavior of ϕ_0 close to the bulk critical temperature, $\phi_0 \simeq [4\pi(K_c - K)]^2$ [Barber and Fisher (1973)][‡], one immediately obtains all previously known results for the critical behavior of the local surface susceptibility:

(a) Neumann boundary conditions ($\omega = 0$)

$$k_B T\, \chi_{1,1}^{(0)}(K) = (2/K)\{\phi_0 + [\phi_0(4 + \phi_0)]^{1/2}\}^{-1}, \quad (7.186)$$

[‡]For comparison with the expressions in [Barber and Fisher (1973)] one has to take into account that our coupling constant K enters into the Hamiltonian with the factor 1/2.

hence, $\gamma_{1,1} = 1$.

(b) Dirichlet boundary conditions ($\omega = 1$; see equation (61) in [Barber (1974)])

$$k_B T\, \chi_{1,1}^{(1)}(K) = (2/K)\{2 + \phi_0 + [\phi_0(4 + \phi_0)]^{1/2}\}^{-1}, \qquad (7.187)$$

hence, $\gamma_{1,1} = -1$.

For $\omega \neq 0,1$ one has the case of mixed boundary conditions. As it is clear from (7.185), $\chi_{1,1}^{(\omega)}(K)$ diverges at $K = K_c$ if and only if $\omega = 0$, i.e. under Neumann boundary conditions.

Finally, by setting

$$\omega = -4\pi(K_c - K), \qquad (7.188)$$

one obtains an effective mean spherical model with the surface critical exponent $\gamma_{1,1} = 2$, which is the same value as for the $O(n)$ model in the limit $n \to \infty$. In that case $\gamma_{1,1}'$ exists too, and $\gamma_{1,1}' = 1$.

7.4.3 *Critical behavior of the local surface susceptibility*

Now we pass to the analysis of the behavior of the local surface susceptibility

$$\chi_{1,1}^{(0)}(K;\rho) \equiv \chi_{1,1}^{(\omega_0)}(K), \quad \text{with} \quad \omega_0 = \omega_0(K;\rho), \qquad (7.189)$$

given by equation (7.185) where $\omega = \omega_0(K;\rho)$ is to be determined as a function of K and ρ from equation (7.174). We confine our analysis to the surface critical regimes that emerge on approaching the bulk critical temperature from above, i.e. when $K = K_c + \Delta K$, where $\Delta K < 0$ and $|\Delta K| \to 0$. Then, as it is well known, the leading asymptotic form of the bulk spherical field follows from the asymptotic expansion

$$\mathcal{W}_3(\phi) = K_c - (4\pi)^{-1}\phi^{1/2} + O(\phi), \quad \phi \to 0^+, \qquad (7.190)$$

and reads [Barber (1974)]

$$\phi = 16\pi^2|\Delta K|^2, \quad \Delta K \to 0^-. \qquad (7.191)$$

From expression (7.185) at $\omega = \omega_0$ it is clear that the local surface susceptibility of the modified mean spherical model may exhibit divergent behavior in two different regimes: (a) when $\omega_0 \to 0^+$, and (b) when $\omega_0 \to \omega_1(\phi)$, $\omega_1(\phi) \to 0^-$. As it is seen from Eq. (7.174) and the above mentioned properties of the function $G_3(\phi,\omega)$, the first regime may occur only when

$K\rho \to K_c\rho_c - 0$, which, in view of our assumption $\Delta K \to 0^-$, requires $\rho = \rho_c$. The second divergent regime of the local surface susceptibility takes place at any fixed $\rho > \rho_c$. Below we derive the leading-order asymptotic solutions for ω_0 in each of the two cases.

Case (a): $\rho = \rho_c$. To obtain an asymptotic expansion of $G_3(\phi,\omega)$ in both arguments $\phi \to 0^+$ and $\omega \to 0^+$, we notice that the integral in the right-hand side of Eq. (7.177) diverges at $\phi = \omega = 0$ and the divergence arises from the integration over the neighborhood of the origin. Therefore, its leading-order asymptotic behavior is given by the small argument expansion of the trigonometric functions which yields

$$G_3^+(\phi,\omega) = K_c\rho_c - (2\pi)^{-1}\phi^{1/2} + (\omega/2\pi)\ln(\phi^{1/2} + \omega) + O(\phi) + O(\omega).$$
$$(7.192)$$

By setting $\rho = \rho_c$, we obtain that $\omega_0 \to 0^+$ obeys the asymptotic equation

$$-(\omega_0/2\pi)\ln(4\pi|\Delta K| + \omega_0) = |\Delta K|/(6K_c) \qquad (\Delta K \to 0^-). \qquad (7.193)$$

The solution which tends to zero from above as $|\Delta K| \to 0$ is

$$\omega_0 \simeq -\frac{\pi|\Delta K|}{3K_c\ln(4\pi|\Delta K|)}. \qquad (7.194)$$

Obviously, this critical regime leads to $\gamma_{1,1} = 1$.

Case (b): $\rho > \rho_c$. The asymptotic behavior of $G_3(\phi,\omega)$ as $\phi \to 0^+$ and $\omega \to 0^-$, so that $\omega > \omega_1(\phi)$, is readily obtained from the exact representation (7.176) and the expansions (7.181) and (7.192):

$$G_3^-(\phi,\omega) = K_c\rho_c - (2\pi)^{-1}\phi^{1/2} - (|\omega|/2\pi)\ln(\phi^{1/2} - |\omega|) + O(\phi) + O(\omega).$$
$$(7.195)$$

At fixed $\Delta\rho \equiv \rho - \rho_c > 0$, the leading-order equation for the surface spherical field becomes

$$-(|\omega_0|/2\pi)\ln(4\pi|\Delta K| - |\omega_0|) = K_c\Delta\rho. \qquad (7.196)$$

Assuming $|\omega_0| = 4\pi|\Delta K| - x$, where $x = o(|\Delta K|)$, one obtains

$$\omega_0 \simeq -4\pi|\Delta K| + \exp\left(-\frac{K_c\Delta\rho}{2|\Delta K|}\right). \qquad (7.197)$$

Therefore, in this critical regime the local surface susceptibility diverges exponentially as the bulk critical temperature is approached from above:

$$k_B T \, \chi_{1,1}^{(0)}(K;\rho) \simeq \frac{1}{K} \exp\left(\frac{K_c \Delta\rho}{2|\Delta K|}\right). \tag{7.198}$$

This behavior reminds us the one of a two dimensional system close to $T = 0$. The fact that the surface is coupled to an infinite three-dimensional system is reflected in the replacement of $T = 0$ by the bulk critical temperature $T = T_c$.

Finally, if $\rho < \rho_c$ it is easy to see that (7.174) has a finite solution $\omega_0(\rho, K)$, where $0 < \omega_0 < 1$, when $\Delta K \to 0^-$. The last actually follows from the inequalities

$$G_3(0, \phi) > \mathcal{W}_3(\phi) > G_3(1, \phi). \tag{7.199}$$

Thus, if $\rho = 1$ the local surface susceptibility $\chi_{1,1}(K_c; 1)$ is finite.

Let us now comment on the critical value ρ_c of the parameter ρ, defined in equation (7.179). By using translation invariance argument, for the mean square length of the spins at the Newmann boundary of the standard spherical model (with one global spherical field ϕ) one obtains in zero magnetic field

$$\langle S^2(1, r_2, r_3)\rangle = \frac{1}{KL_2L_3} \sum_{\mathbf{k}\in\Lambda} |u_L^{(\omega)}(1, k_1)|^2$$

$$\times \left\{\phi + 2[1 - \cos(\pi(k_1 - 1)/L_3)] + 2\sum_{\nu=2}^{3}[1 - \cos(2\pi k_\nu/L_\nu)]\right\}^{-1} \tag{7.200}$$

In the limit of an infinite film geometry this equation yields ($L_1 = L$ is kept finite)

$$\lim_{L_2,L_3\to\infty} \langle S^2(1, r_2, r_3)\rangle = \frac{1}{K}[2\mathcal{W}_3(\phi) - 1/6 + \phi\mathcal{W}_3(\phi)/6 + \mathcal{W}_2(\phi)/L]. \tag{7.201}$$

Hence, at the critical point $K = K_c$ of the infinite system ($L = \infty$), by taking into account the bulk spherical constraint at $\phi_0 = 0$, namely $\mathcal{W}_3(0) = K_c$, one obtains that the mean square length ρ_0 of the spins at the Neumann boundary of the standard shperical model equals precisely the critical value ρ_c for the surface spins in the modified spherical model.

7.4.4 Critical behavior of the local middle-layer susceptibility

The local susceptibility of the l-th layer, defined by Eq. (7.160), reads

$$k_B T \, \chi_{l,l}^{(0)}(K;\rho) = \frac{1}{K} \lim_{L \to \infty} \sum_{k=1}^{L} \frac{|u_L^{(\omega)}(l,k)|^2}{\phi + 2[1 - \cos \varphi_L^{(\omega)}(k)]}. \tag{7.202}$$

With the aid of Eq. (7.24) one explicitly obtains

$$k_B T \, \chi_{l,l}^{(0)}(K;\rho) =$$

$$\frac{2}{K} \lim_{L \to \infty} \frac{1}{L} \sum_{k=1}^{L} \frac{\left\{ \sin[l\varphi_L^{(\omega)}(k)] - (1-\omega)\sin[(l-1)\varphi_L^{(\omega)}(k)] \right\}^2}{\phi + 2[1 - \cos \varphi_L^{(\omega)}(k)]}$$

$$\times \left\{ 1 - 2(1-\omega)\cos\varphi_L^{(\omega)}(k) + (1-\omega)^2 + [1 - (1-\omega)^2]/L \right\}^{-1}. \tag{7.203}$$

We will be interested only in the behavior of this quantity around the middle of the system. Let us set $l = (L+1)/2$. Then, from (7.203) it follows that

$$k_B T \, \chi_{l,l}^{(0)}(K;\rho) = \frac{2}{K} \lim_{L \to \infty} \frac{1}{L} \sideset{}{'}\sum_{k} \left\{ \phi + 2[1 - \cos \varphi_L^{(\omega)}(k)] \right\}^{-1}$$

$$\times \left\{ 1 - \frac{1 - (1-\omega)^2}{L\left[1 - 2(1-\omega)\cos\varphi_L^{(\omega)}(k) + (1-\omega)^2 \right] + 1 - (1-\omega)^2} \right\}, \tag{7.204}$$

where the primed summation is only over the roots of equation (7.26). Having in mind the properties of the roots $\varphi_L^{(\omega)}(k)$, it is readily seen that in the limit $L \to \infty$ this equation leads to

$$k_B T \, \chi_{\infty,\infty}^{(0)}(K;\rho) = \frac{1}{K} \mathcal{W}_1(\phi_0) \tag{7.205}$$

for any ω. The above result shows that the behavior of $\chi_{\infty,\infty}^{(0)}(K;\rho)$ does not actually depend on the boundary conditions. From the temperature dependence of ϕ_0 close to the bulk critical temperature $\phi_0 \simeq [4\pi(K_c - K)]^2$ and the expansion of $\mathcal{W}_1(\phi)$ for small values of the argument [Barber and Fisher (1973)], $\mathcal{W}_1(\phi) = \phi^{-1/2}/2 + O(\phi^{1/2})$, we obtain $\gamma_{\infty,\infty} = 1$. It is clear that the same will be true for any layer at finite distance from the middle of the system. The above result has been derived in [Barber (1974)] for the standard spherical model under Dirichlet boundary conditions (see Eq. (82) in [Barber (1974)]). Here we have shown that it does not depend on the boundary conditions if they are identical at both the boundaries: just due to the symmetry the system models by itself an analogue of the Neumann boundary at the middle layers.

7.5 Conclusions

The main results of Section 7.3, given by Eq. (7.104) for Newmann boundary conditions, and by Eqs. (7.131), (7.140), for Newmann-Dirichlet boundary conditions, explicitly verify the Privman-Fisher finite-size scaling hypotesis for the singular part of the free energy density of the standard mean spherical model with surface fields. These results imply the known surface critical exponent $\Delta_1^{\text{o}} = 1/2$ for the ordinary surface phase transition at a Dirichlet boundary, and the emergence of a new surface critical exponent $\Delta_1^{\text{sb}} = 3/2$, characterizing the Neumann boundary. It has been conjectured [Danchev et. al (1997a)] that the latter critical exponent corresponds to the special (surface-bulk) phase transition within the mean-spherical model. The last is in consistence with the general expectation for the finite-size scaling form of the free energy for this type of phase transitions if one accepts also that the crossover exponent $\vartheta = 0$, as it is for the three-dimensional $O(n)$ models ([Barber (1983)]). From Eqs. (7.57) and (7.106) it immediately follows that the critical exponent $\gamma_{1,1}^{\text{sb}}$ for the local surface susceptibility $\chi_{1,1}$ at the Newmann boundary is $\gamma_{1,1}^{\text{sb}} = 1$. The same result has been obtained for the standard mean spherical model with enhanced surface couplings under Dirichlet boundary conditions [Barber et. al (1974)]. Unfortunately, in that case the model quite unphysically predicts that in the presence of a sufficiently large enhancements the surface orders at some temperature above the bulk critical temperature even for $d = 3$.

When the external magnetic field is applied at the l-th layer under Neumann-Dirichlet boundary conditions, a family of l-dependent critical exponent $\Delta_1(l)$ appears in the limit $L \to \infty$. These exponents change continuously from $\Delta_1^{\text{sb}} = 3/2$ (at the Neumann surface) to $\Delta_1^{\text{o}} = 1/2$ (at the Dirichlet surface), see Section 7.3.4. It is important to note that $\Delta_1(l)$ depends actually on the exponent which scales l with respect to L. For any layer at a finite distance apart from the nearest boundary $\Delta_1(l)$ is the same as for that boundary. This is in full conformity with the situation observed in [Barber (1974)] for the case of Dirichlet boundary conditions. Only close to the Dirichlet boundary, when $l = L - \rho L^\alpha$, with $0 < \alpha < 1$, the exponent $\Delta_1(l)$ depends continuously on α.

It should be emphasized that the spherical model under nonperiodic boundary conditions is not in the same surface universality class as the corresponding $O(n)$ model in the limit $n \to \infty$, in contrast with the bulk universality classes. For example $\Delta_1^{\circ} = 1$ and $\Delta_1^{sb} = 2$ for the $O(\infty)$ model, but $\Delta_1^{\circ} = 1/2$ and $\Delta_1^{sb} = 3/2$ for the spherical model.

In Section 7.4 we have studied the surface critical behavior of a three-dimensional modified mean spherical model with a film geometry under Neumann boundary conditions. The modification of the standard spherical model consists in the introduction of a second spherical field which fixes the mean square value of the spins in the surface layers at some parameter $\rho > 0$. It has been shown that the thus modified model is equivalent with the standard mean spherical model with mixed boundary conditions instead of the Newmann ones. A special feature of these mixed boundary conditions is that their parameter ω depends on the other parameters of the model, such as K, ρ, external fields, through the set of spherical constraints. By considereing ω as an independent free parameter, we have rederived in an uniform way the previously known critical properties of the local surface susceptibility in the standard mean spherical model. They follow directly from equation (7.185) at $\omega = 0$ (for Neumann boundary conditions), and $\omega = 1$ (for Dirichlet boundary conditions). Eq. (7.185) at $\omega \neq 0, 1$ gives the behavior of the local surface susceptibility at mixed boundary conditions. From the analysis of the latter we conclude that $\chi_{1,1}^{(\omega)}(K)$ diverges at $K = K_c$ only under Neumann boundary conditions. Finally, an effective Hamiltonian, given by equation (7.161) with $v = -2\pi K(K_c - K)$, see (7.188), describes an exactly solved mean spherical model with $\gamma_{1,1} = 2$, the value for the $n \to \infty$ limit of the corresponding $O(n)$ model.

We have derived also the critical behavior (on approaching T_c from above) of the local surface susceptibility $\chi_{1,1}^{(0)}(K;\rho)$, see Section 7.4.3, and the local layer susceptibility $\chi_{l,l}^{(0)}(K;\rho)$ for layers close to the middle of the system, see Section 7.4.4. Thcoc quantities have been evaluated exactly in the thermodynamic limit, see Eqs. (7.185) and (7.205), respectively. It has been shown that the critical behavior of $\chi_{1,1}^{(0)}(K;\rho)$ depends crucially on the parameter ρ. If $\rho < \rho_c$ then $\chi_{1,1}^{(0)}(K;\rho)$ has the behavior characteristic of the spherical model with Dirichlet boundary conditions, i.e. $\gamma_{1,1} = \gamma_{1,1}^{\circ} = -1$. At $\rho = 1$ we have found that $\chi_{1,1}^{(0)}(K;\rho)$ is finite at the bulk critical temperature T_c, in contrast with the value $\gamma_{1,1} = 1$ in the case of just one global spherical constraint. The result $\gamma_{1,1} = 1$ is recovered only if $\rho = \rho_c = 2 - (6K_c)^{-1}$, where K_c is the dimensionless critical coupling. When $\rho > \rho_c$, the local surface susceptibility $\chi_{1,1}^{(0)}$ diverges exponentially as $T \to T_c^+$, see equation (7.198). The calculation of the mean square value ρ_0 of the spins at the Neumann boundary of the standard spherical model elucidates the appearance of the critical value of $\rho = \rho_c = 1.34053\ldots$: it turns out that at the bulk critical point $\rho_c = \rho_0$, see equation (7.201) at $K = K_c$, $\phi = 0$ and $L = \infty$. As it is expected, the behavior of the local middle-layer susceptibility $\chi_{\infty,\infty}(K;\rho)$ turns out to be independent of the boundary conditions if they are the same at both boundaries. Note that for any value v of the surface spherical field, just due to the symmetry which arises from the identical boundary conditions and fields ($h_1 = h_L$) at the opposite surfaces, the system models by itself an analogue of a Neumann boundary at the middle layers. Therefore, as one would expect, the critical exponent for l around the middle of the system equals that at the Neumann boundary, $\gamma_{\infty,\infty} = 1$. The last is obviously true even if the system is with Dirichlet boundary conditions at the surfaces of the film.

The results presented above show that the properties of the modified mean spherical model are improved by introducing a second spherical constraint in the sense that they are closer, in a certain way, to the corresponding ones for the $O(n \to \infty)$ model. It seems clear that in order to obtain "correct" surface critical properties, one has to impose a separate spherical constraint on each layer paralell to the surface.

Chapter 8

Finite-Size Scaling at First Order Transitions

8.1 Introduction

The study of rounding effects at first order phase transitions in systems of finite size, initiated by [Imry (1980)], has drawn a considerable attention and soon led to the creation of a corresponding finite-size scaling theory, see, e.g., [Fisher and Berker (1982)], [Privman and Fisher (1983)], [Binder and Landau (1984)], [Fisher and Privman (1985)]. In this theory nontrivial critical exponents do *not* appear, and its results for systems with a discrete symmetry of the order parameter can be simply explained in the framework of the standard thermodynamic fluctuation theory. For example, let us consider a d-dimensional spin system of the Ising-model type with finite geometry L^d. Following [Binder and Landau (1984)], consider the probability density $P_L(m)$ of finding the system in a macroscopic state with order parameter (magnetization per spin) equal to m,

$$P_L(m) \propto \exp\left[-L^d \beta f_L(m)\right]. \tag{8.1}$$

Furhter, assume that the free energy density $f_L(m)$ of the finite system satisfies the Landau expansion for the corresponding bulk free energy density $f_\infty(m)$, see Section 1.5.3. Then, in the presence of an external magnetic field H one has

$$f_L(m) \simeq f_\infty(m) = f_0 - mH + \frac{1}{2}t\,m^2 + \frac{1}{4}u\,m^4 + O(m^6)$$

$$= \tilde{f}_0 - mH + \frac{1}{8m_0^2\chi_\infty}\left(m^2 - m_0^2\right)^2 + O(m^6). \tag{8.2}$$

265

Here t and u are temperature-dependent parameters, such that $t = 0$ at the critical temperature T_c, and $t < 0$, $u > 0$ for $T < T_c$. The function $f_\infty(m)$ at $H = 0$ has two minima with respect to the fluctuating order parameter m which are attained at the points $m = \pm m_0(t, u)$ corresponding to the two pure phases; the quantity $\chi_\infty = \chi_\infty(t, u)$ is the bulk susceptibility per spin in a pure phase. Since in the neighborhood of each of the pure phases one has a leading-order expansion of the form

$$f_L(m) \;\simeq\; \tilde{f}_0 - mH + \frac{1}{2\chi_\infty}(m - m_0)^2, \quad \text{for} \quad m \simeq m_0$$

$$f_L(m) \;\simeq\; \tilde{f}_0 - mH + \frac{1}{2\chi_\infty}(m + m_0)^2, \quad \text{for} \quad m \simeq -m_0, \qquad (8.3)$$

then the probability density (8.1) can be approximated by a sum of two Gaussians [Binder and Landau (1984)]:

$$P_L(m) \;\simeq\; \exp\left(m_0 h L^d\right) \exp\left[-\frac{(m - m_0 - H\chi_\infty)^2 L^d}{2k_{\mathrm{B}}T\chi_\infty}\right]$$

$$+ \;\; \exp\left(-m_0 h L^d\right) \exp\left[-\frac{(m + m_0 - H\chi_\infty)^2 L^d}{2k_{\mathrm{B}}T\chi_\infty}\right]. \qquad (8.4)$$

Here $h = \beta H$ is the dimensionless external field. By using the above expression one can readily calculate the moments of the fluctuationg order parameter m, hence, the susceptibility per spin for a finite-size sysytem:

$$\chi_L \simeq \chi_\infty + \frac{\beta m_0^2 L^d}{\cosh^2(m_0 h L^d)}. \qquad (8.5)$$

Thus one obtains that the smoothening of a first order phase transition with respect to the magnetic field occurs on the scale $h = O(L^{-d})$; the magnetic susceptibility per spin at the transition point grows unboundedly like L^d as $L \to \infty$. The above results will be rigorously derived for the mean spherical model with power-law interaction in Section 8.4, and for the class of spin models with equivalent-neighbor interactions in Section 8.5.

Let us note that the finite-size scaling behavior at first order phase transitions can be derived within the renormalization group approach suggested in [Nienhuis and Nauenberg (1975)]. According to the general scheme, a point of first order phase transition in a bulk system, at which there are p coexisting phases, also corresponds to a fixed point of the renormalization group transformation which is located in the strong-coupling domain. The representation of p different phases in the parameter space requires p independent sets of thermodynamic densities; the corresponding fixed point is characterized by the appearance of a p-fold degenerate trivial eigenvalue b^d, where b is the spatial rescaling factor, and d is the real-space dimensionality. This renormalization group approach has been generalized to the case of finite-size systems in [Fisher and Berker (1982)]. A review on the theory of first order phase transitions can by found in [Binder (1987b)] and [Landau (1990)].

8.2 Choice of appropriate scaling variable

The choice of scaled field variable appropriate for the description of finite-size scaling at first order phase transitions for Ising-like models has been extensively discussed in the literature, see [Privman and Fisher (1983)] and references therein. For $O(n)$ models with $n \geq 2$ in the low-temperature region $T < T_c$, it has been shown [Fisher and Privman (1985)] that a system confined to block geometry behaves in leading order as a single, fully magnetized domain which is free to orient along an external field. As a result, the only basic scaled field variable z_V is the ratio of the total magnetic energy in the volume L^d of the system to the thermal energy per degree of freedom. In the case of an Ising system confined to cylinder geometry $L^{d-1} \times \infty$, it has been argued [Privman and Fisher (1983)] that the dominant spin configurations represent a succession of correlated regions of "up" and "down" spins, extending along the axis of the cylinder over a characteristic length $\xi_{||}(T; L)$ each. The longitudinal correlation length $\xi_{||}$ is a measure of the scale of decay of the correlations in direction parallel to the axis of an infinitely long cylinder of finite cross-section L^{d-1}. This observation suggests that for such a geometry the proper scaled field variable is the ratio of the magnetic energy in the correlated volume $L^{d-1}\xi_{||}$ to the thermal energy per degree of freedom. The same scaled field variable remains for $O(n)$ type models with $n \geq 2$, provided the spin-wave contribution appears only as a correction.

Indeed, it has been shown that the bulk spin-wave singularities arising in systems with an $O(n)$, $n \geq 2$, symmetry, broken by a first order phase transition, are rounded off on a scale different from that for the first order jump in the bulk magnetization [Fisher and Privman (1985)]. In the unified finite-size scaling formulation, which embodies both effects, the spin-wave terms appear as "corrections to scaling". The description of these corrections in the case of block geometry involves the field-independent scale $(d > 2$, short-range interactions)

$$u_L = \frac{L^{-d+2}}{\beta \Upsilon Y(T)},$$
(8.6)

where $\Upsilon Y(T)$ is the helicity modulus, see its definition in Eq. (7.48).

Summarizing, the following finite-size scaling field variables have been suggested:

(a) In the case of block geometry $(\Lambda = L^d)$

$$z_V = m_0(T)h\,L^d,$$
(8.7)

where $m_0(T)$ is the spontaneous magnetization density in the bulk system.

(b) In the case of cylinder geometry $(\Lambda = L^{d-1} \times \infty)$

$$z_A = m_0(T)h\,L^{d-1}\xi_{||}(T;L).$$
(8.8)

Guided by the above heuristic arguments, one can extend the definition of the scaled field variable to the case of general $L^{d-d'} \times \infty^{d'}$ geometry and long-range interactions, by assuming that for any real $d' < d_l$, d_l being the lower critical dimension,

$$z = m_0(T)h\,L^{d-d'}\lambda_L^{d'}(T,0),$$
(8.9)

where $\lambda_L(T,0)$ is the finite-size characteristic length, see Section 4.5, at $h = 0$. Thus the variable z preserves its physical meaning as the ratio of the magnetic energy in the finite-size characteristic volume,

$$V_c^{d,d'}(T,0) = L^{d-d'}\lambda_L^{d'}(T,0),$$
(8.10)

to the thermal energy per degree of freedom.

8.3   Basic hypothesis

The basic hypothesis is that the singular part of the free-energy density of a system of geometry $L^{d-d'} \times \infty^{d'}$, where $d' < d_l$, in the low-temperature region $T < T_c$ takes asymptotically (as $L \to \infty$) the finite-size scaling form

$$\beta f_{L,\text{sing}}(T,H) \simeq \left[V_c^{d,d'}(T,0)\right]^{-1} W_{d'}(z).$$
(8.11)

Here the variable z and the characteristic volume $V_c^{d,d'}$ are defined in Eqs. (8.9) and (8.10), respectively, $W_{d'}(\cdot)$ is some universal function which depends on the geometry through the number of infinite dimensions $d' < d_l$, and may depend also on the type of the interaction potential (short-range or long-range; if long-range power-law, on the decay exponent).

For a comprehensive exposition on finite-size effects at first order phase transitions, both theory and numerical results of Monte Carlo simulations, the reader may refer to [Binder (1987a)].

To illustrate some essential features of finite-size scaling at first order phase transitions, we consider in the next section the mean spherical model with power-law interaction. The class of exactly solved spin models with equivalent-neighbor interactions admits an independent check of the above hypothesis, see Section 8.5.

8.4   Results for the mean spherical model

Here we consider the mean spherical model with a ferromagnetic interaction potential $J(r)$, see Eq. (3.19), which has the power-law asymptotic behavior $r^{-d-\sigma}$ at large distances $r \gg 1$, where σ is a positive parameter. We demonstrate that by considering the model in a geometry of finite size L, and by choosing the external homogeneous field h to vanish in a suitable way as $L \to \infty$, one can explicitly calculate the finite-size scaling functions in the neighborhood of the first order phase transition.

Let us first note that in the case of a fully finite geometry $\Lambda = L^d$, and in the absence of external magnetic field, $h = 0$, the general expression for the singular part of the finite-size free energy density (3.22) takes the simple form

$$\beta f^{m.s.}_{\Lambda,\text{sing}}(K,0) = \frac{1}{2}\sup_{\phi>0}\left\{U_\Lambda^{d,\sigma}(\phi) - K\phi\right\}, \tag{8.12}$$

where

$$U_\Lambda^{d,\sigma}(\phi) \equiv \frac{1}{|\Lambda|}\sum_{\mathbf{k}\in\mathcal{B}_\Lambda}\ln(\phi + |\mathbf{k}|^\sigma). \tag{8.13}$$

From the concavity of the function $U_\Lambda^{d,\sigma}(\phi)$ for all $\phi \geq 0$ it follows that the supremum problem in Eq. (8.12) has a unique solution $\phi = \phi_L(K,0)$. By taking into account the bounds on the first derivative,

$$0 < \frac{\mathrm{d}}{\mathrm{d}\phi} U_\Lambda^{d,\sigma}(\phi) = \Sigma_\Lambda^{d,\sigma}(\phi) \leq \frac{1}{L^d \phi} + K_c + \varepsilon_L, \qquad \phi \geq 0, \tag{8.14}$$

see Eqs. (5.44) and (5.45), where according to Eq. (5.57) $\varepsilon_L \to 0$ as $L \to \infty$, we conclude that at fixed $K > K_c$ the solution $\phi_L(K,0)$ tends to zero as $L \to \infty$.

Consider next the general case of $\Lambda = L^{d-d'} \times \infty^{d'}$ with $0 \leq d' < \sigma$. By using the asymptotic form of the mean spherical constraint (5.85) at $\phi \ll 1$ and $h = 0$, we obtain that the finite-size spherical field $\phi_L(K,0)$ asymptotically (as $L \to \infty$) satisfies the equation

$$K_c - K + \frac{D_{d',\sigma}}{L^{d-d'} \phi^{1-d'/\sigma}} = 0. \tag{8.15}$$

Hence, the explicit asymptotic form of the zero-field solution, as $L \to \infty$ at fixed $K > K_c$, is

$$\phi_L(K,0) \simeq \left(\frac{D_{d',\sigma}}{K - K_c} \right)^{\frac{\sigma}{\sigma - d'}} L^{-\frac{\sigma(d-d')}{\sigma-d'}}, \tag{8.16}$$

and the finite-size characteristic length near a first order phase transition is given by

$$\lambda_L(K,0) \simeq \left(\frac{K - K_c}{D_{d',\sigma}} \right)^{\frac{1}{\sigma-d'}} L^{\frac{d-d'}{\sigma-d'}}. \tag{8.17}$$

Now we switch on a field $h_L = x_2 K^{1/2} L^{-a}$, with some positive exponent $a > 0$, and consider the corresponding supremum problem for the free energy density,

$$\beta f_{\Lambda,\mathrm{sing}}^{m.s.}(K,h) = \frac{1}{2} \sup_{\phi>0} \left\{ U_\Lambda^{d,\sigma}(\phi) - K\phi - \frac{x_2^2}{\phi L^{2a}} \right\}. \tag{8.18}$$

Note that the field term in the right-hand side preserves the concavity of the function in the braces with respect to ϕ, hence the supremum problem has a unique solution. Since the solution tends again to zero as $L \to \infty$ at fixed x_2, we can use the asymptotic form of the corresponding stationarity condition,

$$K_c - K + \frac{D_{d',\sigma}}{L^{d-d'} \phi^{1-d'/\sigma}} + \frac{x_2^2}{\phi^2 L^{2a}} = 0. \tag{8.19}$$

An essential feature of the above equation is that ϕ appears in two scaled combinations, namely $L^a\phi$ and $L^b\phi$, where

$$b = \frac{\sigma(d - d')}{\sigma - d'}.\tag{8.20}$$

Let us analyze the possible behavior of the solution $\phi = \phi_L(K, h_L)$ of Eq. (8.19) by taking into account that the last two terms in its left-hand side, with ϕ replaced by $\phi_L(K, h_L)$, must be bounded in the limit $L \to \infty$.

(1) If $a > b$, then $\phi_L(K, h_L)$ should behave like L^{-b}, which implies that $\phi_L L^a \to \infty$ as $L \to \infty$, and one regains asymptotically the zero-field case described by equation (8.15) and its solution (8.16).

(2) If $a < b$, then the solution $\phi_L(K, h_L)$ of Eq. (8.19) should behave like L^{-a}, which implies $\phi_L L^b \to \infty$ as $L \to \infty$, and one obtains asymptotically the bulk zero-field solution for the magnetization, see Eq. (4.119),

$$\lim_{L \to \infty} \frac{h_L}{K\phi_L(K, h_L)} = \text{sign}(h)m_0(K),\tag{8.21}$$

where $m_0(K) = (1 - K_c/K)^{1/2}$ is the spontaneous magnetization. We may conclude that in this case we have a slowly vanishing field which singles out one of the pure phases, see Chapter 9, quite in the same way as the standard procedure of switching off the field after the thermodynamic limit is taken.

(3) A non-trivial situation occurs if $a = b$, when Eq. (8.19) takes the form

$$K_c - K + \frac{D_{d',\sigma}}{(L^b\phi)^{1-d'/\sigma}} + \left(\frac{x_2}{L^b\phi}\right)^2 = 0.\tag{8.22}$$

Let us introduce the finite-size scaling variable

$$z \equiv m_0(K)L^{d-d'}\lambda_L^{d'}(K, 0)h_L = (K - K_c)^{\frac{\sigma+d'}{2(\sigma-d')}}(D_{d',\sigma})^{-\frac{d'}{\sigma-d'}}x_2,\tag{8.23}$$

which has the physical meaning of the ratio of the magnetic energy in the finite-size characteristic volume (8.10) to the thermal energy per degree of freedom. In terms of that variable Eq. (8.22) can be rewritten as an equation for the unknown ratio of the finite-size magnetization to the spontaneous one,

$$Y \equiv \frac{h_L}{K\phi\, m_0(K)} = \frac{x_2}{L^b\phi\sqrt{K - K_c}},\tag{8.24}$$

in the following universal form

$$Y(1 - Y^2)^{-\sigma/(\sigma-d')} = \frac{z}{D_{d',\sigma}}. \tag{8.25}$$

This is a direct generalization of Eq. (4.27) in [Fisher and Privman (1986)] to the case of real values of $d' < \sigma$. Hence, the finite-size scaling form of the magnetization near a first order phase transition is

$$m_L(K, h) \simeq m_0(K) Y_{d',\sigma}(z), \tag{8.26}$$

where the universal function $Y_{d',\sigma}(\cdot)$ is given by the solution of Eq. (8.25). One may readily see that $Y_{d',\sigma}(z) \to \pm 1$ as $z \to \pm\infty$, and the magnetization tends to the bulk value in the corresponding pure phase.

Note that from the definitions (8.23) and (8.24) we have identically

$$\frac{z}{Y_{d',\sigma}(z)} \equiv (K - K_c)^{\frac{\sigma}{\sigma-d'}} (D_{d',\sigma})^{-\frac{d'}{\sigma-d'}} L^b \phi_L(K, h_L). \tag{8.27}$$

Hence the solution for $\phi = \phi_L(K, h_L)$ can be written in the finite-size scaling form

$$\phi_L(K, h_L) \simeq \phi_L(K, 0) \frac{z}{D_{d',\sigma} Y_{d',\sigma}(z)}, \tag{8.28}$$

where Eq. (8.16) has been taken into account as well.

Remarkably, at $d' = 0$ one obtains an explicit expression for the universal finite-size scaling function for the magnetization,

$$Y_{0,\sigma}(z) = \frac{2z}{1 + \sqrt{1 + 4z^2}}, \tag{8.29}$$

which is actually independent of the interaction decay exponent and, therefore, holds in the limit $\sigma \to 0$ as well, see next section.

Let us consider now expression (8.18) for the singular part of the finite-size free energy density. For small ϕ we can use Eq. (5.98) for $U_\Lambda^{d,\sigma}(\phi)$ and, by neglecting terms of order higher than $O(\phi)$, obtain

$$U_\Lambda^{d,\sigma}(\phi) \simeq U_{d,\sigma}(0) + K_c\phi - \frac{1}{L^d}\Psi_{d,d',\sigma}(L\phi^{1/\sigma}). \tag{8.30}$$

In view of Eqs. (8.16) and (8.28) we have

$$L\phi_L^{1/\sigma}(K, h_L) \sim L^{-\frac{d-\sigma}{\sigma-d'}} \to 0, \qquad \text{as} \quad L \to \infty, \tag{8.31}$$

so that we can make use of the asymptotic form (6.34) of $\Psi_{d,d',\sigma}(y)$ as $y \to 0$. By substitution of (8.30) and (6.34) into the right-hand side of Eq. (8.18), and taking into account that the supremum is attained at $\phi = \phi_L(K, h)$, we obtain the following finite-size scaling form of the singular part of the free energy density near a first order phase transition:

$$\beta f^{m.s.}_{\Lambda,\text{sing}}(K, h) \simeq \frac{1}{2}\left\{ -(K - K_c)\phi_L - \frac{x_2^2}{\phi_L L^{2b}} - \frac{1}{L^d}\Psi_{d,d',\sigma}(L\phi_L^{1/\sigma}) \right\}$$

$$= \frac{1}{2}\left\{ -\frac{K - K_c}{D_{d',\sigma}}\phi_L(K, 0)\left[\frac{z}{Y_{d',\sigma}(z)} + zY_{d',\sigma}(z) \right] \right.$$

$$\left. + \frac{1}{L^d}\frac{\sigma}{d'}D_{d',\sigma}\left[L^{d'}\phi_L^{d'/\sigma}(K, 0)\left(\frac{z}{D_{d',\sigma}Y_{d',\sigma}(z)} \right)^{d'/\sigma} - 1 \right] \right\} . \quad (8.32)$$

Since, as follows from Eqs. (8.16), (8.17) and (8.10),

$$\frac{K - K_c}{D_{d',\sigma}}\phi_L(K, 0) = \frac{1}{L^d}L^{d'}\phi_L^{d'/\sigma}(K, 0) = \frac{1}{V_c^{d,d'}(K, 0)}, \quad (8.33)$$

we can rewrite Eq. (8.32) for $0 < d' < \sigma$ in the universal form

$$\beta f^{m.s.}_{\Lambda,\text{sing}}(K, h) \simeq \frac{1}{V_c^{d,d'}(K, 0)}\frac{1}{2}\left\{ -\frac{z}{Y_{d',\sigma}(z)} - zY_{d',\sigma}(z) \right.$$

$$\left. + \frac{\sigma}{d'}D_{d',\sigma}\left(\frac{z}{D_{d',\sigma}Y_{d',\sigma}(z)} \right)^{d'/\sigma} \right\}, \quad (8.34)$$

where we have omitted the constant proportional to $(\sigma/d')L^{-d}$.

The case of a fully finite geometry can be obtained by taking the limit $d' \to 0^+$ in Eq. (8.32) and by using the explicit form (8.29) of the function $Y_{0,\sigma}(\cdot)$, which yields

$$\beta f^{m.s.}_{\Lambda,\text{sing}}(K, h) \simeq -\frac{1}{2}(d - \sigma)L^{-d}\ln L$$

$$+ L^{-d}\frac{1}{2}\left\{ -\sqrt{1 + 4z^2} + \ln(1 + \sqrt{1 + 4z^2}) \right\} . \quad (8.35)$$

Here the scaled magnetic field variable z, defined in Eq. (8.23), takes the simple form $z = m_0(K)L^d h_L$, in conformity with Eq. (8.7). This completes the verification of the basic hypothesis about the scaled field variable and the finite-size scaling form of the free energy density near a first order phase transition.

Note that in the case of a fully finite system ($d' = 0$), the finite-size scaling function for the free energy (8.35) is independent of the interaction decay exponent and, therefore, it coincides with the corresponding function for the mean spherical model with equivalent-neighbors interaction.

Finally, it has been found [Brankov and Danchev (1991)] that if the field-dependent finite-size characteristic volume,

$$V_c^{d,d'}(K, h_L) = L^{d-d'} \lambda_L^{d'}(K, h_L), \tag{8.36}$$

is used instead of the zero-field one given by Eq. (8.10), then the finite-size scaling expressions for the mean spherical model near a first order phase transition simplify greatly. Thus, if instead of the scaled field variable z we use the nonlinear (in the field) one

$$\tilde{z} = m_0(K) \, V_c^{d,d'}(K, h_L) \, h_L, \tag{8.37}$$

then the free energy density for $0 < d' < \sigma$ takes the following universal finite-size scaling form

$$\beta f_{\Lambda,\mathrm{sing}}^{m.s.}(K, h) \simeq \frac{1}{V_c^{d,d'}(K, h_L)} \frac{1}{2} \left\{ -\sqrt{D_{d',\sigma}^2 + 4\tilde{z}^2} \right.$$
$$\left. + \frac{\sigma}{d'} D_{d',\sigma} \left[1 - L^{-d} V_c^{d,d'}(K, h_L) \right] \right\}. \tag{8.38}$$

Note that for a fully finite system $\tilde{z} = z$ and in the limit $d' \to 0^+$ one recovers Eq. (8.35).

8.5 Results for equivalent-neighbors models

Here, following [Brankov (1990a)], we study the class of equivalent-neighbors models defined in Section 6.4.1. We remind the reader that in Section 6.4.2 we have formally introduced a "time" variable $t = K - K_c$, where $K_c = 1$ is the dimensionless critical coupling, and a "space" coordinate $x = -h$. Then we have shown that the finite-size magnetization (6.62) of these models satisfies exactly the Burgers equation (6.63) with initial condition (6.64). Our aim now is to show how the above facts can be used for the derivation of finite-size scaling near the first order phase transition. To this end we start with the integral representation (6.67), (6.68) for the finite-size magnetization per spin of the Husimi-Temperley model.

In the *low-temperature regime*, $t \in (0, \infty)$, where the first order phase transition takes place, the function

$$G(\eta; t, x) = -\ln(\cosh \eta) + \frac{(x - \eta)^2}{2(1 + t)} \qquad (8.39)$$

has one or three extrema with respect to $\eta \in \mathbf{R}$ which satisfy the stationarity condition

$$\frac{\eta - x}{1 + t} = \tanh \eta. \qquad (8.40)$$

To find the minima of $G(\eta; t, x)$, we consider its second derivative with respect to η, see Eq. (6.72), and define the point $\eta = \eta_c(t) > 0$ at which the latter vanishes:

$$\cosh^2 \eta_c(t) := 1 + t \qquad (t > 0). \qquad (8.41)$$

Let us now set

$$x_c(t) := (1 + t) \tanh \eta_c(t) - \eta_c(t) = \frac{1}{2} \sinh 2\eta_c(t) - \eta_c(t) > 0. \qquad (8.42)$$

The last inequality follows from the series expansion of $\sinh(\cdot)$ for positive arguments. One easily verifies that at $|x| \geq x_c(t)$ Eq. (8.40) has only one root, $\eta = \bar{\eta}(t, x)$, which provides the minimum of $G(\eta; t, x)$. At $|x| < x_c(t)$ there are three roots, $\eta_1(t, x) < \eta_2(t, x) < \eta_3(t, x)$, and we set

$$\bar{\eta}(t, x) = \left\{ \begin{array}{ll} \eta_3(t, x), & \text{if} \quad 0 < x < x_c(t) \\ \eta_1(t, x), & \text{if} \quad -x_c(t) < x < 0. \end{array} \right. \qquad (8.43)$$

An elementary analysis shows that for all $t \in (0, \infty)$, the function $G(\eta; t, x)$ has two local minima at $\eta = \eta_1(t, x)$ and $\eta = \eta_3(t, x)$, the global minimum being given by Eq. (8.43), and a local maximum at $\eta = \eta_2(t, x)$.

Consider first the zero-field case $x = 0$, when $\eta_1(t, 0) = -\eta_3(t, 0)$ and the two points $\eta = \eta_{1,3}(t, 0)$ correspond to global minima of $G(\eta; t, 0)$. Since the second derivative of that function with respect to η takes equal non-vanishing values at the two global minima, by evaluating the contribution from each of them we find that $m_N(t, 0) = 0$ and

$$\frac{\partial}{\partial x} m_N(t, x) \bigg|_{x=0} \simeq -\frac{N}{(1 + t)^2} \bar{\eta}^2(t, 0). \qquad (8.44)$$

Note that from Eq. (8.40) it follows that the spontaneous magnetization of the bulk system is

$$m_0(K) = -\tanh\bar\eta(t,0) = -\frac{\bar\eta(t,0)}{1+t}. \tag{8.45}$$

Hence, Eq. (8.44) yields the leading-order behavior of the zero-field magnetic susceptibility per spin of the finite system in the low-temperature regime,

$$\chi_N(K,0) = -\beta\frac{\partial}{\partial x}m_N(t,x)\bigg|_{x=0} \simeq \beta m_0^2(K)N, \tag{8.46}$$

which is in full conformity with Eq. (8.5) at $h = 0$.

The finite-size scaling function for the free energy density can be explicitly calculated by using the simple identity

$$\exp\left(\frac{1}{2}aA^2\right) = (2\pi a)^{-1/2}\int_{-\infty}^{\infty} dy\, \exp\left(-\frac{1}{2a}y^2 + yA\right), \tag{8.47}$$

valid for $\mathrm{Re}\,a > 0$. By setting $a = K/N$ and $A = \sum_{i=1}^{N} S_i$, the Hamiltonian (6.56) in the expression for the partition function of the Husimi-Temperley model can be linearized, and one readily obtains

$$Z_N(K,h) \equiv \sum_{\{S_i=\pm1\}_{i=1}^{N}} \exp\left[-\beta\mathcal{H}_N(\{S_i\}_{i=1}^{N}|J,H)\right]$$

$$= \left(\frac{N}{2\pi K}\right)^{1/2}\int_{-\infty}^{\infty} dy\, \exp\left(-\frac{N}{2K}y^2\right)\prod_{i=1}^{N}\left\{\sum_{S_i=\pm1}\exp[(y+h)S_i]\right\}$$

$$= 2^N\left(\frac{N}{2\pi K}\right)^{1/2}\int_{-\infty}^{\infty} dy\, \exp[-N\tilde{G}(y;K,h)]. \tag{8.48}$$

Here the function

$$\tilde{G}(y;K,h) = \frac{y^2}{2K} - \ln\cosh(y+h) \tag{8.49}$$

is closely related to the function $G(\eta;t,x)$ defined in Eq. (8.39), namely,

$$\tilde{G}(x-\eta;1+t,-x) \equiv G(\eta;t,x).$$

Thus, for the singular (in the thermodynamic limit) with respect to the magnetic field part of the free energy density we obtain ($K = 1+t, h - -x$)

$$\beta f_{N,\mathrm{sing}}(K, h) \simeq -N^{-1} \ln \int_{-\infty}^{\infty} d\eta \, \exp[-NG(\eta; t, x)]. \qquad (8.50)$$

By differentiation with respect to the magnetic field we recover the independently obtained expression (6.67) for the finite-size magnetization. The leading-order asymptotic form (as $N \to \infty$) of the Laplace-type integral in the right-hand side of Eq. (8.50) is given by the expansion of the function $G(\eta; t, x)$ around its minima at $\eta = \eta_{1,3}(t, x)$:

$$G(\eta_{1,3}(t, x); t, x) = -\ln \cosh[(1 + t)m_0] + \frac{1}{2}(1 + t)m_0^2 \pm m_0 x + O(x^2). \qquad (8.51)$$

Here we have used the fact that for small $x = -z/(m_0 N) = O(N^{-1})$,

$$-\eta_{1,3}(t, x) = \pm(1 + t)m_0 - \frac{x}{(1 + t)m_0^2 - t} + O(x^2). \qquad (8.52)$$

Hence, for the leading-order field-dependent part of the free energy density (8.50) we obtain

$$\beta f_{N,\mathrm{sing}}(K, h) \simeq -N^{-1} \ln \cosh(m_0 h N). \qquad (8.53)$$

On the other hand, for a fully finite system of $L^d = N$ spins the finite-size scaling variable (8.7) equals $m_0 h N$, and the characteristic volume (8.10) is simply $V_c^{d,0} = N$. Therefore, according to the hypothesis (8.11), the singular part of the free energy density near the first order phase transition has the form

$$\beta f_{N,\mathrm{sing}}(K, h) \simeq N^{-1} W_0(m_0(K) h N). \qquad (8.54)$$

By comparing Eqs. (8.53) and (8.54) we obtain

$$W_0(z) = -\ln \cosh(z). \qquad (8.55)$$

Thus we have verified that the finite-size scaling hypothesis at first order phase transitions holds for the Husimi-Temperley model. The same model has been considered in the context of nonsymmetric first order transitions and boundary effects in [Privman and Rudnick (1990)].

Analogy with a shock wave

On the basis of the Burgers equation (6.63) with the initial condition (6.66), an analogy can be drawn between the appearance in the thermodynamic limit of a critical point at $K = K_c = 1$, followed by a first order phase transition with respect to the field variable x at temperatures $K > K_c$, and the development of a shock wave at times $t \geq 0$ [Brankov and Zagrebnov (1983)]. Indeed, from the above analysis it is clear that the steepness of the front of $m_N(t, x)$ as a function of the "space coordinate" x reaches a maximum at the point $x = 0$, where it is proportional to the initial magnetic susceptibility of the system, see Eq. (8.46). The evolution of the initial condition is such that the maximum steepness of the front increases with "time" but remains finite at any finite t and N due to the presence of diffusion which smooths down the front. The situation drastically changes when the limit $N \to \infty$ in the right-hand side of Eq. (6.63) is taken. The diffusion-free evolution of the initial condition for times $t \in [-1, 0)$ is such that the steepness of the front is still finite, but, due to the non-linearity, increases unboundedly as $t \to 0$. At $t = 0$ the front obtains an infinite slope, and then a shock wave appears, which propagates with increasing jump at the front for all $t > 0$; the magnitude of the jump is bounded above by the saturation magnetization per spin. The flat front of the shock wave corresponds to the phase coexistence in the first order phase transition with respect to the magnetic field.

Thermodynamic scaling for the magnetization

Here we show that the scaling law for the bulk magnetization follows from the existence of self-similar solutions of the Cauchy problem (6.63),(6.64) in the limit $N \to \infty$. In the low-temperature regime $t \in (0, \infty)$, we look for self-similar solutions of Eq. (6.79) of the form

$$m = t^{\beta} v_-(y), \qquad y = x t^{-\Delta}, \tag{8.56}$$

with arbitrary positive exponents β and Δ. By inserting (8.56) into (6.79) we obtain an ordinary differential equation for the unknown function $v_-(\cdot)$

$$\left(t^{\beta+1-\Delta} v_- - \Delta y \right) v'_- + \beta v_- = 0. \tag{8.57}$$

The self-similarity condition is again $\Delta = \beta + 1$, hence $\gamma = 1$. The function $v_- = v_-(y)$ is defined as an implicit function of y by the equation

$$|v_-| = A|v_- - y|^{\beta/(\beta+1)}, \tag{8.58}$$

where $A > 0$ is an arbitrary integration constant.

Since $t = 0$ is a point of singularity, a Cauchy problem emerges with a natural initial (under "time" inversion) condition at $t = \infty$. The latter limit corresponds to an infinitely large interaction constant, or to zero temperature. In this limit the magnetization per spin should approach the step function

$$\lim_{t \to \infty} m_\infty(t, x) = -\text{sign}(x)\, m_0(\infty), \tag{8.59}$$

where $m_0(\infty)$ is the saturation magnetization per spin. From Eqs. (8.56) and (8.59) it follows that when $y \to 0^\pm$ the function $v_-(y)$ should have a singular behavior of the form

$$v_-(y) = -\text{sign}(y)\, m_0(\infty)|y|^{\beta/(\beta+1)}. \tag{8.60}$$

However, functions with such asymptotic behavior cannot satisfy Eq. (8.58) at small y, i.e., we find again that a globally self-similar solution of the Cauchy problem (6.79), (8.59) does not exist. Obviously, the restriction of our consideration to the neighborhood of the line $\{0 < t \le \infty, x = 0\}$ does not suffice. To satisfy the self-similarity constraints, we have to set the initial condition closer to the critical point $t = 0^+$, $x = 0$. This, in turn, necessitates the use of phenomenological argument about its shape. By taking into account the existence of spontaneous magnetization $m_\infty(t, x \to \pm 0) = \mp m_0(t)$ and initial magnetic susceptibility $\chi_\infty(t, x \to \pm 0) = \chi_0(t)$ when $t > 0$, we assume that for sufficiently small t_0 the initial condition has the form

$$m_\infty(t_0, -H/k_B T_0) = -\text{sign}(H)\, m_0(t_0) + \chi_0(t_0)\, H + O(H^3), \tag{8.61}$$

where $k_B T_0 = J/(K_c + t_0)$. Note that when $t_0 \to 0$,

$$m_0(t_0) \sim t_0^\beta \to 0, \qquad \chi_0(t_0) \sim t_0^{-\gamma} \to \infty, \tag{8.62}$$

in accordance with the definition of the critical exponents $\beta > 0$ and $\gamma > 0$, see Section 1.5.2. Next, Eq. (8.58) can be rewritten as an equation of state for the magnetization (8.56):

$$|t\,m - x| = |m/A|^{(\beta+1)/\beta}. \tag{8.63}$$

By inserting the initial condition (8.61) into Eq. (8.63) at $t = t_0$ and $x \to 0$, and by comparing terms of the same order of magnitude in the field H, we obtain

$$m_0(t_0) \simeq A^{\beta+1} t_0^{\beta}, \qquad (k_B T_0)\chi_0(t_0) \simeq \beta\, t_0^{-1}. \tag{8.64}$$

Therefore, the self-similar solution (8.56) satisfies the local initial condition (8.61) at arbitrary positive values $\beta > 0$ of the critical exponent for the spontaneous magnetization and the mean-field value $\gamma = 1$ of the initial susceptibility exponent. But to comply with the thermodynamics of the mean-field models we have to set

$$\beta = \frac{1}{2}, \qquad A = \left(\lim_{t\to 0} m_0^2(t)/t\right)^{1/3}. \tag{8.65}$$

Then expressions (8.64) for the spontaneous magnetization and the initial susceptibility take the asymptotic form predicted by the mean field theory.

8.6   Asymmetric transitions and boundary effects

So far we have considered first order phase transitions between symmetric low-temperature phases, i.e., phases which map one onto another under the change of sign of all the spins in the system. Moreover, we have tacitly assumed periodic boundary conditions, with the only exception of Section 8.5, thus ignoring the effect of surface contributions into the free energy density of the finite-size system. However, both asymmetry of the phases and bounary effects are important for fitting experimental data and results of Monte Carlo simulations. For example, asymmetric first order phase transition may appear even in Ising-type systems under nonperiodic boundary conditions which involve surface fields.

To illustrate the main features of the problem, we shall consider here the simple case of ferromagnetic systems with two-phase coexistence. Let the first order phase transition be driven by the external magnetic field h, and let each of the coexisting at $h = 0$ phases, distinguished by the value the scalar order parameter m_α, $\alpha = 1, 2$, have a bulk free energy density $f_\alpha(h)$. Of course, at the coexistence point one has $f_1(0) = f_2(0)$. For simplicity of notation we shall omit the dependence on the dimentionless coupling $K > K_c$. Following the phenomenological approach suggested in [Privman and Rudnick (1990)], we shall consider systems without soft modes, which have finite correlation length ξ_α in each of the pure low-temperature phases near and at the first order transition point $h = 0$. The system is assumed to have the fully finite geometry L^d, where the size L is large in the sense that $L \gg \xi_\alpha$, $\alpha = 1, 2$. The crucial phenomenological assumption about the partition function is that it can be written in the form

$$Z_L(h) \simeq \sum_{\alpha=1,2} \exp\left[-L^d \beta f_{L,\alpha}(h)\right], \qquad (8.66)$$

where $f_{L,\alpha}(h)$ is the finite-size "metastable" free energ density of phase α. Such an expression is expected to hold when the "single-phase" configurations are dominating the partition function. Eq. (8.66) admits an obvious extension to any finite number of coexisting phases.

Naturally, it is assumed that $\lim_{L\to\infty} f_{L,\alpha}(h) = f_\alpha(h)$. Strictly speaking, each of the bulk free energy densities $f_\alpha(h)$ is defined only in the domain ($h \geq 0$ or $h \leq 0$) in which the corresponding phase α is the stable one. Its analytical continuation to the complementary domain of metastability is known to be hampered by essential singularities. Nevertheless, Fisher and Rudnick argue that for the finite-size scaling analysis one needs only the first terms in the asymptotic expansions near $h = 0$,

$$f_\alpha(h) = f(0) - m_\alpha h + \frac{1}{2}\chi_\alpha h^2 + \cdots \qquad (\alpha = 1, 2), \qquad (8.67)$$

where $f(0)$ stands for $f_1(0) = f_2(0)$, and χ_α is the zero-field susceptibility of phase α. We note that there is a more systematic approach to the construction of suitable "metastable" free energies $f_\alpha(h)$, based on the explicit definition of truncated partition functions [Borgs and Kotecký (1992)]. This approach allows for a rigorous verification of the leading-order correctness of representaions of the type (8.66) for models to which the contour technique of Pirogov – Sinai is applicable, see, e.g. [Borgs (1992)], [Borgs and Kotecký (1993)] and references therein. Typically, such are spin models with discrete spins, q-state Potts models, lattice field theories with double well potential, etc.

Turning back to the phenomenological theory based on Eqs. (8.66) and (8.67), we need a conjecture about the finite-size behavior of the free energy densities $f_{L,\alpha}(h)$, $\alpha = 1, 2$. Away from the phase transition point, within the domain of stability of a given phase α, one can account for nonperiodic boundary conditions by assuming an expansion of the form (4.2), i.e.,

$$f_{L,\alpha}(h) = f_\alpha(h) + L^{-1}f_{s,\alpha}(h) + O(L^{-2}). \qquad (8.68)$$

Here, provided the boundary conditions do not introduce interfaces in the system, $f_{s,\alpha}(h)$ is proportional to the surface free energy density of the system. The next crucial assumption of the phenomenological finite-size scaling theory of nonsymmetric first order phase transitions is that Eq. (8.68) holds also near and at $h = 0$, hence, it can be written for each of the phases $\alpha = 1, 2$. In combination with Eq. (8.67), this amounts to the assumption that the functions $f_{L,\alpha}(h)$ in the partition function (8.66) can be approximated by

$$f_{L,\alpha}(h) \simeq f(0) - m_\alpha h + L^{-1}f_{s,\alpha}(0). \qquad (8.69)$$

Without loss of generality, one can assume that $m_2 > m_1$, so that the phase 2 is the stable one for $h > 0$. Finally, by introducing the notation

$$\mu = \frac{1}{2}(m_2 - m_1), \quad m = \frac{1}{2}(m_2 + m_1), \quad h_{L,c} = \frac{f_{s,2}(0) - f_{s,1}(0)}{2L\mu}, \qquad (8.70)$$

the free energy density of the system near the first order phase transition point takes the form [Privman and Rudnick (1990)]

$$f_L(h) := -\frac{1}{\beta L^d}Z_L(h) \simeq f(0) - mh + \frac{f_{s,2}(0) + f_{s,1}(0)}{2L}$$
$$-L^{-d}\ln\left\{2\cosh\left[\mu\left(h - h_{L,c}\right)L^d\right]\right\}. \qquad (8.71)$$

The main qualitative difference from the case of periodic boundaries (when $h_{L,c} = 0$) is the surface-induced shift in the critical value of the external field, $h_{L,c} = O(L^{-1})$. The scale $O(L^{-d})$ and shape of the rounding are the same as in the symmetric case, compare with Eq. (8.53).

There is a number of papers where the theoretical predictions are compared with simulation results, see, e.g., [Landau (1990)], [Billoire et. al. (1993)], [Vollmayr et. al. (1993)]. These works confirm the conclusion that the rounding of the first order transition vanishes as L^{-d} while the response functions (like the susceptibility) diverge as L^d.

Finally, we mention only that there is a set of results on temperature driven first order transitions, as in the q-state Potts model (where q ordered low temperature phases coexist with one disordered high temperature phase), see, e.g. [Borgs and Kotecký (1993)] and references therein.

Chapter 9

Limit Gibbs States and Finite-Size Scaling

So far we have considered mainly the analytical aspects of thermodynamic phases and phase transitions. We have seen that phase transitions manifest themselves as points (or sets of higher topological dimension) of singularity in the dependence of thermodynamic functions on some (relevant) parameters. Here we consider another fundamental aspect of phases and phase transitions, namely the existence and uniqueness of the thermodynamic limit for Gibbs probability distributions. It is demonstrated that the limit Gibbs distributions of block-spin variables, arising in some special ways of taking the thermodynamic limit, are related to the finite-size scaling functions both at criticality and a first order phase transition.

9.1   Preliminaries

The limit probability distributions for the random field $\{S_i, i \in Z^d\}$ are of principle importance for statistical mecanics, since they give a complete description of the equilibrium properties of a system in the thermodynamic limit. These distributions, called *limit Gibbs states* (LGS) can be obtained from the uniquely defined conditional Gibbs distributions for finite systems by using some limit procedures. In general form they have been introduced in [Dobrushin (1968)] and [Lanford and Ruelle (1969)].

There are two basic procedures for sampling the set of limit Gibbs states, which differ in the way the thermodynamic limit is taken. (i) One may apply different boundary conditions at fixed external parameters (temperature, magnetic fields, etc); in this case the state of the system is perturbed by the different interparticle interaction energies across the boundaries of the

283

system. (ii) One may consider size-dependent external parameters tending to some limit values *simultaneously* with the unbounded increase of the system size under fixed boundary conditions; for example, one may perturb the state of the system by an external field (one-spin Hamiltonian perurbation) which vanishes in the thermodynamic limit.

9.1.1 *Pure and mixed phases*

One of the fundamental problems of statistical mechanics is the description of all the LGS corresponding to a given Hamiltonian $\mathcal{H}$. This problem has been solved completely only in some exceptional cases [Sinai (1982)], [Georgii (1988)]. The first question that arises is, of course, the existence of at least one LGS for the given $\mathcal{H}$. The affirmative answer has been given for quite a large class of systems, including the models we study in this book. The second question concerns the uniqueness of the LGS. Under certain general regularity conditions, which are normally fulfilled for systems with short-range interactions, the set $\mathcal{G}(\mathcal{H})$ of all LGS for a given Hamiltonian $\mathcal{H}$ is nonempty, convex, compact set in the space of all probability distributions on the configuration space of the infinite system. The extremal points of $\mathcal{G}(\mathcal{H})$ correspond to *pure phases*. The *mixed phases* are nontrivial linear combinations of pure phases.

According to the general theory, see, e.g., [Ruelle (1969)], [Sinai (1982)], [Georgii (1988)], a LGS is defined by the probability measure, or the corresponding linear functional, which determines the average values $\langle \cdot \rangle$ in that state. One of the characteristic features of a pure LGS is the translation invariance of the averages of all the local observables. In the context of lattice spin systems this property implies that in a pure phase $\langle S(\mathbf{r}) \rangle$ is independent of the lattice site $\mathbf{r} \in \mathbf{Z}^d$. Next, an important necessary (and often sufficient) condition for a pure phase is that the two-point net correlation function (1.70) decays to zero at infinite separation of the spins,

$$\lim_{|\mathbf{r}_i - \mathbf{r}_j| \to \infty} G(\mathbf{r}_i, \mathbf{r}_j; T, H) = 0. \tag{9.1}$$

The general form of this condition is known as the *strong clustering condition*. To formulate it, let us define for any finite set $A \subset \mathbf{Z}^d$ the $|A|$-spin observable

$$S(A) := \prod_{\mathbf{r} \in A} S(\mathbf{r}). \tag{9.2}$$

Let $\tau_t A$ denote the set obtained under translation of A by a vector $\mathbf{t} \in \mathbf{Z}^d$. A LGS $\langle \cdot \rangle$ is said to be translation invariant if $\langle S(\tau_t A) \rangle = \langle S(A) \rangle$ for all finite A and $\mathbf{t}$. A translation-invariant LGS $\langle \cdot \rangle$ is said to have the *strong clustering property* if for all finite sets $A, B \subset \mathbf{Z}^d$

$$\lim_{|\mathbf{t}| \to \infty} \langle S(A \cup \tau_t B) \rangle = \langle S(A) \rangle \langle S(B) \rangle. \tag{9.3}$$

Consider now the case when the set of limit Gibbs states $\mathcal{G}(\mathcal{H})$ has just two extremal points (pure phases), say, $\langle \cdot \rangle_+$ and $\langle \cdot \rangle_-$. Then, in a mixed translation-invariant LGS $\langle \cdot \rangle$ there always exists a number $\alpha \in (0,1)$, such that for all finite sets $A \subset \mathbf{Z}^d$

$$\langle S(A) \rangle = \alpha \langle S(A) \rangle_+ + (1 - \alpha) \langle S(A) \rangle_-, \tag{9.4}$$

where α is independent of A. The above equality means that the pure phase $\langle \cdot \rangle_+$, respectively $\langle \cdot \rangle_-$, is found with probability α, respectively $1 - \alpha$, and that the probability of A crossing the interface between the pure phases vanishes. Assume in addition that the pure phases $\langle \cdot \rangle_\pm$ are related by spin-flip transformation which implies $\langle S(\mathbf{r}) \rangle_\pm = \pm m_0$, where m_0 is the spontaneous magnetization. Then, by setting $A = \{\mathbf{r}\}$ in Eq. (9.4), one finds that in a mixed phase

$$\langle S(\mathbf{r}) \rangle = (2\alpha - 1) m_0. \tag{9.5}$$

Hence, by taking into account the clustering property (9.3) of the pure phases, one obtains that the pair correlation function in a mixed phase does not vanish at infinite separations, see Chapter 5 in [Glimm and Jaffe (1981)],

$$\lim_{|\mathbf{r}_i - \mathbf{r}_j| \to \infty} [\langle S(\mathbf{r}_i) S(\mathbf{r}_j) \rangle - \langle S(\mathbf{r}_i) \rangle \langle S(\mathbf{r}_j) \rangle] = 4\alpha(1 - \alpha) m_0. \tag{9.6}$$

For more general and precise definitions we refer the reader to [Ruelle (1969)], [Sinai (1982)], [Georgii (1988)].

9.1.2 *Vanishing external fields: Generalized quasiaverages*

One of the fundamental ideas developed by N.N. Bogolyubov in statistical mechanics is the use of source terms for breaking the symmetry of the Hamiltonian, and thus obtaining pure Gibbs phases below the critical point [Bogolyubov (1970)]. This procedure, in which the external field h is set to zero after taking the thermodynamic limit, allows one to calculate average values, known as *quasiaverages*, of the dynamical observables in pure phases with broken symmetry.

A generalized version of Bogolyubov's quasiaverages has been suggested in [Angelescu and Zagrebnov (1985)], [Brankov et. al. (1986)]. It makes use of symmetry-breaking fields with an amplitude h_L which depends on the size L of the system and tends to zero *simultaneously with the passage to the thermodynamic limit*. For example, one may set $h_L \sim L^{-a}$ as $L \to \infty$, with some exponent $a > 0$. The resulting limit Gibbs distributions are characterized by average values called *generalized quasiaverages*. They have been used for exploring the set of zero-field limit Gibbs states of some exactly solved ferromagnetic models: the Curie-Weiss-Ising model [Brankov et. al. (1986)], the n-vector Curie-Weiss model [Angelescu and Zagrebnov (1985)], the spherical models with nearest-neighbor [Brankov and Danchev (1987)], [Patrick (1993)] and long-range power-law interactions [Brankov and Danchev (1990)].

Inhomogeneous fields, switched off after the thermodynamic limit is taken, appear to be a useful tool for investigating surface and interface phenomena, as well as phase separation. Interfaces between domains of opposite mean spin orientation can be induced by imposing inhomogeneous fields and/or appropriate boundary conditions at the opposite faces of a system with layer geometry of finite thickness L. For example, a step-like $(+-)$ bulk field, which breaks the translation invariance in one space dimension, has been applied in a study of phase separation in the spherical model [Abraham and Robert (1980)]. It has been shown that the interface is diffuse (i.e., a nontrivial magnetization profile exists only on the macroscopic scale proportional to L) at all temperatures. It has been proved [Angelescu et. al. (1981)] that the interface is diffuse also in the case of a generalized spherical model, in which the overall spherical constraint is replaced by a set of layer mean spherical constraints. However, the latter model has been treated under the simplifying assumption of a Kac-Helfand in-layer interaction.

The case of vanishing in the thermodynamic limit inhomogeneous field perturbations (at fixed temperature) has been considered in detail by Patrick [Patrick (1993)], [Patrick (1994)]. In [Patrick (1993)] the setting of [Abraham and Robert (1980)] has been generalized to amplitudes h_L of the step-like $(+-)$ field which vanish as inverse power of L as $L \to \infty$. A new field-induced critical point has been discovered in the low-temperature moderate-field regime when $h_L = h_0 L^{-2}$. At the new critical temperature $\tilde{T}_c(h_0)$ the leading-order asymptotic form of the free enrgy is analytic, and the next-to-leading correction to it has a singularity. The layer magnetization has been studied and it has been found that there is a regime below $\tilde{T}_c(h_0)$ where a nontrivial on the scale of L frozen (temperature independent) profile exists. The paper [Patrick (1994)] offers a systematic study of the influence of surface fields on some statistical properties of the spherical model. The behaviour of different thermodynamic functions of the spherical and mean spherical models under inhomogeneous external fields has been studied also in the critical finite-size scaling regime, under periodic, antiperiodic, "free" (Neumann), and "fixed" (zero Dirichlet) boundary conditions, see [Barber (1974)], [Danchev et. al (1997a)], [Danchev (1993)] and [Brankov and Tonchev (1994)]. In the three-dimensional mean spherical model with a $(+-)$ field perturbation of amplitude $h_L = O(L^{-5/2})$, an extended, coordinate-dependent finite-size scaling has been found to hold [Brankov and Tonchev (1994)] for the (proportional to $L^{-1/2}$) magnetization profile.

The analysis of the shape of the magnetization profile in the spherical models has lead to some interesting conclusions too. The fact that in a bulk $(+-)$ field the macroscopic magnetization profile near the Dirichlet boundary is exactly the same as the one near the central layer of zero magnetization has been interpreted in [Abraham and Robert (1980)] as a decoupling effect. A similar effect has been noticed [Brankov and Tonchev (1994)] in the critical finite-size scaling regime, with a vanishing $(+-)$ field perturbation. However, it has been shown [Amin (1997)] that in the case of an even number of layers L the decoupling effect takes place only asymptotically, on the scale of the large bulk correlation length close to the critical point. In general, the decoupling hypothesis breaks down since the mean square length of the spins at the Dirichlet (respectively, Neumann) boundary is different from the one at the central layers of a system in a step-like (respectively, uniform) field. Only when the temperature is fixed below the critical one, and $h_L = h_0 L^{-2}$ as $L \to \infty$, the magnetization profile is finite (nonvanishing) and decouples exactly on the macroscopic scale L. It has been established [Amin and Brankov (1998)] that the field-induced critical temperature $\tilde{T}_c(h_0)$, found in the case of periodic boundary conditions [Patrick (1993)], [Patrick (1994)], persists under Neumann-Dirichlet boundary conditions at the opposite surfaces of the layer, when the step-like $(+-)$ external field perturbation changes sign at distance $L/3$ from the Neumann boundary.

We emphasize that all the results obtained for the magnetization profile agree upon its diffuseness, which implies the general conclusion that the limit Gibbs states of the spherical model are translation invariant at arbitrary high space dimensionality.

9.2　Limit Gibbs states of the spherical models with nearest-neighbor interaction

Here we study the effect of vanishing uniform external magnetic field $H_L = H_1 L^{-a}$, where the amplitude H_1 and the exponent $a > 0$ are parameters, on the limit Gibbs states of the spherical and mean spherical models with nearest-neighbor interaction. The Hamiltonian of the system in a finite region $\Lambda \in Z^d$ consisting of $N \equiv |\Lambda| = L^d$ sites is defined by the equation

$$\beta \mathcal{H}_\Lambda(S_\Lambda | J, H_\Lambda) = -\frac{1}{2} K \sum_{\langle \mathbf{r}, \mathbf{r}' \rangle} S(\mathbf{r}) S(\mathbf{r}') - h_1 L^{-a} \sum_{\mathbf{r} \in \Lambda} S(\mathbf{r}), \qquad (9.7)$$

where the first sum in the right-hand side is taken over all pairs of sites $\langle \mathbf{r}, \mathbf{r}' \rangle$ which are nearest neighbors in Λ under periodic boundary conditions.

As in Section 3.1.3, we assume that the spherical constraint on the allowed spin configurations is

$$\sum_{\mathbf{r} \in \Lambda} S^2(\mathbf{r}) = \zeta L^d, \qquad (9.8)$$

where $\zeta > 0$ is a free parameter. The corresponding mean spherical constraint has the form

$$\sum_{\mathbf{r} \in \Lambda} \langle S^2(\mathbf{r}) \rangle_\Lambda^{G \cdot} (K, h_\Lambda, s) = \zeta L^d. \qquad (9.9)$$

Here $h_\Lambda = \{ h_1 L^{-a}, \mathbf{r} \in \Lambda \}$, and the average value $\langle \cdots \rangle_\Lambda^{G \cdot}$, taken with the Gaussian Hamiltonian

$$\beta \mathcal{H}_\Lambda^{G \cdot}(S_\Lambda | J, H_\Lambda, s) = \beta \mathcal{H}_\Lambda(S_\Lambda | J, H_\Lambda) + s \sum_{\mathbf{r} \in \Lambda} S^2(\mathbf{r}), \qquad (9.10)$$

depends on the parameters K, h_Λ and s. The unique solution of Eq. (9.9) for the parameter s we denote by $\bar{s}_\Lambda = \bar{s}_\Lambda(K, h_\Lambda, \zeta)$.

The probability distribution for the spin configurations S_Λ in the spherical model is defined by

$$d\mu_\Lambda^{s\cdot}(S_\Lambda\,|\,K, h_\Lambda, \zeta) = \frac{\exp\left[-\beta\mathcal{H}_\Lambda(S_\Lambda\,|\,J, H_\Lambda)\right]}{Z_\Lambda^{s\cdot}(K, h_\Lambda, \zeta)}$$

$$\times\,\delta\left(\zeta L^d - \sum_{\mathbf{r}\in\Lambda} S^2(\mathbf{r})\right)\prod_{\mathbf{r}\in\Lambda} dS(\mathbf{r}), \qquad (9.11)$$

where the partition function $Z_\Lambda^{s\cdot}(K, h_\Lambda, \zeta)$ is determined by the probability normalization condition. The probability distribution for the mean spherical model is the Gaussian one,

$$d\mu_\Lambda^{m.s\cdot}(S_\Lambda\,|\,K, h_\Lambda, \zeta) = \frac{\exp\left[-\beta\mathcal{H}_\Lambda^{G\cdot}(S_\Lambda\,|\,J, H_\Lambda, \bar{s}_\Lambda)\right]}{Z_\Lambda^{G\cdot}(K, h_\Lambda, \bar{s}_\Lambda)}\prod_{\mathbf{r}\in\Lambda} dS(\mathbf{r}), \qquad (9.12)$$

where $\bar{s}_\Lambda = \bar{s}_\Lambda(K, h_\Lambda, \zeta)$ is the solution of Eq. (9.9).

The LGS will be specified by the thermodynamic limit(s) of finite-size characteristic functions of the form

$$\varphi_\Lambda(t_A\,|\,K, h_\Lambda, \zeta) = \left\langle \exp\left(\mathrm{i}\sum_{\mathbf{r}\in A} t(\mathbf{r})S(\mathbf{r})\right)\right\rangle_\Lambda (K, h_\Lambda, \zeta). \qquad (9.13)$$

Here A is an arbitrary finite subset of Z^d, $t_A = \{t(\mathbf{r}),\ \mathbf{r}\in A\}$ is the finite set of arguments. The thermodynamic limit will be taken at a fixed set A over a sequence of cubic regions $\{\Lambda_L = L^d,\ L \geq L_A\}$, such that $\Lambda_L \supset A$ for all $L \geq L_A$. We remind the reader that the cases of the spherical and mean spherical models are distinguished by the superscripts $s.$ and $m.s.$, respectively.

The thermodynamic limit of the characteristic functions (9.13) defines a family of *finite-dimensional limit characteristic functions*

$$\lim_{L\to\infty} \varphi_\Lambda(t_A\,|\,K, h_\Lambda, \zeta) = \varphi(t_A\,|\,K, h_1, a; \zeta), \qquad (9.14)$$

parametrized by all finite sets $A \subset Z^d$. Provided this family is self-consistent, by a fundamental theorem due to Kolmogorov, see [Billingsley (1968)], it uniquely defines a probability distribution on the σ-algebra of measurable subsets of the configuration space of the infinite system, generated by all the finite-dimensional "cyllindric" sets.

9.2.1 *Limit Gibbs states of the mean spherical model*

By taking into account the Gaussian probability distribution (9.12) for the mean spherical model, we write the finite-size characteristic function (9.13) in the form

$$\varphi_\Lambda^{m.s.}(t_A|\,K, h_\Lambda, \zeta) = \frac{Z_\Lambda^{G.}(K, h_\Lambda + \mathrm{i}t_A, \bar{s}_\Lambda)}{Z_\Lambda^{G.}(K, h_\Lambda, \bar{s}_\Lambda)}. \tag{9.15}$$

Here we have used the fact that the set of arguments t_A formally enters into the expression for the average value (9.13) as inhomogeneous *imaginary* field acting only on the spins in region A:

$$h_\Lambda + \mathrm{i}t_A = \begin{cases} h_1 L^{-a} + \mathrm{i}t(\mathbf{r}), & \text{if } \mathbf{r} \in A, \\ h_1 L^{-a}, & \text{if } \mathbf{r} \in \Lambda \setminus A. \end{cases} \tag{9.16}$$

Since the spin-spin interaction part of the Hamiltonian (9.7) is a real symmetric quadratic form, it can be diagonalized by a real orthogonal transformation of the spin variables, with a matrix constructed from the corresponding eigenvectors. In Section 7.1 the eigenvectors for periodic boundary conditions are given in a complex form, see Eqs. (7.19) and (7.32). Here we remind the reader how a real orthogonal transformation can be introduced when the region Λ is chosen as a parallelepiped (for simplicity, we shall take a cube) with odd number of sites in each spatial dimension. Then, in the complex Fourier transformation

$$S(\mathbf{r}) = \sum_{\mathbf{k} \in \mathcal{B}_\Lambda} u_\Lambda^{(p)}(\mathbf{r}, \mathbf{k}) x_\Lambda(\mathbf{k}), \qquad u_\Lambda^{(p)}(\mathbf{r}, \mathbf{k}) = \frac{1}{L^{d/2}} e^{\mathbf{r} \cdot \mathbf{k}}, \tag{9.17}$$

where $\mathbf{k}$ is a d-dimensional vector taking values in the Brillouin zone (3.15), only the variable $x_\Lambda(\mathbf{0})$ is real. The variables $x_\Lambda(\mathbf{k})$ with $\mathbf{k} \neq \mathbf{0}$ are complex, and $x_\Lambda(-\mathbf{k})$ equals the complex conjugate of $x_\Lambda(\mathbf{k})$. To obtain a real orthogonal transformation we split the Brillouin zone in three disjoint sets, $\mathcal{B}_\Lambda = \mathcal{B}_\Lambda^+ \cup \mathcal{B}_\Lambda^- \cup \{\mathbf{0}\}$, such that $\mathbf{k} \in \mathcal{B}_\Lambda^+$ implies $-\mathbf{k} \in \mathcal{B}_\Lambda^-$. Then, for each $\mathbf{k} \in \mathcal{B}_\Lambda^+$ we introduce two independent real variables $\xi(\mathbf{k})$ and $\eta(\mathbf{k})$ by setting

$$x_\Lambda(\mathbf{k}) = \frac{1}{\sqrt{2}}[\xi_\Lambda(\mathbf{k}) + \mathrm{i}\,\eta_\Lambda(\mathbf{k})], \quad x_\Lambda(-\mathbf{k}) = \frac{1}{\sqrt{2}}[\xi_\Lambda(\mathbf{k}) - \mathrm{i}\,\eta_\Lambda(\mathbf{k})]. \tag{9.18}$$

Now the transformation (9.17) can be rewritten in the real form

$$S(\mathbf{r}) = \frac{1}{L^{d/2}}x_\Lambda(0) + \frac{\sqrt{2}}{L^{d/2}} \sum_{\mathbf{k}\in\mathcal{B}_\Lambda^+} [\cos(\mathbf{r}\cdot\mathbf{k})\xi_\Lambda(\mathbf{k}) + \sin(\mathbf{r}\cdot\mathbf{k})\eta_\Lambda(\mathbf{k})]. \quad (9.19)$$

As one can readily check, the above transformation is orthogonal, since

$$\sum_{\mathbf{r}\in\Lambda} S^2(\mathbf{r}) = x_\Lambda^2(0) + \sum_{\mathbf{k}\in\mathcal{B}_\Lambda^+} [\xi_\Lambda^2(\mathbf{k}) + \eta_\Lambda^2(\mathbf{k})]. \quad (9.20)$$

After performing the transformation (9.19), for the numerator in the right-hand side of Eq. (9.15) we obtain

$$Z_\Lambda^{G\cdot}(K, h_\Lambda + \mathrm{i}t_A, s) = \int_{R^{L^d}} \mathrm{d}x(0) \left(\prod_{\mathbf{k}\in\mathcal{B}_\Lambda^+} \mathrm{d}\xi(\mathbf{k})\mathrm{d}\eta(\mathbf{k})\right)$$

$$\times \exp\left\{-(s-dK)x^2(0) - \sum_{\mathbf{k}\in\mathcal{B}_\Lambda^+}\left[s - \frac{1}{2}\beta\hat{J}^{n.n.}(\mathbf{k})\right][\xi^2(\mathbf{k}) + \eta^2(\mathbf{k})]\right.$$

$$\left. + \left[L^{d/2-a}h_1 + \mathrm{i}L^{-d/2}\sum_{\mathbf{r}\in A}t(\mathbf{r})\right]x(0)\right.$$

$$\left. + \sqrt{2}\,\mathrm{i}L^{-d/2}\sum_{\mathbf{k}\in\mathcal{B}_\Lambda^+}\sum_{\mathbf{r}\in A}t(\mathbf{r})[\cos(\mathbf{r}\cdot\mathbf{k})\xi(\mathbf{k}) + \sin(\mathbf{r}\cdot\mathbf{k})\eta(\mathbf{k})]\right\}. \quad (9.21)$$

Here $\hat{J}^{n.n.}(\mathbf{k})$ is the Fourier transform of the nearest-neighbor spin-spin interaction, see Eq.(3.18). The integration in the above expression can be easily performed. Thus we obtain the exact Gaussian partition function

$$Z_\Lambda^{G\cdot}(K, h_\Lambda, s) = \pi^{N/2}\exp\left[\frac{L^{d-2a}h_1^2}{4(s-dK)}\right]\prod_{\mathbf{k}\in\mathcal{B}_\Lambda}\left[s - \frac{1}{2}\beta\hat{J}^{n.n.}(\mathbf{k})\right]^{-1/2},$$

$$(9.22)$$

and the characteristic function for the finite-size mean spherical model

$$\varphi_\Lambda^{m.s.}(t_A| K, h_\Lambda, \zeta) = \exp\left\{ -\frac{1}{2} \sum_{\mathbf{r},\mathbf{r}'\in A} G_\Lambda(\mathbf{r} - \mathbf{r}'; K, h_\Lambda, \zeta)t(\mathbf{r})t(\mathbf{r}') \right.$$

$$\left. + \frac{ih_1}{2L^a(\bar{s}_\Lambda - dK)} \sum_{\mathbf{r}\in A} t(\mathbf{r}) \right\}. \quad (9.23)$$

Here

$$G_\Lambda(\mathbf{r} - \mathbf{r}'; K, h_\Lambda, \zeta) = \frac{1}{KL^d} \sum_{\mathbf{k}\in\mathcal{B}_\Lambda} \frac{\cos[(\mathbf{r} - \mathbf{r}') \cdot \mathbf{k}]}{\bar{\phi}_\Lambda + 2\sum_{\nu=1}^{d}(1 - \cos k_\nu)} \quad (9.24)$$

is the pair correlation function for the spins in region A, see Eq. (3.17). The dependence on the parameter ζ enters through the solution $\bar{\phi}_\Lambda = 2\bar{s}_\Lambda/K - 2d$ of the mean spherical constraint

$$\left(\frac{h_1}{KL^a\phi} \right)^2 + \frac{1}{KL^d} \sum_{\mathbf{k}\in\mathcal{B}_\Lambda} \frac{1}{\phi + 2\sum_{\nu=1}^{d}(1 - \cos k_\nu)} = \zeta. \quad (9.25)$$

The behavior of the solution $\phi = \bar{\phi}_\Lambda(K, h_\Lambda, \zeta)$ of this equation, when $L \to \infty$ at fixed ζ and finite non-zero h_1, depends on the dimensionality d of the system, the exponent a and the dimensionless coupling K. All the qualitatively different regimes are described below. Their derivation is based on the observations that: (a) As $L \to \infty$ at fixed $\phi > 0$, the second term in the left-hand side of Eq. (9.25) converges *exponentially fast* in L, see [Barber and Fisher (1973)], to $K^{-1}\mathcal{W}_d(\phi)$, where $\mathcal{W}_d(z)$ is the generalized Watson function (7.74). (b) When $\phi \to 0$ as $L \to \infty$, the second term in the left-hand side of Eq. (9.25) has the same leading-order asymptotic properties as its long-wavelength approximation $K^{-1}\Sigma_\Lambda^{d,2}(\phi)$, see Eqs. (4.121) and (4.126),

$$\frac{K_c}{K} + \frac{1}{KL^d\phi} + \frac{1}{KL^{d-2}} \left[C_{d,2} - S_{d,0,2}(\phi L^2) \right]. \quad (9.26)$$

Hence, one obtains the following results [Brankov and Danchev (1987)]:

(1) If $d \le d_l = 2$ (for any K), or if $d > d_l$ and $K < K_c(\zeta)$, where $K_c(\zeta) = \zeta^{-1}K_c$, and $K_c = \mathcal{W}_d(0)$ is the critical coupling at $\zeta = 1$, then the solution $\bar{\phi}_\Lambda$ tends to a finite limit,

$$\lim_{L\to\infty} \bar{\phi}_\Lambda(K, h_\Lambda, \zeta) = \bar{\phi}_\infty(K, 0, \zeta), \quad (9.27)$$

which satisfies the zero-field bulk spherical constraint. In this case the thermodynamic limit of the characteristic function (9.23) is

$$\varphi^{m.s.}(t_A|\,K,h_1,a;\zeta) = \exp\left\{-\frac{1}{2}\sum_{\mathbf{r},\mathbf{r}'\in A} G(\mathbf{r}-\mathbf{r}';K,0,\zeta)t(\mathbf{r})t(\mathbf{r}')\right\},$$
(9.28)

where

$$G(\mathbf{r}-\mathbf{r}';K,0,\zeta) = \frac{1}{K(2\pi)^d}\int_{-\pi}^{\pi}\mathrm{d}x_1\cdots\int_{-\pi}^{\pi}\mathrm{d}x_d$$
$$\times\frac{\cos[(\mathbf{r}-\mathbf{r}')\cdot\mathbf{x}]}{\bar{\phi}_\infty(K,0,\zeta)+2\sum_{\nu=1}^{d}(1-\cos x_\nu)}$$
(9.29)

is the bulk zero-field spin-spin correlation function. Remarkably, it is simply related, see [Joyce (1972)], to the generating function $P_d(\mathbf{r};z)$ of all random walks on the d-dimensional (hyper)cubic lattice Z^d, which start at the origin $\mathbf{0}$ and end at site $\mathbf{r}$:

$$G(\mathbf{r};K,0,\zeta) = \frac{1}{2\bar{s}_\infty}P_d\left(\mathbf{r};\frac{dK}{\bar{s}_\infty}\right),\quad \bar{s}_\infty = dK + \frac{1}{2}K\bar{\phi}_\infty(K,0,\zeta).\quad (9.30)$$

We remind the reader that the coefficients $P_{d,n}(\mathbf{r})$ in the series

$$P_d(\mathbf{r};z) = \sum_{n=0}^{\infty} P_{d,n}(\mathbf{r})z^n,\qquad |z|<1,$$
(9.31)

equal the probability of a n-step random walk, starting at the origin $\mathbf{0}$, to reach site $\mathbf{r}$ not necessarily for the first time.

(2) If $d > d_l = 2$ and $K > K_c(\zeta)$, then $\bar{\phi}_\infty(K,0,\zeta) = 0$, and due to Eq. (9.30) the bulk pair correlation function (9.29) simplifies to

$$G(\mathbf{r}-\mathbf{r}';K,0,\zeta) = \frac{1}{2dK}P_d(\mathbf{r}-\mathbf{r}';1).$$
(9.32)

It is important to note that $P_d(\mathbf{r};1)$ at $d>2$ decays to zero as $|\mathbf{r}|\to\infty$ by following asymptotically the power-law, see [Joyce (1972)],

$$P_d(\mathbf{r};1)\sim|\mathbf{r}|^{-d+2},\qquad |\mathbf{r}|\to\infty.$$
(9.33)

In this case, the thermodynamic limit of the finite-size characteristic function (9.23) depends on the rate at which $\bar{\phi}_\Lambda(K,h_\Lambda,\zeta)$ vanishes as $L\to\infty$, hence, on the value of the exponent a:

(2a) If $0 < a < d$, then the leading-order asymptotic form of $\bar{\phi}_\Lambda$ is

$$\bar{\phi}_\Lambda(K, h_\Lambda, \zeta) \simeq \frac{h_1 L^{-a}}{K(\zeta - K_c/K)^{1/2}}, \tag{9.34}$$

and the thermodynamic limit of the characteristic function (9.23) is

$$\varphi^{m.s.}(t_A \,|\, K, h_1, a; \zeta) = \exp\left\{ -\frac{1}{2} \sum_{\mathbf{r,r'} \in A} \frac{1}{2dK} P_d(\mathbf{r} - \mathbf{r}'; 1) t(\mathbf{r}) t(\mathbf{r}') \right.$$

$$\left. + i\,\mathrm{sign}(h_1) m_0(K, \zeta) \sum_{\mathbf{r} \in A} t(\mathbf{r}) \right\}, \tag{9.35}$$

where $m_0(K, \zeta) = (\zeta - K_c/K)^{1/2}$ is the spontaneous magnetization of the mean spherical model with constraint parameter ζ. The above form of the characteristic function implies that the average magnetization of the spins in A equals $\pm m_0(K, \zeta)$, where the sign coincides with that of the amplitude h_1 of the vanishing external field. Obviously, due to Eq. (9.33) the pair correlation function (9.32) obeys the characteristic of a pure phase clustering condition (9.1).

(2b) If $a = d$, then the leading-order asymptotic form of the solution is

$$\bar{\phi}_\Lambda(K, h_\Lambda, \zeta) \simeq \left[K L^d y(K, h_1, \zeta) \right]^{-1}, \tag{9.36}$$

where

$$y(K, h_1, \zeta) = \frac{2 m_0^2(K, \zeta)}{[1 + 4 h_1^2 m_0^2(K, \zeta)]^{1/2} + 1}. \tag{9.37}$$

Now the thermodynamic limit of the characteristic function (9.23) is

$$\varphi^{m.s.}(t_A \,|\, K, h_1, d; \zeta) =$$

$$\exp\left\{ -\frac{1}{2} \sum_{\mathbf{r,r'} \in A} \left[\frac{1}{2dK} P_d(\mathbf{r} - \mathbf{r}'; 1) + y(K, h_1, \zeta) \right] t(\mathbf{r}) t(\mathbf{r}') \right.$$

$$\left. + i\, h_1 y(K, h_1, \zeta) \sum_{\mathbf{r} \in A} t(\mathbf{r}) \right\}. \tag{9.38}$$

The above expression implies that the pair correlation function changes to, compare with Eq. (9.32),

$$G(\mathbf{r} - \mathbf{r}'; K, h_1, \zeta) = \frac{1}{2dK} P_d(\mathbf{r} - \mathbf{r}'; 1) + y(K, h_1, \zeta), \qquad (9.39)$$

and the average magnetization of the spins in A is

$$m(K, h_1, \zeta) = h_1 y(K, h_1, \zeta). \qquad (9.40)$$

Indeed, from the general expression (4.119) for the finite-size magnetization of the mean spherical model it follows that

$$\lim_{L \to \infty} m_\Lambda(K, h_1 L^{-d}, \zeta) = \lim_{L \to \infty} \frac{h_1 L^{-d}}{K \bar{\phi}_\Lambda(K, h_\Lambda, \zeta)} = h_1 y(K, h_1, \zeta). \qquad (9.41)$$

The term $y(K, h_1, \zeta)$ in the right-hand side of expression (9.39) for the bulk correlation function originates from the mode with $\mathbf{k} = \mathbf{0}$ in the finite-size correlation function (9.24). Note that this term violates the clustering condition (9.1), and due to Eq. (9.6) may indicate that the phase is mixed.

(2c) If $a > d$, then the leading-order asymptotic form of the solution is the same as in the zero-field case,

$$\bar{\phi}_\Lambda(K, h_\Lambda, \zeta) \simeq \left[KL^d(\zeta - K_c/K) \right]^{-1}, \qquad (9.42)$$

and the result for thermodynamic limit of the characteristic function (9.23) is correspondingly

$$\varphi_d^{m.s.}(t_A | K, h_1, a; \zeta) =$$

$$\exp \left\{ -\frac{1}{2} \sum_{\mathbf{r}, \mathbf{r}' \in A} \left[\frac{1}{2dK} P_d(\mathbf{r} - \mathbf{r}'; 1) + m_0^2(K, \zeta) \right] t(\mathbf{r}) t(\mathbf{r}') \right\}. \qquad (9.43)$$

Note that here the term which modifies the zero-field bulk correlation function follows from Eq. (9.37) for $y(K, h_1, \zeta)$ in the limit $h_1 \to 0$. The form of the bulk correlation function is characteristic of a mixed phase, composed of two equally represented ($\alpha = 1/2$) pure phases with order parameter $\pm m_0(K, \zeta)$, see Eq. (9.6).

Summarizing our findings, we expect that the LGS obtained when $a < d$ is a pure phase, and that the cases $a = d$ and $a > d$ correspond to mixed phases. We will return to this question after the study of the set of zero-field LGS of the spherical model.

9.2.2   *Transformation kernel*

As already mentioned in Section 3.1.1, the spherical model due to [Berlin and Kac (1952)], and the mean spherical model due to [Lewis and Wannier (1952)], see also [Lewis and Wannier (1953)], can be considered as one model in two different statistical ensembles: microcanonical and grand canonical, respectively, with regard to the obserable $\sum_{\mathbf{r}} S^2(\mathbf{r})$. Therefore, at least for finite systems, the probability of some event in the mean spherical model with fixed constraint parameter, say $\zeta = 1$, can be expressed as a product of the conditional probability of that event in the spherical model with constraint parameter ζ and the probability of finding the given value ζ of the random variable $\sum_{\mathbf{r}} S^2(\mathbf{r})$ in the ensemble of the mean spherical model (with $\zeta = 1$). Then, the thermodynamic limit can be taken in the sense of weak convergence of probability distributions for finite sets of random variables to some limit distribution. From mathematical point of view, one of the problems consists in proving that the thermodynamic limit can be interchanged with the integration over ζ. In [Kac and Thompson (1977)] it has been shown that for a certain class of functions $g(S_\Lambda)$ on the configuration space of the finite system, their average values in the zero-field limit Gibbs states of the mean spherical and the spherical models are related by the equation

$$\text{t-}\lim \langle g(S_\Lambda) \rangle^{m.s.}(K,0,1) = \int_0^\infty \mathrm{d}\zeta\, \mathcal{K}(\zeta|K,0)\, \text{t-}\lim \langle g(S_\Lambda) \rangle^{s.}(K,0,\zeta).$$

$$(9.44)$$

Here $\text{t-}\lim \langle \cdot \rangle (K,0,\zeta)$ denotes the thermodynamic limit of the corresponding average value in the finite-system ensemble with coupling constant K and external field $h = 0$; $\mathcal{K}(\zeta|K,0)$ is the Kac-Thompson transformation kernal (in the zero-field case), defined as

$$\mathcal{K}(\zeta|K,0)\,\mathrm{d}\zeta = \text{t-}\lim \mathrm{Prob}\left\{ \zeta \le |\Lambda|^{-1} \sum_{\mathbf{r}\in\Lambda} S^2(\mathbf{r}) < \zeta + \mathrm{d}\zeta \right\}_\Lambda^{m.s.} (K,0,1).$$

$$(9.45)$$

The probability in the right-hand side of the above equation is calculated for a sequence of mean spherical models with fixed K, $h = 0$ and $\zeta = 1$, defined in finite regions Λ which expand to infinity in all space dimensions. Therefore, the Kac-Thompson transformation kernel has the meaning of probability density of the limit probability distribution for the normalized random variable $|\Lambda|^{-1} \sum_{\mathbf{r}} S^2(\mathbf{r})$ in the mean spherical model with constraint parameter $\zeta = 1$.

In the case $d > 2$, when the critical coupling K_c is finite, it has been found [Kac and Thompson (1977)] that the kernal (9.45) is δ-function like only in the high-temperature regime $K < K_c$:

$$\mathcal{K}(\zeta|K, 0) = \delta(\zeta - 1). \tag{9.46}$$

In the low-temperature regime, when $K > K_c$, the kernel has the non-trivial explicit form

$$\mathcal{K}(\zeta|K, 0) = \frac{1}{[2\pi(\zeta - K_c/K)(1 - K_c/K)]^{1/2}} \exp\left(-\frac{\zeta - K_c/K}{2(1 - K_c/K)}\right) \tag{9.47}$$

for $\zeta > K_c/K$, and $\mathcal{K}(\zeta|K, 0) \equiv 0$ for $\zeta < K_c/K$.

Since the weak convergence of probability distributions is equivalent to convergence of characteristic functions, we use the analogue of Eq. (9.44) for the finite-dimensional characteristic functions (9.14) under vanishing magnetic field perturbation:

$$\varphi^{m.s.}(t_A|K, h_1, a; 1) = \int_0^\infty \mathrm{d}\zeta\, \mathcal{K}(\zeta|K, h_1, a)\varphi^{s.}(t_A|K, h_1, a; \zeta). \tag{9.48}$$

In this section we calculate the transformation kernel $\mathcal{K}(\zeta|K, h_1, a)$ in the case when the Hamiltonian of the finite system has the form (9.7). Then, in the next section, we solve the resulting integral equation (9.48) for the characteristic function $\varphi^{s.}(t_A|K, h_1, a; \zeta)$ of the spherical model.

The kernel $\mathcal{K}(\zeta|K, h_1, a)$ is most readily obtained by calculating first the characteristic function $\psi_\Lambda(\lambda|K, h_\Lambda, 1)$ of the normalized random variable $L^{-d} \sum_{\mathbf{r}} S^2(\mathbf{r})$ in the finite-size mean spherical model with Hamiltonian (9.7) and constraint parameter $\zeta = 1$:

$$\begin{aligned}
\psi_\Lambda^{m.s.}(\lambda|K, h_\Lambda, 1) &= \left\langle \exp\left(\mathrm{i}\lambda L^{-d} \sum_{\mathbf{r}\in\Lambda} S^2(\mathbf{r})\right) \right\rangle_\Lambda^{m.s.} (K, h_\Lambda, 1) \\
&= \frac{Z_\Lambda^{G.}(K, h_\Lambda, \bar{s}_\Lambda - \mathrm{i}\lambda L^{-d})}{Z_\Lambda^{G.}(K, h_\Lambda, \bar{s}_\Lambda)}.
\end{aligned} \tag{9.49}$$

Here $\bar{s}_\Lambda = \frac{1}{2}K(\bar{\phi}_\Lambda + 2d)$, where $\bar{\phi}_\Lambda = \bar{\phi}_\Lambda(K, h_1, a; 1)$ is the solution of the mean spherical constraint (9.25) at $\zeta = 1$. From the explicit expression

(9.22) for the Gaussian partition function one readily obtains

$$
\psi_\Lambda^{m.s.}(\lambda|\,K, h_\Lambda, 1) = \exp\left[\frac{i\lambda h_1^2}{K\,L^{2a}\bar\phi_\Lambda^2}\left(1 - \frac{2i\lambda}{K\,L^d\bar\phi_\Lambda}\right)^{-1}\right]
$$

$$
\times \exp\left(-\frac{1}{2}\sum_{\mathbf{k}\in\mathcal{B}_\Lambda}\ln\left\{1 - \frac{2i\lambda}{K\,L^d\left[\bar\phi_\Lambda + 2\sum_{\nu=1}^d(1-\cos k_\nu)\right]}\right\}\right). \tag{9.50}
$$

In the different regimes, considered in the previous section, we obtain the following results.

(1) If $d \le d_l = 2$ (for any K), or if $d > d_l$ and $K < K_c$, then the solution $\bar\phi_\Lambda(K, h_1, a; 1)$ tends to a finite limit $\bar\phi_\infty(K, 0, 1)$ which satisfies the zero-field bulk spherical constraint $\mathcal{W}_d(\phi) = K\zeta$ with $\zeta = 1$. In this case we note that

$$
\lim_{L\to\infty}\sum_{\mathbf{k}\in\mathcal{B}_\Lambda}\ln\left\{1 - \frac{2i\lambda}{KL^d\left[\bar\phi_\Lambda + 2\sum_{\nu=1}^d(1-\cos k_\nu)\right]}\right\}
$$

$$
= -\frac{2i\lambda}{K}\mathcal{W}_d\left(\bar\phi_\infty(K, 0, 1)\right) = -2i\lambda. \tag{9.51}
$$

Hence, the thermodynamic limit of the characteristic function (9.50) is

$$
\lim_{L\to\infty}\psi_\Lambda^{m.s.}(\lambda|\,K, h_\Lambda, 1) = e^{-i\lambda}. \tag{9.52}
$$

Now the transformation from a characteristic function $\psi(\lambda)$ to the corresponding probability density $p(\zeta)$,

$$
p(\zeta) = \frac{1}{2\pi}\int_{-\infty}^\infty d\lambda\, e^{-i\lambda\zeta}\psi(\lambda), \tag{9.53}
$$

yields the result (9.46).

(2) If $d > d_l = 2$ and $K > K_c$, then the solution $\bar\phi_\Lambda(K, h_1, a; 1)$ vanishes as $L \to \infty$ at a rate depending on the value of the exponent a:

(2a) If $0 < a < d$, the leading-order asymptotic form of $\bar\phi_\Lambda(K, h_1, a; 1)$ is given by Eq. (9.34). By insering the latter in Eq. (9.50) and taking the thermodynamic limit, one obtains again the results (9.52) and (9.46).

(2b) If $a = d$, the leading-order asymptotic form of $\bar\phi_\Lambda(K, h_1, a; 1)$ is given by Eq.(9.36). In this case the thermodynamic limit of the characteristic function (9.50) is

$$\lim_{L\to\infty} \psi_\Lambda^{m.s.}(\lambda|\,K, h_\Lambda, 1) = [1 - 2i\lambda y(K, h_1, 1)]^{-1/2}$$
$$\times \exp\left\{ i\lambda \left[\frac{K_c}{K} + \frac{1 - K_c/K - y(K, h_1, 1)}{1 - 2i\lambda\, y(K, h_1, 1)} \right] \right\}. \tag{9.54}$$

The above expression implies the following non-trivial generalization of the Kac-Thompson kernel (9.47):

$$\mathcal{K}(\zeta|\,K, h_1, d) = \frac{1}{2m_0(K;\zeta)[2\pi y(K, h_1, 1)]^{1/2}}$$
$$\times \left[\exp\left(-\frac{[m_0(K;\zeta) + h_1 y(K, h_1, 1)]^2}{2y(K, h_1, 1)} \right) \right.$$
$$\left. + \exp\left(-\frac{[m_0(K;\zeta) - h_1 y(K, h_1, 1)]^2}{2y(K, h_1, 1)} \right) \right] \tag{9.55}$$

for $\zeta > K_c/K$, and $\mathcal{K}(\zeta|K, h_1, d) \equiv 0$ for $\zeta < K_c/K$.

(2c) If $a > d$, then the leading-order asymptotic form of $\bar\phi_\Lambda(K, h_1, a; 1)$ is the same as in the zero-field case, see Eq. (9.42). The result for thermodynamic limit of the characteristic function (9.23) is correspondingly

$$\lim_{L\to\infty} \psi_\Lambda^{m.s.}(\lambda|\,K, h_\Lambda, 1) = [1 - 2i\lambda(1 - K_c/K)]^{-1/2} \exp(i\lambda K_c/K). \tag{9.56}$$

This characteristic function corresponds to the zero-field Kac-Thompson kernel (9.47) in the low-temperature regime.

We remark here that cases (2a) and (2c) correspond to the limits $h_1 \to \pm\infty$ and $h_1 \to 0$, respectively, of case (2b). Expression (9.55) interpolates between the δ-function and the zero-field Kac-Thompson kernel (9.47). Threfore, without loss of generality, in the remainder we will confine ourselves to the case $a = d$.

9.2.3   *Limit Gibbs states of the spherical model*

Let us turn now to Eq. (9.48) which relates the limit finite-dimensional characteristic functions in the mean spherical and spherical models. As we have shown in the previous section, in cases (1) and (2a) the transformation kernel is $\mathcal{K}_d(\zeta|\,K, h_1, a) = \delta(\zeta - 1)$. Therefore, whenever $d \leq 2$, or when $d > 2$ and either $K < K_c$, or $K > K_c$ but $0 < a < d$, the characteristic functions of the two models coincide in the thermodynamic limit:

$$\varphi^{m.s.}(t_A|K, h_1, a; 1) = \varphi^{s.}(t_A|K, h_1, a; 1). \tag{9.57}$$

Consider next the case (2b), when $d > 2$ and $K > K_c$, $a = d$. By rescaling the integration variables $S(\mathbf{r}) \to \zeta^{1/2} x(\mathbf{r})$ in the definition of the finite-volume characteristic function in the spherical model,

$$\varphi_\Lambda^{s.}(t_A|\,K, h_\Lambda, \zeta) = \int_{R^{L^d}} \left(\prod_{\mathbf{r}\in\Lambda} dS(\mathbf{r})\right) \delta\left(\zeta L^d - \sum_{\mathbf{r}\in\Lambda} S^2(\mathbf{r})\right)$$
$$\times \frac{1}{Z_\Lambda^{s.}(K, h_\Lambda, \zeta)} \exp\left[-\beta\mathcal{H}_\Lambda(S_\Lambda|\,J, H_\Lambda) + i\sum_{\mathbf{r}\in A} t(\mathbf{r})S(\mathbf{r})\right], \tag{9.58}$$

we obtain the identity

$$\varphi_\Lambda^{s.}(t_A|\,K, h_\Lambda, \zeta) = \varphi_\Lambda^{s.}(\zeta^{1/2} t_A|\,\zeta K, \zeta^{1/2} h_\Lambda, 1) \tag{9.59}$$

for all finite regions $\Lambda \subset Z^d$, such that $\Lambda \supset A$, A fixed. Hence, the above identity remains valid in the thermodynamic limit (9.14). By applying it to the finite-dimensional limit characteristic function which enters into the integrand in the right-hand side of Eq. (9.48), we obtain at $a = d$:

$$\varphi^{m.s.}(t_A|\,K, h_1, d; 1) =$$
$$\int_{K_c/K}^{\infty} d\,\zeta\, \mathcal{K}_d(\zeta|\,K, h_1, d)\, \varphi_d^{s.}(\zeta^{1/2} t_A|\,\zeta K; \zeta^{1/2} h_1, d; 1)$$
$$= \kappa \int_0^{\infty} dz\, \mathcal{K}(\kappa(z+1)|K, h_1, d)$$
$$\times \varphi^{s.}(\sqrt{\kappa(z+1)}\,t_A|\,K_c(z+1), \sqrt{\kappa(z+1)}\,h_1, d; 1). \tag{9.60}$$

In deriving the second equality we have set $\kappa = K_c/K$ and changed the integration variable $\zeta \to \kappa(z+1)$.

Our aim now is to cast Eq. (9.60) in the form of a Laplace transformation, which can be inverted to obtain $\varphi^{s.}(t_A|\,K, h_1, d; 1)$. To this end, we rescale first the variables $t_A \to \kappa^{-1/2} t_A$ and the field amplitude $h_1 \to \kappa^{-1/2} h_1$. Then we note that the equation

$$\frac{\kappa}{2y(K, \kappa^{-1/2} h_1, 1)} = s \tag{9.61}$$

with $s \geq 0$ has a unique solution for κ,

$$\kappa = \frac{4s^2}{4s^2 + 2s + h_1^2} \equiv \kappa(s). \tag{9.62}$$

Hence, from the explicit expression (9.38) for the characteristic function in the mean spherical model we obtain

$$\varphi^{m.s.}\left(\kappa^{-1/2}t_A \,\middle|\, K, \kappa^{-1/2}h_1, d; 1\right)\Big|_{\kappa=\kappa(s)} =$$

$$\exp\left\{-\frac{1}{2}\sum_{\mathbf{r},\mathbf{r}'\in A}\left[\frac{1}{2dK_c}P_d(\mathbf{r}-\mathbf{r}';1) + \frac{1}{2s}\right]t(\mathbf{r})t(\mathbf{r}') + i\frac{h_1}{2s}\sum_{\mathbf{r}\in A}t(\mathbf{r})\right\}, \tag{9.63}$$

and from expression (9.55) for the transformation kernel we have

$$\mathcal{K}\left(\kappa(z+1)\,\middle|\, K, \kappa^{-1/2}h_1, d\right)\Big|_{\kappa=\kappa(s)} =$$

$$\frac{1}{\kappa(s)}\left(\frac{s}{\pi z}\right)^{1/2}\exp(-sz - h_1^2/4s)\cosh(h_1 z^{1/2}). \tag{9.64}$$

Thus, the relationship between the limit characteristic functions in the mean spherical and spherical models can be cast in the form of a Laplace transformation that can be inverted by using the equation

$$\frac{1}{\sqrt{s}}\exp\left(\frac{k}{s}\right) = \int_0^\infty dz\, e^{-sz}\frac{1}{\sqrt{\pi z}}\cosh(2\sqrt{kz}). \tag{9.65}$$

The original variables of the characteristic function can be restored by setting $K_c(z+1) = K$, hence $z = \kappa^{-1} - 1$, and rescaling $t_A \to \kappa^{1/2}t_A$, $h_1 \to \kappa^{1/2}h_1$. By taking into account that $z^{1/2}\kappa^{1/2} = m_0(K,1)$ is the spontaneous magnetization, the finite-dimensional characteristic function of the spherical model can be written in the form [Brankov and Danchev (1987)]:

$$\varphi^{s.}\left(t_A \,\middle|\, K, h_1, d; 1\right) = \alpha(h_1 m_0)\varphi^{s.+}(t_A \,|\, K, 0, 1)$$

$$+ [1 - \alpha(h_1 m_0)]\varphi^{s.-}(t_A \,|\, K, 0, 1), \tag{9.66}$$

where

$$\alpha(h_1 m_0) = \frac{\exp(h_1 m_0)}{\exp(h_1 m_0) + \exp(-h_1 m_0)}, \tag{9.67}$$

and

$$\varphi^{s.\pm}(t_A|\,K,0,1) = \exp\left\{-\frac{1}{2}\sum_{\mathbf{r},\mathbf{r}'\in A} G(\mathbf{r}-\mathbf{r}';K,0,1)t(\mathbf{r})t(\mathbf{r}') \pm \mathrm{i}\,m_0(K,1)\sum_{\mathbf{r}\in A} t(\mathbf{r})\right\} \qquad (9.68)$$

are the characteristic functions of the two pure zero-field, low-temperature phases of the spherical model. A comparison with Eq. (9.35) at $\zeta = 1$ shows that the pure phases in the spherical and mean spherical models are the same. The mixed phase described by the characteristic function (9.66) reduces to one of the pure phases in the limits $h_1 \to +\infty$, when $\alpha(h_1 m_0) \to 1$, and $h_1 \to -\infty$, when $\alpha(h_1 m_0) \to 0$. Alternatively, the pure zero-field phases can be obtained under external field which vanishes slower than L^{-d}. In the particular case of $h_1 = 0$ one has $\alpha(0) = 1/2$ and the results of [Molchanov and Sudarev (1975)] follow.

Conclusions

On the basis of the above results we can make the following conclusions and conjectures.

(1) Below the bulk critical temperature the spherical model has exactly two pure translation-invariant zero-field limit Gibbs states $\varphi^{s.\pm}$, see Eq. (9.68). Any low-temperature zero-field limit Gibbs state $\varphi^{s.}$ is a linear combinations of these two extreme points:

$$\varphi^{s.} = \alpha\varphi^{s.+} + (1-\alpha)\varphi^{s.-} \qquad (0 \le \alpha \le 1). \qquad (9.69)$$

(2) Below the bulk critical temperature all the zero-field limit Gibbs states of the mean spherical model $\varphi^{m.s.}$ (with constraint parameter $\zeta = 1$) have the form

$$\varphi^{m.s.} = \int_{K_c/K}^{\infty} \left\{\alpha(\zeta)\varphi^{s.+}(\zeta) + [1-\alpha(\zeta)]\varphi^{s.-}(\zeta)\right\}\,\mathrm{d}\nu(\zeta), \qquad (9.70)$$

where $\nu(\zeta)$ is a probability measure on the interval $[K_c/K, \infty)$, $\varphi^{s.\pm}(\zeta)$ are the pure limit Gibbs states of the spherical model with constraint parameter ζ, and $0 \le \alpha(\zeta) \le 1$. The measure ν describing the mixed limit Gibbs states which appear under the action of vanishing uniform external field

$h_\Lambda = h_1 L^{-a}$ with parameters $h_1 \in (-\infty, \infty)$ and $a > 0$ is given by

$$d\nu(\zeta|K, h_1, a) = \mathcal{K}(\zeta|K, h_1, a)d\zeta, \qquad (9.71)$$

where the probability density $\mathcal{K}(\zeta|K, h_1, a)$ is actually the Kac-Thompson transformation kernel explicitly obtained in Section 9.2.2.

(3) In the phase coexistence region $K > K_c$, $h = 0$, the pure limit Gibbs states of both spherical models can be singled out by vanishing external field perturbation $h_\Lambda = h_1 L^{-a}$ with $0 < a < d$, where d is the dimensionality.

(4) As generally expected, vanishing of the bulk pair correlation function $G(\mathbf{r} - \mathbf{r}')$ in the limit $|\mathbf{r} - \mathbf{r}'| \to \infty$ is equivalent to pureness of the limit Gibbs states.

(5) Statistical inequivalence of the mean spherical and spherical models takes place only in mixed states; on the set of pure limit Gibbs states these models are statistically equivalent.

9.3   Probabilistic view on finite-size scaling

A probabilistic description of the critical behavior of a lattice system of spin variables $\{S_i \in R, i = 1, \ldots, n\}$ can be given in terms of limit probability distributions of block-spin random variables of the form [Cassandro and Jona-Lasinio (1978)], [Ellis and Newman (1978a)],

$$\eta_n = B_n^{-1} \left\{ \sum_{i=1}^{n} S_i - E\left(\sum_{i=1}^{n} S_i\right) \right\}, \qquad (9.72)$$

where $E(\cdot)$ denotes the expectation (i.e., average) value, and B_n is a normalization constant, such that $B_n \to \infty$ when $n \to \infty$. For independent random variables $\{S_i\}$ with finite variance the central limit theorem implies that the probability distribution of η_n with $B_n \sim \sqrt{n}$ converges weakly as $n \to \infty$ to the normal (Gaussian) distribution. In the case when $\{S_i\}$ are independent but have infinite variance, under appropriate assumptions on the tails of their distribution, the *non-central limit theorem* is valid: a stable non-Gaussian limit distribution of η_n exists for $B_n \sim n^a$ with some exponent $a \in (1/2, 1)$.

When the distribution of S_i depends on n, i.e., when one actually has a *triangular array of random variables* $\{S_i^{(n)}, i = 1, \ldots, n; n = 1, 2, \ldots\}$, then the limit distribution (if it exists) of the sum $\sum_{i=1}^{n} S_i^{(n)}$ belongs to the class of *infinitely divisible distributions*. The above facts are standard results of probability theory, see, e.g., [Feller (1966)].

Consider a triangular array of dependent random variables which at any fixed n obey the Gibbs joint probability distribution with a n-particle Hamiltonian. Let the system exhibit in the thermodynamic limit $n \to \infty$ a second order phase transition at a critical point $K = K_c$, $h = 0$. In the high-temperature region $K \in (0, K_c)$, due to the weak dependence among the random variables $S_i^{(n)}$, $i = 1, \ldots, n$, one expects the result of the standard central limit theorem to be valid. Arguments have been advanced [Cassandro and Jona-Lasinio (1978)], [Ellis and Newman (1978a)], that criticality at $K = K_c$, $h = 0$ can be identified with the case when the implications of the law of large numbers *and the non-central limit theorem* hold. Then, for appropriate $B_n = C\, n^a$, where $C > 0$ and $a \in (1/2, 1)$, the block-spin variable $\eta_n^{(n)}$ has a smooth limit distribution. Note that in some special cases $B_n \sim l(n)n^a$, where $l(n)$ is a slowly varying function of n.

One may also consider Gibbs probability distributions with parameters approaching the critical point simultaneously with $n \to \infty$. For more precise formulations we follow [Ellis and Newman (1978a)], [Ellis and Newman (1978c)]: for each $n = 1, 2, \ldots$ consider a set of random variables $\{S_i^n(\rho), i \in \Lambda_n\}$ with joint probability distribution

$$dP_n(x_1, \ldots, x_n; \rho) = Z_n^{-1} \exp\left(\sum_{i,j \in \Lambda_n} J_{i,j}^{(n)}\, x_i x_j\right) \prod_{i \in \Lambda_n} d\rho(x_i), \qquad (9.73)$$

where Λ_n are finite subsets of the d-dimensional lattice Z^d which tend in some appropriate sense to Z^d as $n \to \infty$, $\{J_{i,j}^{(n)}\}$ are some constants which may depend on n, ρ is a finite measure on the real axis R, and

$$Z_n = \int_{R^n} \exp\left(\sum_{i,j \in \Lambda_n} J_{i,j}^{(n)}\, x_i x_j\right) \prod_{i \in \Lambda_n} d\rho(x_i). \qquad (9.74)$$

All the thermodynamic parameters are included here in the ρ-dependence: we consider single-spin probability measures ρ which depend on a number of relevant parameters, say, $K_1, \ldots, K_q$, $q \in \{1, 2, \ldots\}$, and assume that the system has a critical point at $K_j = K_{j,c}$, $j = 1, \ldots, q$. By analogy with the extension of classical results of probability theory to the case of triangular arrays of random variables, one may now investigate the asymptotic distributions of block spins $\eta_n(\rho_n)$ defined in terms of random variables

$\{S_i^{(n)}(\rho_n)\}$ which have a joint probability distribution of the form (9.73) with ρ replaced by ρ_n. We are interested in the distributions which arise when ρ_n approaches (as $n \to \infty$) a critical point, i.e., a critical measure $\rho = \rho(K_{1,c}, \ldots, K_{q,c})$. In the case when $J_{i,j}^{(n)} = (2n)^{-1}$ and conditions (6.59) on the measure ρ hold, these distributions have been studied in [Ellis and Newman (1978c)]. The results show that equivalent-neighbors models may give rise to q critical exponents $\nu_1, \ldots, \nu_q$ and scaling constants $b_1, \ldots, b_q$, such that the block spins

$$\eta_n(\rho_n) = B_n^{-1} \sum_{i=1}^{n} S_i^{(n)}(K_{1,c} + b_1 n^{-1/\nu_1}, \ldots, K_{q,c} + b_q n^{-1/\nu_q}) \qquad (9.75)$$

have a well-defined asymptotic behavior as $n \to \infty$. The family of limit distributions is parametrized by the amplitudes $b_1, \ldots, b_q$ and describes an infinitesimal neighborhood of the critical measure. One may expect $\eta_n(\rho_n)$ to have a similar behavior in more complicated models [Ellis and Newman (1978a)]. There are two natural choices of the block-spin variable which have been considered in the literature: one of them corresponds to the so called *short long-range order*, and the other to *long long-range order*, notions introduced in [Schultz et. al. (1964)]. They are distinguished by whether the thermodynamic limit is taken prior to or together with the long-range order limit:

$$\eta_n^{(s)} = B_n^{-1} \sum_{i=1}^{n} S_i^{(\infty)}, \qquad \eta_n^{(l)} = B_n^{-1} \sum_{i=1}^{n} S_i^{(n)}. \qquad (9.76)$$

In the first case one considers the "short" block-spin, where $\{S_i^{(\infty)}, i = 1, \ldots, n\}$, for n finite, are random variables obeying the limit Gibbs distribution; in the second case one considers "long" block-spin, where $\{S_i^{(n)}, i = 1, \ldots, n\}$ are defined as random variables with joint probability distribution given by the Gibbs distribution for a finite system of n particles. In the class of equivalent-neighbors models, the "short" block spins possess a trivial statistics but this is not the case for the "long" block spins [Ellis and Newman (1978a)], [Ellis and Newman (1978c)]. Because of this fact, in the remainder we will consider only "long long-range" blocks of random variables and the superscript (l) will be omitted.

Below we consider the probability distributions in the equivalent-neighbor versions of the spherical and mean spherical models, see Section 3.1.3, of two different block variables: magnetization,

$$\eta(a_m, N) = N^{-a_m} \sum_{i=1}^{N} S_i^{(N)}, \tag{9.77}$$

and interaction energy,

$$\xi(a_e, N) = -\frac{1}{2} K_c N^{2a_m - a_e - 1} \eta^2(a_m, N), \tag{9.78}$$

written in terms of the block-spin variable (9.77).

9.3.1   *Block-spin distribution in the equivalent-neighbors spherical model*

The equivalent-neighbors spherical model belongs to the class of models studied in Section 6.4; it is defined by the non-product type "free" measure μ_0 of the form (6.60). As in Section 9.2, we consider a family of measures which confine the allowed spin configurations to the N-dimensional hypersphere of radius $(\zeta N)^{1/2}$. Then, the density (with respect to the Lebesgue measure on R^N) of the joint probability distribution of the set of random variables $\{S_i \in R, \ i = 1, \ldots, N\}$ is given by

$$p_N^s(x_1, \ldots, x_N | K, h, \zeta) = [Z_N^s(K, h, \zeta)]^{-1}$$

$$\times \exp\left[\frac{K}{2N}\left(\sum_{i=1}^{N} x_i\right)^2 + h \sum_{i=1}^{N} x_i\right] \delta\left(\sum_{i=1}^{N} x_i^2 - \zeta N\right), \tag{9.79}$$

where $Z_N^s(K, h, \zeta)$ is the partition function (3.64).

Note that the corresponding grand canonical ensemble with respect to the random variable $\sum_{i=1}^{N} S_i^2$ leads to the equivalent-neighbors Gaussian model, which is defined by the joint probability density

$$p_N^G(x_1, \ldots, x_N | K, h, s) = [Z_N^G(K, h, s)]^{-1}$$

$$\times \exp\left[\frac{K}{2N}\left(\sum_{i=1}^{N} x_i\right)^2 + h \sum_{i=1}^{N} x_i - s \sum_{i=1}^{N} x_i^2\right]. \tag{9.80}$$

The partition function (3.69) of the equivalent-neighbors Gaussian model,

$$Z_N^G(K, h, s) = \left(\frac{2\pi}{K}\right)^{N/2} \phi^{-1/2}(1 + \phi)^{-(N-1)/2} \exp\left[\frac{Nh^2}{2K\phi}\right], \tag{9.81}$$

is defined for values of the spherical fiels $s > K/2$, or $\phi = 2s/K - 1 > 0$. The condition for thermodynamic equivalence of the two ensembles leads to the mean spherical constraint (3.70).

For the sake of brevity, if ξ_n, $n = 1, 2, \ldots$, is a sequence of random variables and μ is a probability distribution, we write $\xi_n \to p(x)$ to denote that the distribution of ξ_n converges weakly as $n \to \infty$ to a distribution proportional to $\mathrm{d}\mu = p(x)\mathrm{d}x$.

Now we are ready to study the probability distribution of the long long-range block-spin variable $\eta(a_m, N)$, defined by Eqs. (9.77) and (9.79). Consider first the probability density corresponding to the block spin with normalization exponent $a_m = 1$:

$$p_{\eta(1,N)}^s(x|K, h, \zeta) = \frac{1}{Z_N^s(K, h, \zeta)} \int_{R^N} \exp\left[\frac{K}{2N}\left(\sum_{i=1}^{N} x_i\right)^2 + h\sum_{i=1}^{N} x_i\right]$$

$$\times \delta\left(\sum_{i=1}^{N} x_i^2 - \zeta N\right) \delta\left(N^{-1}\sum_{i=1}^{N} x_i - x\right) \mathrm{d}x_1 \cdots \mathrm{d}x_N. \tag{9.82}$$

As in the calculation of the partition function performed in Section 3.1.3, we make use of an orthogonal change of variables $(x_1, \ldots, x_N) \to (y_1, \ldots, y_N)$, such that conditions (3.63) are fulfilled. Then, integrating over $y_1, \ldots, y_N$, we obtain

$$p_{\eta(1,N)}^s(x|K, h, \zeta) = \frac{\mathcal{A}_{N-1}(N^{1/2})}{2Z_N^s(K, h, \zeta)}(\zeta - x^2)^{-3/2}$$

$$\times \exp[Ng(x|K, h, \zeta)] \, I_{[-\zeta^{1/2}, \zeta^{1/2}]}(x), \tag{9.83}$$

where $\mathcal{A}_d(r)$ is the constant (3.57), the function $g(x|K, h, \zeta)$ is defined in Eq. (3.65), and $I_{[a,b]}(x)$ is the indicator function which equals unity if $x \in [a, b]$ and zero otherwise. The exact expression for the probability density $p_{\eta(a,N)}^s$ for the block spin with normalization exponent $a > 0$ can be directly obtained from Eq. (9.83) by using the relation

$$p_{\eta(a,N)}^s(x|K, h, \zeta) = N^{a-1}p_{\eta(1,N)}^s(xN^{a-1}|K, h, \zeta). \tag{9.84}$$

By taking into account the extremal properties of the function $g(x|K, h, \zeta)$ with respect to $x \in (-\zeta^{1/2}, \zeta^{1/2})$, see Section 3.1.3, and by standard evaluation of Laplace type integrals (as $N \to \infty$), we obtain the following results [Brankov and Danchev (1989)]:

Low-temperature regime. Let $K > \zeta^{-1}$ and $h = h_1 N^{-1}$. Then

$$\eta(1, N) \to \alpha(h_1 m_0)\delta(x - m_0) + [1 - \alpha(h_1 m_0)]\delta(x + m_0), \tag{9.85}$$

where $m_0 = m_0(K, \zeta)$ is the spantaneous magnetization (3.66), and $\alpha(h_1 m_0)$ is given by Eq. (9.67). The interesting point to note here is that the same result holds for the spherical model with nearest-neighbor interaction, see Eq. (9.66).

Bulk critical point. Let $K = \zeta^{-1}$ and $h = 0$. Then

$$\eta(1, N) \;\to\; \delta(x);$$
$$\eta(3/4, N) \;\to\; \left(\frac{2}{\zeta}\right)^{1/2} [\Gamma(1/4)]^{-1} \exp\left(-\frac{1}{4}\zeta^{-2}x^4\right). \tag{9.86}$$

As mentioned in Section 3.1.3, the model has a line of critical points, $K = \zeta^{-1}$, if $\zeta > 0$ is considered as a thermodynamic parameter.

High-temperature regime. Let $K < \zeta^{-1}$ and $h = 0$. Then, starting from Eq. (9.83) and taking into account (9.84), one easily obtains the normal distribution

$$\eta(1/2, N) \to [(\zeta^{-1} - K)/2\pi]^{1/2} \exp\left[-\frac{1}{2}(\zeta^{-1} - K)x^2\right]. \tag{9.87}$$

The study of the *finite-size critical regime* requires a somewhat more refined analysis. Let g be a bounded continuous function defined on the real axis. Then from Eqs. (9.83) and (9.84), by expanding $\ln[1 - x^2(\zeta N^{1/2})^{-1}]$ around the point $x = 0$, we obtain

$$\int_R dx\, g(x)\, p^s_{\eta(3/4,N)}(x|K,h,\zeta) = \frac{A_{N-1}(N^{1/2})}{Z^s_N(K,h,\zeta)} \frac{\zeta^{N/2}}{2N^{1/4}}$$

$$\times \int_{-(\zeta^2 N)^{1/4}}^{(\zeta^2 N)^{1/4}} \frac{dx\, g(x)}{(\zeta - x^2 N^{-1/2})^{3/2}} \exp\left[\frac{1}{2}N^{1/2}(K - \zeta^{-1})x^2\right.$$

$$\left. + N^{3/4}hx - \frac{x^4}{4\zeta^2} - \frac{(1 - \theta)^2 x^6}{2N^{1/2}\zeta^3}\left(1 - \frac{\theta x^2}{\zeta N^{1/2}}\right)^{-3}\right]. \tag{9.88}$$

where $\theta \in (0, 1)$. It is clear that if one sets

$$K = \zeta^{-1}\left(1 - x_1^* N^{-1/2}\right), \qquad h = \zeta^{-1/2}x_2^* N^{-3/4}, \tag{9.89}$$

with some fixed x_1^* and x_2^*, then a well defined *smooth* limit distribution of $\eta(3/4, N)$ exists, which is parametrized by x_1^* and x_1^*. In this case the first three terms in the argument of the exponential function in the right-hand side of Eq. (9.88) are of the same order of magnitude and the limiting distribution depends on the largest possible number of parameters. By substituting K and h from Eq. (9.89) into Eq. (9.88), and by extending the range of integration to the whole real axis, we obtain in the limit $N \to \infty$

$$\zeta^{-1/2}\eta(3/4, N) \to \frac{\exp\left(-\frac{1}{2}x_1^* x^2 + x_2^* x - \frac{1}{4}x^4\right)}{C(x_1^*, x_2^*)} \equiv p_\eta^{s \cdot}(x | x_1^*, x_2^*), \qquad (9.90)$$

where

$$C(x_1^*, x_2^*) = \int_{-\infty}^{\infty} dx \, \exp\left(-\frac{1}{2}x_1^* x^2 + x_2^* x - \frac{1}{4}x^4\right). \qquad (9.91)$$

Note that in conformity with the modified finite-size scaling at criticality, see Chapter 6, the finite-size scaling temperature and field variables are, see Eq. (6.46),

$$x_1^* = \zeta K t N^{1/2}, \qquad x_2^* = \zeta^{1/2} h N^{3/4}, \qquad t = (\zeta K)^{-1} - 1. \qquad (9.92)$$

Of course, at the critical point $x_1^* = x_2^* = 0$, Eq. (9.90) reduces to Eq. (9.86).

Now we turn to the probability distribution of the interaction energy (9.78) with $K_c = \zeta^{-1}$. Taking into account that when $a_m = 3/4$ the random variable $\eta(a_m, N)$ has a well defined smooth limit distribution, depending on both x_1^* and x_2^*, it is clear that the same will be true for $\xi(a_e, N)$ if and only if $a_e = 1/2$. Then, by using standard relations for transformation of densities of random variables, see, e.g., [Feller (1966)], one can easily obtain the explicit expression for the limit probability density of $\xi(1/2, N)$.

In the equivalent-neighbors spherical model one can explicitly evaluate the large-N asymptotic form of the n-particle correlation function at zero field, $h = 0$, and $x_1^* = O(1)$. To this end we make use of the leading-order asymptotic equality as $N \to \infty$

$$E_{K,h,\zeta}\left(\prod_{j=1}^{n} S_j^{(N)}\right) \simeq E_{K,h,\zeta}\left(\left[N^{-1}\sum_{i=1}^{N} S_i^{(N)}\right]^n\right), \qquad (9.93)$$

where $E_{K,h,\zeta}(\cdot)$ denotes the expectation value with respect to the joint probability density (9.79), in combination with the result

$$
E_{K,0,\zeta}\left(\left[N^{-1}\sum_{i=1}^{N}S_i^{(N)}\right]^n\right)\Bigg|_{K=\zeta^{-1}(1-x_1^* N^{-1/2})} \simeq
$$

$$
N^{-n/4}(\sqrt{2}\,\zeta)^{n/2}\sqrt{\pi}\,\Gamma\left(\frac{n+1}{2}\right)\frac{D_{-(n+1)/2}(x_1^*/\sqrt{2})}{D_{-1/2}(x_1^*/\sqrt{2})}. \qquad (9.94)
$$

Here $D_\nu(\cdot)$ is the Weber parabolic cylinder function of order ν, which for negative values of the order has the following integral representation:

$$
D_{-p}(z) = \frac{\exp(-z^2/4)}{\Gamma(p)}\int_0^\infty dx\, x^{p-1}\exp(-zx-x^2/2)\quad (p>0). \qquad (9.95)
$$

Therefore, in the finite-size critical region the n-particle zero-field correlation function decays to zero as $N^{-n/4}$. In the limit $N\to\infty$ we obtain that at the critical point the spins $S_i^{(\infty)}$, $i=1,2,\ldots$, are independent random variables.

9.3.2   *Block-spin distribution in the equivalent-neighbors mean spherical model*

By direct calculations starting from the finite-size joint probability distribution (9.80) and partition function (9.81) of the equivalent-neighbors Gaussian model, one can easily derive the exact expression for the probability density of the block-spin variable (9.77):

$$
p_{\eta(a,N)}^{G.}(x|K,h,s) = \left[\frac{K}{2\pi}\phi N^{2a-1}\right]^{1/2}\exp\left\{-\frac{1}{2}K\phi N^{2a-1}\left[x-\frac{N^{1-a}h}{K\phi}\right]^2\right\}.
$$

$$
\qquad (9.96)
$$

We remind the reader that the corresponding probability density for the mean spherical model is simply related to the above one,

$$
p_{\eta(a,N)}^{m.s.}(x|K,h,\zeta) = p_{\eta(a,N)}^{G.}(x|K,h,s)\Big|_{s=(K/2)[1+\bar{\phi}_N(K,h,\zeta)]}, \qquad (9.97)
$$

where $\bar{\phi}_N(K,h,\zeta)$ is the solution for $\phi = 2s/K-1$ of the exact mean spherical constraint (3.73).

In the high-temperature regime, $K < \zeta^{-1}$, or in the presence of a magnetic field $h\neq 0$, the solution $\bar{\phi}_N(K,h,\zeta)$ of the mean spherical constraint is strictly positive, as well as its bulk limit $\bar{\phi}_\infty(K,h,\zeta)$. Then, in order to obtain a limit distribution with a finite variance, one has to choose in Eq. (9.96) $a = 1/2$. In the zero-field case ($h = 0$) this choice leads to a Gaussian limit distribution for the block spin. However, when $h \neq 0$ the condition of a finite variance, $a = 1/2$, implies infinite expectation value, while the condition for a finite expectation value, $a = 1$, implies vanishing variance (the law of large numbers).

The situation changes when the point (K, h) approaches the line $\{K \geq \zeta^{-1}, h = 0\}$, since then $\bar{\phi}_\infty(K, h, \zeta) \to 0$ and one has to choose $a > 1/2$ in order to obtain a well defined smooth limit distribution for $\eta(a, N)$. One easily obtains that the most general, two-parameter family of solutions of the mean spherical constraint results under the parametrization (9.89), which correspond to the modified finite-size scaling. Then,

$$\bar{\phi}_N(K, h, \zeta) = \bar{X}(x_1^*, x_2^*)N^{-1/2} + O(N^{-1}), \tag{9.98}$$

where $\bar{X}(x_1^*, x_2^*)$ is the positive solution for X of the equation

$$X^{-1} - X + (x_2^*)^2 X^{-2} + x_1^* = 0. \tag{9.99}$$

By substitution of $\bar{\phi}_N(K, h, \zeta)$ from Eq. (9.98) into the probability density (9.97), one obtains that the block spin in the mean spherical model has a limit Gaussian distribution under the choice $a = 3/4$:

$$\zeta^{-1/2}\eta(3/4, N) \to \left(\frac{\bar{X}}{2\pi}\right)^{1/2} \exp\left[-\frac{1}{2}\bar{X}\left(x - x_2^*\bar{X}^{-1}\right)^2\right]$$
$$\equiv p_\eta^{m.s.}(x|x_1^*, x_2^*). \tag{9.100}$$

Consider now the case when the line of first order phase transitions of the bulk system $\{K > \zeta^{-1}, h = 0\}$ is approached at a fixed $K > \zeta^{-1}$ with external field vanishing as $h = h_1 N^{-1}$. Then the solution of the mean spherical constraint (3.73) has the same leading-order asymptotic form, given by Eqs. (9.36) and (9.37) at $K_c = 1$, as in the case of nearest-neighbor interaction. Now a smooth one-parameter probability distribution has the total spin per site ($a = 1$):

$$\eta(1, N) \to (2\pi y)^{-1/2} \exp\left[-(x - h_1 y)^2/2y\right], \tag{9.101}$$

where $y = y(K, h_1, \zeta)$ is defined by Eq. (9.37) with $m_0(K, \zeta) = (\zeta - K^{-1})^{1/2}$. Hence it follows that the average magnetization per spin in the limit Gibbs state,

$$m_\infty(K, h_1, \zeta) = m_0(K, \zeta) \frac{2z}{\sqrt{1 + 4z^2} + 1}, \qquad z \equiv h_1 m_0(K, \zeta), \qquad (9.102)$$

interpolates between $-m_0$ and $+m_0$ as the amplitude h_1 of the vanishing field changes from $-\infty$ to $+\infty$. This indicates that the obtained limit Gibbs state is a mixed phase. Note that the finite-size scaling field variable z and the scaling function for the magnetization are the same as for the mean spherical model with power-law interaction near the first order transition, see Eq. (8.29).

In the mean spherical model the joint probability distribution of the spin variables $S_i^{(N)}$, $i = 1, \ldots, n$, $n < N$, is Gaussian. The dependence among the spin variables is induced by two-particle correlations only. For the interaction energy block variable $\xi(a_e, N)$, due to the relationship (9.78), we have again $a_e = 1/2$. Hence, taking into account Eq. (9.93), we see that in the modified finite-size critical region, defined by Eq. (9.89) with bounded x_1^*, x_2^*, the pair correlations decay like $N^{-1/2}$ as $N \to \infty$. Therefore, in the bulk limit $\{S_i^{(\infty)}\}$ are independent random variables at the critical point.

The next interesting object to study is the Kac-Thompson type transformation kernel, which connects the limit probability distributions for the block spins in the spherical and mean spherical models.

9.3.3   *Transformation kernel*

From the definition of the joint spin probability distributions in the equivalent-neighbors spherical, Eq. (9.79), and Gaussian, Eq. (9.80), models, it follows that they are related by the integral equation

$$p_N^{G\cdot}(x_1, \cdots, x_N | K, h, s) =$$
$$\int_{-\infty}^{\infty} d\zeta \, \mathcal{K}_N^{G\cdot}(\zeta | K, h, s) p_N^{s\cdot}(x_1, \cdots, x_N | K, h, \zeta), \qquad (9.103)$$

where

$$\mathcal{K}_N^{G\cdot}(\zeta | K, h, s) = \frac{N \exp(-s\zeta N) \, Z_N^{s\cdot}(K, h, \zeta)}{Z_N^{G\cdot}(K, h, s)} \qquad (9.104)$$

is the transformation kernel between the two ensembles. When the mean spherical constraint (3.70) with $\zeta = 1$ is imposed on the conjugate spherical field s, one obtains the transformation kernel between the spherical model with parameter $\zeta > 0$ and the mean spherical model (with $\zeta = 1$):

$$p_N^{m.s.}(x_1, \cdots, x_N | K, h, 1) =$$
$$\int_{-\infty}^{\infty} d\zeta \, \mathcal{K}_N(\zeta | K, h) p_N^{s.}(x_1, \cdots, x_N | K, h, \zeta), \quad (9.105)$$

where

$$\mathcal{K}_N(\zeta | K, h) = \mathcal{K}_N^G(\zeta | K, h, \bar{s}_N(K, h, 1)), \quad (9.106)$$

and $\bar{s}_N(K, h, 1)$ is the solution for the spherical field $s = K(1 + \phi)/2$ obtained from Eq. (3.70) at $\zeta = 1$. Therefore, a statistical equivalence of the spherical and mean spherical models in the thermodynamic limit takes place if

$$\mathcal{K}(\zeta | K, h) \equiv \lim_{N \to \infty} \mathcal{K}_N(\zeta | K, h, \bar{s}_N(K, h, 1)) = \delta(\zeta - 1), \quad (9.107)$$

Evidently, Eq. (9.105) implies relationships between the probability distributions for the block variables $\eta(a_m, N)$ and $\xi(a_e, N)$ in the two models.

For the sake of simplicity, we confine ourselves to the derivation of an explicit expression for the kernel (9.106) in the neighborhood of the zero-field bulk critical point. At $h = 0$ the partition function of the spherical equivalent-neighbors model (3.64) takes the simple form

$$Z_N^{s.}(K, 0, \zeta) = \frac{1}{2} A_{N-1}(N^{1/2}) \int_{-\zeta^{1/2}}^{\zeta^{1/2}} dx \, (\zeta - x^2)^{(N-3)/2} \exp(Kx^2/2).$$
$$(9.108)$$

The asymptotic evaluation of the above integral at $\zeta = 1 + xN^{-1/2}$ in the neighborhood of the critical point defined by Eq. (9.89) with fixed $|x_1^*| < \infty$ and $x_2^* = 0$, yields

$$Z_N^{s.}(1 - x_1^* N^{-1/2}, 0, 1 + xN^{-1/2}) \simeq$$
$$\frac{\sqrt{\pi}}{2} A_{N-1}(N^{1/2}) \left(\frac{2}{N}\right)^{1/4} \exp\left[\frac{1}{8}(x_1^* - x)^2\right] D_{-1/2}\left(\frac{x_1^* - x}{\sqrt{2}}\right). \quad (9.109)$$

By taking into account that the solution of the mean spherical constraint is $\bar{\phi}_N(K, 0, 1) \simeq \bar{X}(x_1^*, 0)N^{-1/2}$, and by using the explicit expression (9.81) for the partition function of the equivalent-neighbors Gaussian model, we obtain that the transformation kernel is smeared over the $O(N^{-1/2})$ scale:

$$\mathcal{K}^{crit}(x|x_1^*,0) \equiv \lim_{N\to\infty} \mathcal{K}_N(1 + xN^{-1/2}|1 - x_1^* N^{-1/2}, 0)\, N^{-1/2}$$

$$= \frac{\exp\left\{-\frac{1}{4}[x + g(x_1^*)]^2 + \frac{1}{8}(x - x_1^*)^2\right\}}{2^{5/4}\pi^{1/2} g^{1/2}(x_1^*)} D_{-1/2}\left(\frac{x_1^* - x}{\sqrt{2}}\right), \quad (9.110)$$

where

$$g(x_1^*) \equiv 1/\bar{X}(x_1^*,0) = \frac{1}{2}[((x_1^*)^2 + 4)^{1/2} - x_1^*]. \quad (9.111)$$

Evidently, $\mathcal{K}^{crit}(x|x_1^*,0)$ is positive for all x, x_1^* and, as one can easily verify, integrated with respect to x over the real axis yields unity. Therefore, $\mathcal{K}^{crit}(x|x_1^*,0)$ is a probability density for all $x \in R$, and the following relationship between the zero-field probability densities $p_\eta^{m.s.}(x|x_1^*,0)$ and $p_\eta^s(x|x_1^*,0)$ of the block spin variable in the two models takes place:

$$p_\eta^{m.s.}(x|x_1^*,0) = \int_{-\infty}^{\infty} dz\, \mathcal{K}^{crit}(z|x_1^*,0)\, p_\eta^s(z|x_1^* + z,0). \quad (9.112)$$

In the high-temperature limit $x_1^* \to \infty$, from the asymptotic expansion of the parabolic cylinder function, $D_{-1/2}(z) \simeq z^{-1/2} e^{-z^2/4}$ as $z \to \infty$, it follows that the kernel $\mathcal{K}^{crit}(z|x_1^*,0)$ becomes Gaussian, centered at the point $z = -(x_1^*)^{-1}$,

$$\mathcal{K}^{crit}(z|x_1^*,0) \simeq \frac{1}{2\sqrt{\pi}} \exp\left[-\frac{1}{4}(z + 1/x_1^*)^2\right]. \quad (9.113)$$

In the low-temperature limit, the maximum of the kernel $\mathcal{K}^{crit}(z|x_1^*,0)$ shifts away to the neighborhood of $z = -|x_1^*| \to -\infty$, where, for $y = O(1)$,

$$\mathcal{K}^{crit}(x_1^* + y|x_1^*,0) \simeq \frac{\exp\left(-y^2/8\right)}{2^{5/4}\sqrt{\pi|x_1^*|}} D_{-1/2}\left(\frac{-y}{\sqrt{2}}\right). \quad (9.114)$$

When $x_1^* \to -\infty$ at a fixed value of z, one has to use the asymptotic expansion of the parabolic cylinder function, $D_{-1/2}(z) \simeq \sqrt{2}|z|^{-1/2} e^{z^2/4}$ as $z \to -\infty$, which yields

$$\mathcal{K}^{crit}(z|x_1^*,0) \simeq \frac{1}{\sqrt{2\pi\, e\, |x_1^*|(|x_1^*| + z)}}. \quad (9.115)$$

Consider now the behavior of the kernel in the low-temperature region $K > \zeta^{-1}$, when $h = h_1 N^{-1}$ as $N \to \infty$. The partition function (3.64) of

the equivalent-neighbors spherical model in such a field is proportional to
the Laplace-type integral,

$$Z_N^s(K, h_1 N^{-1}, \zeta) = \frac{1}{2} \mathcal{A}_{N-1}(N^{1/2})$$

$$\times \int_{-\zeta^{1/2}}^{\zeta^{1/2}} dx \frac{e^{h_1 x}}{(\zeta - x^2)^{3/2}} \exp\left\{\frac{1}{2} N[Kx^2 + \ln(\zeta - x^2)]\right\}, \quad (9.116)$$

which can be asymptotically evaluated as $N \to \infty$ by taking into account
the contribution from the stationary points: $x = \pm(\zeta - K^{-1})^{1/2}$, if $\zeta > K^{-1}$,
and $x = 0$, if $\zeta < K^{-1}$. Thus we obtain that when $\zeta > K^{-1}$,

$$Z_N^s(K, h_1 N^{-1}, \zeta) \simeq \frac{\sqrt{\pi} \mathcal{A}_{N-1}(N^{1/2})}{N^{1/2} K^{(N-1)/2}}$$

$$\times \frac{\cosh[h_1 m_0(K, \zeta)]}{m_0(K, \zeta)} \exp\left[\frac{1}{2} N K m_0^2(K, \zeta)\right], \quad (9.117)$$

and when $\zeta < K^{-1}$,

$$Z_N^s(K, h_1 N^{-1}, \zeta) \simeq \left(\frac{\pi}{2}\right)^{1/2} \mathcal{A}_{N-1}(N^{1/2}) \zeta^{(N-3)/2} [N(\zeta^{-1} - K)]^{-1/2}.$$

$$(9.118)$$

Next we need to evaluate the partition function (9.81) of the equivalent-
neighbors Gaussian model at the solution $\bar{\phi}_N(K, h_1 N^{-1}, 1)$ of the mean
spherical constraint (3.73) with $\zeta = 1$. Since the leading-order asymptotic
form of $\bar{\phi}_N(K, h_1 N^{-1}, 1)$ is given by Eqs. (9.36) and (9.37) at $\zeta = 1$, we
obtain

$$Z_N^G\left(K, h_1 N^{-1}, \bar{s}_N(K, h_1 N^{-1}, 1)\right) =$$

$$\left(\frac{2\pi}{K}\right)^{N/2} [KNy(K, h_1, 1)]^{1/2} \exp\left\{\frac{h_1^2 y^2(K, h_1, 1) - K^{-1}}{2y(K, h_1, 1)}\right\}. \quad (9.119)$$

By substituting Eqs. (9.118) and (9.119) in Eq. (9.104), from the definition
(9.106) of the transformation kernel we finally obtain that

$$\lim_{N \to \infty} \mathcal{K}_N(\zeta | K, h_1 N^{-1}) = \mathcal{K}(\zeta | K, h_1, d), \quad (9.120)$$

where $\mathcal{K}(\zeta | K, h_1, d)$ is the Kac-Thompson kernel (with $K_c = 1$) derived in
Section 9.2.2 for the spherical models with nearest-neighbors interaction,
see Eq. (9.55). The coincidence of the two kernels is in conformity with the

hypothesis that the interaction range (finite or infinite) becomes irrelevant on the line of first order phase transitions.

Finally, we pass to the derivation of the finite-size scaling functions.

9.3.4　*Finite-size scaling functions*

From Eq. (3.64) for the partition function of the equivalent-neighbors spherical model, one obtains that the corresponding free energy density in the modified finite-size critical region (9.89) is given by

$$\beta f_N^s(K, h, \zeta) = -\frac{1}{2}\ln(2\pi e\,\zeta) + N^{-1}\ln(2^{3/2}\pi\zeta N^{1/4})$$
$$-N^{-1}\ln\int_\infty^\infty dx \exp\left(-\frac{1}{2}x_1^* x^2 + x_2^* x - \frac{1}{4}x^4\right) + o(N^{-1}), \qquad (9.121)$$

which is in full conformity with the modified finite-size scaling prediction, see Eqs. (6.43) - (6.46) at $d' = 0$ and $L^d = N$.

Thus we have obtained that the finite-size scaling function for the free energy density of the spherical model close to criticality is

$$X_f^s(x_1^*, x_2^*) = -\ln\int_\infty^\infty dx \exp\left(-\frac{1}{2}x_1^* x^2 + x_2^* x - \frac{1}{4}x^4\right). \qquad (9.122)$$

To evaluate the asymptotic form of the finite-size free energy density of the equivalent-neighbors mean spherical model, defined by Eqs. (3.71) and (3.72), we take into account that the solution of the exact mean spherical constraint (3.73) in the modified finite-size scaling regime (9.89) is given by expression (9.98). Thus, the following result easily follows:

$$\beta f_N^{m.s.}(K, h, \zeta) = -\frac{1}{2}\ln(2\pi e\zeta) - \frac{\ln N}{4N} + N^{-1}X_f^{m.s.}(x_1^*, x_2^*) + o(N^{-1}),$$
$$(9.123)$$

where the finite-size scaling function is

$$X_f^{m.s.}(x_1^*, x_2^*) = \frac{1}{2}\ln\bar{X}(x_1^*, x_2^*) - \frac{1}{4}\left[\bar{X}(x_1^*, x_2^*) - x_1^*\right]^2 - \frac{1}{2}\frac{(x_2^*)^2}{\bar{X}(x_1^*, x_2^*)},$$
$$(9.124)$$

and $\bar{X}(x_1^*, x_2^*)$ is the positive solution for X of Eq. (9.99).

It is interesting to note that in the case under consideration the zero-field bulk free energy consists of two analytical branches, so that the only analytical background one may naturally single out from the total finite-size free energy (9.123) is a trivial constant term. That is why the finite-size scaling functions, given by Eqs. (9.122) and (9.124), contain the complete temperature and field dependence of the free energy density.

The finite-size scaling behavior of the magnetization per spin in either of the models can be obtained by differentiating the asymptotic form of the corresponding free energy density with respect to the field variable:

$$m_N(K, h, \zeta) \simeq -N^{-1/4}\zeta^{1/2}\frac{\partial}{\partial x_2^*}X_f(x_1^*, x_2^*). \qquad (9.125)$$

On the other hand, from the statistical definition of the magnetization we have

$$m_N(K, h, \zeta) = N^{-1/4}\zeta^{1/2}E_{K,h,\zeta}\left(\zeta^{-1/2}\eta(3/4, N)\right). \qquad (9.126)$$

Due to the weak convergence of $\zeta^{-1/2}\eta(3/4, N)$ as $N \to \infty$ to a random variable with limit distribution given by the right-hand side of Eq. (9.90), in the case of the spherical model, or Eq. (9.100), in the case of the mean spherical model, we can write the the asymptotic form of the finite-size magnetization as follows

$$m_N^s(K, h, \zeta) \quad \simeq \quad N^{-1/4}\zeta^{1/2}\frac{\partial}{\partial x_2^*}\ln C(x_1^*, x_2^*), \qquad (9.127)$$

$$m_N^{m.s.}(K, h, \zeta) \quad \simeq \quad N^{-1/4}\zeta^{1/2}\,\frac{x_2^*}{\bar{X}(x_1^*, x_2^*)}. \qquad (9.128)$$

Similarly, for the finite-size internal energy density we have

$$\beta u_N(K, h, \zeta) \simeq -N^{-1/2}\frac{\partial}{\partial x_1^*}\bar{X}_f(x_1^*, x_2^*), \qquad (9.129)$$

and from its statistical definition it follows that

$$\beta u_N(K, h, \zeta) =$$
$$-\frac{1}{2}KN^{-1/2}\left\{V_{K,h,\zeta}(\eta(3/4, N)) + [E_{K,h,\zeta}(\eta(3/4, N))]^2\right\}, \qquad (9.130)$$

where $V_{K,h,\zeta}(\eta)$ is the variance of the random variable η. Thus, from the weak convergence of the random variable $\eta(3/4, N)$, see Eqs. (9.90), (9.91) for the spherical model, and Eqs. (9.99), (9.100) for the mean spherical one, we obtain the asymptotic form of the finite-size internal energy

$$\beta u_N^{s\cdot}(K,h,\zeta) \;\simeq\; N^{-1/2}\frac{\partial}{\partial x_1^*}\ln C(x_1^*,x_2^*), \tag{9.131}$$

$$\beta u_N^{m\cdot s\cdot}(K,h,\zeta) \;\simeq\; -\frac{1}{2}N^{-1/2}[\bar{X}(x_1^*,x_2^*) - x_1^*]. \tag{9.132}$$

The above results show that one may derive the explicit form of the finite-size scaling function for the free energy density near criticality from the smooth limit distributions found for the appropriate triangular arrays of normalized block random variables. Most straightforward is the case of the spherical model. In this case the integrability of the system of first order partial differential equations for the unknown function $X_f^{s\cdot}$,

$$\frac{\partial}{\partial x_1^*}X_f^{s\cdot} = -\frac{\partial}{\partial x_1^*}\ln C(x_1^*,x_2^*), \qquad \frac{\partial}{\partial x_2^*}X_f^{s\cdot} = -\frac{\partial}{\partial x_2^*}\ln C(x_1^*,x_2^*), \tag{9.133}$$

is trivially fulfilled due to the differentiability of the function $C(x_1,x_2)$, see Eq. (9.91). Hence expression (9.122) follows.

In the case of the mean spherical model the integrability of the corresponding system of first order partial differential equations for the unknown function $X_f^{m\cdot s\cdot}$,

$$\frac{\partial}{\partial x_1^*}X_f^{m\cdot s\cdot} = \frac{1}{2}\left[\bar{X}(x_1^*,x_2^*) - x_1^*\right], \qquad \frac{\partial}{\partial x_2^*}X_f^{m\cdot s\cdot} = -\frac{x_2^*}{\bar{X}(x_1^*,x_2^*)}, \tag{9.134}$$

is fulfilled due to the mean spherical constraint (9.99). Hence expression (9.124) follows.

We see that different ensembles produce different limit distributions of the block random variables and hence different finite-size scaling functions for the free energy density near criticality. On the other hand, the thermodynamic equivalence of the Gibbs ensembles requires that these different scaling functions have the same leading-order asymptotic behavior when the parameters leave the infinitesimal neighborhood of the critical point in either direction. This crossover to bulk behavior can be easily derived from the explicit expressions for the functions $C(x_1^*,x_2^*)$ and $\bar{X}(x_1^*,x_2^*)$.

Near the first order phase transition, when $K > \zeta^{-1}$ and $h = h_1 N^{-1} \to 0$ as $N \to \infty$, the finite-size scaling behavior of the magnetization for the spherical model follows from Eqs. (9.85) and (9.67):

$$m_N^{s\cdot}(K,h,\zeta) \simeq m_0(K,\zeta)\tanh z, \tag{9.135}$$

and for the mean spherical model – from Eqs. (9.37) and (9.101):

$$m_N^{m.s.}(K, h, \zeta) \simeq m_0(K, \zeta) \frac{2z}{\sqrt{1 + 4z^2} + 1}, \qquad (9.136)$$

where the relevant scaled field variable $z = m_0(K, \zeta)hN$ is the ratio of the magnetic energy in the correlated volume (in our case - the entire system) to the thermal energy per degree of freedom. Note that the finite-size scaling function for the magnetization in the equivalent-neighbors mean spherical model, see Eq. (9.136), is the same as the one obtained in the case of nearest-neighbors interaction, see Eqs. (9.40), (9.37), as well as long-range power-law interactions, see Eqs. (8.26), (8.29).

Since the finite-size scaling function for the magnetization is proportional to the derivative of the finite-size scaling function $W_0(z)$ for the free energy density, see Eq. (8.54), from Eqs. (9.135) and (9.136) we obtain

$$-\frac{\partial}{\partial z} W_0^{s.}(z) = \tanh z, \qquad -\frac{\partial}{\partial z} W_0^{m.s.}(z) = \frac{2z}{\sqrt{1 + 4z^2} + 1}. \qquad (9.137)$$

The integration of the above equations yields, apart from a constant term, the finite-size scaling functions for the free energy density at a first order phase transition:

$$
\begin{aligned}
W_0^{s.}(z) &= -\ln(\cosh z), \\
W_0^{m.s.}(z) &= \frac{1}{2}\{-\sqrt{1 + 4z^2} + \ln(1 + \sqrt{1 + 4z^2})\}.
\end{aligned}
\qquad (9.138)
$$

As expected, the result for the finite-size function for the free energy density of the equivalent-neighbors mean spherical model near a first order phase transition coincides with the one obtained in the case of a long-range power-law interaction, see Eq. (8.35).

One may conjecture that the universality classes with respect to the behavior near a first order phase transition are completely independent of the interaction range, be it finite, infinite, or even of the equivalent-neighbors type. However, as we have shown, a given system may have different finite-size scaling functions in different statistical ensembles.

9.4 General conjecture

We have seen that in the case of the equivalent-neighbors spherical models the derivatives of the finite-size scaling function for the free energy are closely related to the limit distributions of triangular arrays of properly normalized block variables. The triangular arrays are defined with the aid of a two-parameter family of Gibbs distributions which approach the critical point simultaneously with the increase in the number of particles in the system. One may conjecture that a similar relationship holds true in the case of more realistic models too. Then one may expect the existence of a special set of values for the normalization exponents of the block random variables and the exponents which determine the rate of convergence of the conjugate thermodynamic parameters to the critical point as the size of the system increases. Given such a set of exponents, well defined smooth limit distributions of the block random variables should exist, parametrized by the amplitudes which determine the deviation from the critical point at fixed size of the system. The same amplitudes would appear, on the other hand, as scaled temperature and field variables which define the finite-size scaling region. The conjecture is that the derivatives of the finite-size scaling function for the free energy are directly related to the moments of the block variables with smooth limit distributions. The conjectured relationship is based on the statistical meaning of the derivatives of the free energy of a finite (although large) system. The large size of the system ensures that the leading asymptotic dependence of the moments of the block variables on the scaled thermodynamic parameters is the same as the dependence of the corresponding moments of the limit distributions. *The existence of finite-size universality classes may be understood now in terms of domains of (partial) attraction of different limit distributions.* On the other hand, given one model system in two Gibbs ensembles which are statistically inequivalent, in the sense that the above mentioned limit distributions are different, then, obviously, the model will have different finite-size scaling functions depending on the ensemble.

Chapter 10

Bulk Quantum Systems

In this book we have already considered the critical behaviour of quantum model systems. Many technical difficulties encountered in the rigorous statistical mechanical theory have been demonstrated in Chapter 2 and Section 3.4. Although the quantum nature of the system causes non-trivial complications, the common wisdom says that it is irrelevant to the critical behavior at nonzero temperatures, in particular to the values of the critical exponents. However, the presence of a zero-temperature phase transition in the system makes the low-temperature critical behavior rich in different regimes. Here, in contrast with the previous considerations, an investigation of critical quantum effects in the context of the finite-size scaling theory will be presented. We shall consider the $O(n)$ quantum φ^4-model in the limit $n \to \infty$ and the quantum mean spherical model. As we shall see bellow, these two models belong to the same universality class and yield a conspicuous possibility to investigate the interplay of quantum and classical fluctuations in an exact manner.

10.1 Finite-size interpretation of quantum effects

To elucidate the statement that near $T_c > 0$ classical fluctuations always dominate, let us compare the thermal energy per degree of freedom, $k_B T$, with the zero-point quantum energy per degree of freedom, $\hbar \nu_0$. Quantum effects are *unimportant* and classical fluctuations are relevant as long as $\hbar \nu_0 \ll k_B T_c$. The frequency of the quantum fluctuations can be evaluated as $\nu_0 \sim 1/\tau_0$, where the related correlation time τ_0 generally diverges as a power low, $\tau_0 \sim (T - T_c)^{-z\nu}$, where ν is the critical exponent of the correlation length and z is the *dynamical critical exponent,*. Therefore, $\hbar\nu_0 \ll k_B T_c$ is always the case sufficiently close to $T_c > 0$, and the *classical critical behavior* dominates in the temperature region close to T_c. However, the width of this region shrinks to zero with the lowering of T_c. In other words, if we have at our disposal a parameter which can be used to decrease T_c, we can attain a regime where the *quantum critical behavior* dominates. Obviously, the quantum critical regime occurs precisely at $0K$ if a zero-temperature phase transition takes place. This is the very situation in almost degenerate systems at energy levels crossing (see, e.g. [Goldenfeld (1992)]), or in low-dimensional systems in which classical fluctuations suppress the critical temperature down to zero. Then the phase transition is accomplished by changing some of the coupling constants in the Hamiltonian instead of the temperature. The special values of these coupling constants identify the position of the *quantum critical point* which is associated with the non-analyticity of some ground state property.

A model system of this type is the reduced Q^4 model, considered in Section 3.4, the critical temperature T_c of which vanishes at a critical value λ_c of the quantum parameter $\lambda = \hbar\nu_0/E_0$, see Fig. 3.2. In this particular case, by considering the self-consistent equation one can see that the quantum effects act in much the same way as an additional component (dimension) of the wavevector space. Although based on a concrete example, the above statement is in fact completely general. This, at least formally, follows from the expression for the partition function of quantum systems as a Feynman path integral in imaginary time $\tau = it$, see e.g. [Parisi (1988)] and [Cardy (1996)],

$$\mathcal{Z} = \int \mathcal{D}[p]\mathcal{D}[q] \exp\left[-\frac{1}{\hbar} \int_0^{\beta\hbar} d\tau \mathcal{L}\{p,q\} \right], \qquad (10.1)$$

where $\mathcal{L}$ is the Lagrangian, and the functional integral is over the degrees of freedom $\{p\}$ and $\{q\}$. The boundary condition in the imaginary time direction is periodic for bosonic and antiperiodic for fermionic degrees of freedom. The Lagrangian itself is a lattice sum (or integral) over a local Lagrangian density and formally can be regarded as a partition function for a $d + 1$-dimensional classical system with slab geometry and thickness $\beta\hbar$. Note that in the limit $\beta\hbar \to \infty$ one obtains a "pure classical" $d + 1$-dimensional system and, hence, the deviation of the temperature from the quantum critical point at $T = 0$ can be considered as a finite-size effect. The opposite limit $\beta\hbar \to \infty$ offer the possibility to write the Hamiltonian as a classical φ^4 model in d dimensions. Below we shall consider the realization of these ideas in the framework of the $O(n)$ quantum φ^4-model and the quantum spherical model.

For details about the relationship between dynamics and thermodynamics, see [Sondhi et. al. (1997)], where the above ideas are illustrated in a different physical context. The reader interested in the theoretical description of quantum effects that are experimentally obseved in quantum paraelectrics may consult [Tosatti and Martoňák (1994)].

10.2 $O(n)$ quantum φ^4-model in the limit $n \to \infty$

The "Hamiltonian" of the quantum φ^4-model we are going to investigate is given by

$$\mathcal{H}\{\varphi\} = \frac{1}{2} \int_0^{1/T} \mathrm{d}\tau \int \mathrm{d}x \left[(\partial_\tau \varphi)^2 + (\nabla\varphi)^2 + r_0\varphi^2 + \frac{u_0}{2n}\varphi^4 \right], \qquad (10.2)$$

where φ is a short-hand notation for the space-time dependent n-component field $\varphi(x,\tau)$, u_0 and r_0 are model constants, T is the temperature. In Eq. (10.2) we have assumed $\hbar = k_B = 1$ and the size scale is measured in units in which the velocity of excitations $c = 1$. Under the considered periodic boundary conditions one has the Fourier transformation

$$\varphi(x,\tau) = \sqrt{\frac{T}{V}} \sum_{\mathbf{k},\omega_l} \varphi(\mathbf{k},\omega_l) \exp\left(\mathrm{i}\mathbf{k}\cdot x - \mathrm{i}\omega_l\tau\right), \qquad (10.3)$$

where $\omega_l = 2\pi T l$ (with $l = 0,\pm 1,\pm 2,\cdots$) are the Matsubara frequencies for bosonic systems, $\mathbf{k}$ is a discrete vector with components $k_i = 2\pi n_i/L$, $n_i = 0,\pm 1,\pm 2,\cdots$ and a cutoff Λ, and $V = L^d$ is the volume of the system.

The partition function for the Hamiltonian (10.2) reads

$$\mathcal{Z} = \int \mathcal{D}[\varphi] \exp\left(-\mathcal{H}\{\varphi\}\right). \qquad (10.4)$$

By using a standard decoupling procedure based on the Stratonovich identity [Stratonovich (1957)] (see also [Hubbard (1959)])

$$\exp\left[-\frac{u_0}{4n}\int_0^{1/T}d\tau\int dx\,(\varphi^2)^2\right]$$

$$= \text{const.}\int \mathcal{D}[\psi]\exp\left[\int_0^{1/T}d\tau\int dx\left(-\frac{1}{2}i\,\psi\varphi^2 - \frac{n}{4u_0}\psi^2\right)\right] \quad (10.5)$$

which introduces an auxiliary field ψ in (10.4), one obtains

$$\mathcal{Z} = C_1\int \mathcal{D}[\psi]\mathcal{D}[\varphi]\exp\left\{-\frac{1}{2}\int_0^{1/T}d\tau\int dx\left[(\partial_\tau\varphi)^2 + (\nabla\varphi)^2 + \right.\right.$$

$$\left.\left. + r_0\varphi^2 + i\,\psi\varphi^2 + \frac{n}{2u_0}\psi^2\right]\right\} \quad (10.6)$$

with some constant C_1. Now the integration by parts with account of the periodicity of φ yields

$$\mathcal{Z} = C_1\int \mathcal{D}[\psi]\mathcal{D}[\varphi]\exp\left\{-\frac{1}{2}\int_0^{1/T}d\tau\int dx\left[(\varphi, A\varphi) + \frac{n}{2u_0}\psi^2\right]\right\},$$

$$(10.7)$$

where $A = (-\partial_\tau^2 - \nabla^2 + r_0^2 + i\,\psi)$, and $(\varphi, A\varphi)$ denotes the inner product in the corresponding functional space. By using the fact that the field φ has n-components, we decompose the integral over φ into a n-dimensional Gaussian integral which can be taken with the aid of the identity

$$\int \mathcal{D}[\varphi]\exp\left[-\frac{1}{2}(\varphi, A\varphi)\right] = C_2(\det A)^{-1/2}. \quad (10.8)$$

Next, by assuming that the field ψ is time and space independent, and by using the identity

$$\det A = \exp(\text{Tr}\ln A) \quad (10.9)$$

the formal expression for $\mathcal{Z}$ can be written in the form

$$\mathcal{Z} = C_3\int \mathcal{D}[\psi]\exp\left[-\frac{n\beta V}{4u_0}\psi^2 - \frac{n}{2}\,\text{Tr}\ln[r_0 + i\,\psi - \partial_\tau^2 - \nabla^2]\right]. \quad (10.10)$$

For large n one can use the saddle-point method to evaluate the integral over ψ.

In the momentum space representation, starting from Eq. (10.10) and performing some straightforward calculations, one finally obtains for the free energy density $f = -(T/nV)\mathrm{Tr}\ln \mathscr{Z}$ (per unit volume and per field component) the following result (up to the irrelevant constant $\mathcal{C}_3$)

$$f = -\frac{1}{4u_0}(\phi - r_0)^2 + \frac{1}{V}\sum_{\mathbf{k}} \ln 2\sinh\left[\frac{1}{2T}\sqrt{\phi + |\mathbf{k}|^2}\right]. \qquad (10.11)$$

Here, for convenience, we have used the shifted field $\phi \equiv i\psi + r_0$ instead of ψ itself; it is understood that ϕ is the solution of the corresponding saddle-point equation. With the aid of the identity

$$\frac{\coth y}{y} = \sum_{m-\infty}^{\infty} \frac{1}{y^2 + \pi^2 m^2}, \qquad (10.12)$$

the saddle-point equation can be written in the form

$$\phi = r_0 + u_0\frac{T}{V}\sum_{\mathbf{k},m} \frac{1}{\phi + (2\pi m T)^2 + |\mathbf{k}|^2}. \qquad (10.13)$$

Eqs. (10.11) and (10.13) are the starting point of the study of finite-size and quantum effects in the model (10.2).

To investigate the bulk quantum critical behaviour of the model at zero temperature, we pass in Eqs. (10.11), (10.13) from summation to integration according to the rules

$$T\sum_{\omega_n} \to \int_{-\infty}^{\infty} \frac{d\omega}{2\pi} \quad \text{and} \quad \frac{1}{V}\sum_{\mathbf{k}} \to \int \frac{d^d\mathbf{k}}{(2\pi)^d},$$

to obtain the ground state energy density

$$f_0 = -\frac{1}{4u_0}(\phi - r_0)^2 + \int \frac{d\omega}{4\pi} \int \frac{d^d\mathbf{k}}{(2\pi)^d} \ln\left[\phi + \omega^2 + |\mathbf{k}|^2\right], \qquad (10.14)$$

and the saddle-point equation

$$\phi = r_0 + u_0 \int \frac{d\omega}{2\pi} \int \frac{d^d\mathbf{k}}{(2\pi)^d} \frac{1}{\phi + \omega^2 + |\mathbf{k}|^2}. \qquad (10.15)$$

These equations contain all the necessary information from which we can extract the critical behaviour at zero temperature. The net quantum effect is the appearance of one extra dimension; the quantum critical point is given by

$$r_{0c} = -\frac{u_0}{2} \int \frac{\mathrm{d}^d \mathbf{k}}{(2\pi)^d} |\mathbf{k}|^{-1}. \tag{10.16}$$

The above integral is infrared convergent only at $d > d_l$, where $d_l = 1$ is the *lower quantum critical dimension*.

If one wants to have a cutoff independent theory in the limit $\Lambda \to \infty$, then one has to regulate the divergent summation over $\mathbf{k}$ in Eq. (10.13). A standard way of doing this is to use the relativistic Pauli - Villars regularization procedure, see [Zinn-Justin (1996)]. Below we sketch this procedure on the nice example of the two-dimensional bulk system, see also [Chubukov et. al. (1994)]. From a mathematical point of view one has to replace Eq. (10.13) by the equation

$$\phi = r_0 + \sum_m \int_{R^2} \frac{\mathrm{d}^2 \mathbf{k}}{(2\pi)^2} \left[\frac{u_0 T}{\phi + (2\pi m T)^2 + |\mathbf{k}|^2} - \frac{u_0 T}{\Lambda^2 + (2\pi m T)^2 + |\mathbf{k}|^2} \right]. \tag{10.17}$$

The momentum integration is now convergent in the ultraviolet limit, and can be performed. The subsequent summation over m is also tractable with the aid of the identity

$$\ln \frac{\sinh b}{\sinh a} = \frac{1}{2} \sum_{m=-\infty}^{\infty} \ln \frac{b^2 + \pi^2 m^2}{a^2 + \pi^2 m^2} \qquad (ab > 0), \tag{10.18}$$

and the result is

$$\phi = r_0 + \frac{u_0 T}{2\pi} \ln \frac{\sinh(\Lambda/2T)}{\sinh(\phi^{1/2}/2T)}. \tag{10.19}$$

In the low-temperature limit $T \ll \Lambda$, assuming $\phi \ll r_0$, one obtains

$$\phi^{1/2} \simeq 2T \operatorname{arcsinh} \left\{ \frac{1}{2} \exp \left[-\frac{2\pi}{T} \left(\frac{r_{0c}}{u_0} - \frac{r_0}{u_0} \right) \right] \right\}. \tag{10.20}$$

Note that the cutoff parameter Λ is absorbed in the value of the quantum critical parameter $r_{0c} = -u_0 \Lambda / 4\pi$. We shall study the interesting $T \to 0^+$ behavior of the above equation in the next section.

All the critical exponents of the model under consideration can be easily calculated in the large n limit by expanding the right-hand side of Eq. (10.15) for small ϕ. Note that the model belongs to the universality class of the reduced Q^4 model, see Section 3.4. Due to the similarity of the expressions for the free energy density of these two models, compare Eqs. (10.11) and (3.98), one can directly apply the method of calculation of the critical exponents used there.

Eq. (10.13) has been used in [Morf et. al. (1977)] for calculation of the critical exponents. For a review on the applicability of the model (10.2) to the study of quantum critical phenomena see [D'Auria et. al.]. A finite-size investigation in different regimes, depending on the relative magnitude of the inverse temperature and the finite size of the system, in the case of both short-range and long-range cases has been presented in [Chamati and Tonchev (2000b)].

In the next section we shall analyze in more detail the lattice counterpart of Eq. (10.13). Equations of this type are characteristic of *closed-form approximations* in the theory of phase transitions. Being relatively simple, equations of the type of the mean spherical constraint (3.89) or the self-consistent gap equation (3.103) are used in the theory of various physical phenomena and, as a rule, lead to a similar critical behavior. The central role of such equations is confirmed by saddle-point calculations, like the one mentioned above, or by special selection of Feynman diagrams, as pointed out in Section 1.6.3.

10.3 Quantum mean spherical model: critical behavior

The definition of the quantum mean spherical model is given in Section 3.3. According to the ideas explained above, one can consider the nonzero temperature as an extra finite-size "spatial" dimension. This permits one to fruitfully utilize the ideas of finite-size scaling theory for studying quantum critical phenomena. The expressions for the different thermodynamic quantities can be refined in finite-size scaling forms which, in conjunction with the quantum to classical dimensional crossover rule, can be used to elucidate the nature of quantum effects.

An attractive feature of the present model is its lattice formulation which seems to be more transparent in the finite-size case, since no ultraviolet regularization is necessary and there are no ambiguities associated with the continuum limit. In addition, there exist important differences in the finite-size behavior of continuum models and their lattice prototypes above the upper critical dimension, see, e.g., [Chen and Dohm (1998a)], [Chen and Dohm (1998b)], where the classical case has been considered. It has even been stated there that above d_u the standard $O(n)$ symmetric ϕ^4 field theory does not correctly describe the leading finite-size effects near the critical point of lattice spin systems under periodic boundary conditions.

10.3.1   *Equation of state at low temperatures*

In spite of the simple structure of Eq. (3.89), it takes a rather subtle and careful analysis to extract the necessary physical information contained in it. Here we present some details about the mathematical techniques which can be used in a variety of contexts. First of all we note that the temperature variable $t = (T - T_c)/T_c$, previously used to measure the deviation from the bulk critical temperature T_c, is inconvenient in the study of quantum critical phenomena at $T_c = 0$. Similarly inconvenient is the use of the reduced magnetic field variable $h = H/k_B T$. That is why, hoping that no confusion will arise, we introduce in this chapter new dimensionless temperature t and magnetic field h variables:

$$t \equiv 1/K = T/J \quad (k_B = 1), \qquad h = H/J,$$

and set in the remainder of this chapter $k_B = 1$.

In the thermodynamic limit the d-fold sum over the discrete momentum vector $\mathbf{k}$ in Eq. (3.89)

$$\frac{1}{N} \sum_{\mathbf{k}} \frac{1}{\sqrt{\phi + 2\sum_{i=1}^{d}(1 - \cos k_i)}} \coth\left(\frac{\lambda}{2t} \sqrt{\phi + 2\sum_{i=1}^{d}(1 - \cos k_i)} \right),$$

$$(10.21)$$

where the quantum parameter $\lambda = \sqrt{g/J}$, turns into a d-fold integral over the first Brillouin zone. Then, with the use of the formula (10.12), Eq. (3.89) for the shifted spherical field ϕ can be written as

$$1 = \frac{t}{(2\pi)^d} \sum_{m=-\infty}^{\infty} \int_{[-\pi,\pi]^d} d^d\mathbf{k} \, \frac{1}{\phi + 2\sum_{i=1}^{d}(1 - \cos k_i) + b^2 m^2} + \frac{h^2}{\phi^2},$$

$$(10.22)$$

where $b = 2\pi t/\lambda$.

Note that the additional "finite-size" dimension in the above expression represents a complication in comparison with the classical case due to the anisotropic form of the denominator in the integrand. In the long-wavelength approximation, for example in the $U(n)$ quantum φ^4-model, the additional dimension looks just like the spatial dimensions and the isotropy is restored, see Eq. (10.13).

By using the identity (4.80) we rewrite the first term in the right-hand side of Eq. (10.22) in the form

$$\mathcal{J}_d(\phi, t) = \frac{t}{(2\pi)^d} \sum_{m=-\infty}^{\infty} \int_{[-\pi,\pi]^d} d^d\mathbf{k} \int_0^{\infty} dx$$

$$\times \exp\left\{-x\left[\phi + 2\sum_i (1 - \cos k_i) + \left(\frac{2\pi t}{\lambda}\right)^2 m^2\right]\right\}. \tag{10.23}$$

Now the integrals over $\mathbf{k}$ can be taken and Eq. (10.23) simplifies to

$$\mathcal{J}_d(\phi, t) = t \sum_{m=\infty}^{\infty} \int_0^{\infty} dx \exp\left\{-x\left[\phi + 2d + \left(\frac{2\pi t}{\lambda}\right)^2 m^2\right]\right\} [I_0(2x)]^d, \tag{10.24}$$

where $I_0(\cdot)$ is the modified Bessel function.

In order to investigate the low-temperature effects in the model system with Hamiltonian (3.85), we apply the Jacobi identity (4.129) to Eq. (10.24) and obtain

$$\mathcal{J}_d(\phi, t) = \lambda \mathcal{J}_d(\phi) + \frac{\lambda}{2\pi^{1/2}} \sum_{m=-\infty}^{\infty}{}' \int_0^{\infty} d\frac{x}{x^{1/2}}$$

$$\times \exp\left[-x(\phi + 2d) - \left(\frac{\lambda}{2tx}\right)^2 m^2\right] [I_0(2x)]^d, \tag{10.25}$$

Here we have introduced the function

$$\mathcal{J}_d(\phi) = \frac{1}{2(2\pi)^d} \int_{[-\pi,\pi]^d} d\mathbf{k} \left(\phi + 2\sum_{i=1}^{d}(1 - \cos k_i)\right)^{-1/2}, \tag{10.26}$$

and the prime means that the term with $m = 0$ has been omitted.

At sufficiently low-temperatures ($\lambda/t \gg 1$) we can use the asymptotic form of the modified Bessel function [Singh and Pathria (1985a)]

$$I_\nu(x) \simeq \frac{e^{x-\nu^2/2x}}{\sqrt{2\pi x}} \left[1 + \frac{1}{8x} + \frac{9 - 32\nu^2}{2!(8x)^2} + \cdots\right], \tag{10.27}$$

which upon substitution in Eq. (10.22) yields

$$1 = \lambda \mathcal{J}_d(\phi) + \frac{\lambda \phi^{(d-1)/2}}{(4\pi)^{(d+1)/2}} \sum_{m=-\infty}^{\infty}{}' \frac{K_{\frac{d-1}{2}}(\phi^{1/2} m\lambda/t)}{(\phi^{1/2} m\lambda/t)^{\frac{d-1}{2}}} + \frac{h^2}{\phi^2}. \tag{10.28}$$

Here $K_\nu(x)$ is the MacDonald function (second modified Bessel function).

The function $\mathcal{J}_d(\phi)$ can be investigated in a way similar to that used in Section 3.1.2 for the Watson type integral $W_d(\phi)$. The result is that when $1 < d < 3$ and $\phi \ll 1$, one can use the approximation

$$\mathcal{J}_d(\phi) \simeq \mathcal{J}_d(0) - (4\pi)^{-(d+1)/2} \left| \Gamma\left(\frac{1-d}{2}\right) \right| \phi^{(d-1)/2}, \tag{10.29}$$

which leads to the conclusion that at zero temperature the system exhibits a phase transition driven by the parameter λ at the quantum critical point:

$$\lambda_c = 1/\mathcal{J}_d(0). \tag{10.30}$$

Thus, according to Eqs. (10.29) and (10.30), Eq. (10.22) takes the final form $(1 < d < 3)$

$$\begin{aligned} \frac{1}{\lambda} - \frac{1}{\lambda_c} &= -\frac{1}{(4\pi)^{(d+1)/2}} \left| \Gamma\left(\frac{1-d}{2}\right) \right| \phi^{(d-1)/2} \\ &+ \frac{2}{(4\pi)^{(d+1)/2}} \phi^{(d-1)/2} \mathcal{K}\left(\frac{d-1}{2}, \frac{\lambda}{2t}\phi^{1/2}\right) + \frac{h^2}{\phi^2}, \end{aligned} \tag{10.31}$$

where the function

$$\mathcal{K}(\nu, y) \equiv \mathcal{K}_1(\nu|1, y) = 2\sum_{m=1}^{\infty} (ym)^{-\nu} K_\nu(2my) \tag{10.32}$$

has been introduced. The asymptotic behavior of the functions $\mathcal{K}_a(\nu|p, y)$ has been considered in [Chamati et. al. (1998)].

Now, attaining one of our main goals, we can readily show that Eq. (10.31) takes the conjectured scaling form. As a consequence, the correlation length $\xi = \phi^{-1/2}$, being a solution of that equation, can be written in a scaling form too:

$$\xi = (\lambda/t) f_\xi\left(\delta\lambda(t/\lambda)^{-1/\nu}, h(t/\lambda)^{-\Delta/\nu}\right), \tag{10.33}$$

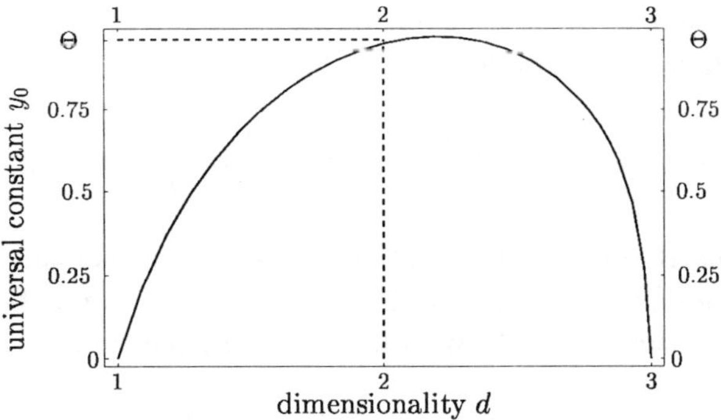

Fig. 10.1   The dependence of the universal value $y_0(d)$ upon the dimensionality d. The constant $\Theta = 0.962424\ldots$ is obtained for the two-dimensional system.

where

$$\delta\lambda = \lambda_c^{-1} - \lambda^{-1}, \qquad \nu = 1/(d-1), \qquad \Delta = \frac{d+3}{2(d-1)}. \tag{10.34}$$

In the remainder of this subsection we shall study the effect of the temperature on the zero-field susceptibility and the equation of state near the quantum critical point $t = 0$, $\lambda = \lambda_c$.

Zero–field susceptibility

From Eq. (10.31) at $h = 0$ we find that the zero-field susceptibility $\chi = (J\phi)^{-1}$ on the line $\lambda = \lambda_c$ $(t \to 0^+)$ is given by

$$\chi = \frac{\lambda_c^2}{Jy_0^2(d)}t^{-2}, \tag{10.35}$$

where $y_0(d)$ is the universal solution of the equation

$$\left| \Gamma\left(\frac{1-d}{2}\right) \right| = 2\mathcal{K}\left(\frac{d-1}{2}, \frac{y}{2}\right). \tag{10.36}$$

The behavior of $y_0(d)$ as a function of the dimensionality d of the system is shown in Fig. 10.1. Note that above the quantum critical point the susceptibility behaves as T^{-2} at low-temperatures.

In what follows we shall investigate Eq. (10.31) between the lower and upper quantum critical dimensions, $1 < d < 3$, and in different regions of the (t, λ) phase diagram. Let us introduce the temperature-shifted critical value of the quantum parameter,

$$\frac{1}{\lambda_c^{\mp}(t)} \simeq \frac{1}{\lambda_c} \mp \frac{1}{2\pi^{(d+1)/2}} \left(\frac{t}{\lambda_c}\right)^{d-1} \Gamma\left(\frac{d-1}{2}\right) |\zeta(d-1)|, \qquad (10.37)$$

where ζ is the Riemann zeta function. One has to take the sign "$-$" for $d < 2$, and the sign "$+$" for $d > 2$. In the first case $(1 < d < 2)$ it is possible to define a *quantum critical region* by the strong inequality

$$\left|\frac{1}{\lambda} - \frac{1}{\lambda_c}\right| \ll \frac{1}{2\pi^{(d+1)/2}} \left(\frac{t}{\lambda_c(t)}\right)^{d-1} \Gamma\left(\frac{d-1}{2}\right) |\zeta(d-1)|. \qquad (10.38)$$

Then, outside the *quantum critical region,* when $\lambda < \lambda_c$ and $y \ll 1$, one has the asymptotic behavior $\mathcal{K}(\nu, y) \sim y^{-1}$. By substituting it into Eq. (10.31), we obtain

$$\chi \simeq J^{-1} \left[\frac{|\Gamma(1 - d/2)|}{(4\pi)^{d/2}\delta\lambda}\right]^{2/(d-2)} t^{2/(d-2)}. \qquad (10.39)$$

From Eq. (10.39) it is clear that the susceptibility diverges according to a power law when the quantum fluctuations are important $(t \to 0^+)$, and that no phase transition driven by λ occurs in a system with dimensionality between 1 and 2.

In the second case $(2 < d < 3)$ one has

$$\chi \simeq J^{-1} \left[\frac{|\Gamma\left(1 - \frac{d}{2}\right)|}{(4\pi)^{d/2}} \frac{\lambda\lambda_c(t)}{\lambda - \lambda_c(t)}\right]^{2/(d-2)} t^{2/(d-2)}, \qquad (10.40)$$

for $\lambda_c(t) < \lambda < \lambda_c$. In obtaining Eq. (10.40), the term of order $y^{-(d-1)/2}$ in the asymptotic expansion of the function $\mathcal{K}(\nu, y)$ has been taken into account in addition to the term $O(y^{-1})$. Now, at finite temperatures there is a phase transition driven by the quantum parameter λ with the critical exponent of the d-dimensional classical spherical model $\gamma = 2/(d - 2)$. Since $d = 2$ is the lower critical dimension in the classical case, this result demonstrates that at $t > 0$ quantum fluctuations are irrelevant to the universal features of the model. However, the validity of the above statement is restricted only to values of λ very close to $\lambda(t)$, when $y \ll 1$. For $\lambda < \lambda_c(t)$ the susceptibility is infinite.

In the region where $\lambda > \lambda_c$ the zero-field susceptibility is given by

$$\chi \simeq J^{-1} \left[\frac{(4\pi)^{(d+1)/2}}{\Gamma\left(\frac{1-d}{2}\right)} \delta\lambda \right]^{2/(1-d)}. \tag{10.41}$$

This result is valid for every d between the lower and upper quantum critical dimensions $(1 < d < 3)$.

The important case $d = 2$ can be solved exactly because of the very simple form of the MacDonald function $K_{1/2}(x) = (\pi/2x)^{1/2} \exp(-x)$. Thus, from Eq. (10.31) we obtain, compare with Eq. (10.20),

$$\phi^{1/2} = \frac{2t}{\lambda} \text{arcsinh} \left[\frac{1}{2} \exp(2\pi\lambda\delta\lambda/t) \right]. \tag{10.42}$$

Eq. (10.42) yields for the susceptibility

$$\chi \simeq J^{-1} \frac{\lambda^2}{t^2} \exp\left(\frac{4\pi\lambda_c}{t} |\delta\lambda| \right), \tag{10.43}$$

when $(2\pi/t)|\lambda/\lambda_c - 1| \gg 1$ and $\lambda < \lambda_c$. At

$$\lambda = \lambda_c = 1/\mathcal{J}_2(0) \simeq 3.1114 \tag{10.44}$$

one obtains

$$\chi = \frac{1}{J\Theta^2} \left(\frac{\lambda_c}{t} \right)^2, \tag{10.45}$$

where the universal constant $\Theta \equiv y_0(2)$ has been first obtained in the framework of the 3-dimensional classical mean spherical model with one finite dimension, see Eq. (4.156). Finally, when $(2\pi/t)(\lambda/\lambda_c - 1) \gg 1$ we obtain

$$\chi \simeq \frac{1}{J(4\pi\delta\lambda)^2} \left\{ 1 + \frac{2t}{\pi\lambda_c\delta\lambda} \exp\left[-\frac{4\pi\lambda_c}{t} |\delta\lambda| \right] \right\}. \tag{10.46}$$

Here the first term in the right-hand side corresponds to the particular case $d = 2$ of Eq. (10.41).

From Eqs. (10.43) - (10.46) one can transparently see the different behavior of the susceptibility $\chi(T)$ in the tree regions: (a) *renormalized classical region* with exponentially divergence as $T \to 0$, (b) *quantum critical region* with $\chi(T) \sim T^{-2}$ and crossover lines $T \sim |\lambda - \lambda_c|$, and (c) *quantum disordered region* with temperature independent susceptibility (up to exponentially small corrections) as $T \to 0$. The location of these regions

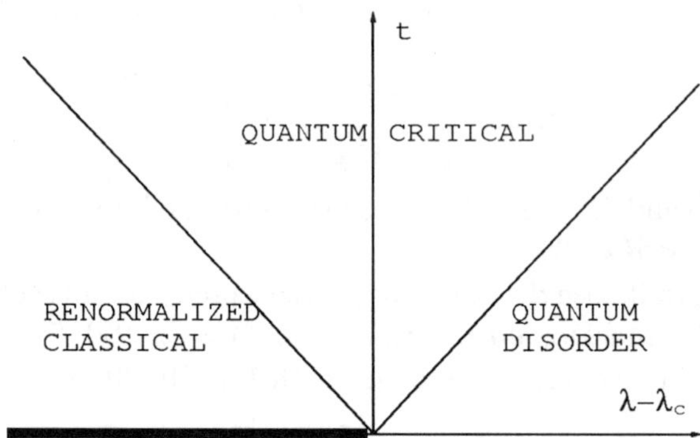

Fig. 10.2   The phase diagram of the two-dimensional model and crossover lines in the plane of the normalized temperature t and quantum parameter λ. One distinguishes renormalized classical, quantum critical and quantum disordered regions. Long-range order is present only at $t = 0$ when $\lambda < \lambda_c$.

is shown in Fig. 10.2. It is useful to clarify the above found bulk critical behavior of χ in the context of finite-size scaling theory. First, from Eq. (10.37) one can see that the exponent ψ of the temperature shift in the quantum parameter λ_c is equal to $\nu^{-1} = d - 1$, in accordance with the finite-size scaling prediction. Second, let us consider the critical behavior of the $(d + 1)$-dimensional classical model with $L_\tau \sim 1/T$ playing the role of a finite-size in the imaginary time direction, i.e., by assuming the slab geometry $\infty^d \times L_\tau$. Finite-size scaling calculations (see, e.g., [Allen and Pathria (1989)]) of the susceptibility at the critical point (in the case under consideration this is $\lambda = \lambda_c$) give:

(i) for $1 < d < 2$,[*]

$$\chi(\lambda_c) \sim L_\tau^{\gamma_\lambda/\nu_\lambda}, \tag{10.47}$$

where $\gamma_\lambda = 2\nu_\lambda = 2/(d - 1)$ and $\chi(\lambda_c) \sim T^{-\gamma_T}$, $\gamma_T = 2$;

(ii) for $2 < d < 3$,

$$\chi(\lambda) \sim L_\tau^{(\gamma_\lambda - \dot{\gamma}_\lambda)/\nu_\lambda} |\lambda - \lambda(L_\tau)|^{-\dot{\gamma}_\lambda}, \tag{10.48}$$

where the exponent $\dot{\gamma}_\lambda = 2/(d - 2)$ pertains to the d infinite-size (spatial) dimensions.

---

[*]The subscripts T and λ mean that the critical exponents are with respect to the temperature and the quantum parameter, respectively.

Note that in the last case Eq. (10.37) yields $\lambda_c - \lambda(L_\tau) \sim t^{d-1}$, which again leads to $\chi(\lambda_c) \sim t^{-\gamma_T}$.

Thus, the different types of critical exponents, depending on λ and T, are related by $\gamma_T = \gamma_\lambda/\nu_\lambda$, or $\gamma_T = \psi\gamma_\lambda$, where $\psi = 1/\nu_\lambda$ is the crossover exponent.

Equation of state

The equation of state of the model system described by Hamiltonian (3.85) can be obtained near the quantum critical point by expressing in Eq. (10.31) the shifted spherical field ϕ in terms of the magnetization density m through the relation $m = h/\phi$. Then the equation of state takes the scaling form

$$1 + \frac{\delta\lambda}{m^{1/\beta}} - \frac{1}{(4\pi)^{(d+1)/2}} \left(\frac{h}{m^\delta}\right)^{1/\gamma} \left[\left|\Gamma\left(\frac{1-d}{2}\right)\right|\right.$$
$$\left. - 2\mathcal{K}\left(\frac{d-1}{2}, \frac{\lambda}{2}\left(\frac{m^{\nu/\beta}}{t}\right)\left(\frac{h}{m^\delta}\right)^\beta\right)\right] = 0. \tag{10.49}$$

Hence we conclude that near the quantum critical point Eq. (10.49) has the general form

$$h = m^\delta f_h(\delta\lambda m^{-1/\beta}, (t/\lambda)^{1/\nu}m^{-1/\beta}), \tag{10.50}$$

where $f_h(x,y)$ is some scaling function. Furthermore, $\gamma = 2/(d-1)$, $\nu = 1/(d-1)$, $\beta = 1/2$ and $\delta = (d+3)/(d-1)$ are the familiar bulk critical exponents for the $(d+1)$–dimensional classical spherical model. E-q. (10.50) provides a direct verification of the finite-size scaling hypothesis, for a system of finite-size $L_\tau \sim 1/T$ in the temporal dimension, in conjunction with the classical-to-quantum dimensional crossover.

Next we aim at obtaining an explicit expression for the scaling function $f_h(x,y)$ in the neighborhood of the quantum critical point. This can be achieved in the case $(t/\lambda)(h/m)^{1/2} \ll 1$ by using the asymptotic form of $\mathcal{K}(\nu,y)$, which yields at $d \neq 2$,

$$f_h(x,y) = \left\{\frac{(4\pi)^{d/2}y^{-\nu}}{\Gamma\left(1-\frac{d}{2}\right)}\left[1 + x + \frac{\zeta(d-1)}{2\pi^{(d+1)/2}}\Gamma\left(\frac{d-1}{2}\right)y\right]\right\}^{2/(d-2)}.$$

In the special case $d = 2$ the scaling function is given by the expression

$$f_h(x, y) = 4y^2 \left\{ \operatorname{arsinh} \left[\frac{1}{2} \exp \left(2\pi \frac{1+x}{y} \right) \right] \right\}^2. \qquad (10.51)$$

At $x = 0$ $(\lambda = \lambda_c)$ and $y \gg 1$ (fixed low temperature and $h \to 0^+$) Eq. (10.51) reduces to

$$f_h(0, y) \simeq \Theta^2 y^2, \qquad (10.52)$$

where the constant Θ is defined in Eq. (4.156). In the region $x < -1$ and $y \ll 1$ (fixed weak field and $t \to 0^+$) the corresponding scaling function has the asymptotic form

$$f_h(x, y) \simeq y^2 \exp \left(-4\pi \frac{|x| - 1}{y} \right). \qquad (10.53)$$

In the region $x > -1$ and $y \ll 1$ we obtain

$$f_h(x, y) \simeq 16\pi^2 (x + 1)^2 \left[1 + \frac{y}{\pi(1 + x)} \exp \left(-4\pi \frac{1 + x}{y} \right) \right]. \qquad (10.54)$$

Note that at zero temperature $(y = 0)$ the scaling function (10.54) becomes identical with that of the three-dimensional classical spherical model, in accordance with the dimensional crossover rule.

10.3.2　*The free energy at low temperatures*

In the preceding subsection we have confined our consideration to the equation for the shifted spherical field (10.22). Here our aim is to obtain closed-form expressions for the free energy density of the model at different spatial dimensions d.

The expression for the free energy can be calculated from Eq. (3.87) by using the identity (10.18), Eq. (4.83) and the Jacobi identity (4.129). After some algebra, in the limit $L \to \infty$ taken at low temperatures $\lambda/t \gg 1$ and zero field $h = 0$, the expression for the free energy density (3.87) can be written in the dimensionless form

$$\tilde{f}_\infty(t, \lambda) := (gJ)^{-1/2} f_\infty(t, \lambda, 0) = \frac{1}{2} \left[a(\phi) - \lambda^{-1}(\phi + 2d) - s(\phi) \right], \qquad (10.55)$$

where $b = 2\pi t/\lambda$,

$$a(\phi) = \frac{1}{2\sqrt{\pi}} \int_0^\infty \frac{dx}{x^{3/2}} e^{-x\phi} \left\{ 1 - \left[e^{-2x} I_0(2x) \right]^d \right\} + \sqrt{\phi}, \qquad (10.56)$$

$$s(\phi, b) = 2 \int_0^\infty \frac{dx}{x} (4\pi x)^{-(d+1)/2} e^{-x\phi} R\left(\frac{\pi^2}{xb^2} \right), \qquad (10.57)$$

and

$$R(x) = \sum_{m=1}^\infty \exp(-xm^2). \qquad (10.58)$$

In Eq. (10.55) the value of ϕ is to be determined from the global minimum condition. The above expressions are valid at *any* d.

In the remainder we shall consider dimensions $d = 1, 2$ and $d = 4$ [†], which represent the three important cases: absence of a critical point, existence of only a quantum critical point, and existence of both quantum and classical critical points, respectively. It can be shown that:

(a) At $d = 1$

$$a(\phi) = \frac{1}{2} \, {}_2F_1\left(-\frac{1}{4}, \frac{1}{4}, 1, \frac{4}{(2+\phi)^2} \right) \sqrt{2+\phi}, \qquad (10.59)$$

$$s(\phi, b) = -\frac{b}{2\pi^2} \sqrt{\phi} \sum_{m=1}^\infty m^{-1} K_1\left(\frac{2\pi m \sqrt{\phi}}{b} \right), \qquad (10.60)$$

where ${}_2F_1$ is the hypergeometric function.

(b) At $d = 2$ and $\phi \ll 1$

$$a(\phi) \simeq a(0, 2) + \mathcal{J}_2(0)\phi - \frac{1}{6\pi} \phi^{3/2} \qquad (10.61)$$

$$s(\phi, b) = -\left(\frac{b}{2\pi} \right)^3 \left[\frac{\phi^{1/2}}{b} \mathrm{Li}_2\left(e^{-2\pi\phi^{1/2}/b} \right) \right.$$

$$\left. + \frac{1}{2\pi} \mathrm{Li}_3\left(e^{-2\pi\phi^{1/2}/b} \right) \right], \qquad (10.62)$$

[†]In both cases $d = 3$ and $d = 4$ the system is characterized by a similar phase diagram, but the mathematical treatment of the $d = 3$ case is more difficult due to the appearance of logarithmic corrections.

where $\mathcal{J}_d(\phi)$ is defined by Eq. (10.26), $\mathcal{J}_2(0) \simeq 0.3214$, and $\mathrm{Li}_p(z)$ are the polylogarithm functions defined by the series

$$\mathrm{Li}_p(z) = \sum_{k=1}^{\infty} \frac{z^k}{k^p}, \tag{10.63}$$

with $|z| \le 1$ for $p \ge 2$ and $-1 \le z < 1$ for $p = 1$.

(c) At $d = 4$ and $\phi \ll 1$

$$a(\phi) \simeq a(0,4) + \mathcal{J}_4(0)\phi - \frac{1}{2}r\phi^2 + \frac{1}{30\pi^{3/2}}\phi^{5/2}, \tag{10.64}$$

$$s(\phi,b) = -\left(\frac{b}{2\pi}\right)^5 \left[\frac{\phi}{b^2}\mathrm{Li}_3\left(e^{-2\pi\phi^{1/2}/b}\right)\right.$$

$$\left. + \frac{3\phi^{1/2}}{2\pi b}\mathrm{Li}_4\left(-e^{-2\pi\phi^{1/2}/b}\right) + \frac{3}{4\pi^2}\mathrm{Li}_5\left(e^{-2\pi\phi^{1/2}/b}\right)\right], \tag{10.65}$$

where $\mathcal{J}_4(0) \simeq 0.1891$ and

$$r = \int_0^{\infty} \sqrt{x}\left[e^{-2x}I_0(2x)\right]^4 \mathrm{d}x \simeq 0.0677. \tag{10.66}$$

Now we have the basic expressions needed to analyze the effect of quantum and classical fluctuations on the phase diagram.

10.4   The finite-temperature C-function

In the remainder of this chapter the interplay of quantum and classical fluctuations will be considered on the basis of a finite-temperature generalization of the C-function.

The famous Zamolodchikov's C-theorem pertains to zero-temperature systems [Zamolodchikov (1986), (1987)]. It establishes the existence of a dimensionless function C of the coupling constants with monotonic properties along the renormalization group trajectories. The assumptions presented in the proof are related with the energy-momentum conservation, the rotational and translational symmetries, and positivity in a two dimensional field theory. The behavior of the C-function reflects the role of quantum fluctuations and it is useful in determining the qualitative futures of the theory away from criticality. At the fixed points it takes the value of the central charge of the corresponding conformal field theory. Since the basic assumptions underlying the C-theorem are not specific to two dimensions

only, a considerable interest exists in generalizations of the Zamolodchikov
result to dimensionalities different from two, as well as to nonzero temperatures [Jack (1992)], [Castro Neto and Fradkin (1993)], [Sachdev (1993)],
[Zabzin (1997)], [Petkou and Vlachos (1999)].

10.4.1 *The definition*

Earlier efforts, see [Jack (1992)] and references therein, have been devoted
to finding a version of the C-theorem valid in four dimensions. The approach has been based on a careful investigation of the form of the trace of
the energy-momentum tensor written in terms of finite local composite operators. Despite the possibility of writing expressions for Zamolodchikov's
equations for the C-function similar to the case of two dimensions, it turned
out inconceivable to demonstrate the monotonicity property. Let us note
also that the $3d$ analog of the central charge [Cardy (1987)] is not equivalent to the universal number characterizing the size dependence of the free
energy at the critical point [Sachdev (1993)] (which is always the case in
$2d$ conformal field theory). This fact indicates that a straightforward generalization of Zamolodchikov's C-theorem is not to be expected for general
value of d. Different approaches to the problem have been offered in [Zabzin
(1997)], and it has been shown that no direct relationships exist between
the finite-temperature C-function and the Zamolodchikov C-function at
zero temperature.

We shall use the approach proposed in [Castro Neto and Fradkin (1993)]
for finding a candidate for the C-function. This approach, in some sense
being a thermodynamic one, gives the possibility to solve the problem in
the field of condensed matter physics. Following [Danchev and Tonchev
(1999)], concrete calculations will be performed on the example of the model
considered in Section 10.3. Since the model is exactly solvable, this can be
done in explicit analytical form.

The following dimensionless function

$$C(T, g) = -\beta^{d+1} \frac{v^d}{n(d)} [f(T, g) - f(0, g)], \qquad (10.67)$$

is considered to be an extension of the Zamolodchikov C-function to nonzero
temperatures.[‡] Here $\beta = 1/T$, g is a set of dimensionful coupling constants,

---

[‡]Note that the definition tacitly assumes that the dynamic critical exponent is $z = 1$.

$f(0, g) \equiv E_0(g)$ is the zero-temperature energy density, $f(T, g)$ is the full free energy density (per unit volume), $n(d) = \Gamma((d+1)/2)\zeta(d+1)/\pi^{(d+1)/2}$ for bosons and $n(d) = \Gamma((d+1)/2)\zeta(d+1)(2 - 2^{1-d})/\pi^{(d+1)/2}$ for fermions, v is the characteristic velocity (e.g., the velocity of quasiparticles) in the system. The function $C(T, g)$ is supposed to be *positive*, and, in the regions where the quantum fluctuations dominate, a *monotonically increasing function of the temperature*. Functions analogous to the one defined in Eq. (10.67) have been discussed also in [Sachdev (1993)], [Zabzin (1997)], [Petkou and Vlachos (1999)]. It has been shown [Castro Neto and Fradkin (1993)], [Zabzin (1997)] that from thermodynamic point of view the positivity of the C-function is related to the positivity of the entropy, hence, it is a quite general property, while for the monotonicity only a sufficient condition has been formulated,

$$\mathrm{d}T\, S(T) \le (d+1)[U(T) - E_0], \qquad (10.68)$$

where U is the internal energy and S is the entropy of the system. This condition bounds the entropy from above whereas the thermodynamic laws bound it only from below, i.e., this condition is *not* a consequence of thermodynamics. Supposing analyticity of the specific heat (no phase transition in the considered system), it can be shown that (10.68) is indeed obeyed, and C is a monotonically increasing function of the temperature in the case of any $d = 1$ system. At a general d, the existence of phase transitions in the system and the interplay of quantum and classical fluctuations make the analysis of the C-function behavior rather difficult. That is why, for any $d \ne 1$ the properties of C have been considered on the example of concrete models: free massive field theories (for any d), the $d = 1$ Ising model in a transverse field, and the $d = 2$ quantum nonlinear σ model (QNLσM) in the limit $N \to \infty$ [Castro Neto and Fradkin (1993)], [Sachdev (1993)]. Recently, the value of the C-function at the critical point of the latter model has been calculated as a function of d, when $1 < d < 3$, in [Petkou and Vlachos (1999)].

10.4.2 *Behavior of the C-function*

Let us explicitly demonstrate the crucial role that the existence of quantum ($T = 0$) and/or classical ($T > 0$) critical points plays in the behavior of C in different regions of the phase diagram in the plane temperature - parameter controlling the quantum fluctuations. The basic expressions needed to analyze the behavior of the finite-temperature C-function as defined by Eq. (10.67) will be obtained in the framework of the quantum mean spherical model (3.85).

Case $d = 1$

From Eqs. (10.55), (10.59) and (10.60) it is easy to see that the only nonanalyticity in the behavior of the free energy exists at $\phi = 0$. In order to clarify the character of this nonanalyticity, let us first consider the zero-temperature system. Then, for $0 < \phi \ll 1$ the free energy density can be written as

$$\tilde{f}_\infty(0, \lambda) = \frac{2}{\pi} - \frac{1}{\lambda} - \frac{1}{8\pi}\phi_0 \ln \phi_0 + \phi_0 \left[\frac{3}{8\pi}\left(2\ln 2 + \frac{1}{3}\right) - \frac{1}{2\lambda}\right], \quad (10.69)$$

where

$$\phi_0 = 64 \exp(-4\pi/\lambda) \tag{10.70}$$

is the solution of the corresponding mean spherical constraint. Such type of solution is well known for different problems, e. g., the one dimensional reduced Q^4 model [Plakida and Tonchev (1986)] and $O(n)$ QNLσM in the large n limit, see [Sénéchal (1993)], [Jolicoeur and Golinelli (1994)]. At nonzero temperatures we are interested in regimes which approach from above the zero-temperature behavior. Thus, for any fixed λ, when $T \to 0^+$, the ratio $\sqrt{\phi}/b$ becomes much greater than 1, and the behavior of the nonzero temperature system is be given by

$$\tilde{f}_\infty(t, \lambda) = \frac{2}{\pi} - \frac{1}{\lambda} - \frac{1}{8\pi}\phi \ln \phi + \phi \left[\frac{3}{8\pi}\left(2\ln 2 + \frac{1}{3}\right) - \frac{1}{2\lambda}\right] -$$
$$- \frac{1}{(2\pi)^2}b^{3/2}\phi^{1/4} \exp\left(-\frac{2\pi\sqrt{\phi}}{b}\right), \tag{10.71}$$

where

$$\phi = \phi_0 \left[1 + 2\sqrt{b}\phi_0^{-1/4} \exp\left(-\frac{2\pi\sqrt{\phi_0}}{b}\right)\right]. \tag{10.72}$$

The solution (10.72) has been obtained and discussed in the framework of the QNLσM in [Jolicoeur and Golinelli (1994)], where "Haldane type" antiferromagnets have been considered. From expressions (10.69) and (10.71)

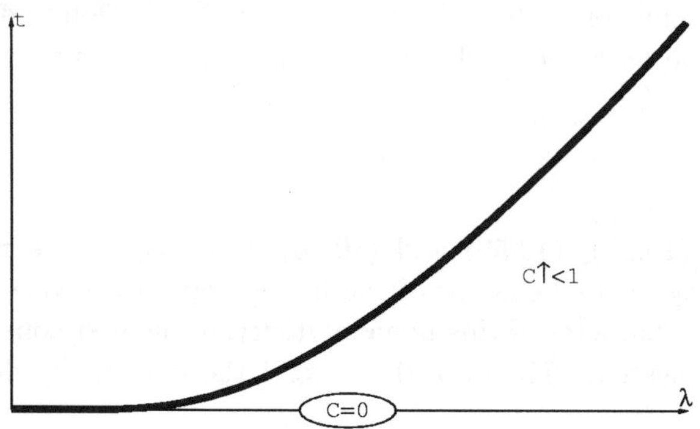

Fig. 10.3 The bold line borders from above the region in the $t-\lambda$ plane where expression (10.73) for the C-function at $d = 1$ is valid. The symbol $C \uparrow$ means that C is a monotonically increasing function of the temperature everywhere that region, starting form $C = 0$ at $t = 0$.

for the free energy, and the definition of the C-function (10.67) we conclude that

$$C(t, \lambda) = \frac{\sqrt{\pi/2}}{6} y_0^{1/2} e^{-y_0},$$ (10.73)

where $y_0 = \lambda \phi_0^{1/2}/t$, and we have set $v = \sqrt{gJ}$. As it is clear from the above expression, C is a positive and monotonically increasing function of the temperature. In fact the C-function obtained here coincides with the C-function of the massive free bosons (at $d = 1$) with mass $\sqrt{\phi_0}$, which is not surprising because of the exponential difference between ϕ and ϕ_0, see Eq. (10.72). Hence, one can consider ϕ as a fixed parameter in Eqs. (10.59) and (10.60). The general case (any d) of free massive bosons actually follows from Eqs. (10.55) - (10.58) by considering ϕ as a fixed parameter related to the mass m of the bosons ($\phi \sim m^2$). In the latter case it is trivial to check that the corresponding C-function is the one obtained in [Castro Neto and Fradkin (1993)].

The results for $d = 1$ are illustrated in Fig. 10.3.

Case $d = 2$

As in the previous case, we are interested in the behavior of the C-function around the critical point only, when $0 < \phi \ll 1$. As it is well known, the critical point is located at $\lambda = \lambda_c$, see Eq. (10.44), and $T = 0$. Then, by setting $n(2) = \zeta(3)/(2\pi)$ and $v = \sqrt{gJ}$ in Eq. (10.67), we obtain from Eqs. (10.61) and (10.62) the C-function

$$C(t,\lambda) = \frac{1}{\zeta(3)} \left[x(y^2 - y_0^2) + \frac{1}{6}(y^3 - y_0^3) + y\mathrm{Li}_2\left(e^{-y}\right) + \mathrm{Li}_3\left(e^{-y}\right) \right],$$

(10.74)

where $x = \pi(1 - \lambda/\lambda_c)/t$; y and y_0 are the solutions of the stationarity equations for the right-hand side of Eq. (10.74) with respect to y and y_0, respectively. Explicitly these solutions are, c.f. Eq. (10.42),

$$y = 2\mathrm{arcsinh}\left(e^{-2x}/2\right)$$

(10.75)

and

$$y_0 = \left\{ \begin{array}{ll} 4|x|, & \lambda > \lambda_c \\ 0, & \lambda \leq \lambda_c \end{array} \right. .$$

(10.76)

The behavior of y in the three regions depicted in Fig. 10.2 is as follows: (i) y tends to zero exponentially fast as a function of x ($x \gg 1$) in the *renormalized classical region*; (ii) $y = O(1)$ ($x = O(1)$) in the *quantum critical region*; and (iii) y diverges as $4x$ for $x \ll -1$ in the *quantum disordered region*. Note that $y \sim (\chi t^2)^{-1/2}$, where χ is the susceptibility of the system. Indeed, the behavior of the C-function also reflects the existence of these three regions.

When $\lambda < \lambda_c$ and $t \to 0$, from Eqs. (10.74) - (10.76) it follows that

$$C(t,\lambda) \simeq 1 - \frac{1}{4\zeta(3)} \exp\left[-4\pi(1 - \lambda/\lambda_c)t^{-1}\right].$$

(10.77)

One explicitly observes the exponentially small corrections to the limit value $C = 1$ at $t = 0$ [Castro Neto and Fradkin (1993)], [Sachdev (1993)], which corresponds to massless bosons in d dimensions.

At $\lambda = \lambda_c$, from Eq. (10.75) one obtains $y = -\ln(2 - \varrho)$, with the "golden mean" $\varrho = (1 + \sqrt{5})/2$, and $C(t,\lambda)$ given by Eq. (10.74) can be significantly simplified. It is easy to show that

$$\exp(-y) = \varrho^{-2} = 2 - \varrho,$$

(10.78)

which reduces the problem to the evaluation of the expression

$$\mathrm{Li}_3(2 - \varrho) - \ln(2 - \varrho)\mathrm{Li}_2(2 - \varrho) - \frac{1}{6}\ln^3(2 - \varrho).$$

(10.79)

Further, by using the Sachdev identity [Sachdev (1993)]

$$\frac{4}{5}\mathrm{Li}_3(1) = \mathrm{Li}_3(2 - \varrho) - \ln(2 - \varrho)\mathrm{Li}_2(2 - \varrho) - \frac{1}{6}\ln^3(2 - \varrho), \qquad (10.80)$$

having in mind that $\mathrm{Li}_3(1) \equiv \zeta(3)$, we conclude that

$$C(t, \lambda_c) = 4/5. \qquad (10.81)$$

This universal *rational* number[§] has been derived for the first time for the $O(n)$ QNLσM in the limit $n \to \infty$ [Sachdev (1993)]. It demonstrates that at the quantum critical point $\lambda = \lambda_c$ the C-function does not depend on the temperature. The discrepancy with the corresponding results in [Castro Neto and Fradkin (1993)] is due to the fact that terms proportional to the difference between y and y_0 in Eq. (10.74) have been neglected there. The above is justified when $y \gg 1$, since then y and y_0 are exponentially close to each other. The analysis of the corresponding equation shows that the last takes place for $x \ll -1$ ($\lambda > \lambda_c$) when $y \sim y_0 \sim 4x$, which is the case of the quantum disorder region. In this case it is easy to see that

$$C(t, \lambda) \simeq \frac{4\pi}{\zeta(3)} \frac{|1 - \lambda/\lambda_c|}{t} \exp\left[4\pi(1 - \lambda/\lambda_c)t^{-1}\right]. \qquad (10.82)$$

The behavior of the C-function in this case is the one for massive free bosons [Castro Neto and Fradkin (1993)]. We observe that C approaches zero exponentially fast in terms of the scaling parameter x. Let us consider now the monotonicity of the C-function. From Eq. (10.74) it follows that

$$\frac{\partial C(t, \lambda)}{\partial t} = -\frac{\pi}{\zeta(3)}(y^2 - y_0^2)(1 - \lambda/\lambda_c)t^{-2}. \qquad (10.83)$$

Since $y > y_0$, we conclude that C is a monotonically increasing function of the temperature when $\lambda > \lambda_c$, and monotonically decreasing function when $\lambda < \lambda_c$. Within exponentially small in the temperature corrections this result coincides with the corresponding one for the QNLσM in the limit $N \to \infty$.

The above results for the behavior of the C-function are illustrated in Fig. 10.4.

[§]Let us note that $\zeta(3)$ is irrational number, as was pointed out by Apéry, 1978 (see [Julia (1990)]). Therefore, none of the intermediate steps suggests that a rational number will be the final result.

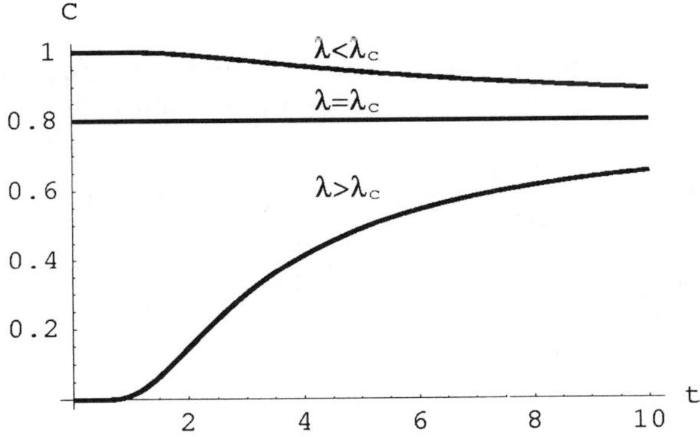

Fig. 10.4   The behavior of the C-function at $d = 2$ is illustrated for $\lambda = 1.5\lambda_c$, $\lambda = \lambda_c$ and $\lambda = 0.5\lambda_c$.

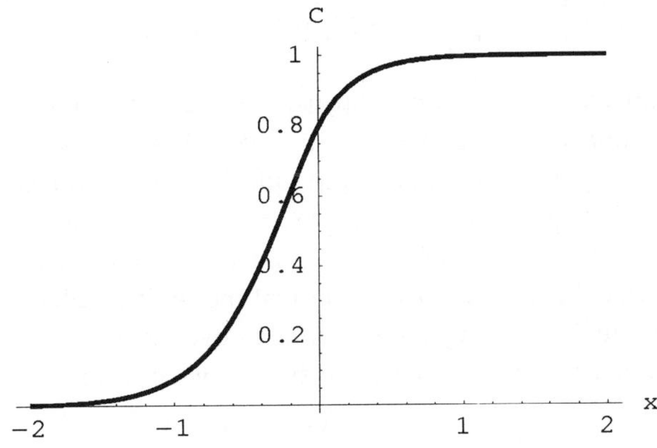

Fig. 10.5   The behavior of C at $d = 2$ as a function of the scaling parameter $x = \pi(1 - \lambda/\lambda_c)/t$.

Finally, it seems worthwhile to mention that, as follows from Eq. (10.74), the C-function is monotonically increasing function of the scaling variable x for any value of t $(\lambda/t \gg 1)$, see Fig. 10.5.

Case $d = 4$

In this case, by setting $n(4) = 3\zeta(5)/(2\pi)^2$ and $v = \sqrt{gJ}$ in Eq. (10.67), from Eqs. (10.64) and (10.65) we obtain the C-function

$$
\begin{aligned}
C(t, \lambda) = \frac{(2\pi)^2}{3\zeta(5)} \Bigg\{ &x(y^2 - y_0^2) + \frac{1}{4} r(y^4 - y_0^4)\frac{\lambda}{t} \\
&+ \frac{1}{(2\pi)^2} \left[y^2 \mathrm{Li}_3\left(e^{-y}\right) + +3y\mathrm{Li}_4\left(e^{-y}\right) + 3\mathrm{Li}_5\left(e^{-y}\right) \right] \Bigg\},
\end{aligned} \quad (10.84)
$$

where

$$
x = \frac{1}{2}\left(\frac{\lambda}{t}\right)^3 \left(\frac{1}{\lambda} - \frac{1}{\lambda_c}\right), \quad (10.85)
$$

and $\lambda_c = 1/\mathcal{J}_4(0) \simeq 5.2882$. Here $y \geq 0$ and $y_0 \geq 0$ are solutions of the stationarity equations for the right-hand side of Eq. (10.84) with respect to y and y_0, respectively. We obtain the following equation

$$
x = -\frac{1}{2}r\frac{\lambda}{t}y^2 + \frac{1}{2(2\pi)^2}\left[y\mathrm{Li}_2\left(e^{-y}\right) + \mathrm{Li}_3\left(e^{-y}\right)\right], \quad (10.86)
$$

for the system at nonzero temperature. For given t and λ the solution of the above equation, if it exists, is unique. The easiest way to see this is simply to verify that its right-hand side is a monotonically decreasing function of y with a maximum $\zeta(3)/(8\pi^2)$ at $y = 0$ for any given t and λ. For $x < \zeta(3)/(8\pi^2)$ the solution formally does not exist, but in a way analogous to that for the classical spherical model it can be shown that the supremum of the free energy density is attained at $y = 0$ in that case, see Section 3.1. For the zero-temperature system we obtain

$$
y_0^2 = \begin{cases} 2t|x|/(r\lambda), & \lambda > \lambda_c \\ 0, & \lambda \leq \lambda_c \end{cases}. \quad (10.87)
$$

We recall that the susceptibility χ of the system is proportional to y^{-2} (if $y > 0$[¶]), which leads to the conclusion that a nonzero-temperature phase transition exists at a given $t_c = t_c(\lambda)$, where $t_c(\lambda)$ is defined by the equation

$$
t_c(\lambda) = \lambda \left[\frac{(2\pi)^2}{\zeta(3)} \left(\frac{1}{\lambda} - \frac{1}{\lambda_c}\right) \right]^{1/3}, \quad \lambda < \lambda_c. \quad (10.88)
$$

[¶]If $y = 0$ the relation between the susceptibility and y is a bit more subtle for dimensionalities above the upper critical one; see, e.g. [Baxter (1982)], Chapter 5.

Note that for $t \leq t_c(\lambda)$ one has $y \equiv 0$. As in the $d = 2$ case, three basically different regimes exist: (i) *renormalized classical region*, where y tends to zero exponentially fast as a function of λ/t; (ii) *quantum critical region*, where y tends to zero algebraically as a function of λ/t, or $y = O(1)$; and (iii) *quantum disordered region*, where $y \gg 1$. Below we analyze in details these three regimes.

(A) Let us first suppose that $y \ll 1$. Then Eq. (10.86) becomes

$$-y^2 \ln(y^2/e) + (4\pi)^2 r \frac{\lambda}{t} y^2 = \theta(t, \lambda). \tag{10.89}$$

where

$$\theta(t, \lambda) = 4\zeta(3) \left[1 + \frac{(2\pi)^2}{\zeta(3)} \left(\frac{\lambda}{t} \right)^3 \delta\lambda \right]. \tag{10.90}$$

Obviously, the above equation may have a positive solution $y \ll 1$ only when $0 < \theta(t, \lambda) \ll 1$. In this region we distinguish between two subregimes: *(a)* when the first term in the left-hand side of Eq. (10.89) dominates, and *(b)* when the second one dominates. In case *(a)* the leading-order solution is

$$y^2 \simeq \frac{\theta(t, \lambda)}{|\ln \theta(t, \lambda)|}, \qquad 0 < \theta(t, \lambda) \ll 1. \tag{10.91}$$

In case *(b)* the solution reads

$$y^2 \simeq \frac{1}{(4\pi)^2 r} \left(\frac{t}{\lambda} \right) \theta(t, \lambda). \tag{10.92}$$

The borderline between these two subregimes:

$$\theta(t, \lambda) = 2(4\pi)^2 r \frac{\lambda}{t} \exp\left[-(4\pi)^2 r \frac{\lambda}{t} + 1 \right], \qquad \lambda/t \gg 1, \tag{10.93}$$

is obtained from the condition that the two terms in the left-hand side of Eq. (10.89) are equal and solve that equation. If $\lambda < \lambda_c$, the function $\theta(t, \lambda)$, given by Eq. (10.90), can be rewritten with the aid of Eq. (10.88) in the form

$$\theta(t, \lambda) \equiv 4\zeta(3) \left[1 - \left(\frac{t_c(\lambda)}{t} \right)^3 \right], \qquad \lambda < \lambda_c. \tag{10.94}$$

One readily sees that the condition $0 < \theta(t, \lambda) \ll 1$ is fulfilled only when t is larger than and very close to $t_c(\lambda)$. Thus, in the $\lambda - t$ plane Eq.

(10.93) determines a line $t = t^*(\lambda)$, $\lambda \le \lambda_c$, which is exponentially close as $\lambda/t \to \infty$ to the line $t = t_c(\lambda)$ and lies above it, see Fig. 10.6. Since $\theta(t, \lambda)$ is a monotonically increasing function of t, one easily verifies that the solution $y(t, \lambda) \ll 1$ of Eq. (10.89) also increases monotonically with t from zero value at $t = t_c(\lambda)$ to

$$y^2(t^*(\lambda), \lambda) = \exp\left[-(4\pi)^2 r \frac{\lambda}{t^*(\lambda)} + 1\right], \qquad \lambda < \lambda_c. \qquad (10.95)$$

at $t = t^*(\lambda)$. Since the leading-order solution (10.91) is exponentially close to zero for parameters in the $\lambda - t$ plane between the lines $t = t_c(\lambda)$ and $t = t^*(\lambda)$, we conclude that this is the *renormalized classical region*.

Consider now the case $\lambda > \lambda_c$ ($\delta\lambda > 0$). The solution given by Eq. (10.92) has been obtained under the conditions that $y \ll 1$ and that the second term in the left-hand side of Eq. (10.89) is much larger than the first term. In view of Eqs. (10.92) and (10.90), the former condition ($y \ll 1$) leads to the strong inequalities

$$t/\lambda \ll 1, \qquad \delta\lambda \ll (t/\lambda)^2, \qquad (10.96)$$

whereas the second condition locates the region of validity of (10.92) above the line $t = t^*(\lambda)$. In this part of the $\lambda - t$ plane the leading-order behavior of y is determined either by the first term in the right-hand side of Eq. (10.90), when $\delta\lambda \ll (t/\lambda)^3$, or by the second one, when $(t/\lambda)^3 \ll \delta\lambda$. In the first case, which includes the line $\lambda = \lambda_c$, from Eqs. (10.92) and (10.90) it follows that $y^2 \sim t$, hence the susceptibility $\chi(t, \lambda) \sim y^{-2}t^{-2}$ diverges as t^{-3} when $t \to 0^+$. In the opposite case, which in view of the second inequality in Eq. (10.96) reads $(t/\lambda)^3 \ll \delta\lambda \ll (t/\lambda)^2$, we obtain $y^2 \sim (\lambda/t)^2 \delta\lambda \ll 1$, hence $\chi(t, \lambda) \gg t^{-2}$. The borderline $t = t_s(\lambda)$ between these two regimes is given by

$$t_s(\lambda) = \lambda \left[\frac{(2\pi)^2}{\zeta(3)} \left(\frac{1}{\lambda_c} - \frac{1}{\lambda}\right)\right]^{1/3}. \qquad (10.97)$$

Note that the line $t = t_s(\lambda)$ is the mirror image of the critical line $t = t_c(\lambda)$ with respect to the axis $\lambda = \lambda_c$, see Fig. 10.6.

(B) Let us now suppose that $y = O(1)$. Then, in the low-temperature region $t \ll 1$, Eq. (10.86) simplifies greatly and within the leading order coincides with the corresponding equation (10.87) for the zero-temperature

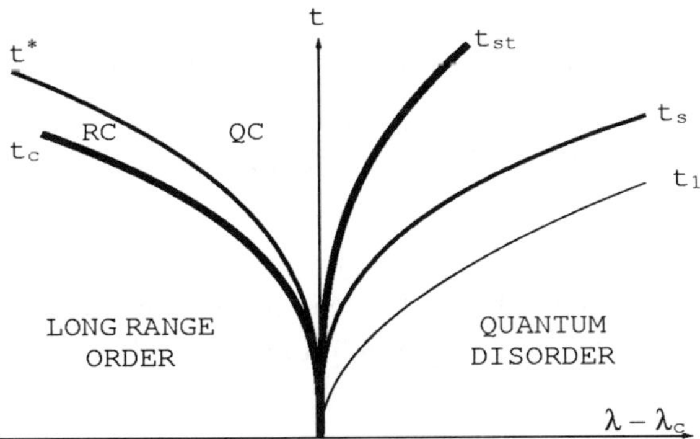

Fig. 10.6   The phase diagram of the model and crossover lines in the case $d = 4$. Long-range order exists below the line $t = t_c(\lambda)$. The line $t = t_{st}(\lambda)$ is the locus of points in the thermodynamic space where $\partial C/\partial t = 0$. The other lines denote crossovers between different regimes described in the text.

system. Its solution is

$$y^2 = \frac{1}{r}\left(\frac{\lambda}{t}\right)^2 \delta\lambda. \tag{10.98}$$

Since $\chi(t,\lambda) \sim (yt)^{-2}$ and $y = O(1)$, one concludes that in this regime $\chi(t,\lambda) \sim t^{-2}$. A formal line in the $\lambda - t$ plane which borders the region in which $y = O(1)$ can be obtained by setting $y = 1$ in Eq. (10.98). Let us denote the so-defined line by $t = t_1(\lambda)$, where $t_1(\lambda) = \lambda(\delta\lambda/r)^{1/2}$, see Fig. 10.6. Summarizing the results for cases *(A)* and *(B)*, we are led to the conclusion that the *quantum critical regime* is observed for values of the parameters t and λ between the lines $t = t^*(\lambda)$ and $t = t_1(\lambda)$.

(C) We are left with the possibility of having $y \gg 1$. In this case one obtains the same solution as the one given by Eq. (10.98). However, now $\chi(t,\lambda) \sim (\delta\lambda)^{-1}$ does not depend on t up to exponentially small in λ/t corrections. We conclude that the region of parameters in the $\lambda - t$ plane below $t_1(\lambda)$ determines the *quantum disordered regime*, see Fig. 10.6.

The existence of the above regions of thermodynamic parameters is reflected by the behavior of the C-function given in Eq. (10.84). We pass now to the investigation of its properties.

First of all we note that in the region of long-range order, when $0 \leq t \leq t_c(\lambda)$, $\lambda \leq \lambda_c$, one has $y = y_0 = 0$, and from Eq. (10.84) one immediately obtains that $C = 1$. Let us emphasize that this result depends only on the existence of long-range order in the system ($y = y_0 = 0$) and not on the dimensionality d. From Eqs. (10.67), (10.55) - (10.57), and the identity

$$\Gamma\left(\frac{s}{2}\right)\pi^{-s/2}\zeta(s) = \int_0^\infty \frac{dx}{x}x^{s/2}R(\pi s), \qquad \mathrm{Re}\, s > 1, \qquad (10.99)$$

it follows that $C = 1$.

Further, we consider the part of the λ - t plane defined by $t > t_c(\lambda)$, where $y > y_0 > 0$. From Eqs. (10.87) and (10.86) it can be easily derived that

$$\frac{\partial C(t,\lambda)}{\partial t} = \frac{2\pi^2}{\zeta(5)}\frac{\lambda}{t^2}(y^2 - y_0^2)\left[\left(\frac{\lambda}{t}\right)^2\delta\lambda - \frac{1}{6}r(y^2 + y_0^2)\right]. \qquad (10.100)$$

The above equation leads us to the conclusion that at any $\lambda < \lambda_c$ the C-function is a *monotonically decreasing function of the temperature*. The same is true also at $\lambda = \lambda_c$ and $t \neq 0$. From the analysis of the solutions of the equations for y and y_0 it is clear that at a fixed λ and $t > t_s(\lambda)$ their *leading-order* asymptotic form is $y = y_0 = (\lambda/t)(\delta\lambda/r)^{1/2}$ (of course, if one takes into account the next-to-the-leading order terms, then $y > y_0$). Setting the above expressions for y and y_0 in the square brackets in Eq. (10.100), we conclude that C is a *monotonically increasing function of t* in the region $t > t_s(\lambda)$, $\lambda > \lambda_c$. It is clear that somewhere between the lines $\lambda = \lambda_c$ and $t = t_s(\lambda)$, $\lambda > \lambda_c$, the derivative of the C-function changes its sign, i.e., somewhere there the C-function has a line $t = t_{st}(\lambda)$ of stationary points. To determine this line we note that from Eqs. (10.85), (10.87) and (10.100) it follows that $\partial C(t,\lambda)/\partial t = 0$ when $y^2 = 5y_0^2 = 5(\lambda/t)^2\delta\lambda/r$. Then, from this relation between y and y_0, by taking into account Eqs. (10.87) and (10.86), we derive that at $t = t_{st}(\lambda)$

$$\frac{(4\pi)^2 r}{5}\frac{\lambda}{t}y^2 = y\mathrm{Li}_2\left(e^{-y}\right) + \mathrm{Li}_3\left(e^{-y}\right). \qquad (10.101)$$

When $t \ll 1$ the above equation reduces to the equation

$$\frac{(4\pi)^2 r}{5}\frac{\lambda}{t}y^2 = \zeta(3) + O(y^2\ln y), \qquad (10.102)$$

which has the leading-order solution $y^2 = 5\zeta(3)t/(16\pi^2 r\lambda)$. Therefore, the

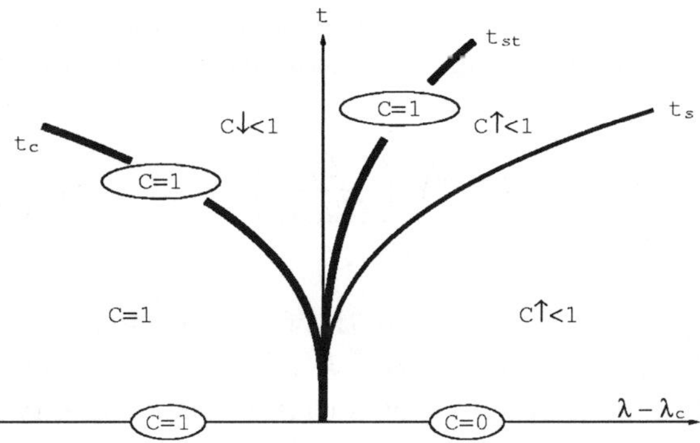

Fig. 10.7 The monotonicity behavior of C as a function of the temperature. The symbols $C \uparrow$ ($C\downarrow$) mean that C is a monotonically increasing (decreasing) function of the temperature. The number in ellipses show the value of C on the corresponding line. In the whole long-range order region $t < t_c(\lambda)$, $\lambda < \lambda_c$, one has $C = 1$.

function $t_{st}(\lambda)$ has the asymptotic form

$$t_{st}(\lambda) \simeq \lambda \left[\frac{(4\pi)^2}{\zeta(3)} \left(\frac{1}{\lambda_c} - \frac{1}{\lambda} \right) \right]^{1/3}. \tag{10.103}$$

Since at the point $\lambda = \lambda_c$, $t = 0$, which lies on the line $t = t_{st}(\lambda)$, one has $y = y_0 = 0$ and $C = 1$, we conclude that $C = 1$ on the whole line $t = t_{st}(\lambda)$, $\lambda > \lambda_c$. It is clear now that C is a *monotonically increasing* function of temperature when $t < t_{st}(\lambda)$, $\lambda > \lambda_c$, and a *monotonically decreasing* function of t when $t > t_{st}(\lambda)$, $\lambda > \lambda_c$, as well as when $t > t_c(\lambda)$, $\lambda < \lambda_c$ (for small t). These results are summarized in Fig. 10.7.

10.4.3   *Finite-size scaling interpretation*

It is interesting to interpret the bulk critical behavior of the C-function in the context of the finite-size scaling theory, see Chapter 4, by introducing the finite "temporal" dimension $L_\tau = \lambda/t$. With this aim we take into account:

(i) The dimensional crossover rule that connects the properties of a given d-dimensional quantum system with those of the corresponding $(d + 1)$-dimensional classical one under the mapping $d \to d + 1$, $L \to L_\tau$, $t \to \lambda$;

(ii) The standard Privman - Fisher hypothesis for the free energy of a finite-size classical system, see Section 4.3, can be used to conjecture

that the free energy density f_∞ of a quantum system with dimensionality $d_l < d < d_u$ (for the model under consideration $d_l = 1$ and $d_u = 3$) should have the form

$$f_\infty(t,\lambda) - f_\infty(0,\lambda) = TL_\tau^{-d} Y\left(\frac{L_\tau}{\xi(0,\lambda)}\right) = T^{1+d}v^{-d}Y\left(\frac{v}{T\xi(0,\lambda)}\right),$$

$$(10.104)$$

where $\xi(0,\lambda)$ is the correlation length of the zero-temperature system, Y is a *universal function* and $v = \sqrt{gJ} = TL_\tau$. As it has been mentioned in Section 10.4.1, one interpretes v as a characteristic velocity in the system. We remind the reader that in order to have no nonuniversal prefactor in front of Y, one considers the dimensionless quantity $\tilde{f}_\infty = \beta f_\infty$ instead of the free energy density f_∞ itself. The normalization of the free energy in Eq. (10.104) simply follows from our choice of L_τ, or, equivalently, v in the system. By inspection of Eqs. (10.74) and (10.104) one concludes that the hypothesis (10.104) is indeed valid for the model under consideration with the standard scaling variable $L_\tau/\xi(\lambda,0) \equiv x = \pi\left(1/\lambda - 1/\lambda_c\right)\lambda/t$, $C(t,\lambda) = X(x) = -Y(x)/n(d)$. It is interesting to note that, despite the lack of hyperscaling at $d_l = 1$, the C-function can again be written as a function of $L_\tau/\xi(\lambda,0)$, see Eq. (10.73), if one identifies $\xi(0,\lambda) = \phi_0^{-1/2}$, where ϕ_0 is given by Eq. (10.70). The case $d = 4 > d_u$ is much more interesting due to the lack of hyperscaling. In the most general case, the C-function cannot be cast in a finite-size scaling form, see Eq. (10.84). That is possible only in an exponentially (in L_τ) narrow neighborhood of the line of finite-temperature phase transitions $t = t_c(\lambda)$, $\lambda < \lambda_c$, where the modified scaling variable is $2x = L_\tau^3/\xi^2(0,\lambda)$, see Eq. (10.85). The standard scaling variable $\tilde{x} = L_\tau/\xi(0,\lambda)$ is restored only for parameters to the right of the curve $t = t^*(\lambda)$ in the $\lambda - t$ plane (see Eq. (10.98) and the comments related to it). This change of scaling variables from x to $\tilde{x}$ is a new point within the finite-size scaling theory. Normally one observes modified finite-size scaling above d_u, see Chapter 6. On the other hand, it has been stated in the context of the $d = 5$ dimensional spherical model with one finite dimension [Barber and Fisher (1973)] that the scaling variable should be the standard one, i.e., $L_\tau/\xi(0,\lambda)$ in our notation. The above results resolve this seeming contradiction: the scaling variable has to be modified very close to the phase boundary, but is the standard one a bit away of it. The physical reasoning for that difference is the existence in the system of a temperature-driven phase transition in addition to the quantum one with respect to λ at $t = 0$. Let us note that all other examples of modified finite-size scaling, considered previously in this book, concern finite-size systems with no sharp phase transition.

10.4.4 *Role of the dimensionality*

The investigation of quantum properties of concrete physical systems, e.g., Kondo lattices, itinerant fermion systems, metallic spin glasses, reveals the existence of a rich set of low-temperature regimes [Sachdev (1996)], [Sondhi et. al. (1997)], [Continentino et. al. (1989)], [Sengupta and Georges (1995)], [Millis (1993)], [Sachdev et. al. (1995)] due to the major role of quantum fluctuations. Despite the simplicity of our model, the corresponding C-function reflects the role that quantum fluctuations play in the system, and its behavior is very rich in possibilities. One generally believes that the monotonicity property of the C-function (to increase with the temperature) holds when the quantum fluctuations "dominate". The real meaning of the term "dominate" turns out to be quite subtle, as we have demonstrated. In fact, we have shown that the region where the C-function remains monotonically increasing as a function of the temperature, and the quantum critical region do essentially intersect but do *not* coincide, see Fig. 10.2. The question of what should be understood as the domination of quantum fluctuations is, indeed, very intriguing. It is a part of the more general problem of a *quantitative* description of the interplay of quantum and critical fluctuations. There exist different views on that issue. The standard one [Sachdev (1996)], [Continentino et. al. (1989)] is based on the ratio of the correlation length to the thermal wavelength of De Broglie. Another possible approach can be based on the behavior of the C-function [Castro Neto and Fradkin (1993)], [Zabzin (1997)]. Furthermore, there is an approach based on the algebra of critical fluctuation operators, developed in [Verbeure and Zagrebnov (1992)], [Verbeure and Zagrebnov (1995)], [Momont et. al. (1997)], where a measure of the "degree of criticality" is introduced in a mathematically rigorous way.

Here we have investigated the behavior of the C-function at $d = 1, 2, 4$. The case $d = 1$ represents the situation with no phase transition and strong quantum fluctuations, $d = 2$ – the one with a quantum critical point at $T = 0$, and $d = 4$ – when there is a line of classical critical points ending up with a zero temperature (quantum) critical point. In fact, these are the most typical cases on which the attention in the literature is focused.

Phase diagrams, qualitatively similar to that in Fig. 10.6, have been obtained for a number of other systems [Sachdev (1996)], [Sondhi et. al. (1997)], [Continentino et. al. (1989)], [Sengupta and Georges (1995)], [Millis (1993)], [Sachdev et. al. (1995)].

Comparing the behavior of the C-function at $d = 1$, $d = 2$ and $d = 4$ we draw the following conclusions:

(a) At $d = 1$ for any fixed λ we have a C-function that is monotonically increasing with the temperature.

(b) At $d = 2$ the above is true only for $\lambda > \lambda_c$.

(c) At $d = 4$ the C-function is a monotonically increasing function of t when $\lambda > \lambda_c$ and t is *small enough*. The monotonicity of the C-function does not change by increasing t *only* at $\lambda = \lambda_c$.

Thus, the region in the parametric space where C remains monotonically increasing with t becomes smaller when d increases.

At any d one can find nontrivial variable(s), function(s) of the temperature and the parameter controlling the quantum fluctuations, in terms of which C is a monotonically increasing function of its variable(s).

We have explicitly demonstrated the crucial role that the dimensionality d and the existence of a phase transition (at high enough d) play in the behavior of the C-function as a function of t. Although we cannot expect a generalization of the Zamolodchikov C-theorem to higher (than $d = 1$) dimensionalities at nonzero temperatures, the investigation of the d-dimensional C-function gives essential information about the balance between the quantum and classical fluctuations: if the quantum fluctuations dominate, it is a monotonically increasing function, and if the classical ones dominate, it is a monotonically decreasing function of the temperature.

Chapter 11

The Casimir Effect

The Casimir effect is a phenomenon common to all systems with interaction forces mediated by fluctuating quantities. This is due to the fact that the geometry of a many-body system, and the properties of its constituent elements, impose some constraints on the fluctuation spectra which make the ground state energy (or the free energy) size and shape dependent. In order to be long-ranged, i.e., to decay with the distance algebraically rather than exponentially, the interaction has to be mediated by massless particles like photons, acoustic phonons, or Goldstone bosons. Posed in the above general form, the Casimir effect is a subject of study in condensed matter physics, quantum electrodynamics, quantum chromodynamics and cosmology.

There exist several recent monographs and reviews on the Casimir effect [Krech (1994)], [Mostepanenko and Trunov (1997)], [Kardar and Golestanian (1999)], [Krech (1999)]. The interested reader may consult them for details of the results in any of the above mentioned fields.

In statistical mechanical systems undergoing a second order phase transition, the fluctuations of the order parameter are correlated over distances set by the correlation length ξ. When a system of finite size L is close enough to the critical point, the correlation length may become of the order of L, which is the main prerequisite for the applicability of the finite-size scaling theory. Usually, the order parameter satisfies some boundary conditions imposed at the surfaces bounding the system, i.e., one has a clear analogy with the above mentioned situation. That is why the analytical theory of the Casimir effect in statistical mechanics is based on the finite-size scaling theory.

11.1   Quantum-mechanical Casimir effect

In 1948 the Dutch physicist H. G. B. Casimir discovers that vacuum fluctuations of the electromagnetic field give rise to an attractive force between
uncharged, perfectly conducting capacitor plates [Casimir (1948)]. What
he found actually is that if the plates are separated by a distance L, they
attract each other with a force per unit area (pressure) $F_{L,\mathrm{Cas}}$ given by
[Casimir (1948)], [Casimir (1953)],

$$F_{L,\mathrm{Cas}} = -\frac{\pi^2}{240}\frac{\hbar c}{L^4}. \tag{11.1}$$

In a recent experiment [Lamoreaux (1997)] this result has been verified
within 5% accuracy*.

The cause of the Casimir effect is in the boundary conditions imposed
by the plates on the zero point fluctuations of the electromagnetic field, as
a consequence of which the vacuum energy of the system is modified in a
distance dependent way. Due to the masslessness of photons, the resulting
interaction has long-range nature.

In a bit more mathematical terms what happens is the following. In
a model without boundary conditions, the eigenvalue of the Hamiltonian
which is associated with the vacuum (or ground) state, known as *zero-point
energy*, is usually discarded because (despite of being infinite) it can be
re-adsorbed in a suitable redefinition of the energy origin. The presence
of neutral conducting plates modifies the energy of the vacuum in such a
way that the difference between the energies with and without boundaries
becomes a measurable quantity. One should take into account that, according to the quantum field theory, the field in the vacuum state does
not really vanish but rather fluctuates. The presence of plates modifies

---

*If L is of the order of 10^{-6} m, $F_{L,\mathrm{Cas}}$ is of the order of 10^{-2} dynes/cm^2. In fact the
measurements have been performed by using conductors in the form of a flat plate and
a sphere, because it turns out very difficult to maintain parallelism with the required
accuracy (10^{-5} rad for plates of 1 cm diameter). If L is the distance of closest approach,
by using the so-called proximity force theorem (for details see [Lamoreaux (1997)] and
references therein) expression (11.1) becomes

$$F_{L,\mathrm{Cas}} = -2\pi R \left(\frac{1}{3}\frac{\pi^2}{240}\frac{\hbar c}{L^3}\right),$$

where R is the radius of the sphere. For more details on Casimir forces in such "curved"
geometries see Section 12.4.

the spectrum of the field, namely, it discretizes the values of the momentum components normal to the plates. As a result the zero-point energy, which is half of the sum of all possible eigenfrequencies compatible with the boundary conditions, changes. Since the vacuum energy density between the plates becomes different from that outside, the plates happen to be under the pressure given by Eq. (11.1). The L-dependence of the force can be inferred on dimensional grounds. The minus sign shows that the force is *attractive*. In general, its sign depends on the exact form of the constraints and on the nature of the fields, and it can be either "minus" (attraction between the plates), as in the considered case, or "plus" (repulsion of the plates). Usually the Casimir force $F_{L,\mathrm{Cas}} \neq 0$ unless special boundary conditions are imposed. Examples of Casimir forces are provided by bag models and by the spherical Einstein Universe.

For more information on the Casimir effect in quantum systems one may consult [Levin and Micha, eds., (1993)] and [Mostepanenko and Trunov (1997)].

11.2 Statistical-mechanical Casimir effect

The confinement of *critical fluctuations* of an order parameter causes long-range forces between the surfaces of a film, which is known as "statistical-mechanical Casimir effect" [Fisher and de Gennes (1978)], [Krech (1994)].

The Casimir force in statistical-mechanical systems is characterized by the excess free energy due to the *finite-size contributions* to the free energy of the system. In the case of a film geometry $L \times \infty^{d-1}$ under given boundary conditions τ imposed in the finite direction, the Casimir force is defined as

$$F_{L,\mathrm{Cas}}^{(\tau)}(T) = -\frac{\partial f_{L,\mathrm{ex}}^{(\tau)}(T)}{\partial L}, \qquad (11.2)$$

where $f_{L,\mathrm{ex}}^{(\tau)}(T)$ is the excess free energy

$$f_{L,\mathrm{ex}}^{(\tau)}(T) = f_{L}^{(\tau)}(T) - L f_{\mathrm{bulk}}(T). \qquad (11.3)$$

Here $f_{L}^{(\tau)}(T)$ is the full free energy per unit area and per $k_B T$, and $f_{\mathrm{bulk}}(T)$ is the corresponding bulk free energy density.

The boundaries influence the system to a depth given by the bulk correlation length $\xi(T)$. When $\xi \ll L$, the Casimir force, being a *fluctuation induced force* between the plates, is negligible. This is always the case above T_c. Note that in statistical-mechanical systems one can turn on and off the Casimir effect merely by changing, e.g., the temperature of the system.

Close to T_c the Casimir effect is long-ranged because of the critical fluctuations, hence it is called *"critical Casimir effect"*. In systems with continuous symmetry below T_c, due to the existence of spin waves excitations (*Goldstone modes*), the correlation length remains infinite and the Goldstone modes give rise to a fluctuation-induced long-range force between the boundaries, i.e., one observes a "noncritical Casimir effect". Fluids with this property are called *correlated fluids*. The most prominent theoretical examples with continuous symmetry are the XY ($O(2)$) and Heisenberg ($O(3)$) models. The most important examples with respect to experimental realizations are liquid ^{4}He below the superfluid transition [Li and Kardar (1991)], [Li and Kardar (1992)] and nematic liquid crystals [Adjari et. al. (1991)], [Adjari et. al. (1992)], [Lyra et. al. (1993)] in the nematically ordered phase[†].

At the bulk critical point T_c, the total free energy of a d-dimensional critical system in the form of a film with thickness L, area A, and boundary conditions (a) and (b) at the opposite surfaces, has the asymptotic form

$$
\begin{aligned}
f_L^{(a,b)}(T_c) \quad \cong \quad & Lf_{\text{bulk}}(T_c) + f_{\text{surf}}^{(a)}(T_c) \\
& + f_{\text{surf}}^{(b)}(T_c) + L^{-(d-1)}\Delta^{(a,b)} + \cdots
\end{aligned}
\qquad (11.4)
$$

as $A \to \infty$, $L \gg 1$. Here f_{surf} is the surface free energy contribution and $\Delta^{(a,b)}$ is *the amplitude of the Casimir interaction*. The L dependence of the Casimir term, the last one in Eq. (11.4), follows from the scale invariance of the free energy; it has been derived in [Fisher and de Gennes (1978)]. The amplitude $\Delta^{(a,b)}$ is *universal*, depending on the bulk universality class and the universality classes of the boundary conditions [Krech (1994)], [Krech and Dietrich (1992)].

For $O(n)$-symmetric model systems ($n \geq 1$), depending on the boundary conditions (a, b), and on the symmetry index n, the excess free energy $f_{L,\text{ex}}^{(a,b)}(T)$ may, or may not, contain contributions independent of L.

[†]In this system the fluctuations of the nematic director are responsible for the long-range nature of the Casimir force. Near the phase transition to the isotropic phase fluctuations of the degree of nematic order and the degree of biaxiality generate short-range corrections to the Casimir force. For more information the interested reader may consult [Ziherl and Žumer (1996)], [Ziherl et. al. (1999)] and references therein.

For Ising-like systems, when $n = 1$, these can be the surface free energies $f_{\text{surf}}^{(a)}(T)$ and $f_{\text{surf}}^{(b)}(T)$, and the interface free energy $\sigma^{(a,b)}(T)$ (for brevity we consider the dependence on the temperature T only). For $O(n)$, $n \geq 2$, models independent of L can be only the contributions stemming from the surface free energies, because the analog of the interface free energy is the helicity modulus $\Upsilon(T)$, see Section 7.2, and the corresponding contribution is of the order $\Upsilon(T)/L$.

Let us consider the *finite-size part* of the excess free energy

$$\Delta f_L^{(a,b)}(T) \equiv f_L^{(a,b)}(T) - L f_{\text{bulk}}(T) - f_{\text{surf}}^{(a)}(T) - f_{\text{surf}}^{(b)}(T).$$

For the "singular part" of $\Delta f_L^{(a,b)}(T)$ one has [Privman (1990a)]

$$\Delta f_{L,\text{sing}}^{(a,b)}(T) = L^{-(d-1)} X_{\text{ex}}^{(a,b)}(a_t t L^{1/\nu}), \tag{11.5}$$

where $t = (T - T_c)/T_c$ is the reduced temperature, a_t is a nonuniversal scaling factor, $X_{\text{ex}}^{(a,b)}$ is *universal* (usually geometry dependent) scaling function, $X_{\text{ex}}^{(a,b)}(0) \equiv \Delta^{(a,b)}$, and ν is the scaling exponent of the correlation length. The nonsingular at T_c part $\Delta f_{L,\text{ns}}^{(a,b)}(T)$ has be assesed as $o(L^{-(d-1)})$; actually, it is supposed to be proportional to L^{-d} [Privman (1990a)]. Summarizing, one has

$$\Delta f_L^{(a,b)}(T) = L^{-(d-1)} X_{\text{ex}}^{(a,b)}(a_t t L^{1/\nu}) + O(L^{-d}), \tag{11.6}$$

when $t L^{1/\nu} = O(1)$. The extension of the above equation to the presence of an external magnetic field is straightforward.

When the system leaves the finite-size scaling region towards high temperatures, $t L^{1/\nu} \to \infty$, one expects that the thermodynamic behavior is approached exponentially fast in L, i.e., it is normally supposed that [Privman (1990a)]

$$\Delta f_L^{(a,b)}(T) = O\left(\exp(-\text{const}\, L/\xi)\right). \tag{11.7}$$

Recently, it has been shown that the leading exponential corrections could stem not only from the singular part of the free energy (when they have to be of the form (11.7)), but from the nonsingular part as well [Chen and Dohm (1999)], when they have a nonuniversal form of the type

$$\Delta f_L^{(a,b)}(T) = O\left(\exp[-c(T)L]\right), \tag{11.8}$$

where $c(T)$ is a nonsingular function.

The situation is more complicated when $tL^{1/\nu} \to -\infty$. In $O(n)$, $n \geq 2$ models, due to the spin wave excitations, below T_c the correlation length remains infinite (in the absence of an external field). It is supposed that if there is no diffuse interface in the system, then [Danchev (1996)], [Danchev (1998)] $\Delta f_L^{(a,b)}(T) = O(L^{-(d-1)})$ is of the same order as in the critical finite-size scaling region. If, due to the boundary conditions an interface is created, then $\Delta f_L^{(a,b)}(T) = \Upsilon(T)L^{-1} + O(L^{-(d-1)})$, where $\Upsilon(T)$ is the helicity modulus. Finally, in the case of an Ising system below T_c, on the ground that the correlation length is finite there, one usually expects all corrections to be exponentially small in L, provided there is no interface in the system. If an interface exists, then some heuristic arguments [Parry et. al. (1991)], [Parry and Evans (1992)] about the form of the *singular* contributions to the excess free energy due to the interface wandering suggest that: when $d < 3$ and $T_w < T \ll T_c$, then $\Delta f_L^{(a,b)}(T) \sim L^{-\tau}$, where $\tau = 2(d-1)/(3-d)$ is the so-called exponent for thermal wandering, and T_w is the wetting transition temperaure [Dietrich (1988)]; when $d \geq 3$, one again expects exponential decay of $\Delta f_L^{(a,b)}(T)$, just as in the case with no interface in the system. The explicit mean field results [Parry and E-vans (1992)] suggest $\Delta f_L^{(a,b)}(T) \sim \exp(-L/2\xi)$. It remains to be seen if the *leading order* finite-size contributions are of universal or nonuniversal character.

Equation (11.4) is valid for both *fluid and magnetic systems at criti-cality*. Prominent examples are: one-component fluid at the liquid-vapor critical point, binary fluid at the consolute point, and liquid ^{4}He at the λ transition point [Krech (1994)].

It should be noted that in contrast with the quantum mechanical Casimir effect, that has been tested experimentally with high accuracy [Lamoreaux (1997)], the critical statistical-mechanical Casimir effect lacks so far a quan-titatively satisfactory experimental verification. For comments on some specific difficulties that the experiment comes across see, [Hanke et. al.]) and the discussion in Section 12.5.

The investigation of the Casimir effect in systems with long-range in-teraction, decreasing at large distances as $r^{-d-\sigma}$, has some peculiarities in comparison with the short-range case. Due to the long-range character of the interaction, an attraction appears between the surfaces bounding the system. In the ordered state one can readily estimate that the increase of the L-dependent part of the excess free energy, due to the direct interparticle interaction, is of the order of $L^{-\sigma+1}$. In the critical region there are still some effects stemming from the direct long-range interaction, on the background of which the fluctuation induced new attraction between the surfaces – the critical Casimir force – develops. In the definition (11.2) used here, which is the common one for systems with short-range interaction, these two effects are superimposed. Therefore, generally speaking, one should expect a crossover from a regime governed by the critical Casimir force (which is of the order of L^{-d}, see Eq. (11.4)), to the one governed by the direct attraction (which of the order of $L^{-\sigma}$); note that if $d = \sigma$ these two effects are of the same order, and each of them will dominate in different temperature region. An interesting case when forces of similar origin are acting simultaneously is provided by wetting: consider a nearly critical wetting layer which intrudes between two noncritical phases, and take into account the effects of long-range correlations and the long-range van der Waals forces [Nightingale and Indekeu (1985)].[‡]

In contrast to temperature driven critical phenomena, quantum phase transitions [Sachdev (1996)], [Sondhi et. al. (1997)], see also Chapter 10, occur at zero temperature with change of some non-thermal control parameter, and the relevant fluctuations are of quantum rather than thermal nature. In such systems one observes a statistical-mechanical Casimir effect in which *critical quantum fluctuations* play essential role.

In quantum systems new features can be observed, since the "temporal direction" formally corresponds to an additional short-range term in the spectrum of their classical analogs, see the discussion following Eq. (10.22). Hence, there is an unavoidable "anisotropy" in the spectrum of a quantum system with long-range interaction. The effect depends on which dimension – the temporal or the spatial one – is the finite-size one. As we shall demonstrate in Section 11.3, when the spatial dimension is finite then the effect is similar to that for classical systems. If the finite dimension is the temporal one, the above effect will not be observed since then the long-range interaction is not affected directly by the "finite size" of the system.

[‡]In this case both finite-size and van der Waals forces give raise to a contribution into the free energy of the wetting layer that depends on its thickness L as L^{-2} at $d = 3$.

11.2.1 *Casimir effect and the C-function: Monotonicity hypotheses*

An interesting point of view on the properties of the excess free energy comes from the finite-temperature generalizations of the Zamolodchikov C-function, considered in Section 10.4. Actually, this function is an *analog of the excess free energy* of the system (11.3).

We recall that according to the standard finite-size scaling theory in the case of periodic boundary conditios, the behavior of $f_{L,\text{ex}}^{(p)}$ close to the critical point $T = T_c$, $h = 0$, is given by

$$f_{L,\text{ex}}^{(p)}(t, h) = L^{-(d-1)} X_{\text{ex}}^{(p)}(a_t t L^{1/\nu}, a_h h L^{\Delta/\nu}). \qquad (11.9)$$

In accordance with the quantum to classical dimensional crossover rule, the statement that the C-function is a positive and a monotonically increasing function of the temperature can be translated in a statement that the universal function $-X_{\text{ex}}^{(p)}$ of the corresponding classical system is positive and a monotonically increasing function of L^{-1}. In Section 10.4 it has been shown that the monotonicity of the C-function is related to the absence of long-range order in the system. The existence of long-range order destroys the general validity of the monotonicity. Hence, we expect the statement formulated for $X_{\text{ex}}^{(p)}$ to be generally valid above T_c for any classical system. Supposing that this is true, and recalling that in the neighborhood of the critical point $X_{\text{ex}}^{(p)}$ is a function of the scaling variables $x_1 = a_t t L^{1/\nu}$ and $x_2 = a_h h L^{\Delta/\nu}$, both of which are monotonically increasing functions of L, we come to the conclusion that *in the vicinity of the critical temperature $(T \geq T_c)$ the excess free energy of a given system is a monotonically increasing function of any of its scaling parameters when the other one is kept fixed*. Since x_1 and x_2 are monotonically increasing functions of the temperature and the magnetic field, respectively, the last implies that $X_{\text{ex}}^{(p)}$ *is a monotonically increasing function of $t > 0$ and h.* It is possible to present some arguments that *the above statement can be extended to the region $t < 0$ for $O(n)$, $n \geq 2$, systems*, in contrast to Ising-like systems. The reasoning about the different behavior of the excess free energy expected in $O(n)$ and Ising-type models is closely related to the well know difference in the behavior of the bulk correlation length $\xi(T)$ in these models: in the Ising model $\xi(T) < \infty$ both below and above the bulk critical temperature, whereas in $O(n)$, $n \geq 2$, models below T_c and in the absence of an external filed $\xi(T)$ is identically infinite due to the existence of soft modes in the system (spin waves). On this basis one expects that away from T_c the function $X_{\text{ex}}^{(p)}$ tends to zero exponentially fast in L (see, e.g., [Privman (1990a)]) for Ising-type models, and, therefore, being of the order of $L^{-(d-1)}$, it cannot be a monotonic function of its scaling parameters in the vicinity of T_c. In $O(n)$, $n \geq 2$, models the finite-size corrections should be essential not only in the vicinity, but also below T_c [Danchev (1996)]. In other words, we expect the monotonicity in the behavior of the correlation length in such models around T_c to be mirrored by a corresponding monotonic behavior of the excess free energy. If an external field $h \neq 0$ is applied, then $\xi(T,h) < \infty$ and, of course, we expect that $X_{\text{ex}}^{(p)} \to 0$ exponentially fast with L again, similarly to the case of Ising-like systems. Since $X_{\text{ex}}^{(p)} < 0$ at any fixed $t < 0$, the last implies that it is a monotonically increasing function of the magnetic field in the sub-critical vicinity of T_c too.

The statements presented above should be considered only as a *plausible hypothesis*, which has to be checked in order to probe the region of its validity. For example, it is questionable whether the monotonicity property of $X_{\text{ex}}^{(p)}$ will still hold if the finite system undergoes a phase transitions of its own. It is reasonable to believe that the hypothesis holds for any $O(n)$, $n \geq 2$, system with $d \leq 3$, when in the finite system with short range interaction there is no sharp phase transition.

In Section 11.2.2 we will show that in the vicinity of T_c the excess free energy scaling function X_{ex} of the three-dimensional mean spherical model is, indeed, a monotonically increasing function of any of its finite-size scaling variables x_1 and x_2 when the other one is kept fixed. The last implies that above T_c the function $X_{\text{ex}}^{(p)}$ monotonically increases with t, h and L; below T_c it monotonically increases with t and h, but decreases with respect L.

Let us turn now to the behavior of $F_{L,\text{Cas}}^{(p)}$. From Eqs. (11.2) and (11.9) it immediately follows that [Danchev (1996)]

$$F_{L,\text{Cas}}^{(p)}(t,h) = L^{-d} X_{\text{Cas}}^{(p)}(x_1, x_2), \tag{11.10}$$

where the finite-size scaling function for the Casimir force is

$$\begin{aligned}
X_{\text{Cas}}^{(p)}(x_1, x_2) &= (d-1)X_{\text{ex}}^{(p)}(x_1, x_2) - \frac{1}{\nu}x_1\frac{\partial}{\partial x_1}X_{\text{ex}}^{(p)}(x_1, x_2) \\
&\quad - \frac{\Delta}{\nu}x_2\frac{\partial}{\partial x_2}X_{\text{ex}}^{(p)}(x_1, x_2).
\end{aligned} \tag{11.11}$$

Note that $X_{\text{Cas}}^{(p)}$ is again a *universal* function of x_1 and x_2. We remind the reader that for finite-size systems this means that $X_{\text{Cas}}^{(p)}$ is the same for all systems of the same universality class, geometry, and boundary conditions. It is believed that if the boundary conditions at the opposite surfaces are the same, then the Casimir force is negative; strictly speaking, for Ising-like systems this is supposed to be true above the wetting transition temperature T_w [Evans and Stecki (1994)], [Parry and Evans (1992)], [Dietrich (1988)], [Evans (1990)]. In the case of a fluid confined between identical walls, the above implies that then the net force between the plates will be attractive at large separations. One of our goals in Section 11.2.2 will be to prove analytically the general expectation that

$$X_{\text{Cas}}^{(p)}(x_1, x_2) < 0 \qquad \text{for any} \quad (x_1, x_2) \in R^2, \tag{11.12}$$

on the example of the three-dimensional mean spherical model. We will also show that *if $T < T_c$ and $H = 0$, the Casimir force is a monotonically increasing function of the temperature.* We believe that these properties are valid for any $O(n)$ model with $n \geq 2$. The existing results for the amplitudes in such models have been obtained by the ϵ-expansion technique up to the first order in ϵ, and in the final expressions ϵ is set equal to one [Krech and Dietrich (1992)]. For periodic boundary conditions [Krech (1994)]

$$\Delta^{(p)} = \Delta_{(1)}^{(p)} \left(1 - \frac{5}{4}\epsilon\frac{n+2}{n+8} \right), \tag{11.13}$$

where $\Delta_{(1)}^{(p)}$ is the one-loop calculation result. As it is clear from Eq. (11.13) at $\epsilon = 1$, for n large enough ($n > 22$) the amplitudes $\Delta^{(p)}$ and $\Delta_{(1)}^{(p)}$ will be even of opposite sign. Therefore it makes no sense to take simply the limit $n \to \infty$ there in order to obtain the spherical model result. So, the direct investigation of the spherical model provides an independent additional information to the field-theoretical results.

11.2.2 *Spherical model results*

Here we consider the only example of a model for which the Casimir force can be investigated exactly as a function of both the temperature and magnetic field scaling variables – the three-dimensional $L \times \infty^2$ mean spherical model with a ferromagnetic short-range interaction, under periodic boundary conditions, see Section 3.1. The main results presented in this section

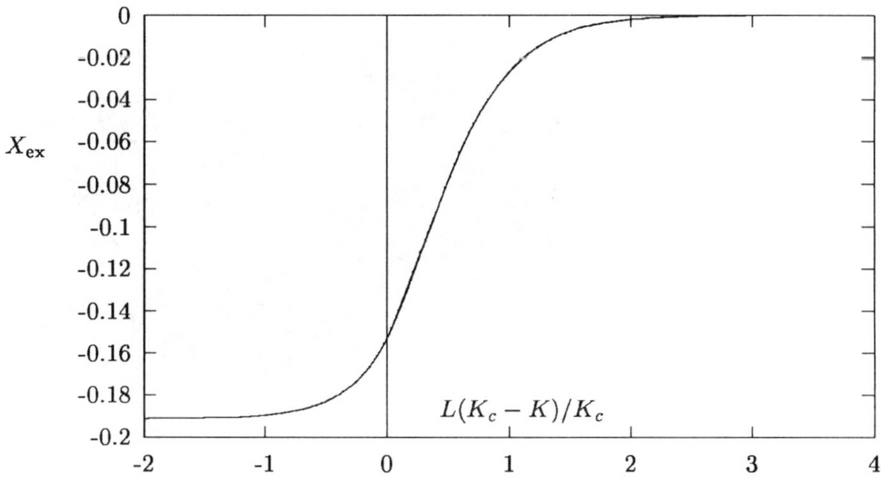

Fig. 11.1 The *universal* finite-size scaling function $X_{ex}(x_1, 0)$ of the zero-field excess free energy as a function of the variable $(1 - K/K_c)L$, where $K_c = 0.25273\ldots$ is the critical coupling.

have been obtained in [Danchev (1996)] and [Danchev (1998)]. For simplicity of notation we omit the superscript (p).

In the case of nearest-neighbor interactions, $d = 3$ and $d' = 2$, we take the finite-size scaling function for the free energy density in the form, compare with Eq. (10.74),

$$X_{ex}(x_1, x_2) = -\frac{1}{2\pi}\left[\frac{1}{6}\left(y_L^3 - y_\infty^3\right) + y_L\, \mathrm{Li}_2\left(e^{-y_L}\right) + \mathrm{Li}_3\left(e^{-y_L}\right)\right]$$
$$+\frac{1}{2}x_2^2(y_\infty^{-2} - y_L^{-2}) - \frac{1}{2}x_1(y_\infty^2 - y_L^2), \quad (11.14)$$

where the finite-size scaling variables x_1 and x_2 are given in Eq. (4.113); $y_L = L/\xi_L$ and $y_\infty = L/\xi_\infty$ are the solutions of the finite-size, Eq. (4.155), and bulk, Eq. (4.117), mean spherical constraints, respectively. The above scaling function $X_{ex}(x_1, x_2)$ can be easily obtained from Eqs. (4.145), (4.148) and Eq. (4.149). Its behavior is illustrated in Fig. 11.1 and Fig. 11.2.

It is convenient to write the finite-size, see Eq. (4.155), and bulk mean sperical constraints in the form

$$x_1 = \frac{1}{2\pi}\ln[2\sinh(y_L/2)] - x_2^2/y_L^4 \equiv g_L(x_2, y_L), \quad (11.15)$$

     The Casimir Effect

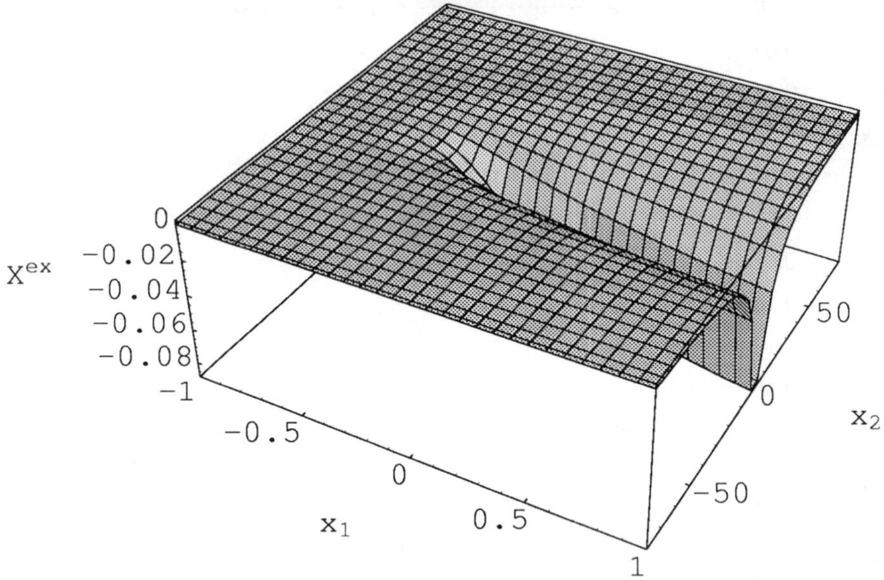

Fig. 11.2   The *universal* finite-size scaling function of the excess free energy $X_{ex}(x_1, x_2)$ as a function of the variables $(K/K_c - 1)L$ and $K_c^{-1/2} h L^{5/2}$. The function monotonically increses with both the temperature T and the magnetic field $|H|$. As a function of the finite-size L of the system it is increasing above and decresing below T_c.

and

$$x_1 = \frac{1}{4\pi} y_\infty - x_2^2/y_\infty^4 \equiv g_\infty(x_2, y_\infty), \tag{11.16}$$

respectively. Note that these equations follow by requiring the first partial derivative of the right-hand side of Eq. (11.14) with respect to the corresponding variable, y_L or y_∞, to vanish. In order to obtain Eq. (11.15) one has to use that $d\mathrm{Li}_p(x)/dx = \mathrm{Li}_{p-1}(x)/x$ and $\mathrm{Li}_1(x) = -\ln(1 - x)$. The main advantage of the above representation of X_{ex} is that it allows one to use the nontrivial Sachdev identity (10.80) for the polylogarithm functions, and thus to obtain a simple expression for the universal constant $\Delta = X_{ex}(0, 0)$.

The finite-size scaling function of the Casimir force immediately follows from Eqs. (11.10), (11.11), (11.14), and the definitions of the scaling

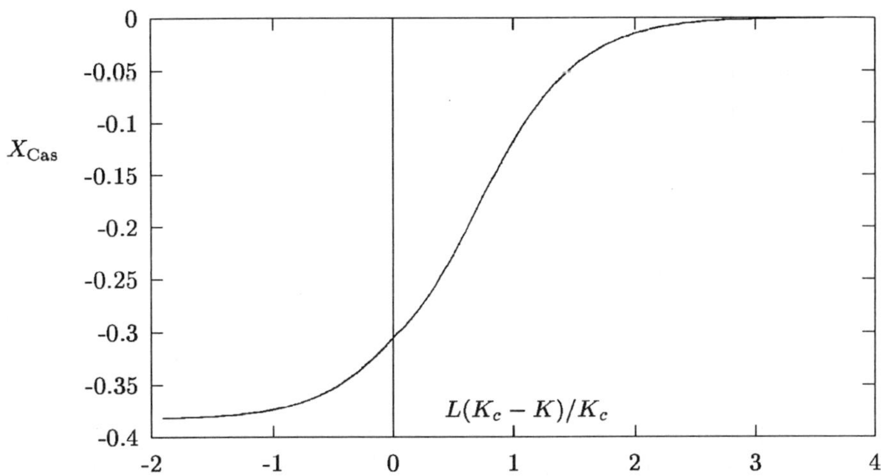

Fig. 11.3   The *universal* finite-size scaling function X_{Cas} of the Casimir force as a function of the variable $(1 - K/K_c)L$.

variables x_1 and x_2:

$$X_{\mathrm{Cas}}(x_1, x_2) = 2X_{\mathrm{ex}}(x_1, x_2) - \frac{5}{2}x_2^2\left(y_\infty^{-2} - y_L^{-2}\right) + \frac{1}{2}x_1(y_\infty^2 - y_L^2). \tag{11.17}$$

Its behavior is shown in Fig. 11.3 and Fig. 11.4).

Verification of the hypotheses

We shall prove analytically that the finite-size scaling function of the excess free energy, given by Eq. (11.14), is a monotonically increasing function of any of its scaling parameters x_1 and x_2 when the other one is kept fixed.

It is easily seen that

$$g_L(x_2, y) = g_\infty(x_2, y) + \frac{1}{2\pi}\ln\left(1 - \mathrm{e}^{-y}\right). \tag{11.18}$$

Hence, having in mind that in Eqs. (11.15) and (11.16) $y_L > 0$, $y \geq 0$, we conclude that $g_L(x_2, y) < g_\infty(x_2, y)$. It is also elementary to verify that $g_L(x_2, y_L)$ and $g_\infty(x_2, y_\infty)$ are monotonically increasing functions of y_L and y_∞, respectively. Let now $y_\infty(x_1, x_2)$ be the solution of Eq. (11.16) for given x_1 and x_2. Then, from the fact that $g_L(x_2, y)$ is a monotonically increasing function of y, and that at the solution $y_L(x_1, x_2)$ of Eq. (11.15)

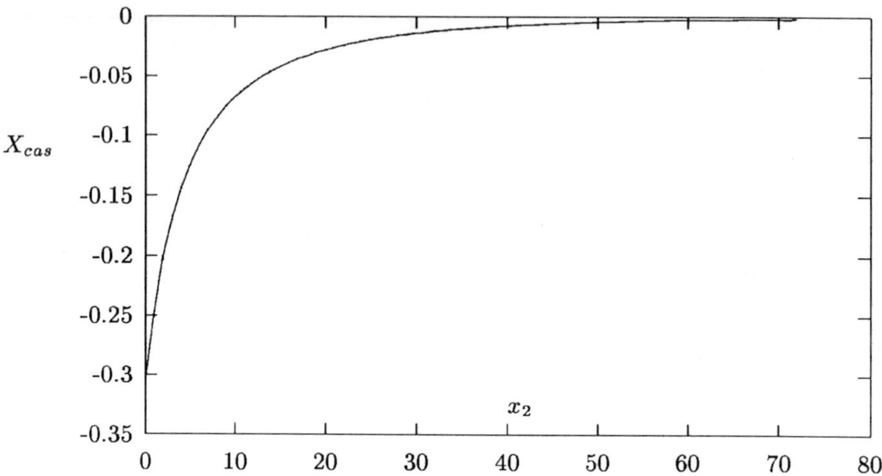

Fig. 11.4 The *universal* finite-size scaling function X_{Cas} of the Casimir force as a function of the variable $K_c^{-1/2} h L^{5/2}$ at $K = K_c$.

one should have $g_L(x_2, y_L) = g_\infty(x_2, y_\infty)$, we obtain

$$y_L(x_1, x_2) > y_\infty(x_1, x_2). \qquad (11.19)$$

We are now ready to deal with the monotonicity properties of the finite-size scaling function for the excess free energy. From Eq. (11.14), having in mind the mean spherical constraints (11.15), (11.16), we derive

$$\frac{\partial X_{\text{ex}}}{\partial x_1} = \frac{1}{2}(y_L^2 - y_\infty^2), \qquad \frac{\partial X_{\text{ex}}}{\partial x_2} = -x_2(y_L^{-2} - y_\infty^{-2}). \qquad (11.20)$$

From these expressions and Eq. (11.19), taking into account the definitions of the scaling variables (4.113), we obtain that the finite-size scaling function for the excess free energy is a monotonically increasing function of both the temperature T and the magnetic field $|H|$. As a function of the finite size L of the system the scaling function is monotonically increasing above and decreasing below T_c, see Fig. 11.2. Finally, it is worth mentioning that from Eqs. (11.14) and (11.19) it follows that in the zero-field case $x_2 = 0$ one has $X_{\text{ex}} < 0$. Numerically this can be checked to hold also for $x_2 \neq 0$.

We turn now to the properties of the Casimir force. Our aim is to show that under periodic boundary conditions this force is negative for all values of T and H. The finite-size behavior of the Casimir force in the vicinity of the critical point is given by Eq. (11.10), where the scaling function is defined in Eq. (11.17). For $T < T_c$, where the long-range Casimir force exists due to the Goldstone-type excitations, the same expressions are valid, although now $x_1 \ll -1$. Here we are not going to discuss whether in this region the above expressions can be simplified further to a function of a given combination of x_1 and x_2, as it is usually the case of first order phase transitions [Fisher and Privman (1985)]. For T above T_c the Casimir force, as it has been shown in [Danchev (1996)], tends to zero exponentially fast with $L \to \infty$, in a full accordance with the general expectations. We are not interested in the explicit form of these exponentially small corrections. Having in mind all these comments, the asymptotic behavior of the Casimir force at any T and H is given by

$$F_{L,\mathrm{Cas}}(t,h) \simeq L^{-3} \left\{ \frac{3}{2} x_2^2 (y_L^{-2} - y_\infty^{-2}) + \frac{1}{2} x_1 (y_L^2 - y_\infty^2) \right.$$
$$\left. - \frac{1}{\pi} \left[\frac{1}{6} (y_L^3 - y_\infty^3) + y_L \, \mathrm{Li}_2 \left(e^{-y_L} \right) + \mathrm{Li}_3 \left(e^{-y_L} \right) \right] \right\}. \qquad (11.21)$$

Since the inequality (11.19) is still valid, from the above expression it follows that $F_{L,\mathrm{Cas}}(t,h) < 0$. Finally, we show that for $T < T_c$ and $h = 0$ ($x_1 < 0$ and $x_2 = 0$) the Casimir force is a monotonically increasing function of the temperature, i.e., monotonically increasing function of x_1. From Eq. (11.21), by taking into account that $y_\infty \equiv 0$ at $T < T_c$, we obtain

$$\frac{\mathrm{d}}{\mathrm{d}x_1} X_{\mathrm{Cas}}(x_1, 0) = \frac{1}{2} y_L^2 - x_1 y_L \frac{\mathrm{d}y_L}{\mathrm{d}x_1} > 0 \qquad (x_1 < 0), \qquad (11.22)$$

since $\mathrm{d}y_L/\mathrm{d}x_1 > 0$ due to Eq. (11.15).

Thus we have completed the verified the hypotheses about the monotonic behavior of the excess free energy and the Casimir force, formulated in the Section 11.2.1.

Some results at the bulk critical point

Here we are interested in the properties of the system at the bulk critical point, $x_1 = x_2 = 0$, where the solutions of the bulk, Eq. (11.16), and finite-size, Eq. (11.15), mean spherical constarints are $y_\infty = 0$ and $y_L \equiv y_{L,c} = \Theta$, see Eq. (4.156), respectively. The problem of obtaining the Casimir amplitude reduces now to the evaluation of the expression

$$\Delta = X_{\mathrm{ex}}(0,0) = -\frac{1}{2\pi}\left[\frac{1}{6}y_{L,c}^3 + y_{L,c}\,\mathrm{Li}_2\left(e^{-y_{L,c}}\right) + \mathrm{Li}_3\left(e^{-y_{L,c}}\right)\right]. \quad (11.23)$$

The expression in the square brackets has been already evaluated in Section 10.4.3, see Eqs. (10.74) – (10.81). Thus we obtain that the Casimir amplitude of the three-dimensional mean spherical model under periodic boundary conditions is

$$\Delta = -\frac{2\zeta(3)}{5\pi} \simeq -0.153051. \quad (11.24)$$

Recalling now that $-X_{\mathrm{ex}}(0,0)$ corresponds to the analog of the C-function for our model, and taking the normalization factor in the form that keeps the C-function of the critical Gaussian model to be $C = 1$ at any d (i.e., setting in Eq. (10.67) $n(3) = \zeta(3)/(2\pi)$), we conclude that *the C-function of the critical three-dimensional mean spherical model under periodic boundary conditions is 4/5.*

From Eqs. (11.21), (11.23) and (11.24) we obtain the remarkably simple result for the Casimir force at the bulk critical point:

$$F_{L,\mathrm{Cas}} = -\frac{4\zeta(3)}{5\pi}\frac{1}{L^3}. \quad (11.25)$$

The exact form of the finite-size scaling function $X(x_1, x_2)$ of the free energy density follows from Eq. (11.14) by omitting the terms depending on y_∞:

$$X(x_1, x_2) = -\frac{1}{2\pi}\left[\frac{1}{6}y_L^3 + y_L\,\mathrm{Li}_2\left(e^{-y_L}\right) + \mathrm{Li}_3\left(e^{-y_L}\right)\right] - \frac{1}{2}\frac{x_2^2}{y_L^2} + \frac{1}{2}x_1 y_L^2. \quad (11.26)$$

Having the exact solution for $y_{L,c}$ from Eq. (10.78), and such a simple form for the free energy density, one can easily determine the exact values of other physically interesting quantities at the critical point. For example it is easy to show that the finite-size specific heat is given by

$$c_L(0,0) = \frac{1}{2} - \frac{8\pi}{\sqrt{5}}\Theta K_c^2 L^{-1}, \quad (11.27)$$

where Θ is the constant (4.156), and that the critical finite-size correlation length $\xi_L(K_c, 0) = L/y_{L,c}$ equals [Singh and Pathria (1986)], [Brankov and

Danchev (1991)],

$$\xi_L(K_c, 0) = L/\Theta. \tag{11.28}$$

For explicit results on the behavior of ξ_L under other geometries, boundary conditions, and for long-range interactions the reader may consult the following works: [Brézin (1982)], [Luck (1985)], [Shapiro and Rudnick (1986)], [Singh and Pathria (1986)], [Singh and Pathria (1987)], [Henkel (1988)], [Allen and Pathria (1989)], [Brankov and Danchev (1991)], [Henkel and Weston (1992)], [Allen and Pathria (1993)], [Allen and Pathria (1994)], [Allen and Pathria (1995)].

Discussion and comments

The results reported here about the mean spherical model are in full agreement with the predictions of the finite-size scaling theory. Eqs. (11.14), (11.15), (11.16), (11.21) and (11.26) give the universal finite-size scaling functions of the excess free energy, Casimir force and free energy density. The universal Casimir amplitude Δ of the mean spherical model has been determined exactly in the simple close form $\Delta = -2\zeta(3)/(5\pi) \simeq -0.153051$. This value is *surprisingly close* (within the error bar) to the Casimir amplitude reported for the Ising model $\Delta = -0.1526 \pm 0.0010$ [Krech (1997)]. In contrast to the Ising-like case, however, the zero-field excess free energy and Casimir force below T_c *do not* tend to zero exponentially fast in L as $L \to \infty$. For example, the finite-size scaling functions of the excess free energy and Casimir force *tend to a constant below* T_c, see Eq. (31) in [Danchev (1996)]. The explanation of this behavior, which, we believe, is common to all $O(n)$, $n \geq 2$, models under periodic boundary conditions, is based on the fact that due to the existing soft modes in the system (spin waves or Goldstone modes), the zero-field correlation length ξ below T_c is identically infinite. If an external field is applied ($h \neq 0$), then $\xi < \infty$, and the excess free energy density $f_{\mathrm{ex}} \to 0$ exponentially fast in L.

Finally, we emphasize that there is a close parallel between the temperature dependence of the C-function, considered in Chapter 10 for a d-dimensional quantum system, and the properties of the scaling function $-X_{\mathrm{ex}}$ for the excess free energy of the corresponding classical system as a function of L^{-1}. We have proposed some arguments that, if the finite system does not exhibit a phase transition, and if the system is equivalent in some sense to a $O(n)$ model with $n \geq 2$, then $-X_{\mathrm{ex}}$ monotonically increases as a function of L^{-1} above T_c and decreases below T_c. We would expect the same to be true for the the C-function of the corresponding quantum system as a function of T close to its quantum critical point. If the classical system is equivalent to some Ising type model, the same type of arguments, with account of the lack of monotonicity of the correlation function in the neighborhood of T_c, lead to the hypothesis that $-X_{\text{ex}}$ should be a monotonic function of L^{-1} both below and above T_c. For the corresponding *C-function of a quantum system, that can be mapped onto a classical Ising model (according to the dimensional crossover rule)*, the above implies that *C is a monotonically increasing function of the temperature both below and above its quantum critical point*. This is indeed the case for the quantum version of the two-dimensional Ising model, see Fig. 2 in [Castro Neto and Fradkin (1993)]. We would like to stress that the relatively simple picture described here should probably change significantly, if the finite system undergoes a phase transition of its own. Such is the case of the Ising model in $4 - \varepsilon$ ($\varepsilon > 0$) dimensions, for which the finite-scaling function of the excess free energy density X_{ex} above T_c is known up to the first order in ε [Krech and Dietrich (1992)]. It has a *minimum* with respect to T *slightly above* T_c. Unfortunately, no results are available for X_{ex} below the shifted critical temperature of the finite system $T_{c,L}$. The above problems can be studied exactly within the spherical model.

11.3   The Casimir effect and quantum critical systems

The investigation of the Casimir effect in a critical quantum system has some peculiarities in comparison with classical systems which are due to an "effective interplay" between the fluctuation induced forces related to the temporal and spatial finite-size dimensions.

11.3.1   *Background ideas*

Let us consider a quantum system with the film geometry $L \times \infty^{d-1} \times L_\tau$, where $L_\tau \sim \hbar/(k_B T)$ is the "finite-size" in the temporal (imaginary time) direction, and let us suppose that *periodic boundary conditions* are imposed across the finite spatial dimension (in the remainder we set $\hbar = k_B = 1$). Let $f_L(T, g, H)$ be the free energy density of the system, where g is the parameter governing the phase transition at $T = 0$, and let g_c is its quantum critical point. Then, according to the dimensional crossover rule, the Privman - Fisher hypothesis for finite classical systems and Eq. (11.3), in the quantum case one could state that

$$L^{-1} f_{L,\text{ex}}(T, g, H) = (T L_\tau) L^{-(d+z)} X_{\text{ex}}(x_1, x_2, \rho), \tag{11.29}$$

with finite-size scaling variables

$$x_1 = L^{1/\nu} \delta g, \quad x_2 = h L^{\Delta/\nu}, \quad \rho = L^z / L_\tau. \tag{11.30}$$

Here Δ and ν are the usual critical exponents of the bulk model, h is a properly normalized external magnetic field H, $\delta g \sim g - g_c$, and X_{ex} is the universal scaling function of the excess free energy. According to the definition (11.2) of the Casimir force, one immediately obtains

$$F_{L,\text{Cas}}(T, g, H) = (T L_\tau) L^{-(d+z)} X_{\text{Cas}}(x_1, x_2, \rho), \tag{11.31}$$

where the *universal* finite-size scaling function $X_{\text{Cas}}(x_1, x_2, \rho)$ of the Casimir force is related to the one of the excess free energy $X_{\text{ex}} \equiv X_{\text{ex}}(x_1, x_2, \rho)$ by the equation

$$\begin{aligned} X_{\text{Cas}}(x_1, x_2, \rho) &= -(d+z) X_{\text{ex}} - \frac{1}{\nu} x_1 \frac{\partial X_{\text{ex}}}{\partial x_1} \\ &\quad + \frac{\Delta}{\nu} x_2 \frac{\partial X_{\text{ex}}}{\partial x_2} + z \rho \frac{\partial X_{\text{ex}}}{\partial \rho}. \end{aligned} \tag{11.32}$$

It follows from Eq. (11.31) that one can consider the general case of Casimir amplitudes depending on the scaling variable ρ:

$$\Delta_{\text{Cas}}(\rho) = X_{\text{Cas}}(0, 0, \rho). \tag{11.33}$$

The classical amplitudes $\Delta^{(a,b)}$ introduced by Eq. (11.4) at $(a, b) \equiv (p)$ (periodic boundary conditions) are particular cases of $\Delta_{\text{Cas}}(\rho)$ at $\rho = 0$, i.e., at $T = 0$. We remind the reader that, according to the dimensional crossover rule, the quantum system is formaly equivalent to a $d + z$ dimen sional classical one.

In addition to the above "usual" excess free energy and Casimir amplitudes, one can define in a completely analogous way "*temporal excess free energy density*" f_{ex}^{t}, c.f. Eq. (11.3),

$$f_{\text{ex}}^{\text{t}}(T, g, H) = f_\infty(T, g, H) - f_\infty(0, g, H) \tag{11.34}$$

and *"temporal Casimir amplitudes"*

$$f_{\mathrm{ex}}^{\mathrm{t}}(T, g_c, 0) = T L_\tau^{-d/z} \Delta_{\mathrm{Cas}}^{\mathrm{t}}. \tag{11.35}$$

Whereas the "usual" amplitudes characterize the leading L corrections to the bulk free energy density at the critical point, the "temporal amplitudes" determine the leading temperature-dependent corrections to the ground state energy of an *infinite* system at its quantum critical point g_c.

When the quantum parameter g is in the vicinity of g_c, one expects

$$f_{\mathrm{ex}}^{\mathrm{t}}(T, g, H) = T L_\tau^{-d/z} X_{\mathrm{ex}}^{\mathrm{t}}(x_1^{\mathrm{t}}, x_2^{\mathrm{t}}), \tag{11.36}$$

i.e. instead of the amplitude $\Delta_{\mathrm{Cas}}^{\mathrm{t}}$ one has $X_{\mathrm{ex}}^{\mathrm{t}}(x_1^{\mathrm{t}}, x_2^{\mathrm{t}})$, a scaling function which is the analog of $X_{\mathrm{ex}}(x_1, x_2, \rho)$. The scaling variables now are

$$x_1^{\mathrm{t}} = L^{1/\nu z} \delta g \qquad x_2^{\mathrm{t}} = h L^{\Delta/\nu z}. \tag{11.37}$$

Obviously

$$\Delta_{\mathrm{Cas}}^{\mathrm{t}} = X_{\mathrm{ex}}^{\mathrm{t}}(0, 0). \tag{11.38}$$

Let us finally note that if $z = 1$, then the temporal excess free energy coincides, up to a (negative) normalization factor, with the definition of the non-zero temperature generalization of the C-function of Zamolodchikov, see Eq.(10.67).

Now we pass to the study of the above introduced quantities on the example of the quantum spherical model with power-law interaction. Since a specific mathematical techniques is required, that has not been presented in the previous chapters of the book, we give here the derivation of the scaling form of the free energy density in some detail. We mention that, up to our knowledge, only few results exist on the Casimir effect in quantum *critical* systems (see, e.g., [Chamati et. al (2000)], [Chamati and Tonchev (2000b)]).

11.3.2 *Film geometry: the free energy density*

The model is described by the Hamiltonian (3.85). Upon setting $s = \beta\mu/2$, the corresponding free energy density for a system in a finite region Λ under periodic boundary conditions has the form

$$\beta f_\Lambda(T, g, H) = \sup_{\mu} \left\{ \frac{1}{N} \sum_{\mathbf{k}} \ln \left[2\sinh\left(\frac{1}{2}\beta\omega(\mathbf{k};\mu)\right) \right] - \frac{\mu}{2}\beta - \frac{\beta g H^2}{2\omega^2(\mathbf{0};\mu)} \right\}.$$

$$(11.39)$$

Here $\omega^2(\mathbf{k};\mu) = g[\mu + U(\mathbf{k})]$ with $U(\mathbf{k}) \simeq J|\mathbf{k}|^\sigma$, $0 < \sigma \le 2$, where $U(\mathbf{k})$ is the Fourier transform of the interaction potential; the energy scale has been fixed so that $U(\mathbf{0}) = 0$. The supremum in Eq. (11.39) is attained at the solutions of the mean-spherical constraint which, by using identity (10.12), can be cast in the form (compare with Eq. (3.89) in the case of nearest-neighbor interaction)

$$1 = \frac{t}{N} \sum_{m=-\infty}^{\infty} \sum_{\mathbf{k}} \frac{1}{\phi + U(\mathbf{k})/J + b^2 m^2} + \frac{h^2}{\phi^2}. \qquad (11.40)$$

Equations similar to (11.39) and (11.40) have already been analyzed in the previous chapter, where the bulk properties of the model have been considered in the case of short-range interaction. Here we remind the reader our notation: $b = (2\pi t)/\lambda$, $\lambda = \sqrt{g/J}$ is the normalized quantum parameter, $t = T/J$ – the normalized temperature, $h = H/J$ – the normalized magnetic field, and $\phi = \mu/J$ – the scaled spherical field.

Below we need expressions specified to the film geometry. First we explain how from Eq. (11.39) one can obtain the full free energy density for a system with a geometry $L \times \infty^{d-1} \times L_\tau$. It is easy to see that in the new notation Eq. (11.39) can be rewritten as

$$f_L(t, \lambda, h|d, \sigma)/J = -\frac{h^2}{2\phi} - \frac{\phi}{2} + A_L(\phi) - t\sum_{n=1}^{\infty} U_L(\phi; n), \qquad (11.41)$$

where

$$A_L(\phi) = \frac{\lambda}{2N} \sum_{\mathbf{k}} \sqrt{\phi + |\mathbf{k}|^\sigma},$$

$$U_L(\phi; n) = \frac{1}{N} \sum_{\mathbf{k}} \exp\left[-n\frac{\lambda}{t}\sqrt{\phi + |\mathbf{k}|^\sigma} \right]. \qquad (11.42)$$

In order to calculate $A(L)$ we use the identity [Chamati (1994)]

$$\sqrt{1 + z^\alpha} = 1 + \frac{\alpha}{2} \int_0^\infty [1 - \exp(-zt)]\, t^{-1-\alpha/2}\, G_{\alpha,1-\frac{\alpha}{2}}(-t^\alpha), \qquad (11.43)$$

where the function $G_{\alpha,\beta}$ is defined in Eq. (5.11). Since we are interested in the geometry $L \times \infty^{d-1} \times L_\tau$, one has by standard arguments

$$\frac{1}{N} \sum_{\mathbf{k}} \rightarrow \frac{1}{L} \sum_{k_1=-(L-1)/2}^{(L-1)/2} \frac{1}{(2\pi)^{d-1}} \int dk_2 \cdots \int dk_d. \tag{11.44}$$

By using the asymptotic expansion (4.90) at $L \gg 1$, we obtain

$$A_L(\phi) = A_\infty(\phi) + \delta A_L(\phi), \tag{11.45}$$

where

$$A_\infty(\phi) = \frac{\lambda}{2} \int_{[-\pi,\pi]^d} d^d\mathbf{k} \sqrt{\phi + |\mathbf{k}|^\sigma} \tag{11.46}$$

and

$$\delta A_L(\phi) = -\frac{\lambda}{4} \frac{\sigma}{(4\pi)^{d/2}} \sum_{l=1}^\infty \int_0^\infty \frac{dx}{x^{\sigma/4+d/2+1}} \exp\left(-\frac{l^2 L^2}{4x}\right) G_{\frac{\sigma}{2},1-\frac{\sigma}{4}}(-x^{\sigma/2}\phi). \tag{11.47}$$

Since the only singularities of $A_\infty(\phi)$ as a function of ϕ are coming from small k's, it is justified to appproximate the Brillouin zone by a shpere of radius x_D, which leads to

$$\begin{aligned}
A_\infty(\phi) &\simeq \frac{\lambda}{2} a_d \int_0^{x_D} \frac{dx}{x^{1-d}} \sqrt{\phi + x^\sigma} \\
&\simeq \frac{\lambda a_d}{2d} x_D^d (x_D^\sigma + \phi)^{1/2} {}_2F_1\left(1, -\frac{1}{2}, 1 + \frac{d}{\sigma}, \frac{x_D^\sigma}{x_D^\sigma + \phi}\right)
\end{aligned} \tag{11.48}$$

Here we have introduced the constant $a_d = (2\pi)^{-d}\mathcal{A}_d(1)$, where $\mathcal{A}_d(1)$ is defined in Eq. (3.57).

Next we turn to the evaluation of the term $U_L(\phi; n)$ which, by taking into account Eq. (11.44), can be written in the form

$$\begin{aligned}
U_L(\phi; n) = {}&\frac{L^{-1}}{(2\pi)^{d-1}} \sum_{k_1=-(L-1)/2}^{(L-1)/2} \int_{-\pi}^{\pi} dk_2 \ldots \int_{-\pi}^{\pi} dk_d \\
&\times \exp\left[-\frac{n\lambda}{t}\sqrt{\phi + (k_1^2 + \ldots + k_d^2)^{\sigma/2}}\right].
\end{aligned} \tag{11.49}$$

By using the Poisson summation formula (4.79) we obtain

$$U_L(\phi; n) = U_\infty(\phi; n) + \delta U_L(\phi; n), \tag{11.50}$$

where

$$U_\infty(\phi; n) = a_d \int_0^{x_D} \mathrm{d}x\, x^{d-1} \exp\left(-\frac{n\lambda}{t}\sqrt{\phi + x^\sigma}\right) \tag{11.51}$$

and

$$\delta U_L(\phi; n) = \frac{2}{(2\pi)^d} \sum_{l=1}^{\infty} \int_{[-\pi,\pi]^d} \mathrm{d}^d\mathbf{k}\, \cos(k_1 lL) \exp\left[-\frac{n\lambda}{t}\sqrt{\phi + |\mathbf{k}|^\sigma}\right]. \tag{11.52}$$

Then, by using the integral representation of the MacDonald function $K_\nu(x)$,

$$K_{\pm\nu}(2\sqrt{zt}) = \frac{1}{2}\left(\frac{z}{t}\right)^{\nu/2} \int_0^\infty \frac{\mathrm{d}x}{x^{\nu+1}} \exp\left(-tx - \frac{z}{x}\right), \tag{11.53}$$

we derive

$$U_\infty(\phi; n) = \frac{\lambda}{t\sigma}\frac{a_d}{\sqrt{\pi}}\Gamma\left(\frac{d}{\sigma}\right) \phi^{\frac{d}{\sigma}+\frac{1}{2}} \left(\frac{n\lambda}{2t}\phi^{\frac{1}{2}}\right)^{-\left(\frac{d}{\sigma}+\frac{1}{2}\right)} K_{\frac{d}{\sigma}+\frac{1}{2}}\left(\frac{n\lambda}{t}\phi^{\frac{1}{2}}\right). \tag{11.54}$$

Now we are left to consider $\delta U_L(\phi; n)$. After sphericalization of the Brillouin zone in (11.52), and performing the integration with the aid of the integral representation for the Bessel function $J_\nu(z)$,

$$\int_{-a}^{a} \mathrm{d}x(a^2 - x^2)^{\beta-1}\mathrm{e}^{\mathrm{i}\lambda x} = \sqrt{\pi}\Gamma(\beta)\left(\frac{2a}{\lambda}\right)^{\beta-1/2} J_{\beta-1/2}(a\lambda),$$

where Re $\beta > 0$, we obtain

$$\delta U_L(\phi; n) = \frac{2L^{-d/2+1}}{(2\pi)^{d/2}} \sum_{l=1}^{\infty} \int_0^{x_D} \frac{\mathrm{d}x\, x^{d/2}}{l^{d/2-1}} J_{d/2-1}(lLx) \exp\left[-n\frac{\lambda}{t}\sqrt{\phi + x^\sigma}\right]. \tag{11.55}$$

In the low-temperature limit the upper limit of integration x_D in the above expressions can be replaced by infinity.

By collecting Eqs. (11.47), (11.48), (11.54) and (11.55), one finally obtains the result [Chamati et. al (2000)]:

$$
\begin{aligned}
f_L(t,\lambda,h|d,\sigma)/J = {}& -\frac{h^2}{2\phi} - \frac{\phi}{2} \\
&+ \frac{\lambda a_d}{2d} x_D^d (x_D^\sigma + \phi)^{\frac{1}{2}} \, {}_2F_1\left(1, -\frac{1}{2}, 1 + \frac{d}{\sigma}, \frac{x_D^\sigma}{x_D^\sigma + \phi}\right) \\
&- \frac{\lambda}{4} \frac{\sigma L^{-(d+\frac{\sigma}{2})}}{(4\pi)^{\frac{d}{2}}} \sum_{n=1}^{\infty} \int_0^{\infty} \frac{dx}{x^{\frac{\sigma}{4}+\frac{d}{2}+1}} \exp\left(-\frac{n^2}{4x}\right) G_{\frac{\sigma}{2},1-\frac{\sigma}{4}}\left(-x^{\sigma/2} L^\sigma \phi\right) \\
&- \frac{\lambda}{\sigma} \frac{a_d}{\sqrt{\pi}} \Gamma\left(\frac{d}{\sigma}\right) \phi^{\frac{d}{\sigma}+\frac{1}{2}} \sum_{m=1}^{\infty} \frac{K_{\frac{d}{\sigma}+\frac{1}{2}}\left(m\frac{\lambda}{t}\phi^{\frac{1}{2}}\right)}{\left(m\frac{\lambda}{2t}\phi^{\frac{1}{2}}\right)^{\frac{d}{\sigma}+\frac{1}{2}}} \\
&- \lambda\sqrt{2}\frac{L^{-(d+\frac{\sigma}{2})}}{(2\pi)^{\frac{d+1}{2}}} \sum_{n=1}^{\infty}\sum_{m=1}^{\infty} \int_0^{\infty} \frac{dz}{m^d z^{\frac{3}{2}}} \mathcal{F}_{\frac{d}{2}-1,\sigma}\left(\frac{z}{m^\sigma}\right) \\
&\qquad\qquad \times \exp\left[-zL^\sigma\phi - \frac{n^2}{4zL^\sigma}\left(\frac{\lambda}{t}\right)^2\right].
\end{aligned}
\tag{11.56}
$$

Here

$$
\mathcal{F}_{\nu,\sigma}(y) = \int_0^{\infty} dx\, x^{\nu+1} J_\nu(x) \exp(-yx^\sigma).
\tag{11.57}
$$

The main advantage of the above expression, despite of its complicated form in comparison with Eq. (11.39), is the simplified dependence on the size L which now enters only via the arguments of some functions. This gives us the possibility to obtain the finite-size scaling functions of the excess free energy and the Casimir force.

11.3.3 *Scaling form of the excess free energy and the Casimir force at low temperatures*

In Eq. (11.56) ϕ is the solution of the corresponding mean spherical constraint that follows by requiring the partial derivative of its right-hand side with respect to ϕ to be zero. The bulk free energy $f_\infty(t,\lambda,h|d,\sigma)$ results from $f_L(t,\lambda,h|d,\sigma)$ by merely taking the limit $L \to \infty$ in it. Let us denote the solution of the bulk mean spherical constraint by ϕ_∞. Then, in full accordance with Eq. (11.29), from Eqs. (11.3) and (11.56), we obtain the finite-size scaling form of the excess free energy density:

$$L^{-1} f_{L,\text{ex}}(t, \lambda, h|d, \sigma) = (TL_\tau) L^{-(d+z)} X_{\text{ex}}(x_1, x_2, \rho|d, \sigma), \qquad (11.58)$$

with scaling variables

$$x_1 = L^{1/\nu}\left(\lambda^{-1} - \lambda_c^{-1}\right), \quad x_2 = hL^{\Delta/\nu}, \quad \rho = L^z/L_\tau, \qquad (11.59)$$

where $L_\tau = \lambda/t$. Here the critical value of $\lambda = \lambda_c$ is

$$\lambda_c^{-1} = \frac{1}{2(2\pi)^d} \int \frac{d^d\mathbf{k}}{\sqrt{U(\mathbf{k})/J}}, \qquad (11.60)$$

and $\nu^{-1} = d - \frac{1}{2}\sigma$, $\Delta/\nu = \frac{1}{2}\left(d + \frac{3}{2}\sigma\right)$, and $z = \frac{1}{2}\sigma$ are the critical exponents of the model. In Eq. (11.58) the *universal* scaling function $X_{\text{ex}}(x_1, x_2, \rho|d, \sigma)$ of the excess free energy has the form

$$
\begin{aligned}
X_{\text{ex}}(x_1, x_2, \rho|d, \sigma) = {}&-\frac{1}{2}x_1(y_0^\sigma - y_\infty^\sigma) - \frac{1}{2}x_2^2(y_0^{-\sigma} - y_\infty^{-\sigma}) \\
&-\frac{a_d}{4\sqrt{\pi}\sigma}\Gamma\left(\frac{d}{\sigma}\right)\Gamma\left(-\frac{d}{\sigma} - \frac{1}{2}\right)\left(y_0^{d+\sigma/2} - y_\infty^{d+\sigma/2}\right) \\
&-\frac{a_d}{\sigma\sqrt{\pi}}\Gamma\left(\frac{d}{\sigma}\right)\sum_{m=1}^{\infty}\left[\frac{(2y_0)^{d+\frac{\sigma}{2}}K_{\frac{d}{\sigma}+\frac{1}{2}}\left(my_0^{\sigma/2}/\rho\right)}{\left(my_0^{\sigma/2}/\rho\right)^{d+\frac{\sigma}{2}}}\right. \\
&\left.\qquad\qquad - \frac{(2y_\infty)^{d+\frac{\sigma}{2}}K_{\frac{d}{\sigma}+\frac{1}{2}}\left(my_\infty^{\sigma/2}/\rho\right)}{\left(my_\infty^{\sigma/2}/\rho\right)^{d+\frac{\sigma}{2}}}\right] \\
&-\frac{1}{4}\frac{\sigma}{(4\pi)^{\frac{d}{2}}}\sum_{n=1}^{\infty}\int_0^{\infty}\frac{dx}{x^{\frac{\sigma}{4}+\frac{d}{2}+1}}e^{-\frac{n^2}{4x}}G_{\frac{\sigma}{2},1-\frac{\sigma}{4}}\left(-x^{\frac{\sigma}{2}}y_0^\sigma\right) \\
&-\frac{\sqrt{2}}{(2\pi)^{\frac{d+1}{2}}}\sum_{n=1}^{\infty}\sum_{m=1}^{\infty}\int_0^{\infty}\frac{dz}{m^d z^{\frac{3}{2}}}e^{-zy_0^\sigma - \frac{n^2}{4z\rho^2}}\mathcal{F}_{\frac{d}{2}-1,\sigma}\left(\frac{z}{m^\sigma}\right), \quad (11.61)
\end{aligned}
$$

where $y_0 = L\phi^{1/\sigma}$ and $y_\infty = L\phi_\infty^{1/\sigma}$.

By direct evaluation of the above expression, one can check that if the finite system below λ_c is in the ordered phase (with $y_0 = y_\infty = 0$), then $X_{\text{ex}} \sim L^{-\sigma}$, which reflects the dominating contribution of the direct spin-spin long-range interaction in that region.[§] As we see from Eq. (11.58), a

---

[§]The requirement the finite nonzero-temperature system to be in ordered state involves $d - 1 > \sigma$, which together with $\sigma/2 < d < 3\sigma/2$ excludes the case $\sigma = 2$.

crossover takes place from $L^{-(d+z)}$ behavior, where the fluctuation-induced interactions dominate, to $L^{-\sigma}$ behavior, where the direct spin-spin interactions become essential.

For the Casimir force we obtain

$$F_{L,\mathrm{Cas}}(T,\lambda,h|d,\sigma) = (TL_\tau)L^{-(d+z)}X_{\mathrm{Cas}}(x_1,x_2,\rho|d,\sigma), \qquad (11.62)$$

where the *universal* finite-size scaling function X_{Cas} of the Casimir force is related to that of the excess free energy X_{ex} by Eq. (11.32).

The above expressions for the scaling functions of the excess free energy (and the Casimir force) are the most general ones, which give one the possibility to analyze issues as: *(i)* the sign of the Casimir force; *(ii)* monotonicity of the Casimir force as a function of the temperature; *(iii)* the relation of the excess free energy scaling function to the corresponding finite-temperature C-function and its monotonicity properties; *(iv)* finite-system generalization of the finite-temperature C-function, etc.

It is clear that just due to the dimensional crossover rule, the size L plays the same role for the finite-size system at $t = 0$, as L_τ does for the corresponding infinite quantum system. Therefore, due to the symmetry that obviously arises when $\sigma = 2$, one should expect that the behavior of the two types of amplitudes ("normal" and "temporal") will be essentially the same. We show that explicitly below.

For the model under consideration one can show that

$$\begin{aligned}
X_{\mathrm{ex}}^{\mathrm t}(x_1^{\mathrm t},x_2^{\mathrm t}|d,\sigma) = &-\frac{1}{2}x_1^{\mathrm t}(y_0^2 - y_\infty^2) - \frac{1}{2}(x_2^{\mathrm t})^2(y_0^{-2} - y_\infty^{-2}) \\
&- \frac{a_d}{4\sqrt{\pi}\sigma}\Gamma\left(\frac{d}{\sigma}\right)\Gamma\left(-\frac{d}{\sigma} - \frac{1}{2}\right)\left(y_0^{d/z+1} - y_\infty^{d/z+1}\right) \\
&- \frac{a_d}{\sigma\sqrt{\pi}}\Gamma\left(\frac{d}{\sigma}\right)(2y_0^2)^{\frac{d}{\sigma}+\frac{1}{2}}\sum_{m=1}^\infty \frac{K_{\frac{d}{\sigma}+\frac{1}{2}}(my_0)}{(my_0)^{\frac{d}{\sigma}+\frac{1}{2}}}.
\end{aligned} \qquad (11.63)$$

where the scaling variables are defined by

$$x_1^{\mathrm t} = L_\tau^{1/\nu z}(1/\lambda - 1/\lambda_c), \qquad x_2^{\mathrm t} = hL_\tau^{\Delta/z\nu}, \qquad (11.64)$$

$y_0 = L_\tau\phi_0^{1/2}$, and $y_\infty = L_\tau\phi_\infty^{1/2}$. Note that y_0 is the solution of the mean spherical constraint for the system at non-zero temperature, whereas y_∞ is the solution for the zero-temperature ("infinite" in the "temporal" dimension) one. The direct evaluation of Eq. (11.63), assuming the finite system in an ordered state, shows that X_{ex}^{t} remains of the same order of magnitude in that region. This is exactly the behavior to be expected, despite the long-range nature of the interactions, since the finite "dimension" of the system is now the temporal one.

Now we are ready to investigate in a bit more detail the behavior of the Casimir amplitudes as a function of d, σ and ρ.

11.3.4 *Evaluation of some Casimir amplitudes*

Here we determine the Casimir amplitudes, both usual and "temporal", of the model in the case of short-range interactions at $d = 2$ and in the special case $d = \sigma$ of long-range interactions.

"Usual" Casimir amplitudes

Two-dimensional system with short-range interactions $(d = \sigma = 2)$

In this case essential simplifications in the expression for the Casimir amplitudes occur, since the functions $G_{\alpha,\beta}$ and $\mathcal{F}_{\alpha,\beta}$ used in the general expression of the free energy (11.56) take the explicit form

$$G_{1,1/2}(z) = \frac{1}{\sqrt{\pi}} \exp(z), \quad \mathcal{F}_{\alpha,2}(z) = (2z)^{-\alpha-1} \exp\left(-\frac{1}{4z}\right).$$

At the quantum critical point (see Eq. (10.44)) $\lambda = \lambda_c$, $h = 0$, this leads to $(y_\infty = 0)$

$$X_{\text{ex}}(0,0,\rho|2,2) = -\frac{y_0^3}{(2\pi)^{3/2}} \sum_{m,n}{}' \frac{K_{\frac{3}{2}}\left(y_0\sqrt{n^2/\rho^2 + m^2}\right)}{\left(y_0\sqrt{n^2/\rho^2 + m^2}\right)^{\frac{3}{2}}} - \frac{y_0^3}{12\pi}, \quad (11.65)$$

where $y_0 = L\phi^{1/2}$; the primed summation over the integers m and n indicates that the term with $m = n = 0$ is excluded.

Since it is not clear how to obtain an explicit analytical solution for y_0 in a film geometry at nonzero temperature, the above expression has to be analyzed numerically (see, e.g., [Chamati et. al. (1998)] for a numerical analysis of the mean spherical constraint). Nevertheless, that expression significantly simplifies at zero temperature when $L_\tau = \infty$, $1 \ll L \ll \infty$. Then, at the quantum critical point $\lambda = \lambda_c$, $h = 0$, of a finite system with the film geometry $L \times \infty \times L_\tau$, Eq. (10.42) yields $y_0 = \Theta$, see Eq. (4.156). By inserting this value of y_0 in Eq. (11.65), taking into account that $K_{3/2}(x) = \sqrt{\pi/(2x)}\exp(-x)(1 + 1/x)$, and using the identity (10.80) we obtain:

$$\Delta_{\text{Cas}}(0|2, 2) = -\frac{2\zeta(3)}{5\pi}, \tag{11.66}$$

which indeed coincides with Eq. (11.24) due to the dimensional crossover rule.

One-dimensional system with long-range interactions ($d = \sigma = 1$)

In this case the functions G and $\mathcal{F}$ become

$$G_{1/2,3/4}(-z) \;=\; \frac{\sqrt{z}}{\pi}\exp(z^2/2)K_{1/4}(z^2/2),$$

$$\mathcal{F}_{\nu,1}(y) \;=\; \frac{2^{\nu+1}}{\sqrt{\pi}}\Gamma(\nu + 3/2)\frac{y}{(1 + y^2)^{\nu+3/2}}. \tag{11.67}$$

In order to obtain explicit results for the amplitudes, numerical evaluations are unavoidable even in the simplest case of zero temperature. Then, from Eq. (11.61) we obtain

$$X_{\text{ex}}^{\text{u}}(0,0,0|1,1) =$$

$$-\frac{y_0^{1/2}}{8\pi^{\frac{3}{2}}}\sum_{n=1}^{\infty}\int_0^{\infty} dx\, x^{-\frac{3}{2}}\exp\left(\frac{y_0^2 x}{2} - \frac{n^2}{4x}\right)K_{\frac{1}{4}}\left(\frac{y_0^2 x}{2}\right) - \frac{1}{3\pi}y_0^{3/2}. \tag{11.68}$$

The numerical solution of the mean spherical constraint yields $y_0 \simeq 0.6248$. After substitution of this solution in Eq. (11.68) we obtain

$$\Delta_{\text{Cas}}(0|1, 1) \simeq -0.3157. \tag{11.69}$$

The above result shows that the value of the Casimir amplitude in the case $\sigma = 1$ is smaller than the one in the case of short-range interaction. One can ask whether this is just a coincidence, or the Casimir amplitudes are increasing function of σ at a given d.

Relation with the Zamolochikov's C-function

In terms of the critical-point value of an analog of the finite-temperature C-function, see (10.67), the result given by Eq. (11.66) can be rewritten in the form

$$\Delta_{\text{Cas}}(0|d,2) = -n(d,2)\tilde{c}(d,2), \tag{11.70}$$

where, by analogy with the case of short-range interaction, we have define the number $\tilde{c}$ via the relation

$$\tilde{c}(d,\sigma) = -\Delta_{\text{Cas}}(0|d,\sigma)/n(d,\sigma). \tag{11.71}$$

As usual, the normalization factor $n(d,\sigma)$ is chosen so that to ensure $\tilde{c} = 1$ for the corresponding Gaussian model:

$$n(d,\sigma) = \frac{2^{\sigma/2}\Gamma\left(\frac{d}{2} + \frac{\sigma}{4}\right)\zeta\left(d + \frac{\sigma}{2}\right)}{\pi^{d/2}\left|\Gamma\left(-\frac{\sigma}{4}\right)\right|}. \tag{11.72}$$

For $\sigma = 2$ we immediately obtain $\tilde{c}(2,2) = 4/5$, and $n(d,2)$ becomes the normalization factor given in [Castro Neto and Fradkin (1993)]. In that way we reproduce the well known result for $\tilde{c}$ due to Sachdev [Sachdev (1993)], see Section 10.4. The d-dependence of the value $\tilde{c}(d,2)$ has been considered in [Petkou and Vlachos (1999)] for a d-dimensional ($2 < d < 4$) conformally invariant field theory.

In the particular case of long-range interaction $d = \sigma = 1$, from Eqs. (11.71) and Eq. (11.69) one obtains $\tilde{c}(1,1) = 0.606\ldots$.

"Temporal" Casimir amplitudes

σ-dimensional system $(d = \sigma)$

Let us note that unlike the case of usual Casimir amplitudes, here it is possible to consider the more general case $d = \sigma$, where $0 < \sigma \leq 2$. From Eq. (11.63) one obtains a general expression for the temporal Casimir amplitudes for a system with geometry $\infty^d \times L_\tau$, at the quantum critical point $\lambda = \lambda_c$, $h = 0$.

$$\Delta_{\text{Cas}}^{t}(d,\upsilon) = -\frac{a_d}{4\sqrt{\pi}\sigma}\Gamma\left(\frac{d}{\sigma}\right)\Gamma\left(-\frac{2d}{\sigma} - \frac{1}{2}\right)y_0^{2(d/z+1)}$$

$$-\frac{a_d}{\sigma\sqrt{\pi}}\Gamma\left(\frac{d}{\sigma}\right)(2y_0^4)^{\frac{d}{\sigma}+\frac{1}{2}}\sum_{m=1}^{\infty}\frac{K_{\frac{d}{\sigma}+\frac{1}{2}}(my_0^2)}{(my_0^2)^{\frac{d}{\sigma}+\frac{1}{2}}}, \tag{11.73}$$

where y_0 is the solution of the corresponding mean spherical constraint. We notice that in the particular case $d/\sigma = 1$ Eq. (11.73) simplifies considerably. In a way similar to that explained in the case of short-range interactions, one obtains $(0 < \sigma \le 2)$

$$\Delta^{\mathrm{t}}_{\mathrm{Cas}}(\sigma, \sigma) = -\frac{16}{5\sigma}\frac{\zeta(3)}{(4\pi)^{\sigma/2}}\frac{1}{\Gamma(\sigma/2)}. \tag{11.74}$$

Note that the "temporal Casimir amplitude" $\Delta^{\mathrm{t}}_{\mathrm{Cas}}(\sigma, \sigma)$ reduces at $\sigma = 2$ to the "normal" Casimir amplitude $\Delta_{\mathrm{Cas}}(0|2, 2)$, given by Eq. (11.66). This reflects the existence of a special symmetry in that case between the "temporal" and spatial dimensionalities of the system.

When $\sigma \ne 2$ it is easy to verify that the following general relation between the temporal amplitudes holds:

$$\frac{\Delta^{\mathrm{t}}_{\mathrm{Cas}}(\sigma, \sigma)}{\Delta^{\mathrm{t}}_{\mathrm{Cas}}(2, 2)} = \frac{8\pi}{\sigma(4\pi)^{\sigma/2}\Gamma(\sigma/2)}. \tag{11.75}$$

The right-hand side of the above equation is a decreasing function of σ.

Relation with the Zamolochikov's C-function

As already mentioned above, when $z = 1$ the temporal excess free energy, Eq. (11.34), coincides, up to a negative normalization constant, with the nonzero-temperature generalization of the C-function. At $z \ne 1$ a straightforward generalization of this definition can be proposed at least in the case of long-range power-low interaction:

$$C(T, g|d, z) = -T^{-(1+d/z)}\frac{v^{d/z}}{n(d, z)}f^{\mathrm{t}}_{\mathrm{ex}}(T, g), \tag{11.76}$$

where $z = \sigma/2$, the nonuniversal constant v in our notation is $v = TL_\tau$, see Eq. (11.36), and the normalization factor is taken to be such that the corresponding Gaussian model would have $C(T, g_c|d, z) \equiv \tilde{c}^{\mathrm{t}}(d, \sigma) = 1$ at its critical point, i.e.

$$n^{\mathrm{t}}(d, \sigma) = \frac{4}{\sigma}\frac{\zeta(1 + 2d/\sigma)}{(4\pi)^{d/2}}\frac{\Gamma(2d/\sigma)}{\Gamma(d/2)}. \tag{11.77}$$

Let us note that the above choice of the normalization constant $n^{\mathrm{t}}(d, z)$ preserves not only the value of $\tilde{c}$ for the Gaussian model, but also for the spherical model if $d = \sigma$. Indeed, in that case from Eq. (11.74) and the

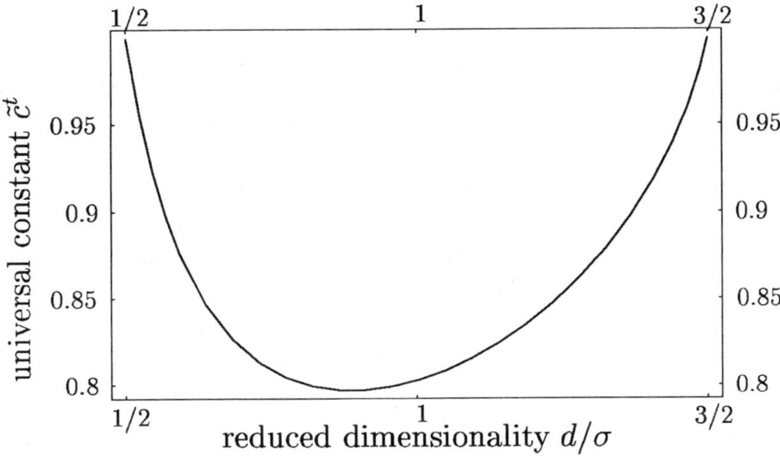

Fig. 11.5   Behaviour of the universal constant $\tilde{c}^t$ as a function of d/σ.

above definitions one again obtains for the spherical model:

$$\tilde{c}^t(\sigma,\sigma) = 4/5. \qquad (11.78)$$

This result is a generalization to the case of long-range interaction of Sachdev's one, see Eqs. (11.24) and (10.81) and the discussion related to them.

The quantity $\tilde{c}^t(d,\sigma)$ is an important universal characteristic of the theory. That is why we have analyzed numerically its behavior for dimensions between the lower critical ($d_l^q = \sigma/2$) and upper critical ($d_u^q = 3\sigma/2$) dimension at arbitrary values of $0 < \sigma \leq 2$. The resulting universal function of d/σ is presented in Fig. 11.5.

11.3.5   *Concluding remarks*

Expression (11.58) represents actually an analog of the Privman – Fisher hypothesis for the finite-size scaling form of the free energy (formulated initially for classical systems, see (4.28) in the case when quantum fluctuations are essential. According to the finite-size scaling hypothesis one has to expect that the temperature multiplying the universal scaling function would appear with exponent $p = 1 + d/z$ [Sachdev (1994)], where the dynamical critical exponent z reflects the anisotropic scaling between spatial and "temporal" ("imaginary-time") directions. Eqs. (11.61) – (11.62) present a general expression for the Casimir force in the *quantum* mean spherical model. In the classical limit ($\lambda = 0$) for a system with short-range interaction it coincides with the corresponding one derived for the classical mean spherical model, see Section 11.2.2. In the short-range case ($\sigma = 2$) the corresponding amplitude is given by Eq. (11.66). This amplitude is equal to the "temporal Casimir amplitude" for the $O(n)$ sigma model in the limit $n \to \infty$ [Sachdev (1993)]. In the short-range case we have demonstrated *explicitly* that the two models, due to the fact that they belong to the same universality class, indeed possess equal Casimir amplitudes as it is to be expected on the basis of the *hypothesis of universality.*

In the long-range case ($d = \sigma \neq 2$), the correction to the ground-state energy of the bulk system due to the nonzero temperature is determined by Eq. (11.74). One observes that in this case $c^t(\sigma,\sigma) = 4/5$ does not depend on σ. This could be understood by noting that the change of σ does not alter the exponent in the spectrum corresponding to the finite "temporal" dimension, see Eq. (11.40). At zero temperature we have evaluated numerically the Casimir amplitude in the particular case $d = \sigma = 1$ of long-range interaction. The result (11.69) shows that the Casimir amplitude is smaller than in the case of short-range interaction ($\sigma = 2$). Furthermore, the usual universal amplitude $c(1,1)$ is no longer σ-independent, because the finite-size part of the spectrum is σ-dependent in this case. In accordance with the general expectations, all the amplitudes are *negative*, see Eqs. (11.65), (11.66), (11.74)) and (11.69).

Finally, let us note that the basic expression Eqs. (11.61) for the finite-size scaling function of the excess free energy can be used as a starting point for generalization of some of the existing results on the C-function to the case of long-range interactions. We have suggested in Eq. (11.76) a generalization of the nonzero-temperature C-function to the case of power-low long-range interactions. For the quantum mean spherical model this definition leads to $\tilde{c} = 4/5$ at any $d = \sigma$, which generalizes to long-range interactions the corresponding result for short-range interactions.

Chapter 12

Survey of Results on the Casimir Effect

In Chapter 11 we have presented analytical and numerical results about the Casimir effect in the context of the spherical approximation. This simple approximation gives the possibility of describing the main features of the phenomenon. Of course, there exist numerous theoretical results on the subject, obtained for more realistic models by using different methods, such as ε-expansion, conformal invariance, and transfer-matrix calculations, in conjunction with a variety of possible boundary conditions. All these results are important for the interpretation of experimental data and Monte Carlo sumulations. We hope that our survey of the field will provide the reader with useful information needed for the better understanding of the phenomenon.

12.1 Ising model results

As expected, the Casimir effect has been most thoroughly studied in the case of the Ising model. Since the correlation length ξ is finite both above and below the critical point ($T = T_c$, $h = 0$), the effect is long-ranged only in the finite-size critical region $tL^{1/\nu} = O(1)$ and $hL^{\Delta/\nu} = O(1)$. Away from that region the Casimir force is supposed to fall down exponentially fast in L, as is the expected behavior of the finite-size corrections.

12.1.1 *Casimir amplitudes*

It turns out that only the type of the surface phase transition is important for the finite-size scaling description. That is why we introduce here the

the superscript (o) for all boundary conditions, e.g., zero Dirichlet, that lead to ordinary surface phase transition, see Chapter 7. Exact analytical results are available for the two-dimensional Ising model under periodic (p), antiperiodic (a), ordinary (o), $(+,+)$ and $(+,-)$ boundary conditions. The results for periodic, $(+,+)$ and $(+,-)$ boundary conditions are obtained via exact solutions, whereas those for antiperiodic and ordinary boundary conditions are derived by conformal invariance methods. The amplitudes available for the $2d$ Ising model are summarized in Table 12.1. We note that in this case the extraordinary and special (surface-bulk) surface phase transitions (respectively, boundary conditions) do not exist. At

$\Delta^{(o,o)}$	$\Delta^{(+,+)}$	$\Delta^{(p)}$	$\Delta^{(+,-)}$	$\Delta^{(a)}$	$\Delta^{(o,+)}$
$-\frac{1}{48}\pi$	$-\frac{1}{48}\pi$	$-\frac{1}{12}\pi$	$\frac{23}{48}\pi$	$\frac{1}{6}\pi$	$\frac{1}{24}\pi$

Table 12.1   The exact Casimir amplitudes for the $d = 2$ Ising model.

$d = 3$ mean-field and/or Monte Carlo results are available under periodic, surface-bulk (sb), ordinary, $(+,+)$ and $(+,-)$ boundary conditions. The numerical values of the Casimir amplitudes for the three-dimensional Ising model are summarized in Table 12.2, where the values in the row denoted by $\varepsilon = 1$ are obtained by extrapolating the ε-expansion to $\varepsilon = 1$ [Krech (1997)]; those in the row PA are obtained from Pade type approximants at $\varepsilon = 1$ [Krech (1997)]; the Monte - Carlo (MC) results are taken from [Krech and Landau (1996)]; the Migdal-Kadanoff real-space renormalization group results (MKRSRG) are from [Indekeu et. al. (1986)]; the last row (LFM) contains results obtained by local-functional methods [Borjan and Upton (1998)]. The analytical results obtained by ε-expansion techniques are:

	$\Delta^{(o,o)}$	$\Delta^{(+,+)}$	$\Delta^{(p)}$	$\Delta^{(sb,+)}$	$\Delta^{(+,-)}$	$\Delta^{(o,+)}$
$\varepsilon = 1$	-0.0139	-0.173	-0.1116	-0.093	1.58	0.165
PA	-0.0164	-0.326	-0.1315		2.39	0.208
MC	-0.0114	-0.345	-0.1526		2.450	0.197
MKRSRG	-0.015	0		0.017	0.279	0.051
LFM		-0.428			3.1	

Table 12.2   The Casimir amplitudes for the $d = 3$ Ising model.

$$\Delta^{(o,o)} = \Delta^{(sb,sb)} = -\frac{\pi^2}{1440}\left\{1 + \varepsilon\left[\ln\left(2\sqrt{\pi}\right) + \frac{\gamma-1}{2} - \frac{\zeta'(4)}{\zeta(4)} - \frac{5}{12}\right]\right\},$$

$$\Delta^{(o,sb)} = \frac{7}{8}\frac{\pi^2}{1440}\left\{1 + \varepsilon\left[\ln\left(2\sqrt{\pi}\right) + \frac{\gamma-1}{2} - \frac{\zeta'(4)}{\zeta(4)} - \frac{\ln 2}{7} + \frac{5}{42}\right]\right\},$$

$$\Delta^{(p)} = -\frac{\pi^2}{90}\left\{1 + \varepsilon\left[\ln\sqrt{\pi} + \frac{\gamma-1}{2} - \frac{\zeta'(4)}{\zeta(4)} - \frac{5}{12}\right]\right\},$$

$$\Delta^{(a)} = \frac{7}{8}\frac{\pi^2}{90}\left\{1 + \varepsilon\left[\ln\sqrt{\pi} + \frac{\gamma-1}{2} - \frac{\zeta'(4)}{\zeta(4)} - \frac{\ln 2}{7} + \frac{5}{42}\right]\right\},$$

where $\gamma \simeq 0.577216$ denotes Euler's constant, $\zeta'(4) \simeq -0.068911$ and $\zeta(4) = \pi^4/90$. The results for $\Delta^{(o,o)}$ are due to [Symanzik (1981)], the others are taken from [Krech and Dietrich (1992)].

12.1.2 *Scaling functions*

Let us consider first the exact results for the finite-size scaling functions of the excess free energy and/or the Casimir force in the two-dimensional case, as well as some analytical mean-field results at $d = 3$.

Exact results in the two-dimensional case

From the results obtained in [Evans and Stecki (1994)], see also [Au-Yang and Fisher (1980)], one can extract that the excess free energy scaling function under $(+, +)$ boundary conditions is, see the solid line in Fig. 12.1,

$$X_{\text{ex}}^{(+,+)}(x) = -\frac{1}{4\pi}\int_0^\infty \ln\left[1 + \frac{x + \sqrt{x^2 + y^2}}{\sqrt{x^2 + y^2} - x}\exp\left(-\sqrt{x^2 + y^2}\right)\right]dy,$$

where

$$x = 8K_c t L, \quad t = (T/T_c - 1), \quad K_c = \ln(1 + \sqrt{2})/2,$$

and $X_{\text{ex}}^{(o,o)}(x) = X_{\text{ex}}^{(+,+)}(-x)$.

At the critical point one obtains the Casimir amplitudes of the model:

$$\begin{aligned}
X_{\text{ex}}^{(+,+)}(0) &= X_{\text{ex}}^{(o,o)}(0) = \Delta^{(+,+)} = \Delta^{(o,o)} \\
&= -\frac{1}{4\pi}\int_0^\infty \ln\left[1 + \exp(-y)\right]dy = -\frac{\pi}{48}.
\end{aligned}$$

Having in mind that for the two-dimensional Ising model $A_\xi^- = (8K_c)^{-1}$ and $A_\xi^-/A_\xi^+ = 2$, one can rewrite the above expression in a form that allows

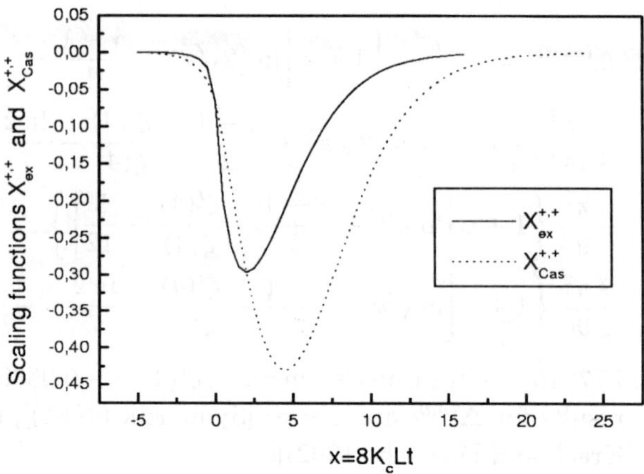

Fig. 12.1   The finite-size scaling functions of the excess free energy $X_{ex}^{(+,+)}$ and the Casimir force $X_{Cas}^{(+,+)}$ under $(+,+)$ boundary conditions.   Note that they are everywhere negative and both have minima *above* T_c: the former attains minimum $X_{ex}^{(+,+)} \simeq -0.2965$ at $x \simeq 2.0$, while the latter attains minimum $X_{Cas}^{(+,+)} \simeq -0.4305$ at $x \simeq 4.4571$.

direct comparison (actually, only at $T > T_c$) with the renormalization group results [Krech and Dietrich (1992)]. Thus, see [Krech (1994)],

$$X_{ex}^{(o,o)}(x_-) = -\frac{1}{4\pi} \int_{x_-}^{\infty} \frac{y}{(y^2 - x_-^2)^{1/2}} \ln\left[1 + \frac{x_- + y}{y - x_-} \exp(-y)\right] dy \, ,$$

where $t < 0$, $x_- = L/\xi(t < 0)$, and

$$X_{ex}^{(o,o)}(x_+) = -\frac{1}{4\pi} \int_{2x_+}^{\infty} \frac{y}{(y^2 - 4x_+^2)^{1/2}} \ln\left[1 + \frac{y - 2x_+}{y + 2x_+} \exp(-y)\right] dy \, ,$$

where $t > 0$, $x_+ = L/\xi(t > 0)$. The behavior of $X_{ex}^{(o,o)}$ is visualized in Fig. 12.2.

Due to the absence of a phase transition in the two-dimensional Ising strip, $X_{ex}^{(+,+)}(x)$ is an analytic function of x for $x > 0$ and $x < 0$. In the

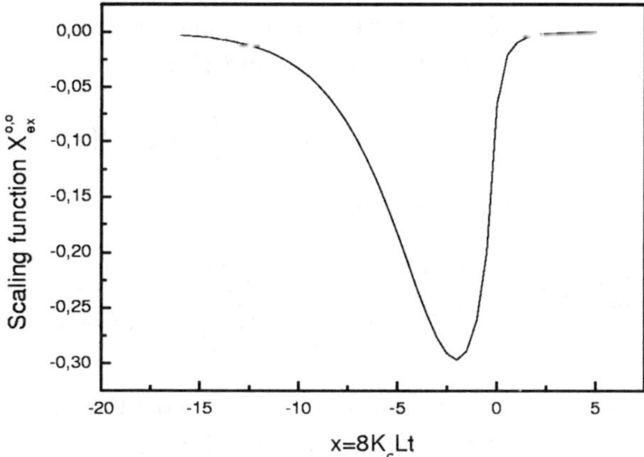

Fig. 12.2   The finite-size scaling function of the excess free energy under (o, o) boundary conditions. It is everywhere negative and attains minimum $X_{ex}^{+,+} \simeq -0.2965$ at $x \simeq -2.0$, i.e. *below* T_c.

limits $x \to 0$ and $x \to \pm\infty$ one has

$$X_{ex}^{(+,+)}(x) = \begin{cases} -\dfrac{1}{16\sqrt{2\pi|x|}}\exp(-|x|), & x \to -\infty \\[2mm] -\dfrac{\pi}{48} + \dfrac{1}{2\pi}x\ln|x| + cx - \dfrac{1}{8\pi}x^2\ln|x|, & x \to 0 \\[2mm] -\dfrac{1}{2}x\exp\left(-x/2\right), & x \to \infty \end{cases} \quad , \quad (12.1)$$

where c is a constant not specified here. The reformulation of the above limits for $X_{ex}^{(o,o)}$ in terms of the scaling variables x_- and x_+ yields:

$$X_{ex}^{(o,o)}(x_-) = \begin{cases} -\dfrac{\pi}{48} + \dfrac{1}{4\pi}x_-\ln x_- + O(x_-), & x_- \to 0, \\[2mm] -\dfrac{1}{2}x_-\exp(-x_-/2), & x_- \to \infty, \end{cases}$$

and

$$X_{ex}^{(o,o)}(x_+) = \begin{cases} -\dfrac{\pi}{48} + \dfrac{1}{2\pi}x_+\ln x_+ + O(x_+), & x_+ \to 0, \\[2mm] -\dfrac{1}{32\sqrt{\pi}}x_+^{-1/2}\exp\left(-2x_+\right), & x_+ \to \infty. \end{cases}$$

It can be shown that $X_{ex}^{(o,o)}(x)$ posesses a well pronounced minimum at $x_- \simeq 2.0$ $(t < 0)$ [Krech (1994)], and has an infinite slope at $x = 0^*$, see

---

*The different expansions around $x = 0$ in terms of $x = x_-$ and $x = x_+$ are due to the different values of the constants A_ξ^+ and A_ξ^-: $A_\xi^-/A_\xi^+ = 2$.

Fig. 12.2.

For the sake of completeness we give also the finite-size scaling function for $X_{\mathrm{Cas}}^{(+,+)}(x) = X_{\mathrm{Cas}}^{(o,o)}(-x)$. Having in mind that

$$X_{\mathrm{Cas}}^{(+,+)}(x) = X_{\mathrm{ex}}^{(+,+)}(x) - x\frac{d}{dx}X_{\mathrm{ex}}^{(+,+)}(x),$$

one obtains [Evans and Stecki (1994)]

$$X_{\mathrm{Cas}}^{(+,+)}(x) = -\frac{1}{\pi}\int_0^\infty \frac{\sqrt{x^2+y^2}}{1+\frac{\sqrt{x^2+y^2}-x}{x+\sqrt{x^2+y^2}}\exp\sqrt{x^2+y^2}}\,dy.$$

This function attains its minimum $X_{\mathrm{Cas}}^{(+,+)} \simeq -0.4305$ at $x \simeq 4.4571$, see the dotted line in Fig. 12.1. Note that the finite-size scaling functions of the Casimir force $X_{\mathrm{Cas}}^{(+,+)}$ and the excess free energy $X_{\mathrm{ex}}^{(+,+)}$ attain minima at quite different values of the scaling argument, though on the same side of T_c.

Furthermore, it is interesting to mention that $X_{\mathrm{Cas}}^{(+,+)}$ has minimum *above* T_c, while $X_{\mathrm{Cas}}^{(o,o)}$ attains its minimum *below* T_c. The asymptotic behavior of $X_{\mathrm{Cas}}^{(+,+)}(x)$ and $X_{\mathrm{Cas}}^{(o,o)}(x)$ readily follows from Eq. (12.1).

Let us consider now the case of a semi-infinite Ising strip under $(+,-)$ boundary conditions, when an interface develops at $T < T_c$. Taking into account that

$$\Delta f_L^{(+,-)}(T) = \Delta f_L^{(+,+)}(T) + \sigma_L^{(+,-)}(T), \tag{12.2}$$

one concludes that the corresponding scaling functions are related by the equation

$$X_{\mathrm{ex}}^{(+,-)}(x) = X_{\mathrm{ex}}^{(+,+)}(x) + \Sigma^{(+,-)}(x),$$

where $\Sigma^{(+,-)}$ is the finite-size scaling function of the *finite-size* interface free energy $\sigma_L^{(+,-)}$. In fact, one can consider Eq. (12.2) as a *definition* of the finite-size interface free energy, see [Privman (1990a)]. The relationship between the Casimir forces now readily follows

$$F_{L,\mathrm{Cas}}^{(+,-)}(T) = F_{L,\mathrm{Cas}}^{(+,+)}(T) - \frac{\partial}{\partial L}\sigma_L^{(+,-)}(T).$$

Hence one obtains

$$X_{\mathrm{Cas}}^{(+,-)}(x) = X_{\mathrm{Cas}}^{(+,+)}(x) + \delta X_{\mathrm{Cas}}^{(+,-)}(x), \tag{12.3}$$

where

$$\delta X_{\mathrm{Cas}}^{(+,-)}(x) = (d-1)\Sigma^{(+,-)}(x) - \frac{x}{\nu}\frac{d}{dx}\Sigma^{(+,-)}(x).$$

At any d, one has $\sigma_L^{(+,-)}(T) \to \sigma(T)$ as $L \to \infty$, where the interface free energy $\sigma(T)$ is independent of the precise choice of the boundary conditions used to create the interface. The following exact result for the $2d$ Ising model is known [Onsager (1944)]:

$$\sigma(T) = 2K - \mathrm{arcsinh}\,[1/\sinh(2K)] \qquad (d=2).$$

It can be shown that $\sigma_L^{(+,-)}(T) > \sigma(T) \geq 0$ [Evans and Stecki (1994)]. At the critical temperature $\sigma(T_c) = 0$ and [Evans and Stecki (1994)], [Stecki (1993)]

$$\sigma^{(+,-)}(T_c) = \frac{\pi}{2}L^{-1} - \sqrt{2}\pi L^{-2} + \frac{\pi}{4}\left(1 - \frac{1}{96}\pi^2\right)L^{-3} + \cdots, \qquad (12.4)$$

hence

$$\Delta^{(+,-)} = \Delta^{(+,+)} + \frac{\pi}{2} = \frac{23}{48}\pi \qquad (d=2).$$

The next-to-the-leading order terms in Eq. (12.4) yield corrections to scaling.

At low temperatures the behavior of $\sigma_L^{(+,-)}(T)$ is [Abraham and Svrakic (1986)], [Stecki (1993)], [Evans and Stecki (1994)]:

$$\sigma_L^{(+,-)}(T) = \sigma(T) + \frac{\pi^2}{2\sinh[\sigma(T)]}L^{-2} + O(L^{-3}),$$

and, therefore,

$$F_{L,\mathrm{Cas}}^{(+,-)}(T) = F_{L,\mathrm{Cas}}^{(+,+)}(T) + \frac{\pi^2}{\sinh[\sigma(T)]}L^{-3} + O(L^{-4}). \qquad (12.5)$$

The scaling function $\Sigma^{(+,-)}(x)$ of $\sigma_L^{(+,-)}(T)$ is also known [Abraham and Svrakic (1986)], [Sachdev (1993)], [Evans and Stecki (1994)]. It is given by the implicit equation $\Sigma^{(+,-)}(x) = y/\sin y$, where $y \cot y = x/2$ and $y \in [0,\pi]$, i.e., $x \in (-\infty, 2]$. At higher temperatures, when $x > 2$, one finds exponential decay of $\sigma_L^{(+,-)}(T)$, namely, it has been shown that [Evans and Stecki (1994)]

$$\Sigma^{(+,-)}(x) \simeq \frac{1}{2}x^2 \exp(-x/2).$$

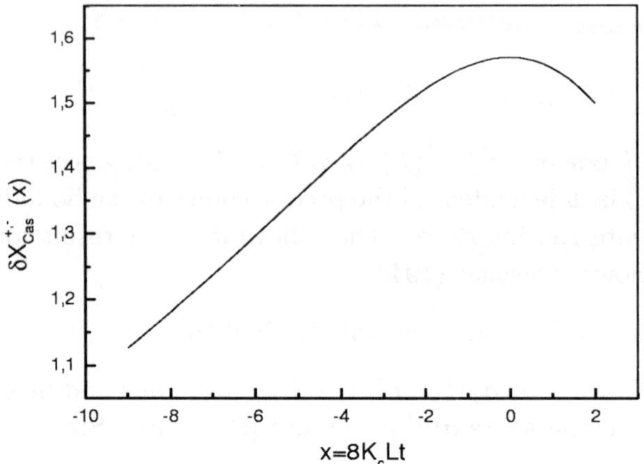

Fig. 12.3 The finite-size scaling function of the contribution to the free energy $\delta X_{\mathrm{Cas}}^{(+,-)}(x)$ from the interface in the system. It has a maximum $\delta X_{\mathrm{Cas}}^{+,-} = \pi/2$ at T_c.

Close to T_c one has in addition ($|x| \ll 1$)

$$\delta X_{\mathrm{Cas}}^{(+,-)}(x) \simeq \frac{\pi}{2} - \frac{\pi^2 - 8}{4\pi^3}x^2 - \frac{5\pi^2 - 48}{3\pi^5}x^3.$$

The analysis shows that $\delta X_{\mathrm{Cas}}^{(+,-)}(x)$ is positive and has maximum at T_c, see Fig. 12.3. Since $F_{L,\mathrm{Cas}}^{(+,-)}(T)$ and $F_{L,\mathrm{Cas}}^{(+,+)}(T)$ are related via the relation (12.3), the former function turns out to have a maximum $X_{\mathrm{Cas}}^{(+,-)} \simeq 1.5331$ below T_c, at $x \simeq -0.9642$, see Fig. 12.4. When $x \to -\infty$,

$$\delta X_{\mathrm{Cas}}^{(+,-)}(x) \simeq -2\pi^2/x + O(x^{-2}),$$

which leads to low-temperature asymptotic form of $F_{L,\mathrm{Cas}}^{(+,-)}(T)$ of the type given by Eq. (12.5).

Finally we note that close to T_c both $F_{L,\mathrm{Cas}}^{(+,-)}(T)$ and $F_{L,\mathrm{Cas}}^{(+,+)}(T)$ obtain $1/L$ nonuniversal corrections to their leading-order scaling behavior [Evans and Stecki (1994)].

Some results for the excess free energy under periodic boundary conditions can be inferred from [Ferdinand and Fisher (1969)]. At the critical point one obtains

$$f_{L,\mathrm{ex}}^{(p)}(T_c) = \frac{\pi}{12}L^{-2} + O(L^{-4}\ln^3 L),$$

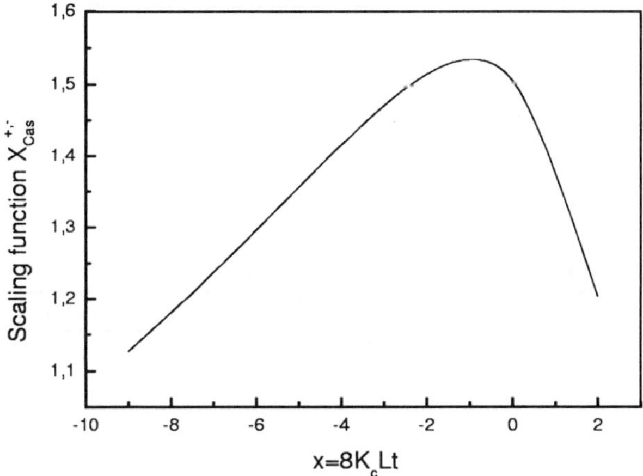

Fig. 12.4   The finite-size scaling function of the Casimir force $X_{\text{Cas}}^{(+,-)}(x)$. Note that the force is *positive*, i.e. it corresponds to a *repulsion* of the layers bounding the system. It attains maximum $X_{\text{Cas}}^{(+,-)} \simeq 1.5331$ at $x = -0.9642$, *below* T_c.

whereas at $x < 1$ the corresponding finite-size scaling function is given by

$$X_{\text{ex}}^{(p)}(x) = \frac{\pi}{12} + \pi \sum_{i=2}^{\infty} \binom{1/2}{i} \left(\frac{x}{2\pi}\right)^{2i} \left(i - 2^{-2i+1}\right) \zeta(2i - 1).$$

At $x \ll 1$ one obtains the expansion

$$X_{\text{ex}}^{(p)}(x) = \frac{\pi}{12} - \frac{7}{4^5 \pi^3} x^4 + O(x^6),$$

i.e., the leading x dependence is of the order x^4, instead of the order $x \ln x$ as for systems with edges (under $(+,+)$, $(+,-)$ or (o,o) boundary conditions).

Exact mean-field results in the three-dimensional case

The finite-size scaling functions for the Casimir force in mean-field approximation, $X_{\text{Cas}}^{(a,b)}(x)$ with $x = tL^2$, have been considered in [Krech (1997)] at $d = 3$. The following results have been obtained.

Under $(+,+)$ boundary conditions:

(i) If $x \geq -\pi^2$, then

$$X_{\text{Cas}}^{(+,+)}(x) = -(2K)^4 k^2 (1 - k^2), \qquad (12.6)$$

where $k = k(x)$ is given by the parametric equation $x = (2K)^2(2k^2 - 1)$;

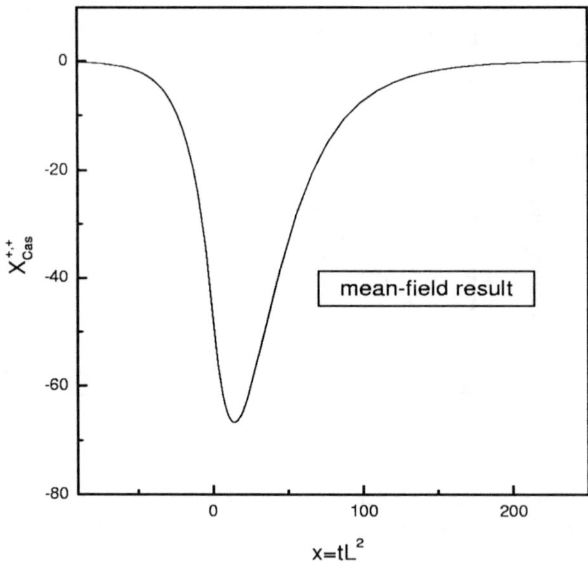

Fig. 12.5   The zero-field finite-size scaling function $X_{\mathrm{Cas}}^{(+,+)}$ of the Casimir force as a function of the scaling variable $x = tL^2$, obtained by mean-field methods for the $d = 3$ Ising model uder $(+, +)$ boundary conditions. The function is given in arbitrary units. Note that it is negative and has a *minimum above* T_c, as in the case of the $2d$ Ising model.

(ii) If $x \leq -\pi^2$, then

$$X_{\mathrm{Cas}}^{(+,+)}(x) = (2K)^4 k^2, \tag{12.7}$$

where $k = k(x)$ is to be determined from $x = -(2K)^2(k^2 + 1)$. In the above equations $K \equiv K(k)$ is the complete elliptic integral of the first kind. The behavior of $X_{\mathrm{Cas}}^{(+,+)}(x)$ is shown in Fig. 12.5.

Under $(+, -)$ boundary conditions:

(i) If $x \geq 2\pi^2$, then

$$X_{\mathrm{Cas}}^{(+,-)}(x) = (2K)^4(1 - k^2)^2, \tag{12.8}$$

where $k(x)$ is given by the equation $x = 2(2K)^2(k^2 + 1)$.

(ii) If $x \leq 2\pi^2$. then

$$X_{\mathrm{Cas}}^{(+,-)}(x) = (2K)^4, \tag{12.9}$$

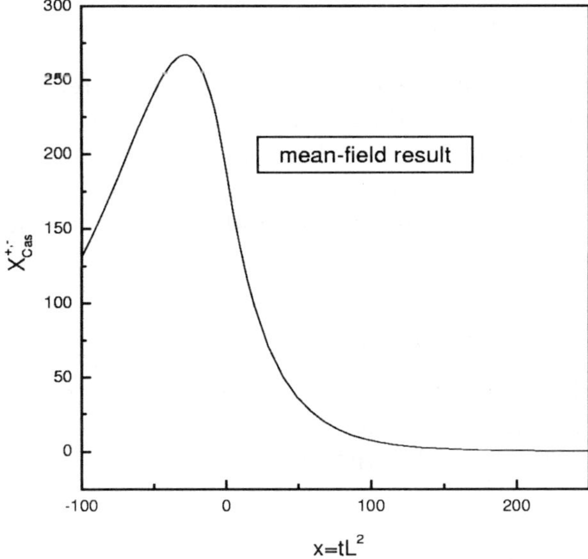

Fig. 12.6   The zero-field finite-size scaling function $X^{(+,-)}_{\mathrm{Cas}}$ of the Casimir force as a function of the scaling variable $x = tL^2$, obtained by mean-field methods for the $d = 3$ Ising model uder $(+, -)$ boundary conditions. The function is given in arbitrary units. Note that it is positive and has a *maximum below* T_c, as in the case of the 2d Ising model.

where $k(x)$ is to be determined from $x = -2(2K)^2(2k^2 - 1)$. The behavior of $X^{(+,-)}_{\mathrm{Cas}}(x)$ is shown in Fig. 12.6.

Under $(o, +)$ boundary conditions:

$$X^{(o,+)}_{\mathrm{Cas}}(x) = \frac{1}{16} X^{(+,-)}_{\mathrm{Cas}}(4x). \tag{12.10}$$

Under $(sb, +)$ boundary conditions:

$$X^{(sb,+)}_{\mathrm{Cas}}(x) = \frac{1}{16} X^{(+,+)}_{\mathrm{Cas}}(4x). \tag{12.11}$$

All the above Casimir force functions are given in arbitrary units; it is their functional form that gives us the important information. Note that both $X^{(+,+)}_{\mathrm{Cas}}$ and $X^{(sb,+)}_{\mathrm{Cas}} > 0$ are positive and have a minimum *above* T_c, whereas $X^{(+,-)}_{\mathrm{Cas}}$ and $X^{(o,+)}_{\mathrm{Cas}}$ exhibit maximum *below* T_c.

12.2   Results for $O(n)$ models

The existing analytical results are due to field-theoretical analysis of the Casimir effect in $d = 4 - \varepsilon$ dimensions of the Ginzburg-Landau Hamiltonian for $O(n)$ symmetric systems. Some quantitative information about the dependence of the Casimir amplitude as a function of the angle between two infinite surface fields, applied to a three-dimensional film, are also available due to mean-field calculations.

12.2.1   *Casimir amplitudes*

The Casimir amplitudes have been obtained to first order in ε in a two-loop approximation, by using dimensional regularization techniques and ε-expansion [Krech and Dietrich (1992)]:

$$
\frac{\Delta^{(o,o)}}{n} = \frac{\Delta^{(sb,sb)}}{n}
$$

$$
= -\frac{\pi^2}{1440}\left\{1 + \varepsilon\left[\ln\left(2\sqrt{\pi}\right) + \frac{\gamma - 1}{2} - \frac{\zeta'(4)}{\zeta(4)} - \frac{5\,n + 2}{4\,n + 8}\right]\right\},
$$

$$
\frac{\Delta^{(o.sb)}}{n} = \frac{7}{8}\frac{\pi^2}{1440}\left\{1 + \varepsilon\left[\ln\left(2\sqrt{\pi}\right) + \frac{\gamma - 1}{2} - \frac{\zeta'(4)}{\zeta(4)} - \frac{\ln 2}{7} + \frac{5}{14}\frac{n + 2}{n + 8}\right]\right\},
$$

$$
\frac{\Delta^{(p)}}{n} = -\frac{\pi^2}{90}\left\{1 + \varepsilon\left[\ln\sqrt{\pi} + \frac{\gamma - 1}{2} - \frac{\zeta'(4)}{\zeta(4)} - \frac{5\,n + 2}{4\,n + 8}\right]\right\},
$$

$$
\frac{\Delta^{(a)}}{n} = \frac{7}{8}\frac{\pi^2}{90}\left\{1 + \varepsilon\left[\ln\sqrt{\pi} + \frac{\gamma - 1}{2} - \frac{\zeta'(4)}{\zeta(4)} - \frac{\ln 2}{7} + \frac{5}{14}\frac{n + 2}{n + 8}\right]\right\}.
$$

Some available mean-field results for $O(n)$ models at $d = 3$ [Krech (1997)] illustrate the dependence of the Casimir amplitude Δ_α on the angle α between two surface vector fields $\mathbf{h}_1$ and $\mathbf{h}_2$ in the limit $|\mathbf{h}_1|, |\mathbf{h}_2| \to \infty$. The amplitude Δ_α is given implicitly as a solution of the set of equations

$$
\alpha = \sqrt{1 + \omega}\int_1^\infty x^{-1}\left[x^3 - 1 + \omega(x - 1)\right]^{-1/2}\mathrm{d}x,
$$

$$
\psi = \int_1^\infty \left[x^3 - 1 + \omega(x - 1)\right]^{-1/2}\mathrm{d}x, \quad \omega = \psi^{-4}\Delta_\alpha,
$$

where $0 \le \alpha \le \pi$. The functional dependence of Δ_α on α is illustrated in Fig. 12.7. Note that Δ increases monotonically from $\alpha = 0$ to $\alpha = \pi$, and vanishes at $\alpha = \pi/3$.

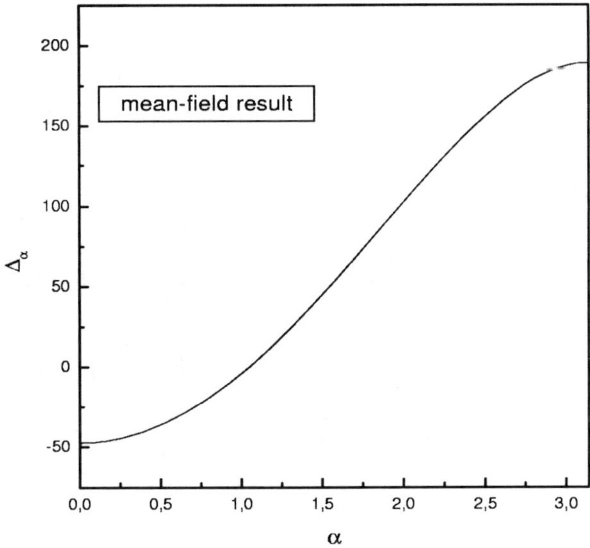

Fig. 12.7 The Casimir amplitude Δ_α (in arbitrary units) for an $O(n)$ model as a function of the angle between the surface vector fields $\mathbf{h_1}$ and $\mathbf{h_2}$ in the limit $|\mathbf{h_1}|, |\mathbf{h_2}| \to \infty$. Note that $\Delta_\alpha = 0$ at $\alpha = \pi/3$.

12.2.2 *Scaling functions*

As for the Casimirs amplitudes, the results available for the universal finite-size scaling function of the excess free energy *above* the bulk critical temperature T_c have been obtained to first order in ε in a two-loop approximation, by using dimensional regularization techniques and ε-expansion [Krech and Dietrich (1992)]:

$$\frac{X_{\text{ex}}^{(o,o)}(x)}{n} = \frac{x^4}{8\pi^2}\left\{\left[-2 + \varepsilon\left(\gamma - \frac{8}{3} + \ln\frac{x^2}{\pi}\right)\right]g_{3/2,0}(x)\right.$$
$$\left. + \varepsilon\left[g_{3/2,1}(x) + \frac{n+2}{n+8}\left(\frac{\pi}{x}g_{1/2,0}(x) + g_{1/2,0}^2(x)\right)\right]\right\},$$

$$\frac{X_{\text{ex}}^{(sb,sb)}(x)}{n} = \frac{x^4}{8\pi^2}\left\{\left[-2 + \varepsilon\left(\gamma - \frac{8}{3} + \ln\frac{x^2}{\pi}\right)\right]g_{3/2,0}(x)\right.$$
$$\left. + \varepsilon\left[g_{3/2,1}(x) + \frac{n+2}{n+8}\left(-\frac{\pi}{x}g_{1/2,0}(x) + g_{1/2,0}^2(x)\right)\right]\right\},$$

$$\frac{X_{\text{ex}}^{(o,sb)}(x)}{n} = \frac{x^4}{8\pi^2}\left\{\left[2 - \varepsilon\left(\gamma - \frac{8}{3} + \ln\frac{x^2}{\pi}\right)\right] h_{3/2,0}(x) \right.$$
$$\left. -\varepsilon\left[h_{3/2,1}(x) - \frac{n+2}{n+8}h_{1/2,0}^2(x)\right]\right\},$$

$$\frac{X_{\text{ex}}^{(p)}(x)}{n} = \frac{x^4}{8\pi^2}\left\{\left[-2 + \varepsilon\left(\gamma - \frac{8}{3} + \ln\frac{x^2}{\pi}\right)\right] g_{3/2,0}(x/2) \right.$$
$$\left. +\varepsilon\left[g_{3/2,1}(x/2) + \frac{n+2}{n+8}g_{1/2,0}^2(x/2)\right]\right\},$$

$$\frac{X_{\text{ex}}^{(a)}(x)}{n} = \frac{x^4}{8\pi^2}\left\{\left[2 - \varepsilon\left(\gamma - \frac{8}{3} + \ln\frac{x^2}{\pi}\right)\right] h_{3/2,0}(x/2) \right.$$
$$\left. -\varepsilon\left[h_{3/2,1}(x/2) - \frac{n+2}{n+8}h_{1/2,0}^2(x/2)\right]\right\},$$

where $x = L/\xi$, and

$$g_{a,b}(y) = \frac{1}{a}\int_1^\infty \frac{(x^2-1)^a}{\exp(2xy)-1}\left[\ln(x^2-1)\right]^b dx,$$
$$h_{a,b}(y) = \frac{1}{a}\int_1^\infty \frac{(x^2-1)^a}{\exp(2xy)+1}\left[\ln(x^2-1)\right]^b dx.$$

The asymptotic expansion of the functions $g_{a,b}(y)$ and $h_{a,b}(y)$ as $y \to \infty$ leads to the following asymptotic behavior of the finite-size scaling functions for the excess free energy:

$$\frac{X_{\text{ex}}^{(o,o)}(x)}{n} \simeq -\frac{1}{16\pi^{3/2}}\left[1 + \varepsilon\left(\ln\left(2\sqrt{\pi}\right) - \frac{n+2}{n+8}\pi\right)\right] x^{(d-1)/2}e^{-2x},$$

$$\frac{X_{\text{ex}}^{(sb,sb)}(x)}{n} \simeq -\frac{1}{16\pi^{3/2}}\left[1 + \varepsilon\left(\ln\left(2\sqrt{\pi}\right) + \frac{n+2}{n+8}\pi\right)\right] x^{(d-1)/2}e^{-2x},$$

$$\frac{X_{\text{ex}}^{(o,sb)}(x)}{n} \simeq \frac{1}{16\pi^{3/2}}\left[1 + \varepsilon\ln\left(2\sqrt{\pi}\right)\right] x^{(d-1)/2}e^{-2x},$$

$$\frac{X_{\text{ex}}^{(p)}(x)}{n} \simeq -\frac{1}{(2\pi)^{3/2}}\left[1 + \frac{\varepsilon}{2}\ln(2\pi)\right] x^{(d-1)/2}e^{-x},$$

$$\frac{X_{\text{ex}}^{(a)}(x)}{n} \simeq \frac{1}{(2\pi)^{3/2}}\left[1 + \frac{\varepsilon}{2}\ln(2\pi)\right] x^{(d-1)/2}e^{-x}.$$

These results show that the system approaches its thermodynamic behavior exponentially fast in L on leaving the finite-size scaling region $L/\xi = O(1)$.

Finaly, we present a comparison of the renormalization group results for $X_{ex}^{(p)}/n$ when $n = 1, 2, 3$, with the exact spherical model results corresponding to $n = \infty$, see Fig. 12.8. Note that, against the general expectation, the finite-size scaling function of the Ising model, rather than that of the Heisenberg one ($n = 3$), lies closest to the finite-size scaling function of the spherical model. Furthermore, one observes that $X_{ex}^{(p)}/n$ is an *increasing* function of n. Having in mind that the spherical model under periodic boundary conditions is a $n \to \infty$ limit of $O(n)$ models, see Section 3.2, one may conclude that the ϵ-expansion results in first order of ϵ are too crude approximation in order to capture the way in which the excess free energy of the $O(n)$ models approaches that of the spherical model. It is worth mentioning also that the scaling function X_{ex} of the spherical model is a *monotonically* increasing function of the scaling variable x, whereas the renormalization group results for the Heisenberg and XY models look quite simmilar to those for the Ising model: their functions $X_{ex}^{(p)}/n$ have a *minimum* above T_c. The Monte Carlo results for the Ising model yield $\Delta = -0.1526 \pm 0.0010$ [Krech (1997)], which is *surprisingly close* to the exact value for the spherical model $\Delta^{(p)} = -2\zeta(3)/(5\pi) \simeq -0.153051$. Note that all the curves practically overlap at $L > 2\xi$.

12.3 Results for two-dimensional strips

In two dimensions the Casimir amplitudes for systems with strip geometry can be obtained by conformal invariance methods. Within conformal field theory in $2d$ the universality class of a critical system is characterized by the value of a single scalar parameter c called *central charge* or *conformal anomaly number*. The value of c cannot be determined from general conformal invariance considerations alone. However, the requirement that the *critical* statistical mechanical system, which one describes as a conformal field theory, has a *unitary transfer matrix* leads to a "quantization" of the possible values of $c < 1$. According to [Friedan et. al. (1984)], one has

$$c = 1 - \frac{6}{m(m+1)},$$

where $m \in \{3, 4, 5, \cdots\}$. The following identifications with critical and tricritical models have been made for given values of m [Friedan et. al. (1984)]

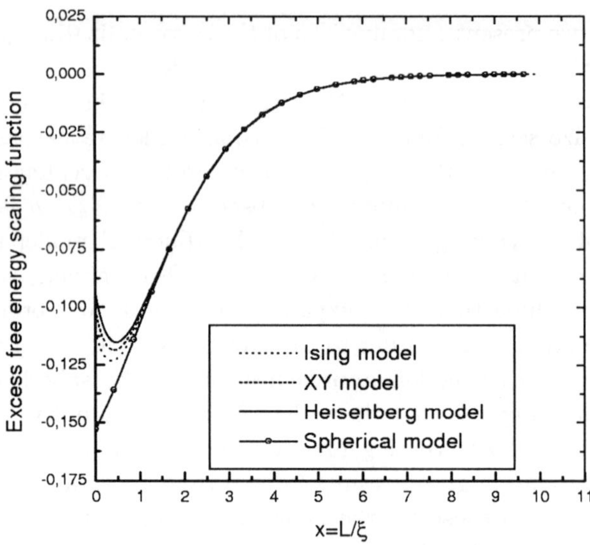

Fig. 12.8 The *universal* zero-field finite-size scaling functions $X_{\mathbf{ex}}$ of the excess free energy as a function of the scaling variable $x = L/\xi(T > T_c)$ for the Ising, XY, Heisenberg and spherical models.

m	model	c
3	critical Ising	1/2
4	tricritical Ising	7/10
5	critical 3-state Potts	4/5
6	tricritical 3-state Potts	6/7
∞	critical Gaussian	1

The universality class of these $2d$ models is uniquely determined by the value of c (or m).

The following general results for the Casimir amplitudes are known [Cardy (1986)], [Blöte et. al. (1986)], [Affleck (1986)]:

a) under periodic boundary conditions

$$\Delta^{(p)} = -\frac{\pi}{6}c$$

b) under ordinary and $(+,+)$ boundary conditions

$$\Delta^{(o,o)} = \Delta^{(+,+)} = -\frac{\pi}{24}c.$$

Under other boundary conditions, the amplitudes have been determined in [Cardy (1986)] for the critical Ising and 3-state Potts models. Note that for the 3-state Potts model one considers also (s, s) boundary conditions, which means that equal surface Potts fields are applied, such that the Potts spins at the boundaries have the same value s. Then $\Delta^{(o,o)} = \Delta^{(s,s)}$ for any Potts state s.

12.4 Results for curved geometries

As already mentioned, it is demanding to maintain parallelism of the two plates with sufficiently high accuracy (10^{-5} rad for 1 cm diameter plates [Lamoreaux (1997)]). Therefore, some theoretical efforts have been made to investigate Casimir forces in geometries like a sphere in front of a planar wall[†] [Eisenriegler and Ritschel (1995)], [Hanke et. al.], or in the case of two spheres [Burkhardt and Eisenriegler (1995)], [Eisenriegler and Ritschel (1995)]. These geometries are much more easy to control experimentally, besides, they are closer to reality in, e.g., colloidal suspensions [Beysens and Esteve].

Most of the results available for the two spheres geometry are due to conformal invariance methods, which are a valuable and powerful tool for investigations at the critical point of the system. Let $f^{(a,b)}(T_c, r, r_1, r_2)$ be the free energy density (per $k_B T$) of an infinite critical fluid in which two spheres with radii r_1 and r_2, at a distance r apart (measured from center to center), are immersed. Boundary conditions (a) and (b) are imposed at the surfaces of the spheres. Let

$$f_{\text{ex}}^{(a,b)}(T_c, r, r_1, r_2) = f^{(a,b)}(T_c, r, r_1, r_2) - f^{(a,b)}(T_c, \infty, r_1, r_2)$$

be the excess free energy of such a system. Scale invariance requires $f_{\text{ex}}^{(a,b)}$ to be a function of two independent, scale-invariant variables, e.g., $r^2/(r_1 r_2)$ and r_1/r_2. Conformal invariance in *general* dimensionality d leads to the result that $f_{\text{ex}}^{(a,b)}$ depends on a single variable $\kappa = (2r_1 r_2)^{-1}|r^2 - r_1^2 - r_2^2|$:

$$f_{\text{ex}}^{(a,b)}(T_c, r, r_1, r_2) = X_{\text{ex}}^{(a,b)}(\kappa), \quad 1 < \kappa < \infty. \tag{12.12}$$

It turns out possible to make definite predictions, on the ground of general reasoning, about the behavior of $X_{\text{ex}}^{(a,b)}(\kappa)$ in the limits $\kappa \to 1$ and

---

[†]Such a geometry is particularly important for interpreting direct force measurements made by atomic force microscope.

$\kappa \to \infty$.

First, let us note that the case of a two spheres with radii r_1 and r_2 (let $r_2 > r_1$ for definitness) in an unbounded critical fluid is conformally equivalent to that of a single sphere of radius r_1 immersed in a critical fluid of radius r_2. The last implies that these two cases can be described by the same universal scaling function $X_{\mathrm{ex}}^{(a,b)}(\kappa)$. Furthermore, it is clear that the asymptotic form of $X_{\mathrm{ex}}^{(a,b)}(\kappa)$ as $\kappa \to 1^+$ is determnined by the Casimir amplitude $\Delta^{(a,b)}$ for the parallel plates geometry. Indeed, considering the case of concentric spheres ($r = 0$) in the limit $r_1, r_2 \to \infty$ at fixed $L = r_2 - r_1$, and comparing Eqs. (12.12) and (11.4), one obtains [Burkhardt and Eisenriegler (1995)]

$$X_{\mathrm{ex}}^{(a,b)}(\kappa) \simeq \mathcal{A}_d(1)\Delta^{(a,b)}[2(\kappa - 1)]^{-(d-1)/2}, \qquad (12.13)$$

where $\mathcal{A}_d(1)$ is the surface area of the unit sphere in R^d, see Eq. (3.57). This expression determines the Casimir interaction of two spheres that nearly touch immersed in an unbounded critical fluid. If $D = r - r_1 - r_2$ is the distance between the closest points of the two spheres, and if $D \ll r_1, r_2,,$ then $\kappa \simeq 1 + D(1/r_1 + 1/r_2)$ and

$$f_{\mathrm{ex}}^{(a,b)}(T_c, r, r_1, r_2) \simeq \mathcal{A}_d(1)\Delta^{(a,b)} \left\{ 2\left[1 + D\left(\frac{1}{r_1} + \frac{1}{r_2}\right)\right] \right\}^{-(d-1)/2} .$$

In the opposite limit $r \gg r_1, r_2$, by using operator product expansion. one derives [Burkhardt and Eisenriegler (1995)] (performing "small-sphere expansion")

$$f_{\mathrm{ex}}^{(a,b)}(T_c, r, r_1, r_2) = -B^{(a,b)} \left(\frac{r_1 r_2}{r^2}\right)^{x_\psi}, \qquad (12.14)$$

where ψ is the order parameter operator ϕ if the two boundary conditions ($a = b = e$) correspond to the extraordinary surface universality class. In this case the scaling exponent is $x_\phi = \beta/\nu$ (for the $d = 3$ Ising model $x_\phi \simeq 0.518$, and for the $d = 3$ XY model $x_\phi \simeq 0.519$). If at least one of the two spheres does not have a symmetry-breaking boundary condition, then the operator ψ corresponds to the local energy density operator ε and the corresponding scaling exponent is $x_\varepsilon = d - 1/\nu$ (for the $d = 3$ Ising model $x_\varepsilon \simeq 1.41$, and for the $d = 3$ XY model $x_\varepsilon \simeq 1.51$). The amplitude

$$B^{(a,b)} = A_\psi^{(a)} A_\psi^{(b)} / B_\psi,$$

where $A_\psi^{(a)}$ and $A_\psi^{(b)}$ are the amplitudes of the critical profiles of the operator ψ in a seminfinite system bounded by a planar surface of type $(\tau) = (a,b)$ in the z direction, i.e., $\langle\psi(z)\rangle_{\text{half space}}^{(\tau)} = A_\psi^{(a)}(2z)^{-x_\psi}$. The amplitude B_ψ is the amplitude of the bulk correlation function $\langle\psi(\mathbf{R}_1)\psi(\mathbf{R}_2)\rangle_{\text{bulk}} = B_\psi R^{-2x_\psi}$. Although neither $A_\psi^{(\tau)}$, nor B_ψ are universal, their combination $B^{(a,b)}$ is universal. Its value for the $d = 2$ Ising model is exactly known (then $x_\phi = 1/8$, $x_\epsilon = 1$): $B^{(+,+)} = \sqrt{2}$ and $B^{(a,b)} = 1$ for all other boundary conditions. In $d = 4 - \epsilon$ dimensions the renormalization group calculations yield [Burkhardt and Eisenriegler (1995)]:

$$
\begin{aligned}
B^{(+,+)} &= 45\epsilon^{-1}\left[1 - \frac{62}{27}\epsilon + O\left(\epsilon^2\right)\right], \\
B^{(o,o)} &= \frac{1}{2}\left[1 + O(\epsilon^2)\right], \\
B^{(sb,sb)} &= \frac{1}{2}\left[1 + \frac{2}{3}\epsilon + O(\epsilon^2)\right].
\end{aligned}
$$

The Casimir interaction of a single sphere with a planar boundary follows from the results for the two-spheres interaction in the limit $r_2 \to \infty$. Denoting the radius of the single sphere by r and the distance from its center to the planar boundary by $r + D$, one obtains $\kappa = 1 + D/r$ and from Eqs. (12.13) and (12.14) it follows that

$$
f_{\text{ex}}^{(a,b)}(T_c, D, r) = \begin{cases} S_d \Delta^{(a,b)}(r/2D)^{(d-1)/2}, & D \ll r, \\ -B^{(a,b)}(r/2D)^{x_\psi}, & D \gg r. \end{cases} \tag{12.15}
$$

From Eqs. (12.13), (12.14) and (12.15) one concludes that the Casimir interaction between two spheres, and between a sphere and a wall, is very long-ranged. For the extraordinary surface universality class the value of x_ϕ in $d = 3$ implies a $r^{-1.04}$ decay of the Casimir potential energy of two widely separated colloidal particles in a one-component fluid at the liquid-vapor critical point, or at the consolute point of a binary mixture. This is the case of the $d = 3$ Ising bulk universality class and $(+, +)$ boundary conditions. Note that the Casimir potential energy decays almost as slowly as the Coulomb interaction or Newtonian gravitation. (The thermodynamics of system controlled by gravitational forces is an extensively studied but still controversal topic.) Then the volume integral $\int d^3r\, f_{\text{ex}}^{(a,b)}(T_c, r, r_1, r_2)$ diverges, i.e. the total potential energy of a homogeneous configuration of colloidal particles is more than extensive. For colloidal particles in Helium at the lambda point (the case of the XY model with (o, o) boundary conditions), the Casimir potential decays as $r^{-3.02}$. Note that in both cases the Casimir force decays much slower than the van der Waals force, and is stronger than it for all r greater and close to $r_1 + r_2$. This implies that even if the van der Waals force alone is not strong enough to produce a phase transition of colloidal particles, they will form a condensed phase at the critical point of the fluid due to the strong and long-ranged Casimir force. Recalling that $\xi \to \infty$ as $T \to T_c$, one expects aggregation or flocculation due to the Casimir force in a temperature range $T_1 < T < T_2$, where $T_1 < T_c < T_2$. Flocculation of colloidal particles in nearly critical fluids has been observed experimentally.

Let us turn now to the full functional form of $X_{\text{ex}}^{(a,b)}(\kappa)$. Here the results are quite scarse and are actually mean field type ones. In [Eisenriegler and Ritschel (1995)] the properties of $X_{\text{ex}}^{(a,b)}(\kappa)$ at d near the upper critical dimension $d_u = 4$ have been derived. Under boundary conditions with *broken order parameter symmetry*, i.e., $(sb, +)$, $(o, +)$, $(+, +)$ and $(+, -)$, the leading behavior in the ϵ expansion follows from the mean-field theory, while under *symmetry preserving* boundary conditions, i.e., (o, o), (o, sb) and (sb, sb), the leading behavior is that of the Gaussian model. Since the Gaussian model is conformally invariant for all $d > 2$, one can obtain the corresponding critical point behavior at any dimensionality d. The results obtained in [Eisenriegler and Ritschel (1995)] are given below.

Under boundary conditions with broken order parameter symmetry one has

$$X_{\text{ex}}^{(a,b)}(\kappa(\rho(x))) = \frac{9S_d}{2^{d+1}\pi^2\epsilon} \int_0^x dy \frac{d\ln(\rho(y))}{dy} y, \qquad (12.16)$$

where $\kappa(\rho) = \frac{1}{2}(\rho + \rho^{-1})$, $\rho = r_1/r_2$, $\epsilon = 4 - d$, and the function $\rho(x)$ is given by:

(a) Under $(+, +)$ boundary conditions

$$\ln \rho(x) = -\psi(x), \qquad (12.17)$$

with

$$\psi(x) = 2\omega^{-1/2}K([(1 + 1/\omega)/2]^{1/2}). \qquad (12.18)$$

Here $K(\kappa) = F(\pi/2, \kappa)$ is the complete elliptic integral and $\omega = \sqrt{1 - x}$.

(b) Under $(sb, +)$ boundary conditions

$$\psi(x) = -\omega^{-1/2} K\left(\left[(1 + 1/\omega)/2\right]^{1/2}\right).$$ (12.19)

(c) Under $(+, -)$ boundary conditions

$$\ln \rho(x) = -\varphi(x),$$ (12.20)

where

$$\varphi(x) = 2^{3/2} x^{-1/4} \frac{2}{\delta} K\left(\frac{2}{\delta} - 1\right)$$ (12.21)

with

$$\delta = 1 + \left[(1 + x^{-1/2})/2\right]^{1/2}.$$ (12.22)

(d) Under $(+, o)$ boundary conditions

$$\ln \rho(x) = -2^{1/2} x^{-1/4} \frac{2}{\delta} K\left(\frac{2}{\delta} - 1\right).$$ (12.23)

From the above expressions it follows that $X_{\text{ex}}^{(a,b)} < 0$ under $(+, +)$ and $(+, sb)$ boundary conditions, which implies that the Casimir force is *attractive*. For $(+, -)$ and $(o, +)$ boundary conditions $X_{\text{ex}}^{(a,b)} > 0$, i.e., the force is *repulsive*.

To the group of symmetry preserving boundary conditions belong the types (o, o), (o, sb) and (sb, sb). Under these boundary conditions the $O(n)$ Gaussian model yields [Eisenriegler and Ritschel (1995)]

$$
\begin{aligned}
X_{\text{ex}}^{(a,b)}(\kappa) &= \frac{N}{d-2} \sum_{l=0}^{\infty} \ln[1 - \delta^{(a,b)} \rho(\kappa)^{2\lambda}] \frac{1}{2\lambda} \\
&\quad \times \binom{d-3+l}{l} \left\{ 2l(d-2+l) + \frac{1}{2}(d-2)^2 \right\},
\end{aligned}
$$ (12.24)

where $\lambda = l + (d-2)/2$. One observes that for all values of κ (or ρ) the interaction is attractive if $(a) = (b)$ and repulsive if $(a) \neq (b)$.

Although the results available are only of mean-field type calculations, their combination with the exact Ising model results at $d = 2$ leads to reliable estimates of $X_{\text{ex}}^{(a,b)}(\kappa)$ within the Ising universality class under $(+, +)$, $(+, -)$, $(o, +)$, and (o, o) boundary conditions.

All the results for the Casimir force presented above have been obtained by using conformal transformations and, therefore, are restricted only to

$T = T_c$, where the correlation length ξ is infinite (if ξ is finite, the conformal invariance no longer holds). Nevertheles, there exist some results for the temperature dependence of the force. This problem has been considered in [Hanke et. al.]. So far, the results are available only for Ising like systems and for the generic case of a spherical particle with radius r located at a distance $D = r - r_1 - r_2$ from a planar wall under $(+, +)$ boundary conditions (when the same of the two coexisting bulk phases is enriched near the wall and sphere surfaces). The particle may be considered as a freely moving colloidal particle but also as a sphere attached to the tip of an atomic force microscope.

Close to T_c the singular contribution of the Casimir force is of the form

$$F_{\text{Cas}}^{(+,+)} = r^{-1} X_{\text{Cas}}^{(+,+)}(D/\xi, D/r) \qquad (T > T_c). \qquad (12.25)$$

The universal function $X_{\text{Cas}}^{(+,+)}$ corresponds to attraction and has (as the parallel plates geometry does) a maximum as a function of T, with D and r fixed, at $T_{max}(D, r)$ *above* T_c. The fluid is described by the standard φ^4 Hamiltonian, see also Section 1.6.1,

$$H\{\varphi\} = \int_V dV \left\{ \frac{1}{2}(\nabla\varphi)^2 + \frac{\tau}{2}\varphi^2 + \frac{u}{24}\varphi^4 - h\varphi \right\}, \qquad (12.26)$$

for a scalar order parameter $\varphi(\mathbf{r})$ in cylindrical coordinates $\mathbf{r} = (\rho, z) \in R^d$ with boundary condition $\varphi = \infty$ at the wall and sphere surfaces (corresponding to the critical adsoption fixed point). The region V is the half-space $z \geq 0$ excluding the volume ocupied by the sphere. The field h is conjugate to the deviation of the concentration from the critical composition. The scaling functions are obtained from mean field evaluation of the stress tensor by using the mean field order parameter profile. The profile diverges near the wall and the sphere surfaces; it is obtained by numerical solution of the Euler-Lagrange equation for Eq. (12.26).

When $r \gg D$ one can use the so-called Derjaguin approximation, in which the presence of the sphere is modeled by a system of parallel plate contributions, the "plates" being of width $d\rho$ and radius ρ, at distance $L(\rho) = D + \rho^2/(2r)$ from the wall. Then, as $D/r \to 0$ at $T > T_c$, the scaling function $X_{\text{Cas}}^{(+,+)}(D/\xi, D/r)$ of the Casimir force *on the sphere* is expressed in terms of $X_{\text{Cas}}^{(+,+)}(L/\xi)$ for the force between two *parallel plates* at distance L. At $d = 3$ and $d = 4$ this procedure ends up with the result

[Hanke et. al.]

$$X_{\text{Cas}}^{(+,+)}(D/\xi, D/r) = \omega(d) \left(\frac{D}{r}\right)^{-(d+1)/2}$$

$$\times \int_0^\infty \frac{\mathrm{d}\alpha\, \alpha^{d-2}}{(1+\alpha^2/2)^d} X_{\text{Cas}}^{(+,+)}((D/\xi)(1+\alpha^2/2)), \qquad (12.27)$$

where $\omega(3) = 4\pi$ and $\omega(4) = 12\pi$. In order to obtain a reliable approximation in the case $d = 3$, one interpolates to $d = 3$ the exact results available for $X_{\text{Cas}}^{(+,+)}(L/\xi)$, $T > T_c$, at $d = 2$ and $d = 4$.

When $r \ll D, \xi$ one can apply the so-called small sphere expansion [Burkhardt and Eisenriegler (1995)], [Eisenriegler and Ritschel (1995)]. The corresponding results for $d = 4 - \varepsilon$ dimensions are quite cumbersome and we refer the reader to [Hanke et. al.], [Diehl and Smock (1993a)], [Diehl and Smock (1993b)]. Here we only quote that at $d = 3$

$$X_{\text{Cas}}^{(+,+)}(x,y) = -\frac{B^{(+,+)}}{c_+} \frac{x^{\beta/\nu+1}}{2^{\beta/\nu}} P_+(x) y^{-\beta/\nu-1}$$

$$+ \left[\frac{B^{(+,+)}}{c_+^2}\right]^2 \frac{x^{2\beta/\nu+1}}{2^{2\beta/\nu}} P_+(x) P_+'(x) y^{-2\beta/\nu-1} + O(y^{-d+1/\nu-1}),$$
$$(12.28)$$

where $P_+(x)$ is the universal scaling function (in self-explained notations)

$$P_+(z/\xi_+) = \langle\varphi\rangle_{\text{half space},t>0}^+ / \langle\varphi\rangle_{\text{bulk},t>0}.$$

The universal amplitude c_+ characterizes the order parameter profile (for $T > T_c$) at critical adsorption on a planar substrate. In three dimensions $B^{(+,+)} \simeq 7.73$ and $c_+ \simeq 0.717$. From these results one concludes that at values typical for atomic force microscope ($r \simeq 10^{-6}$ m, $D \simeq 10^{-8}$ m) and $T_c \simeq 300$ K, the singular part of the Casimir force should be of the order of 10^{-10} N, while the corresponding van der Waals force is of the order 10^{-11} N. Therefore, near T_c the critical Casimir force *dominates* the background dispersion forces. Finally, it is worth mentioning that, within the small sphere expansion, the Casimir force between two spherical particles turns out to be nonsymmetric with respect the to deviations from the critical concentration of a critical binary liquid mixture in such a way, that that the Casimir force is *enhanced* when the concentration of the component preferably adsorbed by the (colloidal) particles is *reduced*. This assymetry is consistent with the shape of experimentally observed flocculation phase diagrams [Beysens and Esteve], [Gallagher and Maher (1992)], [Jayalakshmi and Kaler (1997)].

The Casimir force in the concentric spheres geometry has been considered also at tricritical points [Ritschel and Gerwinski (1997)]. Results at $T = T_c$ are still not available.

12.5   Experimental results

The quantitative experimental investigation of the critical Casimir effect is an extremely challenging task. One faces problems like high temperature resolution, and temperature stabilization of the sample and the apparatus, which have to be prepared with great care since high accuracy data are required. Nevertheless, the results of some recent experiments are in a qualitative agreement with the existing theory. These results can be divided in two groups: *(i)* based on wetting effects and *(ii)* based on direct force measurements by atomic force microscopes. At present, the results in the second group are at exploratory stage and we refer the interested reader to [Krech (1999)] for a brief review. Here we shall give only some details about the results based on wetting effects [Mukhodaphyay and Law (1999)], [Garcia and Chan (1999)]. In [Mukhodaphyay and Law (1999)] one considers the critical Casimir effect in binary wetting films, whereas in [Garcia and Chan (1999)] the effect is observed in ^{4}He films near the superfluid transition.

In [Mukhodaphyay and Law (1999)] the authors present the first experimental determination of the universal function $X_{\text{ex}}^{(+,-)}$ for critical binary liquid mixtures ($n = 1$, $d = 3$) adsorbing from the vapor onto a noncritical substrate where opposite boundary conditions hold at the substrate-liquid and liquid-vapor surfaces of the adsorbed film. They studied the vapor adsorption behavior of two different critical binary liquid mixtures, methanol + hexane (MH) and 2-methoxy-ethanol + methylcyclohexane (MM), onto a Si wafer placed at a height H above the liquid-vapor surface of the critical liquid mixture. In a MH mixture hexane adsorbs at the vapor surface while methanol adsorbs at the Si surface. In a MM mixture methylcyclohexane adsorbs at the vapor surface while 2-methoxy-ethanol adsorbs at the Si surface. In both cases the film can be modeled as a slab of thickness L with constant dielectric constant. For such a film the total free energy per unit area $\mathcal{F}_L^{(+,-)}(t,H)$ can be written as

$$\mathcal{F}_L^{(+,-)}(t,H) = \sigma_{sl} + \sigma_{lv} + \rho gLH + \omega_{nc} + k_B T_c L^{-2} X_{\text{ex}}^{(+,-)}(L/\xi_\pm),$$
$$(12.29)$$

where σ_{sl} and σ_{lv} are the surface tensions between the substrate and the liquid, and between the liquid and the vapor, respectively; ρgLH is the gravitational contribution to the free energy, ρ being the film density; ω_{nc} is the noncritical contribution that can be written in the form

$$\omega_{nc} = W/L^2 - A\exp(-L/\delta) \qquad (12.30)$$

Here W is the Hamaker constant characterizing the dispersion interaction, while $A\exp(-L/\delta)$ models the effect of a hard wall (substrate) that structures the fluid over a (molecular) distance δ. Considering L as a variational parameter (as in mean field theory) one determines the equilibrium thickness L by minimizing $\mathcal{F}_L^{(+,-)}(t,H)$ with respect to L. This leads to

$$k_B T_c X_{\text{Cas}}^{(+,-)}(L/\xi_\pm) = \rho gHL^3 - 2W + \frac{AL^3}{\delta}\exp(-L/\delta), \qquad (12.31)$$

where the universal function $X_{\text{Cas}}^{(+,-)}(x)$ of the critical Casimir force, is related to $X_{\text{ex}}^{(+,-)}(L/\xi_\pm)$ via

$$X_{\text{Cas}}^{(+,-)}(x) = 2X_{\text{ex}}^{(+,-)}(x) - x\frac{d}{dx}X_{\text{ex}}^{(+,-)}(x).$$

Having determined the parameters W, A and δ, one can evaluate also $X_{\text{Cas}}^{(+,-)}(x)$ from the t and H dependence of the thickness L. The data for the two mixtures are similar. In MH, for T sufficiently far from $T_c \simeq 307K$, the thickness L has approximately the same values above and below T_c, being relatively temperature independent; L *increases* sharply as $T \to T_c^+$ and exhibits a maximum below T_c, at $T \simeq 294K$, in agreement with the expectations based on the $2d$ Ising model, see Fig. 12.4 and the mean field theory [Krech (1997)]. It should be noted that the pure components (hexane and methanol) do not exhibit unusual behavior in the vicinity of T_c, thus the behavior of the mixture around T_c is indeed primarily due to the Casimir force. At $T \leq 295\,K$ the thickness of the pure films increases sharply, which is believed to demonstrate the existence of a surface freezing transition. The one-phase measurements allow the authors to deduce the universal scaling function $X_{\text{Cas}}^{(+,-)}(x)$. This function turns out to be the same, in a quite good approximation, for the two mixtures. At the critical point it has been evaluated as $X_{\text{Cas}}^{(+,-)}(0) = 2\Delta^{(+,-)} \sim 0.0053$, which is considerably lower than the theoretical predictions ranging in the interval from 0.279 to 3.1. One may ask if that can be due to a small off-critical composition of the mixtures, since the average composition has not been controlled during the experiment.

A similar experiment has been performed [Garcia and Chan (1999)], in which the thinning of a ^{4}He film adsorbed on a Cu surface has been investigated near the superfluid transition. The effect is due to the critical fluctuations which develop in the system in this temperature regime. In this case Eq. (12.29) has to be modified. First, the boundary conditions at the two interfaces of the wetting layer are believed to be well approximated by Dirichlet boundary conditions which belong to the (o) surface universality class. Second, one can take into account the change in ω_{nc} due to corrections in the van der Waals energy by setting

$$\omega_{nc} = \frac{W}{L^2}(1 + L/l)^{-1},$$

where $l \simeq 193\text{Å}$ denotes the crossover length to retardation [Cheng and Cole (1988)]. Near the lambda point this leads to

$$k_B T_\lambda X_{\text{Cas}}^{(o,o)}(tL^{1/\nu}) = \rho g H L^3 - 2W(1 + L/l)^{-1}, \qquad (12.32)$$

where $T_\lambda = 2.1768K$ and $\nu = 0.6704$. Note that away from T_λ one has the equation

$$\rho g H L^3 = 2W(1 + L/l)^{-1}, \qquad (12.33)$$

which is a simplified version of the theory of Dzyaloshinskii, Lifshitz and Pitaevskii (DLP) [Dzyaloshinskii et. al. (1961)]. This equation simply describes the effect of the competition between the gravitational potential and the van der Waals attraction of the substrate on the thickness L of the film. It should be emphasized that the DLP theory often shows discrepancies with the experiment [Garcia and Chan (1999)], particularly severe in the case of ^{4}He films on metal substrate. The latter discrepancies are attributed to surface roughness, presence of dust particles, scratches or other imperfections of the substrate, or to excitations effects in the film (e.g., capillary waves on the free liquid-vapor interface). Further, since $X_{\text{Cas}}^{(o,o)}(tL^{1/\nu}) < 0$, one expects a critical *thinning* of the wetting layer thickness. At the λ

point, taking into account that $\Delta^{(o,o)} \simeq -0.024$, one obtains that the thinning is $\sim 0.3\%$ for standard substrates like copper. In the experimental realization one measures the capacitance of 5 copper capacitors at different elevation h in a cell containing liquid ^{4}He at the bottom. The films are formed on the surfaces of the capacitors. Measuring their capacitance at different temperatures one obtains the thickness L of the film on them as a function of T. Finally, one subtracts a temperature independent constant L_0 from L, where L_0 is determined from Eq. (12.33) for each capacitor. The so obtained thickness is used in Eq. (12.32). One finds that both the magnitude of the thinning ΔL and the temperature of the minimum decrese systematically with inceasing h. If one uses the corrected thickness $L - L_0$, the minimum of the film thickness turns out to be below T_λ, at the same value $x_m = -9.2 \pm 0.2$ of $x = tL^{1/\nu}$ for all the films. This value coincides with the minimum of $X_{\mathrm{Cas}}^{(o,o)}(x)$. However, the experimental estimate of $X_{\mathrm{Cas}}^{(o,o)}(x)$ does not show the expected data collapse. The magnitude of $X_{\mathrm{Cas}}^{(o,o)}$ increases almost linearly with h, but there is no clear explanation of this effect. If one assumes off-coexistence effects, i.e., a second scaling variable $y = hL^{\Delta/\nu}$, with $\Delta/\nu \simeq 2.47$, the resulting deviation in the data is reduced tenfold. Equivalently good improvement is obtained if a correction factor accounting for the roughness of the Cu surface is introduced. Much better is the situation above T_λ where the data collapse seems to work, and there is a reasonable agreement between the experiment and the theoretical predictions [Krech and Dietrich (1992)], [Krech (1994)]. Finally, it is worth mentioning that away from the transition region the superfluid film is noticeably thinner than the normal fluid film, as one would expect from the presence of Goldstone modes [Li and Kardar (1991)], [Danchev (1998)]. The overall behavior of $X_{\mathrm{Cas}}^{(o,o)}(x)$ is definitely nonmonotonic, in contrast with the monotonic behavior of $X_{\mathrm{Cas}}^{(p)}(x)$ for the spherical model.

Bibliography

Abe R. (1972) "Expansion of a critical exponent in a inverse power of spin dimensionality", *Prog. Theor. Phys.* **48**, 1414; *ibid* **49**, 113.

Abe R. and Hikami S. (1973) "Critical exponents and scaling scaling relations in $1/n$ expansion", *Prog. Theor. Phys.* **49**, 442.

Abraham D. B. and Robert M. A. (1980) "Phase separation in the spherical model", *J. Phys. A* **13**, 2229.

Abraham D. B. and Svrakic N. M. (1986) "Exact finite-size effects in surface tension", *Phys. Rev. Lett.* **56**, 1172.

Adjari A., Peliti L. and Prost J. (1991) "Fluctuation-induced long-range forces in liquid crystals", *Phys. Rev. Lett.* **66**, 1481.

Adjari A., Duplantier B., Hone D., Peliti L. and Prost J. (1992) "Pseudo-Casimir effect in liquid crystals", *J. Phys. (France)* **2**, 487.

Affleck I. (1986) "Universal term in the free energy at a critical point and the conformal anomaly", *Phys. Rev. Lett.* **56**, 746.

Ahlers G. (1980) "Critical phenomena at low temperature", *Rev. Mod. Phys.* **52**, 489.

Aksenov V. L., Stamenković S. and Plakida N. M. *Neutron Scattering by Ferroelectrics* (World Scientific, Singapore, 1989).

Allen S. and Pathria R. K. (1989) "Finite-size scaling of O(n) models with singular behaviour", *Can. J. Phys.* **67**, 952.

Allen S. and Pathria R. K. (1993) "Explicit results for the correlation length of the finite-sized spherical model of ferromagnetism", *J. Phys. A* **26**, 6797.

Allen S. and Pathria R. K. (1994) "Spin-spin correlations in a finite-sized spherical model with under twisted boundary conditions", *Phys. Rev. B* **50**, 6765.

Allen S. and Pathria R. K. (1995) "Finite-size corrections to the correlation function of the spherical model at $d \geq 4$", *Phys. Rev. B* **52**, 15925.

Amin M. E. (1997) "On the decoupling effects in the spherical model", *Physica A* **255**, 137.

Amin M. E. and Brankov J. G. (1998) "New effects in the magnetization profile of the spherical model in a step-like field", *J. Phys. A* **31**, 4821.

Anders R. and Stierstadt K. (1981) "Experimental determination of the critical exponent η for nickel", *Solid State Commun.* **39**, 185.

Angelescu N., Bundaru M. and Costache G. (1981) "On phase separation in the generalized spherical model", *J. Phys. A* **14**, L533.

Angelescu N. and Zagrebnov V. A. (1985) "A generalized quasiaverage approach to the description of the limit states of the n-vector Curie-Weiss ferromagnet", *J. Stat. Phys.* **41**, 323.

Au-Yang H. and Fisher M. E. (1980) "Wall effects in critical systems :Scaling in Ising model strips", *Phys. Rev. B* **21**, 3956.

Bally D., Grabchev B., Popovici M., Totia M. and Lungu A. M. (1968) "Critical scattering of thermal neutrons in some ferromagnetic metals ", *J. Appl. Phys.* **39**, 459.

Barber M. N. (1974) "Critical behavior of a spherical model with a free surface", *J. Stat. Phys.* **10**, 59.

Barber M. N. (1983), "Finite-size Scaling" in *Phase Transitions and Critical Phenomena*, Vol. **8**, edited by Domb C. and Lebowitz J. L. (Academic Press, London), 145.

Barber M. N. and Fisher M. E. (1973) "Critical Phenomena in Systems of Finite Thickness. I The Spherical Model", *Annals of Phys.* **77**, 1.

Barber M. N., Jasnow D., Singh S. and Weiner R. A. (1974) "Critical behavior of the spherical model with enhanced surface exchange", *J. Phys. C.: Solid State Phys.* **8**, 3408.

Bardeen, J., Cooper, L.N. and Schrieffer, J.R. (1957) "Theory of superconductivity", *Phys. Rev.* **108**, 1175-1204.

Bateman H. end Erdélyi A. (1955) "Higher Transcendental Functions", Vol. 3 (McGraw-Hill, New York, 1955).

Baxter R. J. (1972) "Partition function of the eight-vertex model", *Ann. Phys. (NY)* **70**, 193.

Baxter R. J. (1982), *Exactly Solved Models in Statistical Mechanics*, (Academic Press, London, 1982).

Bayong E. and Diep H. T. (1999) "Effect of long-range interactions on the critical behavior of the continuous Ising model", *Phys. Rev. B* **59** (1999), 11919.

Belokolos E. D. and Petrina D. Ya. (1984) "On a relation between the methods of approximating hamiltonian and finite zone integration", *Teor. i Mat. Fiz.* **58**, 61.

Berlin T. H. and Kac M. (1952) "The spherical model of a ferromagnet", *Phys. Rev.* **86**, 821.

Beysens D. and Esteve D. (1985) "Adsorption phenomena at the surface of silica spheres in a binary liquid mixture", *Phys. Rev. Lett.* **54**, 2123.

Billingsley P (1968), *Convergence of Probability Measures*, (Wiley, New York, 1968).

Billoire A., Neuhaus T. and Berg B. (1993) "Observation of FSS for a first-order phase transition", *Nucl. Phys. B* **396**, 779.

Binder K. (1983) "Critical Behaviour at Surfaces", in *Phase Transitions and Critical Phenomena*, Vol. 6, edited by Domb C. and Lebowitz J. L. (Academic Press, New York).

Binder K. (1987a) "Finite size effects on phase transitions", *Ferroelectrics* **73**, 43.

Binder K. (1987b) "Theory of first-order phase transitions", *Rep. Progr. Phys.* **50**, 783-859.

Binder K. (1992) "Finite size effects at phase transitions", in *Lecture Notes in Phys.*, Vol. 409 Computational Methods in Field Theory, edited by Gausterer H. and Lang C.B. (Springer-Verlag).

Binder K. and Landau D. P. (1984) "Finite-size scaling at first-order phase transitions", *Phys. Rev. B* **30**, 1477.

Binder K., Nauenberg M., Privman V. and Young A. P. (1985) "Finite-size tests of hyperscaling", *Phys. Rev. B* **31**, 1498.

Binney J.J.,Dowrick N.J., Fisher A.J. and Newman M.E.J. *The Theory of Critical Phenomena, An Introduction to the Renormalization Group*, (Clarendon Press, Oxford 1992).

Blöte H. W., Cardy J. L. and Nightingale M. P. (1986) "Conformal invariance, the central charge, and universal finite-size amplitudes at criticality", *Phys. Rev. Lett.* **56**, 742.

Bogolyubov, N.N. (1947) "On the Theory of Superfluidity", *J. Phys.* **11**, No. 1, 23-32.

Bogolyubov, N.N., Zubarev, D.N. and Tserkovnokov, Yu.A. (1957) "On the Theory of Phase Transition", *Dokl. Akad. Nauk SSSR* **117**, No. 5, 788-791.

Bogolyubov, N.N. (1960) "On Some Problems of the Theory of Superconductivity", *Physica Suppl.* **26**, S1-S16.

Bogolyubov, N.N. (1961) "Quasiaverages in problems of statistical mechanics", Dubna, 1961, preprint JINR No. D-781.

Bogolyubov Jr., N.N. (1966) "On model dynamical systems in statistical mechanics", *Physica* **32**, 933-944.

Bogoliubov N. N. (1970) "Quasi-Averages", in *Lectures on Quantum Statistics*, Vol. 2 (Gordon and Breach, New York).

Bogolyubov Jr., N.N. (1970) "On a minimax principle for some model problems of statistical physics", *Soviet J. Nuclear Phys.* **10**, 243-245.

Bogolyubov Jr., N.N. (1972) *A Method for Studying Model Hamiltonians*, Pergamon Press, Oxford.

Bogolyubov Jr., N.N. and Sadovnikov, B.I. (1975) *Some problems of statistical mechanics* (in Russian) (Izdat. Vysshaya Shkola, Moscow, 1975), Pt. II.

Bogolyubov Jr., N.N., Brankov, J.G., Zagrebnov, V.A., Kurbatov, A.M. and Tonchev, N.S. (1981) *The Approximating Hamiltonian Method in Statistical Physics* (in Russian), Publ. House Bulg. Acad. Sci., Sofia.

Bogolyubov Jr., N.N., Brankov, J.G., Zagrebnov, V.A., Kurbatov, A.M. and Tonchev, N.S. (1984) "Some classes of exactly soluble models of problems in quantum statistical mechanics", *Russian Math. Surveys* **39**, No. 6, 1-50.

Borgs C. (1992) "A microscopic theory of finite-size scaling", *Int. J. Mod. Phys.*

C **3**, 897.

Borgs C. and Kotecký R. (1992) "Finite-Size Effects at Asymmetric First-Order Phase Transitions", *Phys. Rev. Lett.* **68**, 1734.

Borgs C. and Kotecký R. (1992) "Finite-size scaling for first-order phase transitions (rigorous results)", *Physica A* **194**, 128.

Borjan Z. and Upton P. J. (1998) "Order-parameter profiles and Casimir amplitudes in critical slabs", *Phys. Rev. Lett.* **81**, 4911.

Botet R., Jullien R. and Pfeuty P. (1982) "Size scaling for infinitely coordinated systems", *Phys. Rev. Lett.* **49**, 478.

Botet R.and Jullien R. (1983) "Large-size critical behavior of infinitely coordinated systems", *Phys. Rev. B* **28**, 3955.

Brankov, J.G., Zagrebnov, V.A. and Tonchev N.S. (1975) "An asymptotically exact solution of the generalized Dicke model", *Theoret. and Math. Phys.* **22**, 13-20.

Brankov, J.G., Tonchev N.S. and Zagrebnov, V.A. (1975) "An Exactly Solvable Model for Metal-Insulator Phase Transition", *Physica* **79A**, 125-127.

Brankov, J.G., Tonchev N.S. and Zagrebnov, V.A. (1977) "A Nonpolynomial Generalization of Exactly Soluble Models in Statistical Mechanics", *Ann. Phys. (N.Y.)* **107**, 82-94.

Brankov, J.G., Tonchev N.S. and Zagrebnov, V.A. (1979) "On a Class of Exactly Soluble Statistical Mechanical Models with Nonpolynomial Interactions", *J. Stat. Phys.* **20**, 317.

Brankov J. G. and Zagrebnov V. A. (1983) "On the description of the phase transition in the Husimi-Temperley model", J. Phys. A **16**, 2217.

Brankov, J.G., Zagrebnov, V.A. and Tonchev, N.S. (1986) "Limit Gibbs states for the Curie - Weiss - Ising model", *Teor. Mat. Fiz.* **66**, 109.

Brankov J. G. and Danchev D. M. (1987) "On the limit Gibbs states of the spherical model", J. Phys. A **20**, 4901.

Brankov J. G. and Tonchev N. S. (1988) "On the finite-size scaling equation for the spherical model", *J. Stat. Phys.* **52**, 143.

Brankov J. G. (1989) "Finite-size scaling for the mean spherical model with inverse power law interaction", *J. Stat. Phys.* **56**, 309.

Brankov J.G. and Danchev D.M. (1989) "A probabilistic view on finite-size scaling in infinitely coordinated spherical models", *Physica A* **158**, 842-863.

Brankov J. G. (1990) "Derivation of finite-size scaling for mean-field models from the Burgers equation", *J. Phys. A* **23**, 5647.

Brankov J. G. (1990) "Finite-size effects in the approximating Hamiltonian method", *Physica A* **168**, 1035.

Brankov J. G. and Danchev D. M. (1990) "Finite-Size Scaling at First-Order Phase Transitions: The Spherical Model with Long-Range Interactions", in *Selected Topics in Statistical Mechanics*, edited by Logunov A. A., Bogolubov Jr., N. N., Kadyshevsky V. G. and Shumovsky A. S. (World Scientific, Singapore), pp. 325-333.

Brankov J. G. and Tonchev N. S. (1990) "An investigation of finite-size scaling for systems with long-range intercation: The spherical model", *J. Stat. Phys.* **59**, 1431.

Brankov J. G. and Danchev D. M. (1991) "Finite-size scaling for the correlation function of the spherical model with long-range interactions", *J. Math. Phys.* **32**, 2543.

Brankov J. G. and Danchev D. M. (1993) "Logarithmic Finite-Size Corrections in the Three-Dimensional Mean Spherical Model", *J. Stat. Phys.* **71**, 775.

Brankov J. G. and Danchev D. M. (1993) "Finite-size logarithmic corrections in the free energy of the mean spherical model", *J. Phys.* **A26**, 4485.

Brankov J. G. and Priezzhev V. B. (1992) "Test of finite-size predictions in three dimensions", *J. Phys.* **A25**, 4297.

Brankov J. G. and Tonchev N. S. (1992) "Finite-size scaling for systems with long-range interactions", *Physica A* **189**, 583.

Brankov J. G. and Tonchev N. S. (1994) "Finite-size scaling in the presence of an inhomogeneous field", *Phys. Rev. B* **50**, 2970.

Bray and Moore (1978) "Critical temperature shifts for finite slabs in the ϵ-expansion", *J. Phys. A* **11**, 715.

Brézin E. (1982) "An investigation of finite size scaling", *J. Phys. (France)* **43**, 15.

Brézin E. (1983) "Finite size scaling: part II", *Ann. N. Y. Acad. Sci* **410**, 339.

Brézin E. and Zinn-Justin J. (1985) "Finite-size effects in phase transitions", *Nucl. Phys.* **B 257**, 867.

Brézin E., Korutcheva E., Jolicoeur Th., and Zinn-Justin J. (1993) "$O(N)$ Vector Model with Twisted Boundary conditions", *J. Stat. Phys.* **70**, 583.

Bruce A. D. and Cowley K. A. *Structural Phase Transitions* (Taylor and Francis, London, 1981).

Burkhardt T. W. and Eisenriegler E. (1995) "Casimir interaction of spheres in a fluid at the critical point", *Phys. Rev. Lett.* **74**, 3189.

Casimir H. B. G. (1948), *Proc. K. Ned. Acad. Wet.* **51**, 793.

Casimir H. B. G. (1953) "Introductory remarks on quantum electrodynamics", *Physica* **19**, 846.

Cassandro M. and Jona-Lasinio G. (1978) "Critical point behaviour and probability theory", *Adv. Phys.* **27**, 913.

Cardy J. L. (1986) "Effect of boundary conditions on the operator content of two-dimensional conformal invariant theories", *Nucl. Phys. B* **275**, 200.

Cardy J. L. (1987) "Anisotropic correction to correlation functions in finite-size systems", *Nucl. Phys. B* **290**, 355.

Cardy J. L., in *Phase Transitions and Critical Phenomena*, Vol.11, edited by Domb C. and Lebowitz J. L. (Academic, New York, 1987).

Cardy J.L., ed. (1988) *A collection of reprints: Finite-Size Scaling*, in *Current Physics: Sources and Comments*, Vol. 2, (North-Holland, Amsterdam, 1988).

Cardy J.L. (1996) *Scaling and Renormalization in Statistical Physics*, (Cambridge University Press, Cambridge, 1996).

Car A. and Zagrebnov V. A. (1994) "Critical fluctuation operators for a quantum model of ferroelectric", *Physica A* **212**, 398.

Castro Neto A. H. and Fradkin E. (1993) "The theromodynamics of quantum systems and generalization of Zamolodchikov's C-theorem", *Nuclear Physics B* **400**[FS], 525.

Chaikin P. M. and Lubensky T. C. (1995), *Principles of Condensed Matter Physics,* (Cambridge University Press, 1995).

Chakravarty S., Halperin B. I. and Nelson D. R. (1989) "Two-dimensional quantum Heisenberg antiferromagnet at low temperatures", *Phys. Rev. B* **39**, 2344.

Chamati H. (1994) "$T = 0$ finite-size scaling for a quantum system with long-range interaction", *Physica A* **212**, 357.

Chamati H., Danchev D. M. and Tonchev N. S. (2000) "Casimir amplitudes in a quantum spherical model with long-range interaction", *Eur. Phys. J. B* **14**, 307.

Chamati H. and Tonchev N. S. (1992) "Long-Range Order of an Exactly Solvable Model of a Quantum Antiferromagnet", *Phys. status solidi (b)* **174**, 505.

Chamati H. and Tonchev N. S. (1994) "Symmetry breaking and long-range order: An n-component model of a structural phase transition", *Phys. Rev. B* **49**, 4311.

Chamati H. and Tonchev N. S. (1996) "Finite-size shift of the critical temperature in the spherical model", *J. Stat. Phys.* **83**, 1211.

Chamati H. and Tonchev N. S. (2000) "Exact results for some Madelung-type constants in the Finite-size scaling theory", *J. Phys. A* **33**, L167.

Chamati H. and Tonchev N. S. (2000) "Finite-size scaling investigations in the quantum ϕ^4-model with long-range interaction", *J. Phys. A* **33**, 873.

Chamati H., Danchev D. M., Pisanova E. S. and Tonchev N. S. (1997) "Low-temperature regimes and finite-size sacling in a quantum spherical model", cond-mat/9707280 ICTP, Trieste, preprint IC/97/82, July 1997.

Chamati H., Pisanova E. S. and Tonchev N. S. (1998) "Theory of a spherical-quantum-rotor model: Low-temperature regime and finite-size scaling", *Phys. Rev. B* **57**, 5798 .

Chang R. F., Barstyn H. and Sengers J. V. (1979) "Correlation function near the critical mixing point of a binary liquid", *Phys. Rev. A* **19**, 866.

Chen X. S. and Dohm V. (1998) "Failure of unuversal finite-size scaling above the upper critical dimension", *Physica A* **251**, 439.

Chen X. S. and Dohm V. (1998) "Finite-size scaling in the ϕ^4 theory above the upper critical dimension", *Eur. Phys. J. B* **5**, 529.

Chen X. S. and Dohm V. (1999) "Violation of finite-size scaling in three dimensions", *Eur. Phys. J. B* **10**, 687.

Cheng E. and Cole M. W. (1988) "Retardation and many-body effects in multilayer film adsorption", *Phys. Rev. B* **38**, 987.

Chubukov A. V., Sachdev S. and Ye J. (1994) "Theory of two-dimensional quantum Heisenberg antiferromagnets with a nearly critical ground state", *Phys.*

Rev. B **49**

Cohen J. D. and Carver T. R. (1977) "New measurment of the critical exponent β for nickel by microwave transmission", *Phys. Rev. B* **15**, 5350.

Collins M. F., Minkiewicz V. J., Nathans R., Passell L. and Shirane G. (1969) "Critical and spin-wave scatering of neutrons from iron", *Phys. Rev.* **179**, 417.

Continentino M. A., Japiassu G. M. and Troper A. (1989) "Critical approach to the coherence transition in Kondo lattices", *Phys. Rev. B* **39**, 9734.

Danchev D. M. (1990) "Classical dipoles on a finite triangular lattice: The spherical model approximation", *Physica A* **163**, 835.

Danchev D. M. (1993) "Finite-size dependence of the helicity modulus within the mean spherical model", *J. Stat. Phys.* **73**, 267.

Danchev D. M. (1996) "Finite-size scaling Casimir force function: Exact spherical model results", *Phys. Rev. E* **53**, 2104.

Danchev D. M. (1998) "Exact three-dimensional Casimir force amplitude, C function, and Binder's cumulant ratio: Spherical model results", *Phys. Rev. E* **58**, 1455.

Danchev D. M., Brankov J. G. and Amin M. E. (1997) "New surface critical exponents in the spherical model", *J. Phys. A* **30**, 1387.

Danchev D. M., Brankov J. G. and Amin M. E. (1997) "Surface critical exponents for a three-dimensional modified spherical model", *J. Phys. A* **30**, 5645.

Danchev D. M. and Tonchev N. S. (1999) "On the finite-temperature generalization of the C-theorem and the interplay between classical and quantum fluctuations", *J. Phys. A*, **32**, 7057.

D'Auria A.C., De Cesare L., Esposito U. and Rabuffo I. (1997) "Temperature-dependent scaling fields and quantum criticality within the Wilson renormalization group approach", *Physica A* **243**, 152.

Diehl H. W. (1986) "Field-theoretical approach to critical behavior of surfaces", in *Phase Transitions and Critical Phenomena*, Vol. **10**, edited by Domb C, and Lebowitz J. L. (New York, Academic), p. 76.

Diehl H. W. (1987) "Finite size effects in critical dynamics and the renormalization group", *Z. Phys. B* **66**, 211.

Diehl H. W. and Smock M. (1993a) "Critical behavior of supercritical surface enhancement: Temperature singularity of surface magnetization and order parameter profile to one-loop order", *Phys. Rev. B* **47**, 5841.

Diehl H. W. and Smock M. (1993b) "Erratum: Critical behavior at extraordinary transition: Temperature singularity of surface magnetization and order parameter profile to one-loop order [Phys. Rev. B47, 5841 (1993)]", *Phys. Rev. B* **48**, 6740.

Dietrich S. (1988), in *Phase Transitions and Critical Phenomena*, Vol.12, edited by Domb C. and Lebowitz J. L. (Academic, New York, 1988).

Dietrich S. (1990) "Critical phenomena at intefaces", *Physica A* **168**, 160.

Dobrushin R. L. (1968) "The problem of uniqueness of the Gibbs random field and the problem of phase transitions", *Funct. Anal. Appl.*, **2**, No. 4, 44.

Domb C. (1996) *The Critical Point: A Historical Introduction to the Modern Theory of Critical Phenomena* (Taylor and Francis, London, 1996).

Domb C. and Hunter D. L. (1965) "On the critical behavior of ferromagnets", *Proc. Phys. Soc. (London)*, **86**, 1147.

Dörre, P., Haug, H., Heise, M. and Jelitto, R.J. (1979) "On Quantum Statistical Variational Principles", *Z. Phys. B* **34**, 117-122.

Durr W., Taborelli M., Paul O., Germar R., Gudat W., Pascia D. and Laudolt M. (1989) "Magnetic Phase Transition in Two-Dimensional Ultrathin Fe Films on Au (100)", *Phys. Rev. Lett.* **62**, 206.

Dutta A., Chakrabarti B. K. and Bhattacharjee J. K. (1997) "Quantum Lifshitz point: ϵ expansion and the spherical limit", *Phys. Rev. B* **55**, 5619.

Dyson F.R., Lieb E. and Simon B. (1978) "Phase transition in quantum spin systems with isotropic and nonisotropic interaction", *Journ. of Stat. Phys.* **18**, 335.

Dzyaloshinskii I. E., Lifshitz E. M. and Pitaevskii L. P. (1961) "The general theory of Van der Waals forces", *Adv. Phys.* **10**, 165.

Eisenriegler E. and Ritschel U. (1995) "Casimir forces between spherical particles in a critical fluid and conformal invariance", *Phys. Rev. B* **51**, 13717.

Ellis R. S. and Newman C. M. (1978) "Fluctuationes in Curie-Weiss Exemplis", in *Lecture Notes in Physics*, ed. Dell'Antonio G., Doplicher S., and Jona-Lasinio G., Vol. 80, pp. 313-324 (Springer-Verlag, Berlin, 1978).

Ellis R. S. and Newman C. M. (1978) "The Statistics of Curie-Weiss Models", *J. Stat. Phys.* **19**, 149.

Ellis, R.S. and Newman, C.M. (1978c) "Limit theorems for sums of dependent random variables occurring in statistical mechanics", *Z. Wahrsch. verw. Geb.* **44**, 117-139.

Evans R., in *Liquids at Interfaces*, Les Houches Session XLVIII, edited by J. Charvolin, J. Joanny and J. Zinn-Justin (Elsevier, Amsterdam, 1990), p. 3.

Evans R. and Stecki J. (1994) "Solvarion force in two-dimensional Ising strips", *Phys. Rev. B* **49**, 8842.

Feller W. (1966) *An Introduction to Probability Theory and its Applications* (Wiley, New York, 1966).

Ferdinand, A.E. and Fisher, M.E. (1969) "Bounded and Inhomogeneous Ising Models. I. Specific-Heat Anomaly of a Finite Lattice", *Phys. Rev.* **185**, 832-846.

Ferrenberg A. M. and Landau D. P. (1991) "Critical behavior of the three-dimensional Ising model: A high-resolution Monte Carlo study", *Phys. Rev. B* **44**, 5081.

Fisher M. E. (1964) "The Free Energy of a Macroscopic System", *Arch. Rat. Mech. Anal.* **17**, 377.

Fisher M.E. (1965) "Correlation Function and the Coexistence of Phases", *J. Math. Phys.* **6**, 1643-1653.

Fisher M. E. (1971) "Theory of critical point singularities", in *Critical Phenomena*, ed. Green M.S., Proc. Enrico Fermi Int. School of Physics, Vol. 51, pp.

1-99 (Academic, New York, 1971).

Fisher M. E. (1974) "The renormalization group in the theory of magnetism", *Rev. Mod. Phys.* **46**, 597.

Fisher M.E. (1983) "Scaling universality and renormalization group theory", in *Critical Phenomena*, Lecture Notes in Physics v.186, ed. F. J. W. Hahne (Springer-Verlag, 1983).

Fisher M. E. (1988) "Condensed Matter Physics: Does quantum mechanics matter", in *Niels Bohr: Physics and the world*, edited by Matsui T. and Oleson A. (Harwood Academic, chur, 1988), 177.

Fisher M. E. (1998) "Renormalization group theory :Its basis and formulation in statiatical physics" *Rev. Mod.Phys.* **70**,653.

Fisher M. E. and Barber M. N. (1972) "Scaling theory for finite-size effects in the critical region", *Phys. Rev. Lett.* **28**, 1516.

Fisher M. E., Barber M. N. and Jasnow D. (1973) "Helicity modulus, superfluidity, and scaling in isotropic systems", *Phys. Rev. A* **8**, 1111.

Fisher M. E. and Berker A. N. (1982) "Scaling for first-order phase transitions in thermodynamic and finite systems", *Phys. Rev. B* **26**, 2507.

Fisher M. E. and Burford R. J. (1967) "Theory of critical-point scattering and correlations. I. The Ising model", *Phys. Rev.* **156**, 583.

Fisher M. E. and de Gennes P. G. (1978) "Phénomènes aux parois dans un mélange binaire critique", *C.R. Acad. Sci. Paris B* **287**, 207.

Fisher M. E., Ma S.-k. and Nikel B. G. (1972), "Critical exponents for long-range interactions" *Phys. Rev. Lett.* **29**, 917.

Fisher M. E. and Privman V. (1985) "First-order transitions breaking $O(n)$ symmetry: Finite-size scaling", *Phys. Rev. B***32**, 447.

Fisher M. E. and Privman V. (1986) "First-order transitions in spherical models: Finite size scaling", *Comm. Math. Phys.***103**, 527.

Friedan D., Qui Z. and Shenker S. (1984) "Conformal invariance, unitarity, and critical exponents in two dimensions", *Phys. Rev. Lett.* **52**, 1575.

Gallagher P. D. and Maher J. V. (1992) "Partitioning of polystyrene latex spheres in immiscible critical mixtures", *Phys. Rev. A* **46**, 2012.

Garanin D. A. (1996) "Bloch-wall phase transition in the spherical model", *J. Phys. A* **29**, 2349.

Garcia R. and Chan M. H. W. (1999) "Critical fluctuation-indiced thinning of ^4He films near the superfluid transition", *Phys. Rev. Lett.***83**, 1187.

Gasparini F. M. and Rhee I. "Critical behavior and scaling of confined ^4He", in *Progress in Low Temperature Physics,* ed. Brewer D. F. (North-Holland, Amsterdam, 1992), Vol. XIII, p. 1.

Gelfand M. P. and Fisher M. E. (1988) "Finite-size effects in surface tension: Thermodynamics and the Gaussian interface model" *Int. J. Thermophys.* **9**, 713.

Gelfand M. P. and Fisher M. E. (1990) "Finite-size effects in fluids interfaces", *Physica A* **166**, 1.

Georgii H.-O. (1988) *Gibbs measures and Phase Transitions* (de Gruyter, Berlin,

1988).

Gibbs J. W., *Elementary Principles in Statistical Mechanics*, (Reprinted by Dover, New York 1960).

Ginibre, J. (1968) "On the Asymptotic Exactness of the Bogoliubov Approximation for Many Boson Systems", *Comm. Math. Phys.* **8**, 26-51.

Glasser M.L. and Zucker I.J. (1980) "Lattice Sums", *Theor. Chem.: Adv. and Persp.* **5**, 67.

Glimm J. and Jaffe A. (1981) *Quantum Physics. A Functional Integral Point of View* (Springer-Verlag, New York, 1981).

Gochev I. G. and Tonchev N. S. (1992) "Long-range order of the Lieb-Mattis model of a quantum antiferromagnet", *Phys. Rev. B* **45**, 480.

Goldenfeld N. (1992) *Lecture on Phase Transitions and the Renormalization Group* in *Frontiers in Physics*, edited by Pines D.(Addison-Wesley, New York, 1982).

Goldschmidt Y. Y. (1987) "Dynamical relaxation in finite size systems", *Nucl. Phys. B* **285**, 519.

Gough J. and Pulé J. V. (1993) "The Spherical Model and Bose-Einstein Condensation", *Helv. Phys. Acta* **66**, 17.

Gradshteyn I. S. and Ryzhik I. H. (1973) *Table of Integrals, Series, and Products* (Academic, New York).

Griffiths, R.B. (1964) "A Proof that the Free Energy of a Spin System is Extensive", *J. Math. Phys.* **5**, 1215.

Griffiths R. B. (1966) "Spontaneous magnetization in idealized ferromagnets", *Phys. Rev.* **152**, 240.

Griffiths R. B. (1967) "Thermodynamic functions for fluids and ferromagnets near the critical point", *Phys. Rev.* **158**, 176.

Hanke A., Schlesener F., Eisenriegler E., Dietrich S. (1998) "Critical Casimir forces between spherical particles in fluids", *Phys. Rev. Lett.* **81**, 1885.

Heller P. (1967) "Experimental investigation of critical phenomena", *Rep. Prog. Phys.* **30**, 731.

Henkel M. (1988) "Amplitude-exponent relation for the correlation length in the spherical model", *J. Phys. A* **21**, L227.

Henkel M. (1999) *Conformal Invariance and Critical Phenomena* (Springer, Berlin, 1999).

Henkel M. and Weston R. A. (1992) "Universal amplitudes in finite-size scaling: the antiperiodic 3D spherical model", *J. Phys. A* **25**, L207.

Hepp K. and Lieb E. H. (1973a) "On the supperradiant phase transition for molecules in a quantized radiation field: the Dicke maser model", *Ann. Phys. (New York)* **76**, 360.

Hepp K. and Lieb E. H. (1973b) "The equilibrium statistical mechanics of matter interacting with quantized radiation field", *Phys. Rev. A* **8**, 2617.

Hertz J. (1976) "Quantum critical phenomena" *Phys. Rev. B* **14**, 1165.

Hiroyoshi H., Hoshi A., Fujimoti H, and Nakagava Y. (1980) "Arrott – Noakes Plots for Magnetization of Some 3d-Transition Metals and Alloys in High

Magnetic Fields", *J. Phys. Soc. Japan* **48**, 830.

Hornreich R. M. and Schuster H. G. (1982) "Thermodynamic properties of the random-field spherical model", Phys. Rev. B **26**, 3929.

Hocken R. and Moldover M. R. (1976) "Ising critical exponents in real fluids: An experiment", *Phys. Rev. Lett.* **37**, 29.

Huang K. (1987) *Statistical Mechanics* (Wiley, New York, 1987).

Hubbard J. (1959) "Calculation of partiticion functions", *Phys. Rev. Lett.* **3**, 22.

Husimi K. (1953) "Statistical mechanics of condensation", in *Proc. Intern. Conf. Theor. Phys.* (Kyoto and Tokyo, 1953), p. 531.

Imry Y. (1980) "Finite-size rounding of a first-order phase transition" *Phys. Rev. B* **21**, 2042.

Indekeu J. O., Nightingale M. P. and Wang W. V. (1986) "Finite-size interaction amplitudes and their universality: exact, mean-field, and renormalization group results", *Phys. Rev. B* **34**, 330.

Jack I., in *Renormailzation Group 91*, edited by D. V. Shirkov and V. B. Priezzhev, (World Scientific, Singapore, 1992), pp. 55-65.

Jagannathan A. and Rudnick J. (1989) "The spherical model for spin glasses revisited", *J. Phys. A* **22**, 5131.

Jayalakshmi Y. and Kaler E. W. (1997) "Phase behavior of colloids in binary liquid mixtures", *Phys. Rev. Lett.* **78**, 1379.

Jolicoeur Th. and Golinelli O. (1994) "σ-model study of Haldane-gap antiferromagnets", *Phys. Rev. B* **50**, 9265.

Joyce G. S. (1966) "Spherical model with long-range ferromagnetic interactions", *Phys. Rev.* **146**, 349.

Joyce G. S. (1969) "Absence of a ferromagnetism or antiferromagnetism in the isotropic Heisenberg model with long-range interactions", *J. Phys. C.* **2**, 1531.

Joyce G. S. (1972) "Critical properties of the spherical model", in *Phase Transitions and Critical Phenomena*, C. Domb and M. S. Green, eds., (Academic, London, 1972), Vol. 2, p. 375.

Julia B. "Statistical theory of numbers", in *Number Theory and Physics*, Springer Proceedings in Physics **47**, edited by Luck J. M., Moussa P. and Waldschmidt M., (Springer, Berlin, 1990), p. 276.

Jund P., Kim S.G. and Tsallis S. (1995) "Crossover from extensive to nonextensive behavior driven by long-range interaction" *Phys. Rev. B* **52**, 50.

Kac M. and Thompson C. J. (1971) "Spherical model and the infinite spin dimensionality limit" *Phys. Norveg.* **5**, 163.

Kac M. and Thompson C. J. (1977) "Correlation functions in the spherical and mean spherical models" *J. Math. Phys.* **18**, 1650.

Kadanoff L. P. (1966) "Scaling laws for Ising models near T_c", *Physics (New York)* **2**, 263.

Kadanoff L. P. (1971), in *Proc. Intern. School of Physics "Enrico Fermi"*, Corso LI, edited by Green M. S. (Academic Press, New York, 1971).

Kadanoff L. P. and Wegner F. J. (1971) "Some critical properties of the eight-vertex model", *Phys. Rev. B* **4**, 3989.

Kaplan T. A., Horsch P. and Von der Linden W. (1989) "Order Parameter in Quantum Antiferromagnets", *Journ. Phys. Soc. Jpn.* **58**, 3894.

Kaplan T. A., Horsch P. and Von der Linden W. (1990) "Spontaneous symmetry breaking in the Lieb-Mattis model of antiferromagnetism", *Phys. Rev. B* **42**, 4663.

Kardar M. and Golestanian R. (1999) "The "friction" of vacuum, and other fluctuation-induced forces", *Rev. Mod. Phys.* **71**, 1233.

Khorunzhy A. M., Khoruzhenko B. A., Pastur L. A. and Shcherbina M. V. (1992) "The large-n limit in statistical mechanics and spectral theory of disordered systems", in *Phase Transitions and Critical Phenomena*, Vol. 15, edited by Domb C. and Lebowitz L. (Academic Press, New York), p. 73.

Klemm A. and Zagrebnov V. A. (1977) "A compressible Dicke model (exact solution)", *Physica* **86** A, 400.

Klemm A. and Zagrebnov V. A. (1978) "On Dicke-type hamiltonian with hidden variables", *Physica* **92** A, 599.

Knops H. J. F. (1973) "Infinite spin dimensionality limit for nontranslationally invariant interactions" *J. Math. Phys.* **14**, 1918.

Kobeissi M. A. (1981) "Mössbauer study of static and dynamics critical behavior in Fe", *Phys. Rev. B* **24**, 2380.

Koma T. and Tasaki H. (1993) "Symmetry breaking and long range order in Heisenberg antiferromagnets", *Phys. Rev. Lett.* **70**, 93.

Kopec T. K. and Pirc R. (1997) "Quantum spherical description of an Ising spin glass in a transversee field", *Phys. Rev. B* **55**, 5623.

Kosterlitz J. M., Thouless D. J. and Jones R. C. (1976) "Spherical model of a spin-glass", *Phys. Rev. Lett.* **36**, 1217.

Krech M. (1994) *The Casimir Effect in Critical Systems* (World Scientific, Singapore).

Krech M. (1997) "Casimir forces in binary liquid mixtures", *Phys. Rev. E* **56**, 1642.

Krech M, (1999) "Fluctuation-induced forces in critical fluids", *J. Phys. Cond. Mat.* **11**, R391.

Krech M. and Dietrich S. (1992) "Free energy and specific heat of critical films and surfaces", *Phys. Rev. A* **46**, 1886.

Krech M. and Dietrich S. (1992b) "Specific heat of critical films, the Casimir force, and wetting films near critical end points", *Phys. Rev. A* **46**, 1922.

Krech M. and Landau D. P. (1996) "Casimir effect in critical systems: A Monte Carlo simulation", *Phys. Rev. E* **53**, 4414.

Kulkarni G. U., Kannan K. R., Arunarkavalli T and Rao C. N. R. (1994) "Particle-size effects on the value of T_c of $MnFe_2O_4$: Evidence for finite-size scaling", *Phys. Rev. B* **49**, 724.

Kumar A., Krishnamurthy H. R. and Gopal E. S. R. (1983) "A Comprehensive review of critical phenomena in fluid systems", *Phys. Rep.* **98**, 57.

Kupiainen A. J. (1980) "On the $1/n$ expansion", *Commun. Math. Phys.* **73**, 273.

Lamoreaux S. K. (1997) "Demonstration of the Casimir force in the 0.6 to 6 μm range", *Phys. Rev. Lett.* **78**, 5.

Landau D. P. (1990) "Monte Carlo studies of finite-size effects at first and second order phase transitions", in *Finite Size Scaling and Numerical Simulations of Statistical Systems*, edited by Privman V. (World Scientific, Singapore, 1990), 223.

Landau L. D. (1971), reprinted in *Collected Papers of L. D. Landau*, ed. ter Haar D. (Pergamon, London, 1965).

Lanford O. E. and Ruelle D. (1969) "Observables at infinity and states with short range correlations in statistical mechanics", *Commun. Math. Phys.* **13**, 194.

Lavis D.A. and Bell G.M. *Statiatical Mechanics of Lattice Systems, Exact, Series and Renormalization Group Methods*, in *Text and Monograph in Physics*, Vol.2 (Springer-Verlag, Berlin, 1999).

Lawrie I.D. (1978) "Quantum crossover in the transverse Ising model", *J. Phys. C* **11**, 1123.

Lawrie I.D. (1978a) "Dimensional crossover in quantal and finite-sized Ising models", *J. Phys. C* **11**, 3857.

Lederman D., Rammos C. A. and Jaccarino V. (1993) "Thermodynamic properties of $(FeF_2)_n(CoF_2)_n$ superlattices", *J. Phys. Cond. Matt.* **5**, 373.

Lederman D., Rammos C. A. and Jaccarino V., Cardy J. L. (1993) "Finite-size sacling in FeF_2/ZnF_2 superlattices", *Phys. Rev. B* **48**, 8365.

Levin F. S. and Micha D. A. (1993), *Theory and recent experiments on atomic systems* (Plenum Press, New York and London, 1993).

Lewis H. W. and Wannier G. H. (1952) "Spherical model of a ferromagnet", *Phys. Rev.* **88**, 682.

Lewis H. W. and Wannier G. H. (1953) "Spherical model of a ferromagnet" (Erratum), *Phys. Rev.* **90**, 1131.

Li Y. and Baberschke K. (1992) "Dimensional crossover in ultrathin Ni(111) films on W(110)", *Phys. Rev. Lett.* **68**, 1208.

Li H. and Kardar M. (1991) "Fluctuation-induced forces between rough surfaces", *Phys. Rev. Lett.* **67**, 3275.

Li H. and Kardar M. (1992) "Fluctuation-induced forces between manifolds immersed in a correlated fluids", *Phys. Rev. A* **46**, 6490.

Luck J. M. (1985) "Corrections to finite-size-scaling laws and convergence of transfer-matrix methods", *Phys. Rev. B* **31**, 3069.

Luijten E., Binder K. and *Blöte* H.W.J. (1999) "Finite-size scaling above the upper critical dimension revisited: the case of the five-dimensional Ising model", *Eur. Phys. J. B* **9**, 289.

Ma S.K. (1973) "Critical exponents above T_c to $O(1/n)$", *Phys.Rev. A* **7**, 2172.

Ma S.K. (1976) *Modern Theory of Critical Phenomena*, Frontiers in Physics **46**, (Benjamin, London, 1976).

Manousakis E. (1991) "The spin 1/2 Heisenberg antyferromagnet on a sguare lattice and its application on the cuprous oxides", *Rev. Mod. Phys.* **63**, 1.

Lyra M. L., Kardar M. and Svaiter N. F. (1993) "Effects of surface enhancement on fluctuation-induced interactions", *Phys. Rev. E* **47**, 3456.

Marshall A. W. and Olkin I. (1979), *Inequalities: Theory of Majorization and Its Applications* (Academic, New York, 1979).

Mattis D. C. and Langer W. D. (1970) "Role of phonons and band structure in metal-insulator phase transition", *Phys. Rev. Lett.* **25**, 376.

McCoy B. M. and Wu T. T. (1973), *The Two-dimensional Ising Model* (Harvard Univ. Press, Cambridge, 1973).

Mermin N. D. and Wagner H. (1966) "Absence of ferromagnetism or antiferromagnetism in one- or two-dimensional isotropic Heisenberg model", *Phys. Rev. Lett.* **17**, 1133.

Michielsen K., Schneider T. and De Raedt H. (1992) "Finite-size effects in layered superconductors", *Z. Phys. B* **85**, 15.

Millis A. J. (1993) "Effect of a nonzero temperature on quantum critical points in itinerant fermion system", *Phys. Rev. B* **48**, 7183.

Molchanov S. A. and Sudarev Ju. N. (1975) "Gibbs states in the spherical model", *Soviet Math. Dokl.* **16**, 1254.

Momont B., Verbeure A. and Zagrebnov V. A. (1997) "Algebraic structure of quantum fluctuations", *J. Stat. Phys.* **89**, 633.

Morf R., Schneider T. and Stoll E. (1977) "Nonuniversal critical behavior and its suppression by quantum fluctuations", *Phys. Rev. B* **16**, 462.

Montroll E. W. (1949) "Continuum models of cooperative phenomenon", *Nuovo Chim. Suppl.* **6** 265.

Mostepanenko V. M. and Trunov N. N. *The Casimir effect and its applications* (Moscow, Energoatomizdat, 1990, in Russian); English version: (New York, Clarendon Press, 1997).

Mukhodaphyay A. and Law B. M. (1999) "Critical Casimir effect in binary liquid wetting films", *Phys. Rev. Lett.* **83**, 772.

Nagaev E. D. (1988) *Magnets with complicated exchange interaction* (Nauka, Moskow, 1988, in Russian).

Niel J. C. and Zinn-Justen J. (1987) "Finite size effects in critical dynamics", *Nucl. Phys. B* **280**, 355.

Nienhuis B. and Nauenberg M. (1875) "First-order phase transition in renormalization group theory", *Phys. Rev. Lett.* **35**, 477.

Nieuwenhuizen Th. M. (1995) "Quantum description of spherical spins", *Phys. Rev. Lett.* **74**, 4293.

Nieuwenhuizen Th. M. and Ritort F. (1998) "Quantum phase transition in spin glasses with multi-spin interaction", *Physica A* **250**, 8.

Nightingale M. P. and Indekeu J. O. (1985) "Effect of criticality on wetting layers", *Phys. Rev. Lett.* **54**, 1824.

Nissen J. A., Chui T. C. P. and J.A. Lipa J. A. (1993) "The Specific Heat of Confined Helium near the Lambda Point", *J. Low Temp. Phys.* **92**, 353.

Obermair G. (1972) "A dynamical spherical model", in *Dynamical Aspects of Critical Phenomena*, eds. Budnick J. I. and Kawatra M. P. (Gordon and Breach, New York, 1972), p. 137.

Oitmaa, J. and Barber, M.N. (1975) "On the critical behaviour of an Ising system with lattice coupling", *J. Phys. C* **8**, 3653.

Onsager L. (1944) "Crystal statistics. I. A two-dimensional model with an order-disorder transition", *Phys. Rev.* **65**, 117.

den Ouden L.W.J., Capel H.W. and Perk J.H.H. (1976) "Systems with separable many-particle interactions. II.", *Physica* **85A**, 425.

Parisi G. (1988) *Statistical Field Theory*, (Addison - Wesley, The Advaced Book Program, New-York, 1988).

Parry A. O. and Evans R. (1992) "Novel phase behavior of a confined fluid or Ising magnet", *Physica A* **181**, 250.

Parry A. O., Evans R. and Nicolaides D. B. (1991) "Long-ranged surface perturbations for confined fluids", *Phys. Rev. Lett.* **67**, 2978.

Pastur L. A and Shcherbuna M. V. (1984) "Infinite range limit for correlation functions of lattice systems", *Teor. i Mat. Fiz.* **61**, 3.

Pathria R. K. (1972) "Bose-Einstein condensation in thin films", *Phys. Rev A* **5**, 1451.

Patrick A. E. (1993) "On phase separation in the spherical model of a ferromagnet: Quasiaverage approach", *J. Stat. Phys.* **72**, 665.

Patrick A. E. (1994) "The influence of external boundary conditions on the spherical model of ferromagnet. I Magnetization profiles", *J. Stat. Phys.* **75**, 253.

Petkou A. C. and Vlachos N. D. (1999) "Finite-size effects and operator product expansions in a CFT for $d > 2$", *Phys. Lett. B* **446**, 306.

Petrina D Ya (1995) *Mathematical Foundations of Quantum Statistical Mechanics. Continuous Systems.* (Kluwer, Dordrecht/Boston/London, 1995).

Pisanova E. S. and Tonchev N. S. (1993) "On the interplay of classical and quantum fluctuations: an exactly solvable model for a structural phase transition", *Physica A* **197**, 301.

Pisanova E. S. and Tonchev N. S. (1995) "Universality amplitudes in finite-size scaling for an anharmonic crystal", *Physica A* **217**, 419.

Plakida N.M. (1981) "A ferroelectric model with strong low-temperature anharmonicity" *Sov. J. Low. Temp. Phys.* **7**, 644.

Plakida N. M. and Tonchev N. S. (1986) "Quantum effects in a d-dimensional exactly solvable model for a structural phase transition", *Physica A* **136**, 176.

Plakida N. M. and Tonchev N. S. (1987) "Equation of state for an exactly solvable model for structural phase transition", *Teor. i Mat. Fiz.* **72**, 269.

Plakida N. M., Radosz A. and Tonchev N. S. (1987) "Exactly soluble model of phase transition with cubic anisotropy", *Physica A* **143**, 227.

Plakida N. M., Radosz A. and Tonchev N. S. (1991) "Exactly solvable model of a structural phase transition", *Phase Transitions* **29**, 179.

Privman V. (1990) "Finite-size scaling theory", in *Finite Size Scaling and Numerical Simulations of Statistical Systems*, edited by Privman V. (World Scientific, Singapore, 1990), 1.

Privman V., ed. (1990) *Finite Size Scaling and Numerical Simulations of Statis-*

tical Systems, (World Scientific, Singapore, 1990).

Privman V. (1990b) "New logarithmic term in the superfluid density scaling in confined geometry", *J. Phys. A.* **23**, L711.

Privman V. and Fisher M. E. (1983) "Finite-Size Effects at First-Order Transitions", *J. Stat. Phys.* **33**, 385.

Privman V. and Fisher M. E. (1984) "Universal critical amplitudes in finite-size scaling", *Phys. Rev. B* **30**, 322.

Privman V and Rudnick J. (1990) "Nonsymmetric First-Order Transitions: Finite-Size Scaling and Test for Infinite-Range Models", J. Stat. Phys. **60**, 551.

Privman V., Hohenberg P. C. and Aharony A. (1991) "Universal critical point amplitude relations", in *Phase Transitions and Critical Phenomena*, edited by Domb C. and Lebowitz J. L. (Academic, New York, 1991), Vol. 14, p.1.

Radosz A. (1988) "Pseudo-Goldstone mode in the exactly solvable model of phase transition", *Phys. Lett. A* **127**, 319.

Radosz A. (1990) "Soft mode in the class of exactly soluble models of phase transition", *Physica A* **168**, 853.

Ravandal F. (1976) "Scaling And Renormalization Groups", (Introductory Lectures given at the Niels Bohr Institute and Nordita, 1976).

Reed M. and Simon B. (1980) "Methods of Modern Mathematical Physics. I: Functional Analysis" (Academic, New York, 1980).

Ritschel U. and Gerwinski M. (1997) "Casimir forces at tricritical points: theory and possible experiments", *Physica A* **243**, 362.

Rocker W., Kohlhaus R. and Schöpgens A. W. (1971) "Magnetocalorischer Effect und kritische Exponenten des Eisens in der Umgebung seiner Curietemperature", *Z. Agnew. Phys.* **32**, 164.

Rowlinson J. S. and Winton F. L. S. (1982) *Liquids and Liquid Mixtures* (Butterworth, London, 1982).

Rudnick J. (1978) "First order transition induced by cubic anisotropy", *Phys. Rev. B* **18**, 1406.

Rudnick J., Hong Guo and Jasnow D. (1985) "Finite-Size Scaling and Renormalization Group" *Journ. Stat. Phys.* **41**, 353.

Rudnick J. (1990), in *Finite Size Scaling and Numerical Simulation of Statistical Systems*, edited by Privman V. (World Scientific, Singapore, 1990), 141.

Ruelle D. (1969), *Statistical Mechanics. Rigorous Results* (Benjamin, New York, 1969).

Sachdev S. (1993) "Polylogarithm identities in a conformal field theory in three dimensions", *Phys. Lett. B* **309**, 285.

Sachdev S. (1994) "Quantum phase transitions and conseverd charges", *Z. Phys. B* **94**, 469.

Sachdev S. (1996) "A quantum critical trio: solvable models of finite temperature crossover near quantum phase transition", in *Strongly Correlated Magnetic and Superconducting Systems*, edited by Sierra G. and Martin-Delgado M. A., (Springer, Berlin, 1996).

Sachdev S., Read N. and Oppermann R. (1995) "Quantum field theory of metallic spin glasses", *Phys. Rev. B* **52**, 10 286.

Sachdev S. and Ye J. (1992) "Universal quantum-critical dynamics of two-dimensional antiferromagnets", *Phys. Rev. Lett.* **69**, 2411.

Saleur H. and Derrida B. (1985) "A combination of Monte Carlo and transfer matrix methods to study 2D and 3D percolation", *J. Physique* **46**, 1043.

Scheiber B. A., Meadows M. R., Mockler R. C. and O'Sullivan W. J. (1979) "Critical Phenomena in Fluid Films: Scaling Crossover and Law of Corresponding States", *Phys. Rev. Lett.* **43**, 590.

Schneider T., Stoll E., and Beck H. (1975) "Exactly soluble models for distortive structural phase transitions", *Physica* **79A**, 201.

Schneider T. (1991) "Dynamical crossover in cuprate superconductors", Z. Phys. B **85**, 187.

Schultz T. D., Mattis D. C. and Lieb E. H. (1964) "Two-Dimensional Ising Model as a Soluble Problem of Many Fermions", *Rev. Mod. Phys.* **36**, 856.

Sénéchal D. (1993) "Mass gap of the nonlinear-σ model through the finite-temperature effective action", *Phys. Rev. B* **47**, 8353.

Sengupta A. M. and Georges A. (1995) "Non-Fermi-liquid behavior near a $T = 0$ spin-glass transition", *Phys. Rev. B* **52**, 10 295.

Shapiro J. and Rudnick J. (1986) "The fully finite spherical model", *J. Stat. Phys.* **43**, 51.

Shcherbina M.V. (1988) "Spherical limit of n-vector model", *Teor. Mat. Fiz.* **77**, 460.

Silver H., Frankel N. E. and Ninham B. W. (1972) "A class of mean field models", *J. Math. Phys.* **13**, 468.

Sinai Ya. G. (1982) *Theory of Phase Transitions. Rigorous Results*, (Akademiai Kiado, Budapest, 1982).

Singh S., Jasnow D. and Barber M. N. (1975) "Critical behaviour of the spherical model with enhanced surface exchange: two spherical fields" *J. Phys. C* **8**, 3408.

Singh S. and Pathria R. K. (1985) "Privman-Fisher hypothesis on finite systems: Verification in the case of the spherical model of ferromagnetism", *Phys. Rev. B* **31**, 4483.

Singh S. and Pathria R. K. (1985) "Finite-size effects in the spherical model of ferromagnetism: Antiperiodic boundary conditions", *Phys. Rev. B* **32**, 4618.

Singh S. and Pathria R. K. (1986) "Spin-spin correlations in finite systems: Scaling hypothesis and corrections to bulk behavior", *Phys. Rev. B* **33**, 672.

Singh S. and Pathria R. K. (1987) "Spin-spin correlations in finite systems with $O(n)$ symmetry: Scaling hypothesis and corrections to bulk behavior", *Phys. Rev. B* **36**, 3769.

Singh S. and Pathria R. K. (1989) "Finite-size scaling of $O(n)$ models with long-range interactions", *Phys. Rev. B* **40**, 9238.

Singh S. and Pathria R. K. (1989) "Analytical evaluation of a class of lattice sums in arbitrary dimensions", *J. Phys. A* **22**, 1883.

Soeffge F. (1980) "Stress-induced crossover effect near the Curie point of nickel I. Experimental details and results", *Phil Mag. B* **42**, 47 (in German); "Stress-induced crossover effect near the Curie point of nickel II. Theory and discussion", *Phil Mag. B* **42**, 63 (in German).

Sondhi S. L., Girvin S. M., Carini J. P. and Shahar D. (1997) "Continuous quantum phase transitions", *Rev. Mod. Phys.* **69**, 315.

Stamenković S., Tonchev N.S. and Zagrebnov V.A. (1987) "Exactly soluble model for structural phase transition with a Gaussian type anharmonicity", *Physica A* **145**, 262.

Stanley H. E. (1968) "Spherical model as the limit of infinite spin dimensionality", *Phys. Rev.* **176**, 718.

Stanley H. E. (1969a) "Critical indices for a system of spins of arbitrary dimensionality situated on a lattice of arbitrary dimensionality", *J. Appl. Phys.* **40**, 1272.

Stanley H. E. (1969b) "Exact solution for a linear chain of isotropically interacting classical spins of arbitrary dimensionality", *Phys. Rev.* **179**, 570.

Stanley H. E. (1971) *Introduction to Phase Transitions and Critical Phenomena* (Oxford University Press, Oxford, 1971).

Stanley H. E. (1999) "Scaling, universality, and renormalization: Three pillar of modern critical phenomena", *Rev. Mod. Phys.* **71**, S358.

Stauffer D., Ferer M. and Wortis M. (1972) "Universality of second order phase transitions: the scale factor for the correlation length", *Phys. Rev. Lett.* **29**, 345.

Stecki J. (1993) "Capillary length of planar interface from low temperatures to the critical point: An Ising $d = 2$ strip", *Phys. Rev. B* **47**, 7519.

Stratonovich R.L. (1957) "A method for calculating quantum distribution functions", *Dokl. Akad. Nauk S.S.S.R* **115**, 1097 (in Russian); (1958) *Sov. Phys. Doklady* **2**, 416, (in English).

Symanzik K. (1981) "Schrödinger representation and Casimir effect in renormalizable quantum field theory", *Nucl. Phys. B* **190**, 1.

Suter R. M. and Hohenemser C. (1978) "Crossover in the dynamic exponent z for three-dimensional ferromagnets", *Phys. Rev. Lett.* **41**, 705.

Suzuki M. (1973) "Critical exponents for long-range interaction I, Dimensionality, symmetry and potential range", *Prog. Theor. Phys* **49**, 424.

Tang Z. X., Sorensen C. M., Klabunde K. J. and Hadjipanayis G. C. (1991) "Size-Dependent Curie Temperature in Nanoscale $MnFe_2O_4$ Particles", *Phys. Rev. Lett.* **67**, 3602.

Temperley H. N. V. (1954) "The Mayer theory of condensation tested against a simple model of imperfect gas", *Proc. Phys. Soc. London, Sect. A* **67**, No. 411, 233.

Theumann W. K. (1970) "Some critical properties of Ornstein-Zernike systems and spherical model", *Phys. Rev. B* **2**, 1396.

Tonchev N. S. and Brankov J. G. (1980) "On the $s - d$ Model for Coexistence of

Ferromagnetism and Superconductivity", *phys. stat. sol. (b)* **102**, 179.

Tonchev N. S. (1991) "On the finite-size scaling in quantum critical phenomena", *Physica A* **171**, 374.

Tosatti E. and Martoňák R. (1994) "Rotational melting in displacive quantum paraelectics", *Solid St. Commun.* **92**, 167.

Toupin R. A. and Lax M. (1957), "Lattice of Partly Permanent Dipoles", *J. Chem. Phys.* **27**, 458.

Tu Y. and Weichman P. B. (1994) "Quantum spherical models for dirty phase transitions", *Phys. Rev. Lett.* **73**, 6.

Van Hove L. (1949) "Quelques propriétés générales de l'intégrale de configuration d'un systéme de particules avec interaction", *Physica* **15**, 951.

van Hemmen L. and Zagrebnov V. A. (1988) "Models of a structural phase transition with general anharmonicity and disorder", *J. Stat. Phys.* **53**, 835.

Verbeure A. and Zagrebnov V. A. (1992) "Phase Transitions and Algebra of Fluctuation Operators in an Exactly Soluble Model of a Quantum Anharmonic Crystal", *J. Stat. Phys.* **69**, 329.

Verbeure A. and Zagrebnov V. A. (1995) "Dynamics of Quantum Fluctuations in an Anharmonic Crystal Model", *J. Stat. Phys.* **79**, 377.

Verbeure A. and Zagrebnov V. A. (1995) "No-go theorem for quantum structural phase transition", *J. Phys. A* **28**, 5415.

Verbeure A. and Zagrebnov V. A. (1995) "Collective excitations in the anharmonic cristal" *Physica A* **215**, 394.

Vojta T. (1993) "Spherical random-field systems with long-range interactions: general results and apllication to Coulomb glass", *J. Phys. A* **26**, 2883.

Vojta T. (1996) "Quantum version of a spherical model: Crossover from quantum to classical critical behavior ", *Phys. Rev. B* **53**, 710.

Vojta T. and Schreiber M. (1994) "Generalised Coulomb gap in the spherical version of a lattice model of disordered and correlated localized particles", *Phys. Rev. B* **49**, 7861.

Vojta T. and Schreiber M. (1994) "Critical correlations and susceptibilities in the random-field spherical model", *Phys. Rev. B* **50**, 1272.

Vojta T. and Schreiber M. (1996) "Critical behavior of a quantum spherical model in a random field", *Phys. Rev. B* **53**, 8211.

Vollmayr K., Reger J. D., Scheucher M. and Binder K. (1993) "Finite size effects at thermally-driven first order phase transitions: a phenomenological theory of the order parameter distribution", *Z. Phys. B* **91**, 113.

Vygovskiy V. P. and Yergin Yu. V. (1972) "Some peculiarities of magnetization of Ni in the temperature region of magnetic transition" (in russian), *Fizika Metallov i Metallovedenia* **34**, 491.

Whitham G. B. (1974) *Linear and Nonlinear Waves* (Wiley, New York, 1974).

Widom B. (1965) "Surface tension and molecular correlations near the critical point", *J. Chem. Phys.* **43**, 3892; *ibid* "Equation of state in the neighbourhood of the critical point", **43**, 3898.

Yan C. C. and Wannier G. H. (1965) "Observations on the spherical model of a ferromagnet", *J. Math. Phys.* **6**, 1833.

Yukhnovskii I. R. (1987) *Phase transition of the second order. Collective variables method.* (World Scientific, Singapore, 1987).

Zabzin M. (1997) "A finite-temperature generalisation of the Zamolodchikov's C-theorem", hep-th/9705015.

Zagrebnov, V.A., Brankov, J.G. and Tonchev N.S. (1976) "A rigorous result on systems interacting with a Boson field", *Soviet Phys. Dokl.* **20**, 754-755.

Zagrebnov V. A. (1984) "The Approximating Hamiltonian Method for an Infinite-Mode Dicke Mazer Model with A^2-Term", *Z. Phys. B* **55**, 75.

Zamolodchikov A. B. (1986) "Irreveresibility of the flux of the renormalization group in a 2D field theory", *JETF Lett.* **43**, 731; (1987) "Renormalization group and perturbation theory about fixed points in two-dimensional field theory" *Sov. J. Nucl. Phys.* **46**, 1090.

Ziff R., Uhlenbeck G. E. and Kac M. (1977) "The Ideal Bose-Einstein Gas, Revisited" *Phys. Rep.* **32**, 169.

Ziherl P. and Žumer S. (1996) "Fluctuations in confined liquid cristals above nematic-isotropic phase transition temperature", *Phys. Rev. Lett.* **78**, 682.

Ziherl P., Podgornik R. and Žumer S. (1999) "Wetting-driven Casimir force in nematic liquid crystals", *Phys. Rev. Lett.* **78**, 682.

Zinn-Justin J. (1996) *Field Theory and Critical Phenomena*, (Clarendon, Oxford, 1996).

Zuk J. A. (1992) "Eigenvalue problem for tridiagonal matrices arising in the scattering-theory analysis of the disordered conductors", *Can. J. Phys.* **70**, 257.

Index